PARALLEL COMPUTATIONAL FLUID DYNAMICS

TRENDS AND APPLICATIONS

PARALLEL COMPUTATIONAL FLUID DYNAMICS

TRENDS AND APPLICATIONS

Proceedings of the Parallel CFD 2000 Conference
Trondheim, Norway (May 22-25, 2000)

Edited by

C.B. JENSSEN
Statoil
Trondheim, Norway

T. KVAMSDAL
SINTEF
Trondheim, Norway

H.I. ANDERSSON
NTNU
Trondheim, Norway

B. PETTERSEN
NTNU
Trondheim, Norway

A. ECER
IUPUI, Indianapolis
Indiana, U.S.A.

J. PERIAUX
Dassault-Aviation
Saint-Cloud, France

N. SATOFUKA
Kyoto Institute of Technology
Kyoto, Japan

Assistant Editor

P. FOX
IUPUI, Indianapolis
Indiana, Japan

N·H

2001
ELSEVIER
Amsterdam – London – New York – Oxford – Paris – Shannon – Tokyo

ELSEVIER SCIENCE B.V.
Sara Burgerhartstraat 25
P.O. Box 211, 1000 AE Amsterdam, The Netherlands

First edition 2001

Library of Congress Cataloging-in-Publication Data

Parallel CFD 2000 Conference (2000 : Trondheim, Norway)
Parallel computational fluid dynamics : trends and applications : proceedings of the Parallel CFD 2000 Conference / edited by C.B. Jenssen ... [et al.].
p. cm.
ISBN 0-444-50673-X (hardcover)
1. Fluid dynamics--Data processing--Congresses 2. Parallel processing (Electronic computers)--Congresses. I. Jenssen, C. B. (Carl B.) II. Title.

QA911 .P35 2000
532'.00285'435--dc21

2001023148

ISBN: 0-444-50673-X

∞ The paper used in this publication meets the requirements of ANSI/NISO Z39.48-1992 (Permanence of Paper).
Transferred to digital printing 2006

PREFACE

Parallel CFD 2000, the twelfth in an international series of meetings featuring computational fluid dynamics research on parallel computers, was held May 22-25, 2000 in Trondheim, Norway, returning to Europe for the first time since 1997. More than 125 participants from 22 countries converged for the conference which featured 9 invited lectures and 70 contributed papers.

Following the trend of the past conferences, areas such as numerical schemes and algorithms, tools and environments, load balancing, as well as interdisciplinary topics and various kinds of industrial applications were all well represented in the work presented. In addition, for the first time in the Parallel CFD conference series, the organizing committee chose to draw special attention to certain subject areas by organizing a number of special sessions. Particularly the special sessions devoted to affordable parallel computing, large eddy simulation, and lattice Boltzmann methods attracted many participants.

We feel the emphasis of the papers presented at the conference reflect the direction of the research within parallel CFD at the beginning of the new millennium. It seems to be a clear tendency towards increased industrial exploitation of parallel CFD. Several presentations also demonstrated how new insight is being achieved from complex simulations, and how powerful parallel computers now make it possible to use CFD within a broader interdisciplinary setting.

Obviously, successful application of parallel CFD still rests on the underlying fundamental principles. Therefore, numerical algorithms, development tools, and parallelization techniques are still as important as when parallel CFD was in is infancy. Furthermore, the novel concepts of affordable parallel computing as well as metacomputing show that exciting developments are still taking place.

As is often pointed out however, the real power of parallel CFD comes from the combination of all the disciplines involved: Physics, mathematics, and computer science. This is probably one of the principal reasons for the continued popularity of the Parallel CFD Conferences series, as well as the inspiration behind much of the excellent work carried out on the subject. We hope that the papers in this book, both on an individual basis and as a whole, will contribute to that inspiration.

The Editors

ACKNOWLEDGMENTS

Parallel CFD 2000 was organized by SINTEF, NTNU, and Statoil, and was sponsored by Computational Dynamics, Compaq, Fluent, Fujitsu, Hitachi, HP, IBM, NEC, Platform, Scali, and SGI.

The local organizers would like to thank the sponsors for their generous financial support and active presence at the conference. We are also grateful for the help and guidance received form Pat Fox and all the other members of the international organizing committee. We would like to especially thank Günther Brenner, Kjell Herfjord, and Isaac Lopez, for proposing and organizing their own special sessions. Last, but not least, we would like to thank the two conference secretaries, Marit Ødegård and Unn Erlien for their professional attitude and devotion to making the conference a success.

Carl B. Jenssen
Chairman, Parallel CFD 2000

INTERNATIONAL SCIENTIFIC ORGANIZING COMMITTEE PARALLEL CFD 2000

LOCAL ORGANIZING GROUP PARALLEL CFD 2000

TABLE OF CONTENTS

1. Invited Papers

2. Affordable Parallel Computing

5. Tools and Environments

6. Numerical Schemes and Algorithms

11. Industrial Applications

12. Multiphase and Reacting Flows

13. Unsteady Flows

1. Invited Papers

Parallel Computational Fluid Dynamics – Trends and Applications
C.B. Jenssen et al. (Editors)

Perspectives and Limits of Parallel Computing for CFD Simulation in the Automotive Industry

H. Echtle, H. Gildein, F. Otto, F. Wirbeleit, F Klimetzek

DaimlerChrysler AG,
HPC E222, D70546 Stuttgart, Germany

1 ABSTRACT

To achieve shorter product development cycles, the engineering process in the automotive industry has been continuously improved over the last years and CAE techniques are widely used in the development departments. The simulation of the product behaviour in the early design phase is essential for the minimisation of design faults and hence a key factor for cost reduction.

Parallel computing is used in automotive industry for complex CFD simulations since years and can be considered as state of the art for all applications with non-moving meshes and a fixed grid topology. The widely used commercial CFD packages (e.g. Fluent, StarCD etc.) show an acceptable performance on massively parallel computer systems.

Even for complex moving mesh models, as they are used for the simulation of flows in internal combustion engines excellent speed-ups were demonstrated recently on MPP systems and a parallel efficiency of 84 % on 96 nodes of a Cray T3E-900 was achieved within the ESPRIT Project 20184 HPSICE.

In the near future parallel computing will allow a nearly instantaneous solution for selected 3d simulation cases. Within the ESPRIT Project 28297 ViSiT Virtual Reality based steering techniques for the simulation are already tested and developed. This allows the intuitive steering of a 3d simulation running on a MPP system through a direct interaction with the simulation model in VR.

2 KEYWORDS

CFD, combustion, spray, grid generation visualisation, HPC, VR, parallel computing, engine simulation, computational steering, MPP

3 PROCESS CHAIN ENGINEERING SIMULATION

Due to the requirements of the market, car manufacturers are currently faced with the situation to develop more and more products for small and profitable niche markets (e.g. sport utility vehicles). This requires the development of hardware in a shorter time. In addition the development costs must be decreased to remain competitive.

In order to achieve these contradictory goals the behaviour of the new product has to be evaluated in the early design phase as precise as possible. The digital simulation of the product in all design stages is a key technology for the rapid evaluation of different designs in the early design phase, where as shown in Figure 1 the largest impact on production costs can be achieved. The costs associated with a design adjustment should be kept small by minimising changes in the pre-production or production phase. Ideally no design changes should be required after job # 1, when the first vehicle leaves the factory.

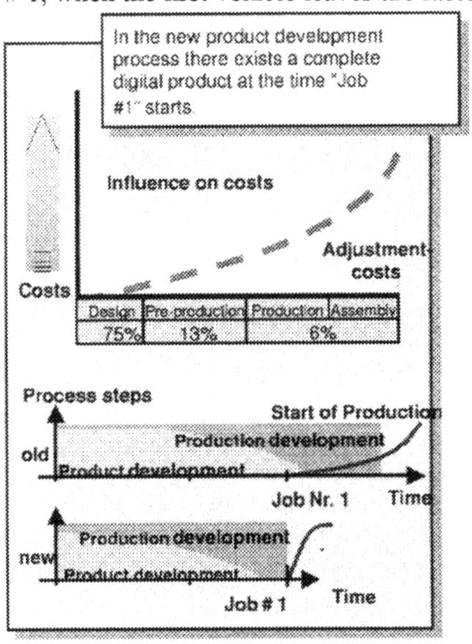

Figure 1:Typical Cost Relationships for Car Development

4 CFD SIMULATION CYCLE

CFD applications are beside crash simulation the most demanding and computationally intensive application in automotive development. CFD is used for a wide range of problems including external aerodynamics, climate systems, underhood flows and the flow and combustion process in engines. In the past the usage of CFD as a regular design tool was limited mainly due to the extremely long CPU time and complex mesh generation.

A typical simulation sequence starting from CAD data and valid for in-cylinder analysis is given in Figure 2. The different steps of the entire engine simulation are depicted including the names of the simulation software used (in grey boxes). STAR-HPC is the parallel version of the numerical simulation code STAR-CD from Computational Dynamics (CD). The programs ProICE and ProSTAR are the pre-processing tools from ADAPCO used for the benchmark results shown in the figures below. Similar tools from other companies e.g. ICEM-CFD are available as well. The visualisation package COVISE is developed at the

University of Stuttgart and is commercialised by VirCinity. To complete such a cycle it typically took 12 only 3 years ago and takes now one week by using advanced mesh generation tools, parallel computers and new post processing techniques.

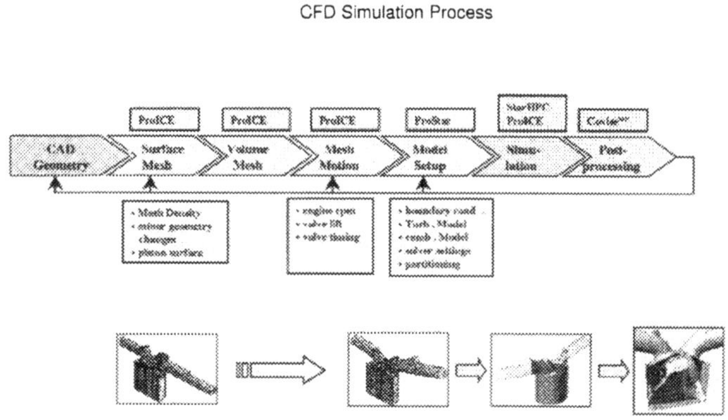

Figure 2: CFD Simulation Cycle

Most commercially available CFD codes are implemented efficiently on MPP systems at least for non moving meshes This reduced the computer time by nearly two orders of magnitude as shown in. Figure 3 for a non moving mesh and Figure 4 for a moving mesh case. Using the implementation strategy for the coupling of StarHPC and ProICE shown in Figure 6 a parallel efficiency of 84 percent on 96 processors for moving grid problems with a reasonable grid size of 600000 cells was demonstrated and a typical simulation can be done within a day or two now, instead of several weeks. Recently similar improvements in the parallelisation of two phase flows with a lagrangian spray simulation could be shown (Figure 5) and parallel computing can be efficiently used for the design of direct injected engines with low fuel consumption as well.

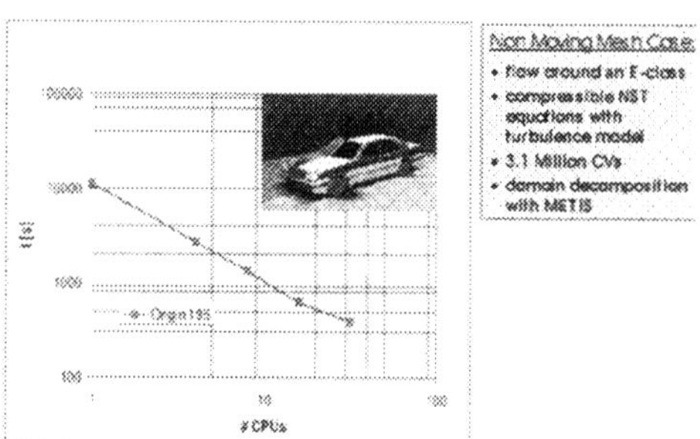

Figure 3: Speed-up steady state, non moving mesh case

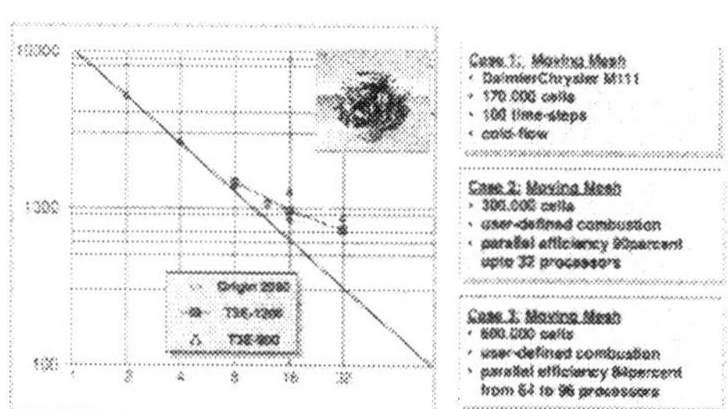

Figure 4: Speed-up transient, moving mesh case

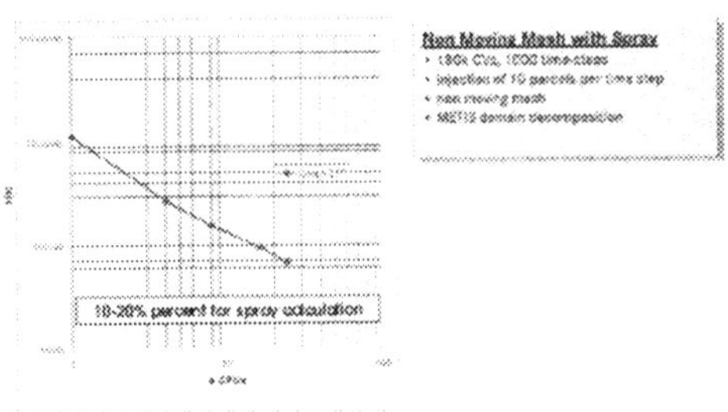

Figure 5: Speed-up transient spray simulation

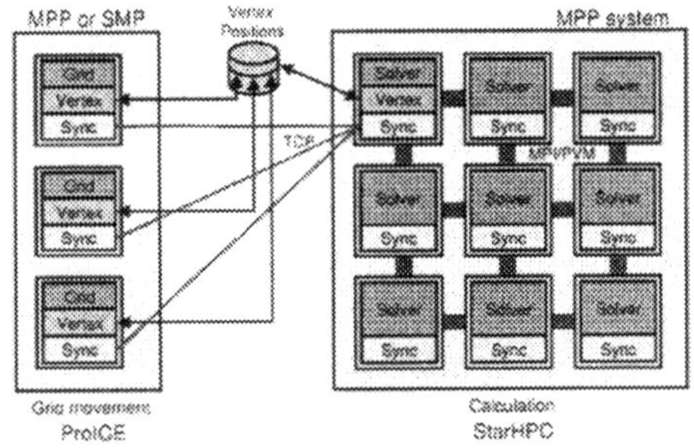

Figure 6: Scalable Implemenation of StarCD for Moving Grid Problems

The speed-up achieved in simulation automatically shifted the bottlenecks in the simulation process to the pre- and post-processing.(Figure 7) Although considerable achievements were made in the pre-processing with semi-automatic mesh generators for moving mesh models, further improvements in this domain and a closer integration with existing CAD packages are required.

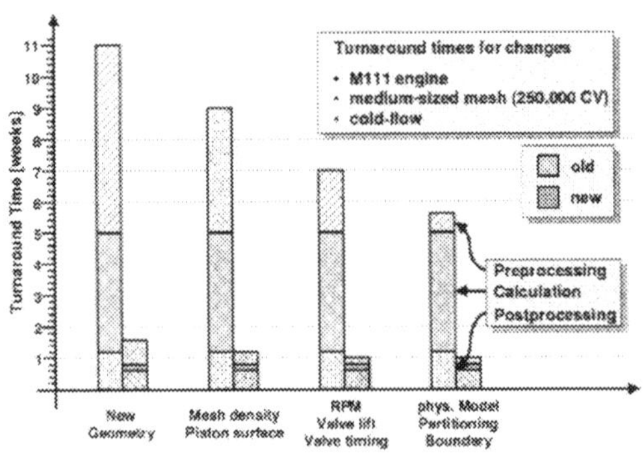

Figure 7: Turnaround time for engine simulation

5 ENGINE SIMULATION

An overview of the physics, which are simulated in a typical spark ignited engine configuration, is shown in Figure 8. Due to the moving valves and piston the number of cells and the mesh structure is changed considerably during a simulation run. Beside the cold flow properties the fuel spray and the combustion process has to be simulated. Spray and fluid are tightly coupled and the correct prediction of mixture formation and wall heat transfer are essential for an accurate combustion simulation. In particular the combustion process and the spray fluid interaction are still a matter of research.

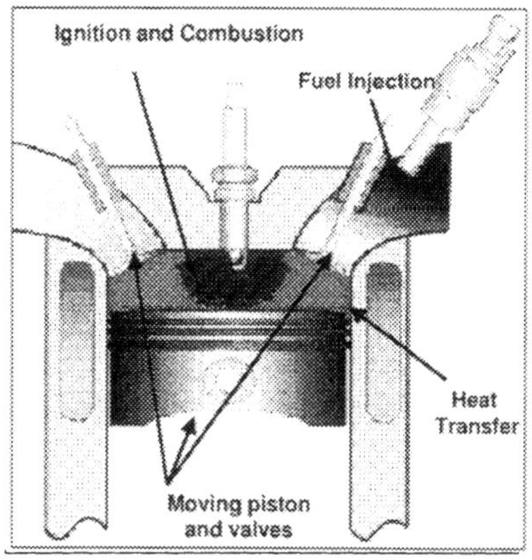

Figure 8: Engine Configuration

5.1 *Mathematical Method and Discretisation*

The implicit finite volume method which is used in STAR-HPC discretises the three dimensional unsteady compressible Navier-Stokes equations describing the behaviour of mass, momentum and energy in space and time.

All results for engines shown here, were done with:

- k-ε turbulence model with a wall function to model the turbulent behaviour of the flow,
- combustion modelling (e.g. premixed version of the 2-equation Weller model),
- several scalar transport equations to track the mixture of fresh and residual gas and the reactants.

The fuel injection is modelled by a large number of droplet parcels formed by droplets of different diameter. The number of parcels has to be large enough to represent the real spray in a statistical sense. An ordinary differential equation for every parcel trajectory has to be solved as a function of the parcel and flow properties (mass, momentum, energy, drag, heat conduction). Each droplet is considered as a sphere and based on this geometric simplification droplet drag and vaporisation rates are evaluated. In addition collision and break-up models for droplet-droplet and droplet-wall interaction are used to describe the spray and its feedback on the flow realistically.

5.2 *Domain Decomposition and Load Balancing*

To get scalability of a parallel application for a high number of processors it is necessary to balance the load and restrict the memory address space locally for each processor. A standard domain decomposition is used for non moving grid problems and the grid is decomposed in different parts. MPI or PVM is used for inter-processor communication in StarHPC.

For moving grid problems with sprays, as in engine simulation, an adapted decomposition strategy is required to account for:

- the number of cells in the grid, changing due to the mesh movement,
- the computational effort, depending on the complexity of physics in a cell (number of droplets, chemical reactions etc.),

Currently this problem is not yet solved in general terms.

5.3 *Results*

Figure 9 shows the mixing process of fresh air (blue) and residual gas (yellow) in a cross section of an engine, which is a typical result of a transient cold flow simulation. It can be seen how the piston is going down from top dead centre (step 1) to bottom dead centre (step 4). The gray surface below the intake valve at the right side is an iso-surface of a constant residual gas concentration. This type of simulation can be used to optimise valve timings or port geometries.

A typical combustion result for a gasoline engine with premixed gas is shown in Figure 10. The development and motion of the theoretically predicted flame front coincides quite well in shape and phase with the experimentally measured flame front.

Figure 11 shows a comparison of the spray formation and flame propagation in a diesel engine compared to an experiment of the soot luminosity. Again the agreement with the experiment is quite good. This examples illustrates the degree of complexity achieved in simulation today.

To achieve these results considerable expertise and tuning of the simulation models is still required and additional research is needed to improve the prediction of these methods.

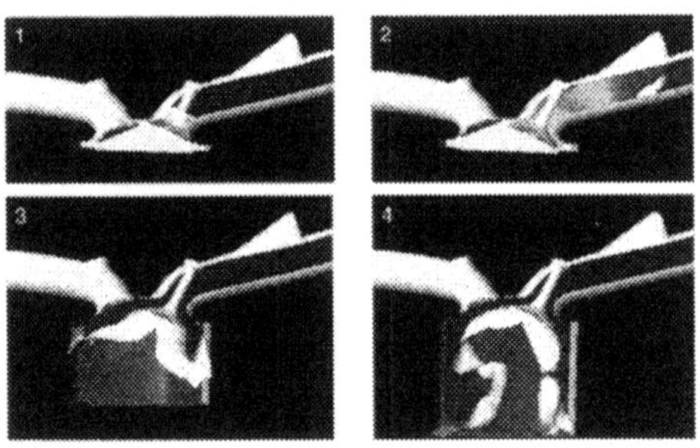

Figure 9: Mixing Process of Fresh Air (blue) and Residual Gas (yellow) in an internal combustion engine

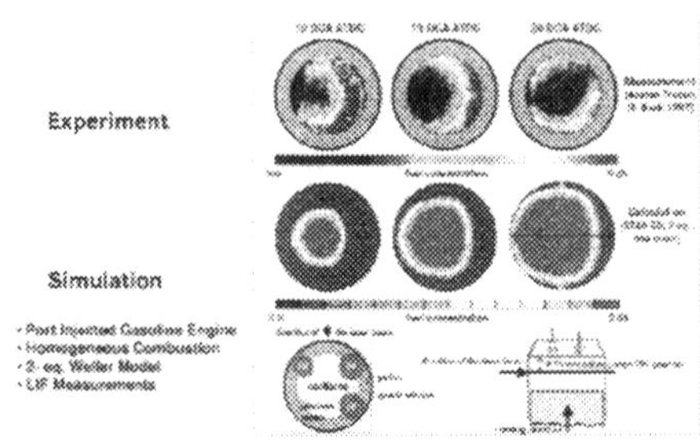

Figure 10: Simulated Flame Propagation, Comparison to Experiment

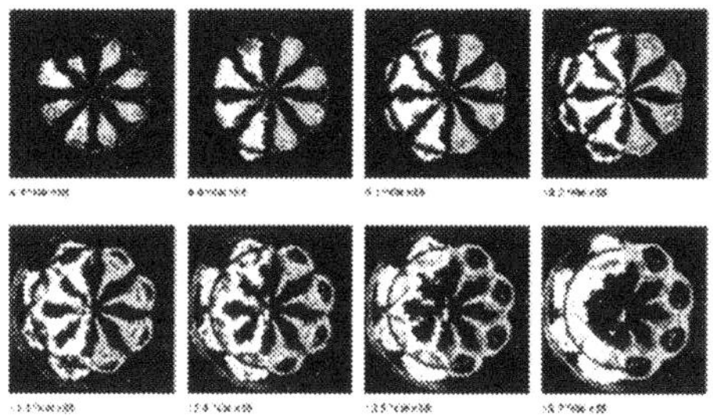

Figure 11: Experimental soot luminosity compared to simulated isosurface of temperature

6 SIMULATION OF HVAC SYSTEMS

The simulation of **H**eating **V**entilation and **A**ir **C**onditioning systems (HVAC) is another domain where CFD is widely used in automotive industry as shown in Figure 12.

This type of simulation typically requires large and complex grids with several million cells. In addition many geometric configurations (passengers, outlets of ducts) etc. has to be taken into account in order to predict the passenger comfort, the system efficiency and energy consumption.

By combining the CFD results with a model for the solar radiation and a thermophysical passenger model the thermal comfort can finally be evaluated as shown in Figure 13.

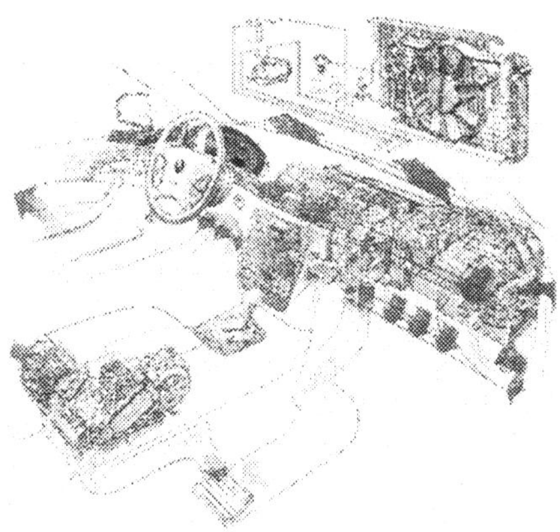

Figure 12: Simulation of HVAC systems in cars.

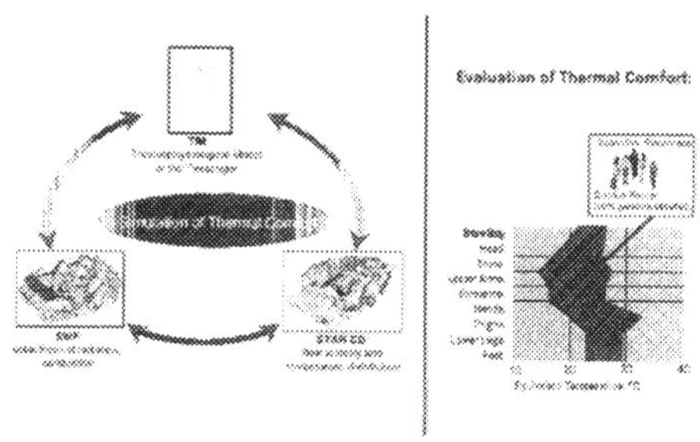

Figure 13:Evaluation of thermal comfort

7 RECENT ACTIVITIES & OUTLOOK

The previous examples have shown the complexity, CFD simulation has reached in automotive industry. The availability of cheap multiprocessor systems in combination with parallel codes within the last few years is considered as a key success factor for the widespread acceptance of these technologies in the development departments.

In addition, parallel computing allowed to increase the model size and physical complexity, which improves the accuracy and reliability of the predicted results. The reduced simulation time allows a faster development of sophisticated physical models, e.g. for combustion and sprays.

For selected 3d simulation cases significant changes in the solution can be observed in under minute. This is an acceptable response time for the interactive steering of the computation, which opens new possibilities for the use of 3d CFD simulation. Within the ESPRIT Project ViSiT Virtual Reality based steering techniques are already tested and developed. In such an environment the user interacts directly with a simulation running on an MPP system as shown in Figure 14.

The scope of interaction with the simulation model within ViSiT ranges from a simple change in the boundary conditions like velocity direction and magnitude of duct openings to a complete interactive exchange of a driver and seat as shown in Figure 15.

Figure 14: Interaction with simulation model in VR

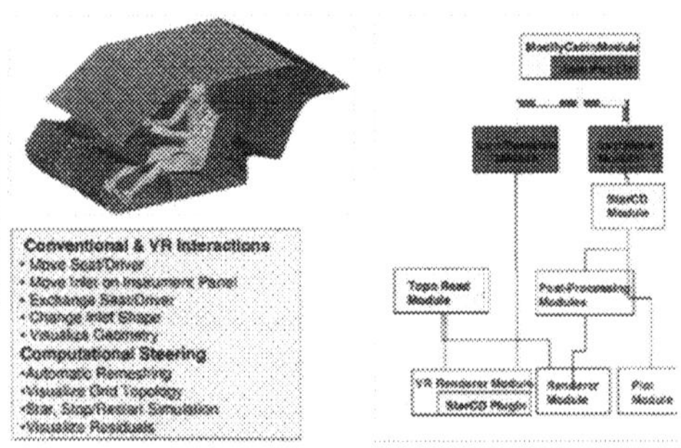

Figure 15: Scope of ViSiT (Virtual interactive Simulation Testbed)

Beside the interactive steering automatic geometry and parameter optimisation is getting feasible for 3d CFD as well with a reasonable response time. Here the combined usage of parametric CAD systems, automatic mesh generation and simulation is required to guarantee a rapid optimisation and the fast feedback of the optimised geometry into the design system. Although all components for such an optimisation are already available now, the integration of these tools for CFD application has to be improved to exploit the potential benefit of such an approach in the design process.

8 CONCLUSIONS

The integration of CFD into the development process of the automotive industry required a reduction in turnaround time by more than an order of magnitude.This reduction was made possible by a combined improvement of mesh generation, simulation and visualisation

Beside the speedup in simulation execution time achieved with high performance computing the short response times stimulate a rapid improvements in physical modelling as needed for a widespread usage of CFD simulation.

VR offers an intuitive way to analyse 3d simulation results and even direct interaction with simulation models in VR can already be demonstrated for selected test cases.

Whereas considerable progress has been achieved in accelerating the simulation process, the integration of CAD and CAE should be improved in the future. The combination of parametric CAD systems with 3d simulation tools and numerical optimisation will be an extremely powerful tool for a rapid product design and HPC is required for the exploitation and integration of these technologies in the design process.

9 ACKNOWLEDGEMENTS

The HPSICE and ViSiT project were funded by the European Commission in the ESPRIT program. The authors would like to thank the project partners for their excellent collaboration

Contact Points:	VirCinity	www.vircinity.com
	CD	www.cd.co.uk
	adapco	www.adapco.com
	sgi	www.sgi.de
	ICEM CFD	www.icemcfd.com
	HLRS	www.hlrs.de

Parallel Computational Fluid Dynamics - Trends and Applications
C.B. Jenssen et al. (Editors)

Application of Navier-Stokes Methods to Predict Vortex-Induced Vibrations of Offshore Structures

Y. Kallinderis[1] K. Schulz [2] W. Jester [3]

Dept. of Aerospace Engineering and Engineering Mechanics
The University of Texas at Austin
Austin, TX 78712

A major issue for the design of offshore structures is calculation of the forces and responses under the action of waves and currents. Use of empirical models has proven to be inadequate especially for deepwater applications. Navier-Stokes simulations have emerged as a powerful tool for predictions of vortex-induced vibrations (VIV) including the highly nonlinear situation of resonance (lock-in) of the structure. A numerical simulator that uses Navier-Stokes solvers and deformable mixed-element grids is presented and validated via comparisons with experiments. Three different levels of approximation are considered: (i) 2-D solutions, (ii) quasi-3D simulations based on a "strip theory" approach, as well as (iii) full 3-D computations. Qualitative and quantitative comparisons with published experimental data are made which show the ability of the present numerical method to capture complex, unsteady flow phenomena. Two special issues related to marine risers that are addressed are (i) the strong interference between different structures, and (ii) VIV suppression devices.

1 INTRODUCTION

A critical issue related to flow-structure interactions at offshore oil installations is the prediction and suppression of vortex-induced vibrations (VIV). Typical such structures are risers and spar platforms which are typically cylindrical in shape and are an essential part of any offshore oil exploration or production. Modeling of the structural aspects of these elements has reached a substantial degree of maturity, but the understanding and prediction of VIV is still a perplexing issue. Although typical amplitudes of vibration for risers undergoing VIV are small, the risers can still fail as a result of the persistent high frequency dynamic stresses causing fatigue. Resonance (*lock-in*) occurs when the natural structural frequency of the cylinder dominates the vortex shedding frequency which can result in large amplitude vibrations of the cylinder.

To address VIV difficulties, the offshore industry typically attempts to infer hydrodynamic loads based on experimental measurements which may be scaled to fit the particular problem of interest. Most all of the current models used to predict VIV response characteristics are derived from databases of experimental results primarily from shallow water

[1] Professor
[2] Postdoctoral fellow
[3] Graduate research assistant

installations. A large scatter of predicted responses has been observed [1]. Data for deepwater installations are very rare.

As numerical methods for solving the Navier-Stokes equations have matured substantially in recent years, an effort to utilize Navier-Stokes technology as a primary VIV analysis tool has been underway. Several two-dimensional Navier-Stokes flow-structure interaction methods have been developed which treat the offshore structures as being rigidly mounted on linear elastic springs (see e.g. Schulz and Kallinderis [2], Meling [3], Dalheim [4], Yeung [5]). However, not all of the pertinent flow physics and geometric characteristics can be correctly modeled with two-dimensional calculations (e.g. oblique shedding and helical strake geometries). Employment of a full three-dimensional Navier-Stokes solver can be prohibitive in terms of computing resources for deepwater cases such as riser calculations. In such cases, a quasi-3D approach which considers 2-D "cuts" of the flowfield and structure can be a practical solution [6]. This is also called the *strip theory* approach and allws the "CFD planes" to be coupled through the three dimensional structure that is considered.

Numerical results based on solution of the Navier-Stokes equations are presented for two classes of offshore problems: *fixed* and *elastically-mounted* structures. The fixed cases correspond to a circular cylinder with roughness in the *supercritical* (high Reynolds number) regime, as well as simulations of two cylinders and their interaction. The elastically-mounted cases focus on the VIV response of a circular cylinder for various Reynolds numbers. The quasi-3D method is applied to a flexible riser. Finally, the VIV results include an investigation of the effectiveness of two different classes of suppression devices: strakes and fairings.

2 NUMERICAL METHOD

Solution of the governing incompressible Navier-Stokes equations are accomplished using a forward Euler marching scheme in time for the momentum equations and a pressure correction formulation to obtain a divergence free velocity field at each time level. This pressure correction method is implemented using a finite-volume spatial integration scheme on non-staggered hybrid grids composed of both quadrilateral and triangular elements. The quadrilateral elements are used near viscous boundaries where they can efficiently capture strong solution gradients, and the triangular elements are used elsewhere allowing for complex geometries to be discretized [7]. In three dimensions, prismatic and tetrahedral elements are employed.

To include turbulence effects for high Reynolds numbers flows, the numerical method is coupled with the Spalart-Allmaras turbulence model [8]. This model is coupled with the solution of the Navier-Stokes equations by providing a local eddy viscosity (μ_t) throughout the flow-field by solving a separate partial differential equation. A more detailed presentation on the specifics of the outlined numerical procedure including the pressure correction formulation, edge-based finite volume discretization, artificial dissipation, and boundary conditions is presented in Ref. [2].

2.1 Elastically mounted structures

To simulate the VIV phenomenon, a structural response is required which dictates the displacement and velocity of each body as they respond to the surrounding flow field. Consequently, the incompressible fluid mechanics solution procedure must be coupled with a rigid body structural response in order to adequately resolve the flow-structure interaction. If each structure is treated as a rigidly mounted elastic body moving in the transverse direction only, the resulting equation of motion is:

$$m\ddot{y} + c\dot{y} + ky = f_y(t) \tag{1}$$

where m is the mass per unit length of the body, c is the damping coefficient, k is the stiffness coefficient, and y denotes the transverse location of the body centroid [9]. The right hand side of equation (1) contains the time-dependent external force, $f(t)$, which is computed directly from the fluid flow field. If the equation of motion is non-dimensionalized using the same parameters as the Navier-Stokes equations (U_∞ and D), the following equation of motion is obtained:

$$\ddot{y} + \left(\frac{4\pi\zeta_s}{U_{red}}\right)\dot{y} + \left(\frac{4\pi^2}{U_{red}^2}\right)y = \left(\frac{\rho_f D^2}{2m}\right)C_L(t) \tag{2}$$

where ζ_s is the non-dimensional damping coefficient, U_{red} is the reduced velocity, ρ_f is the fluid density, and C_L is the lift coefficient. The reduced velocity is an important parameter relating the structural vibration frequency to the characteristic length and free-stream fluid velocity. The reduced velocity for a circular cylinder of diameter D is defined by:

$$U_{red} = \frac{U_\infty}{f_n D} \tag{3}$$

where f_n is the natural structural frequency of the cylinder. Another important non-dimensional parameter arriving out of the above normalization is the mass ratio. The mass ratio for a circular cylinder is defined as:

$$n = \frac{m}{\rho_f D^2} \, . \tag{4}$$

The mass ratio is useful in categorizing the lock-in range that exists for a cylinder undergoing vortex-induced vibrations. Note that in general, low mass ratio cylinders have a much broader lock-in range than do cylinders with high mass ratios [10].

To obtain flow-structure solutions, the two problems are coupled via the hydrodynamic force coefficients acting on each body in the domain (C_L and C_D) which are the forcing functions in the equation of motion for each body. Note that equation (2) considers only transverse motion, but an identical equation of motion can be constructed for the in-line direction in terms of the normalized drag coefficient (C_D). Consequently, the present approach uses superposition of the two responses to obtain arbitrary two-dimensional motions. The overall solution procedure for marching forward one global time step is outlined as follows:

- Obtain pressure and velocity fields at the current time level using the numerical pressure correction algorithm.
- Compute the lift and drag coefficients acting on each body from the pressure and velocity fields.
- Compute the new centroid displacement and velocity of each body using a standard 4th-order Runge Kutta integration for equation (2).
- Deform the mesh and update grid velocities accordingly to match the new body displacements and velocities.

Additionally, note that if *multiple* bodies are moving within a single domain, then a deforming computational mesh is required in order to accommodate arbitrary motions of each body. Specific details on how this mesh deformation is accomplished are discussed in Ref. [2].

3 APPLICATIONS

All three levels of approximation (2-D, quasi-3D , and 3-D) are employed for different applications. Numerical results are presented for both *fixed* and *elastically-mounted* structures.

3.1 Fixed Cylinder with Roughness in two dimensions

This section considers flow about a fixed cylinder in a steady current with various roughness coefficient values. Surface roughness is an important concern for offshore applications since structures in the marine environment are often augmented by the addition of marine growth. For these applications, the roughness coefficients were chosen to match the experimental results of Achenback and Heinecke [11]. Three roughness coefficient values were considered along with a smooth circular cylinder which provides a baseline for the roughness results. Note that the Reynolds number presented in the experiments and used in all of the numerical simulations was $Re = 4 \times 10^6$ which corresponds to flow in the *supercritical* regime.

A uniform roughness was achieved in the experimental setup by placing pyramids with predefined heights onto the surface of an otherwise smooth cylinder. An analogous setup was utilized for the two-dimensional numerical simulations using triangular roughness elements on the cylinder surface. Two of the resulting surface roughness geometries for the numerical results are illustrated in Figure 1. Figure 1(a) corresponds to a roughness parameter of $k_s/D = 0.03$ while Figure 1(b) corresponds to a value of $k_s/D = 0.009$. The roughness coefficient simply characterizes the magnitude of the roughness with k_s referring to the nominal height of the roughness element and D to the smooth cylinder diameter.

Comparisons between the experimental and numerical results are presented in Figure 2 which shows the drag coefficient of a fixed cylinder as a function of surface roughness. The numerical results are in excellent agreement with the experimental measurements

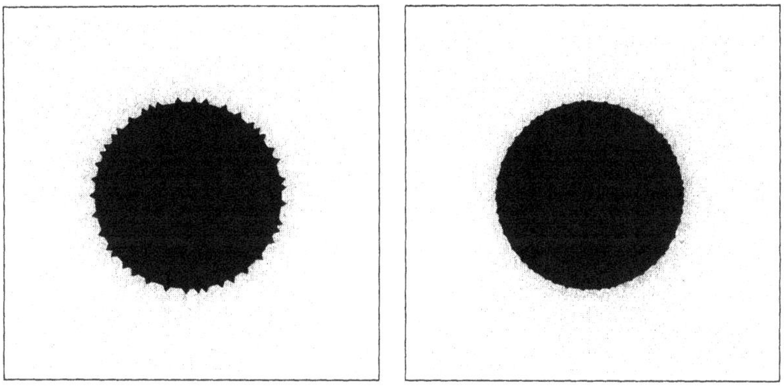

(a) $k_s/D = 0.03$ (b) $k_s/D = 0.009$

Figure 1: Illustration of surface roughness geometries

and capture several important physical phenomenon. In particular, the experimental measurements indicate that the cylinders with larger surface roughness values have larger drag coefficient values. However, the results from the two highest surface roughness cylinders yielded almost identical drag values. This similarity was also observed in the numerical results. In addition, the smooth cylinder results for $k_s/D = 0.0$ agree reasonably well and indicate the applicability of the method to flow configurations in the supercritical regime.

3.2 Flow about Fixed Cylinder Pairs

This section considers uniform flow about a pair of circular cylinders in both a tandem and side-by-side arrangement. Experimental results summarized by Zdravkovich [12] and Chen [13] indicate a wide variety of interference effects depending on the orientation and spacing of the cylinders. The orientation of the cylinders is measured by the longitudinal spacing (L/D) and transverse spacing (T/D) relative to the flow. Results for a pair of tandem cylinders in a *bi-stable* transition regime with $L/D = 2.15$ and a pair of side-by-side cylinders in the *biased gap* regime with $T/D = 2.5$ are presented below.

3.2.1 Tandem Orientation: Transition Region

For certain tandem separations between $L/D = 2$ and $L/D = 2.5$, the exierimtally observed *bistable* nature of the flow has been observed numerically. For $L/D \approx 2.15$, it is possible to drive the flow into either the *Reattachment* or *Two Vortex Streets* regimes by selecting the initial conditions.

To achieve the *Reattachment* regime, a steady solution at $Re = 100$ is first obtained. This lower Reynolds number result establishes the steady recirculation region between the cylinders. The Reynolds number is then slowly increased to $Re = 1000$. The resulting flow pattern shown in Figure 3(a) indicates the *Reattachment* regime observed in experiments. In this regime, the shear layer separating from the upstream cylinder reattaches to the

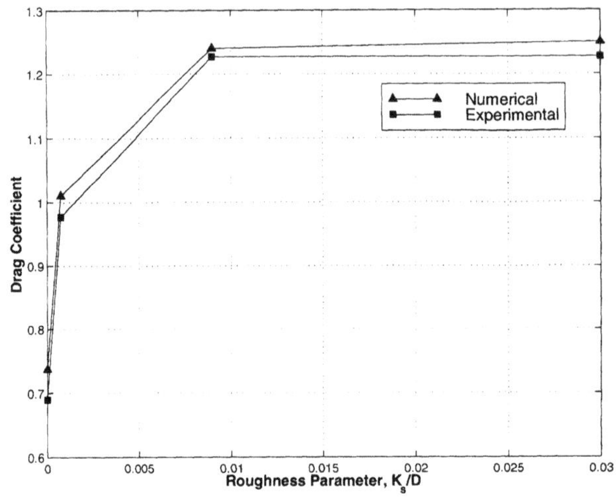

Figure 2: Drag coefficient of a fixed cylinder as a function of surface roughness, $Re = 4 \times 10^6$ (*supercritical*). A roughness parameter of $k_s/D = 0.0$ indicates a smooth cylinder with no roughness. Experimental results from Achenback and Heinecke [11].

downstream cylinder. A steady recirculation region exists in the gap between the cylinders with no vortex shedding occurring behind the upstream cylinder. This state was observed to be stable in the sense of persisting for over 1000 periods of vortex shedding.

To achieve the *Two Vortex Streets* regime, the flow is impulsively started at $Re = 1000$. In this case, the small asymmetry in the mesh is sufficient to cause vortex shedding from the upstream cylinder to begin before the steady recirculation region can be fully established. The final flow pattern, shown in Figure 3(b), indicates the *Two Vortex Streets* regime in which a vortex street is formed behind each cylinder. As before, this state persisted for over 1000 periods of vortex shedding.

3.2.2 Side-by-Side Configuration: Biased gap regime

For intermediate transverse spacings of side-by-side cylinders ($1.2 < T/D < 2.0$), an asymmetric *biased gap* flowfield is observed [12, 13]. In this regime, the flow in the gap between the cylinders is deflected towards one of the cylinders. Thus, two distinctive near wakes are formed, one wide wake and one narrow. The particular direction of the bias will intermittently change, indicating another *bistable* state. In the present study, the *Biased Gap* flow regime have been simulated and analyzed at $Re = 1000$. Qualitative comparisons with experimental observations are excellent.

Particle traces for the biased gap regime ($T/D = 1.5$) are shown on Figure 4. This figure shows four snapshots with the gap flow biased downwards. Each bias tends to persist for between five and ten periods of vortex shedding, then a transition to the other bias will tend to occur. The *flopping* between states occurs at time intervals roughly two orders of magnitude shorter than those reported in experimental results by Kim and

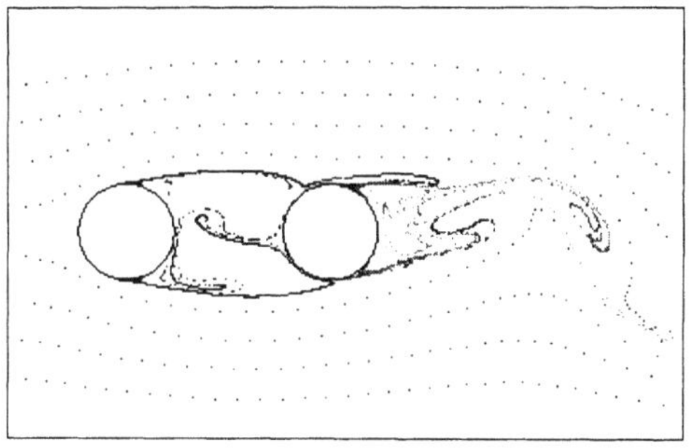

(a) Reattachment Regime

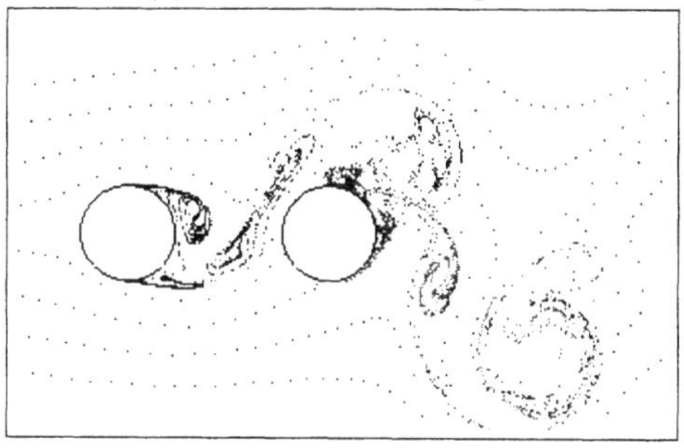

(b) Two Vortex Streets Regime

Figure 3: Particle traces in bistable region, $Re = 1000$, $L/D = 2.15$.

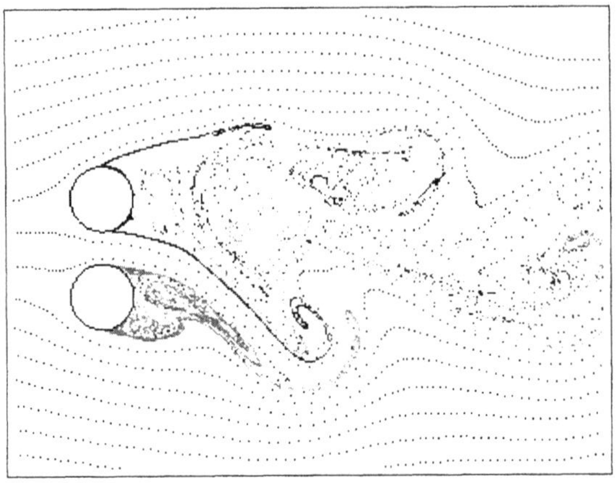

Figure 4: Particle traces in biased-gap region, $Re = 1000$, $T/D = 1.5$.

Durbin [14] at $Re = 3500$ and $T/D = 1.75$, although they are consistent with other numerical results of Chang and Song [15]. The reason for this discrepancy is not clear.

3.3 VIV and the Reynolds number

The speed of the current has a significant effect on the VIV response of the structure. The extend of the resonance (*lock-in*) region, as well as the amplitudes and frequencies of the response of the structure depend on the Reynolds number of the flow to a large degree. To demonstrate the fluid-structure coupling present during VIV, several series of different VIV simulations are presented combined with sample displacement histories and frequency responses. The first set corresponds to *low* Reynolds number tests ($90 \leq Re \leq 140$), while the second set refers to *moderate* Reynolds number tests ($6.83 \times 10^3 \leq Re \leq 1.85 \times 10^4$). A final set of cases considers *high* Reynolds numbers near the critical regime ($2.25 \times 10^5 \leq Re \leq 4.75 \times 10^5$). All three cases exhibit the *lock-in* phenomenon associated with VIV indicating that the natural frequency of the structure dominates the vortex shedding frequency.

3.3.1 Low Reynolds Number VIV

For the first VIV series, the Reynolds numbers, reduced velocities, and parameters characterizing the cylinder structural properties were chosen to match the experimental setup presented in Ref. [16]. The cylinder was constrained to move in the transverse direction only, and the surrounding fluid was assumed to be water as per the experiment. The cylinder was not allowed to move until the flow was fully developed for the fixed cylinder geometry at the desired Reynolds number. Figure 5 shows the evolution of the cylinder displacement for $U_{red} = 6.13$ ($Re = 110$). This reduced velocity falls within the lock-in regime, and the cylinder displacement is seen to increase monotonically until it reaches a peak amplitude of over 40% of the cylinder diameter.

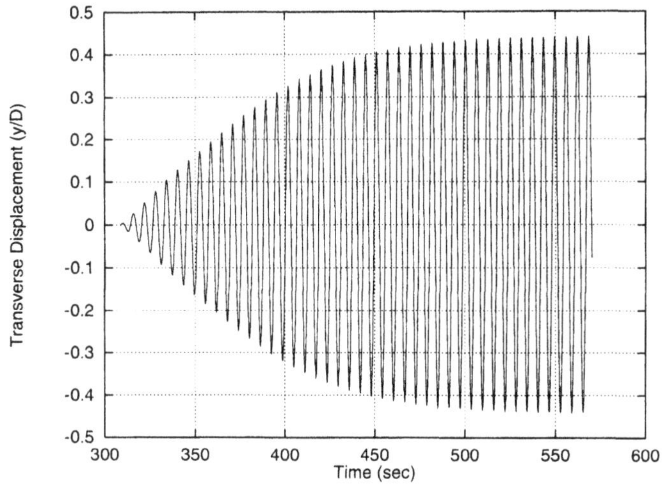

Figure 5: Displacement of an elastically mounted cylinder vibrating freely in the transverse direction (inside lock-in regime), $U_{red} = 6.13$, $Re = 110$.

The remaining results for this set of VIV simulations are presented in Figure 6 which includes the experimental results from Ref [16]. Figure 6 shows two series of results: the maximum transverse displacement of the cylinder at each Reynolds number and the ratio of the vortex shedding frequency (f) to the natural structural frequency (f_n). Frequency ratios of $f/f_n = 1.0$ indicate that the vortex shedding frequency has locked-in to the structural frequency. The Reynolds number of the numerical results ranges from $Re = 90$ to $Re = 140$ which corresponds to reduced velocities in the range from $U_{red} = 5.02$ to $U_{red} = 7.81$. The numerical results are seen to agree reasonably well with the experimental results and capture the large displacement amplitudes that occur within the lock-in regime. The numerical results are also able to detect the beginning and end of the lock-in regime although the beginning of the numerical lock-in regime occurs at about a 10% lower Reynolds number than does the experimental lock-in regime. Additionally, the end of the numerical lock-in regime occurs at a lower Reynolds number than its experimental counterpart. The authors in Ref. [16] note that no end plates were used on the cylinder during the experiments which can introduce some heightened three-dimensional effects not present in this strictly two-dimensional numerical study. This difference might account for the small shift between the two lock-in regions.

An additional point to note is that both the numerical and experimental maximum vibration amplitudes occur near the lower limit of the lock-in region. The authors of the experimental data point out that this result is in contrast to several other VIV experiments [17, 18] at higher Reynolds numbers where the maximum amplitude is observed near the middle of the lock-in region.

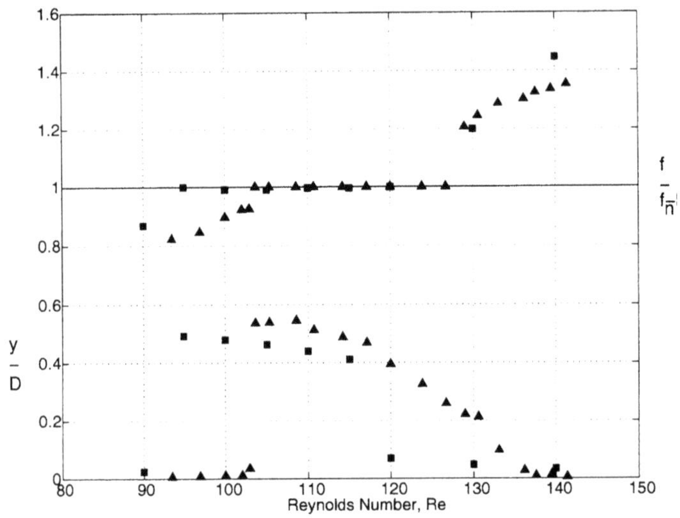

Figure 6: Maximum transverse displacements ($\frac{y}{D}$) and shedding frequency ratios ($\frac{f}{f_n}$) of an elastically mounted cylinder throughout the lock-in regime, Low Reynolds Number Cases.
▲ Experimental results from Ref. [16]
■ Present numerical results

3.3.2 Moderate Reynolds Number VIV

The second set of results for the moderate Reynolds number cases are shown in Figure 7. The structural properties, Reynolds numbers, and reduced velocities were chosen to match the experimental results presented in Ref. [19], and motions were again constrained to be in the transverse direction. Note that at these higher Reynolds numbers, the maximum transverse displacement is higher than the lower Reynolds number cases. The maximum displacement predicted by the simulation is over 75% of the cylinder diameter for these cases versus a maximum displacement of under 50% for the lower Reynolds number cases. The higher Reynolds number results also exhibit the more standard *bell* response shape typical of most VIV experiments. Note that the maximum numerical displacement is approximately 15% lower than its experimental counterpart and occurs at a higher Reynolds number. The numerical results also predict the lock-in range to occur slightly below the experimental results though the end of the lock-in range is quite similar. Despite these differences, the numerical results offer reasonable agreement with the experimental values in spite of the difficulties in reproducing the experiment exactly. For the numerical tests, the flow field at each Reynolds number is developed independently, but there is no guarantee that the experimental water tunnel results were developed in a similar fashion. In particular, experimental results are likely obtained by increasing the free-stream velocity while the cylinder continues to be in motion. In contrast, the flexibility afforded by the numerical simulation allows the solution at each free-stream velocity (and resulting

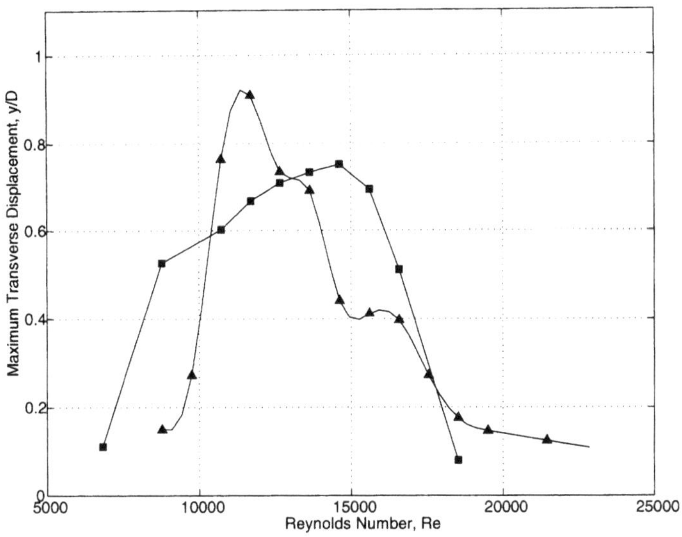

Figure 7: Maximum transverse displacements of an elastically mounted cylinder throughout the lock-in regime, Moderate Reynolds Number Cases.
▲ Experimental results from Moe et al. [19]
■ Present numerical results

Reynolds number) to be developed independently prior to letting the cylinder move freely.

In any case, the numerical results do seem to present a tractable option for approaching VIV problems and are seen to adequately resolve the beginning and end of the lock-in regime. In water experiments, typical excitation regions for transverse vibrations occur with reduced velocities in the range $4.5 \leq U_{red} \leq 10$ with the maximum transverse amplitude falling within the range of $6.5 \leq U_{red} \leq 8$ [20]. The numerical results concur with these observations with the excitation range occurring for $4.5 \leq U_{red} \leq 8.5$ and the maximum transverse amplitude attained at $U_{red} = 7.5$.

3.3.3 High Reynolds Number VIV

For these VIV cases, the Reynolds numbers, reduced velocities, and parameters characterizing the cylinder structural properties were chosen to match the experimental setup presented by Allen and Henning [21]. The Reynolds number in these experiments ranged from $2.25 \times 10^5 \leq Re \leq 4.75 \times 10^5$ which corresponds to reduced velocities in the range $4.86 \leq U_{red} \leq 10.26$. The cylinder was constrained to move in the transverse direction only and the solution at each Reynolds number was obtained by continuing with the already developed VIV solution from the next lower Reynolds number. The development of the solution in this manner is believed to be more representative of the experimental results, which were obtained from tow tests in a large model basin.

Note that at these higher Reynolds numbers, the maximum transverse displacement is higher than the lower Reynolds number cases. The maximum numerical displacement is over 75% of the cylinder diameter for these cases versus a maximum displacement of under 50% for the lower Reynolds number cases. The higher Reynolds number results also exhibit the more standard *bell* response shape typical of many VIV experiments. Note that the maximum numerical displacement is approximately 15% lower than its experimental counterpart and occurs at a higher Reynolds number. The numerical results also predict the lock-in range to occur slightly below the experimental results though the end of the lock-in range is quite similar. Despite these differences, the numerical results offer reasonable agreement with the experimental values in spite of the difficulties in reproducing the experiment exactly. For the numerical tests, the flow field at each Reynolds number and corresponding reduced velocity is developed independently but there is no guarantee that the experimental water tunnel results were developed in a similar fashion. However, the numerical results do seem to present a tractable option for approaching VIV problems and are seen to adequately resolve the beginning and end of the lock-in regime. In water experiments, typical excitation regions for transverse vibrations occur with reduced velocities in the range $4.5 \leq U_{red} \leq 10$ with the maximum transverse amplitude falling within the range of $6.5 \leq U_{red} \leq 8$ [20]. The numerical results concur with these observations with the excitation range occurring for $4.5 \leq U_{red} \leq 8.5$ and the maximum transverse amplitude attained at $U_{red} = 7.5$.

An example of the cylinder displacement for a reduced velocity inside the lock-in regime is presented in Figure 8. This figure shows the time evolution of the transverse cylinder displacement for $U_{red} = 7.02$ ($Re = 3.25 \times 10^5$) once it is allowed to respond freely to the surrounding flow-field. This reduced velocity falls within the lock-in regime and the cylinder displacement is seen to oscillate to amplitudes of over 80% of the cylinder diameter. Previous experimental and numerical results at very low Reynolds numbers ($Re \approx 100$) have shown transverse VIV displacement histories which start small and then increase monotonically until the cylinder reaches a peak vibration amplitude [16, 2]. In contrast, the high Reynolds number vibrations presented in Figure 8 are not at all monotonic and additional low frequency oscillations are evident. This difference illustrates the more chaotic influence of turbulence and non-linear flow-structure interaction that is present in high Reynolds number flows.

A summary of the high Reynolds number results are presented in Figure 9, which shows the transverse RMS displacements of both the numerical and experimental results as a function of Reynolds number. Note that freely-published controlled VIV experiments of a flexible cylinder in water near the critical regime are extremely scarce and the usual VIV generalizations may not apply at these high Reynolds numbers. In particular, the experimental results of Figure 9 do not depict the standard *bell* response typical of many experiments at lower Reynolds numbers [20]. The numerical results are seen to support this behavior and offer reasonable agreement with the experimental data.

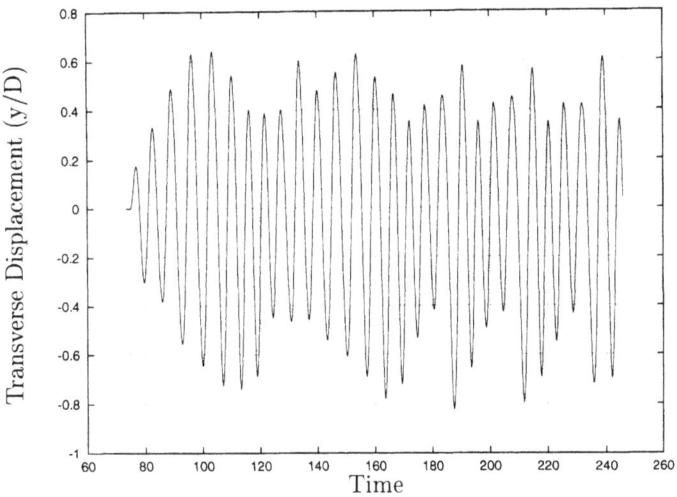

Figure 8: Transverse displacement history of an elastically mounted cylinder undergoing VIV (High Reynolds Number Case). The displacement pattern for this case differs significantly from the low Reynolds example of Figure 5. $U_{red} = 7.02$, $Re = 3.25 \times 10^5$.

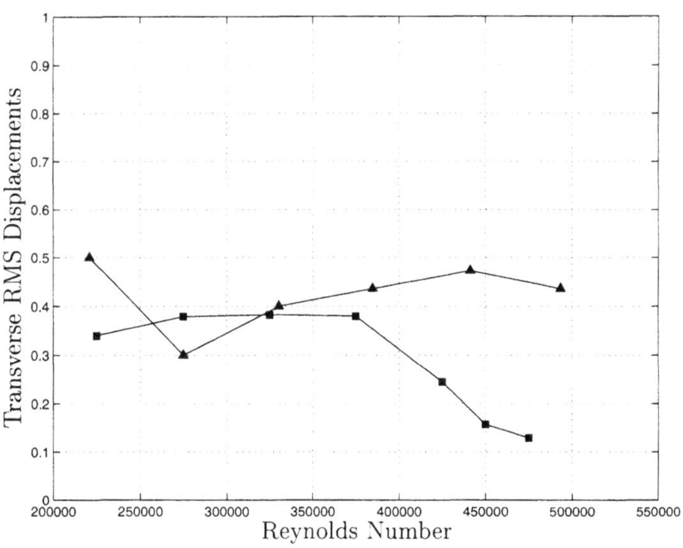

Figure 9: Transverse RMS displacements of an elastically mounted cylinder throughout the lock-in regime, High Reynolds Number Cases (numerical solutions developed *independently)*.

▲ Experimental results from Allen and Henning [21]

■ Present numerical results

3.4 Use of the quasi-3D method for flexing risers

The Hydroelastic method was used to compute the VIV displacements of a long flexible tube subjected to a fixed free-stream current. The physical parameters were taken from an experiment performed with the Marintek Rotating Rig; the conditions are as follows:

- $L/D = 574$
- $D = 0.2m$
- $U_{freestream} = 0.22m/s$ (low) and $0.42m/s$ (medium)
- Total Mass $= 5.3kg$
- Axial Tension $= 712N$
- 8 CFD Planes
- Tube is Pinned at both ends

Figure 10 shows the RMS VIV displacements for the riser subjected to the low and medium speed currents compared with experimental data on the Marintek Rotating Rig. This figure shows a simple 1st-bending mode excitation for the low current speed and a complex multi-mode excitation for the medium current speed.

3.5 The 3-D method applied to risers with strakes and fairings

Not all modeling requirements can be carried out in two dimensions. For example, the simulation of helical strakes using a two-dimensional model is an oversimplification. In reality, suppression devices like helical strakes are effective because they disturb the pattern of vortex-shedding along the spanwise direction of the cylinder – a phenomenon which requires 3D simulations to capture. Consequently, this section presents the simulation of a bare riser geometry compared to a riser with two types of suppression devices attached: strakes and fairings. The motivation for performing these cases was to demonstrate the ability of the method to predict the substantial VIV mitigation which is observed experimentally when attaching helical strakes. For example, experimental results with helical strakes have observed reductions in VIV response by 70-90% of the bare cylinder response [22].

The three geometries modeled all had an aspect ratio of $L/D = 12$ which corresponds to one full wrap of the three helical strakes. Examples of each geometry are shown on the left side of Figure 11. The simulations were carried out at a Reynolds number of $Re = 1.27 \times 10^4$ and a reduced velocity of $U_{red} = 6.5$, which corresponds to the middle of the lock-in regime. Note that each structure was allowed to move in the transverse direction only. The resulting structural displacements for each of the three geometries are shown on the right side of Figure 11. The bare cylinder without any suppression devices experienced peak VIV displacements of approximately 50% of the cylinder diameter. In contrast, the fairing geometry experienced displacements of less than 20% of the diameter while the helical strake geometry experienced displacements of less than 3% of the cylinder diameter.

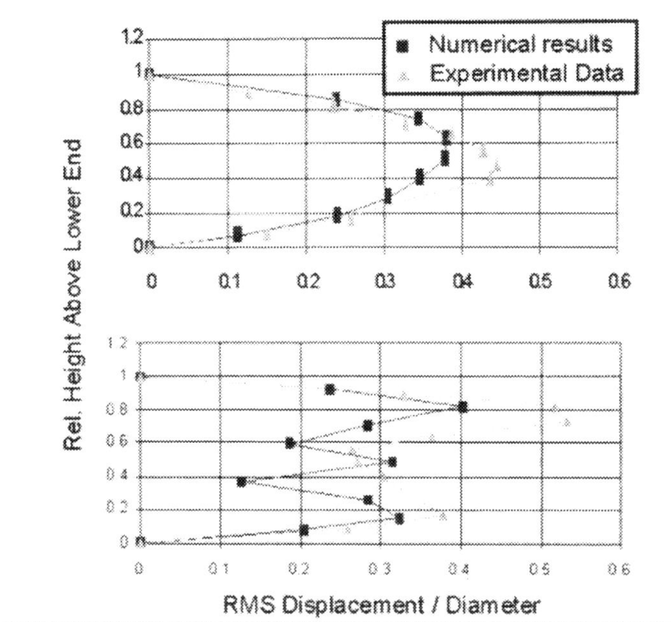

Figure 10: RMS VIV displacement for a flexible riser subjected to low and medium speed currents. Comparison between the present numerical results and the experimental data from the Marintek Rotating Rig

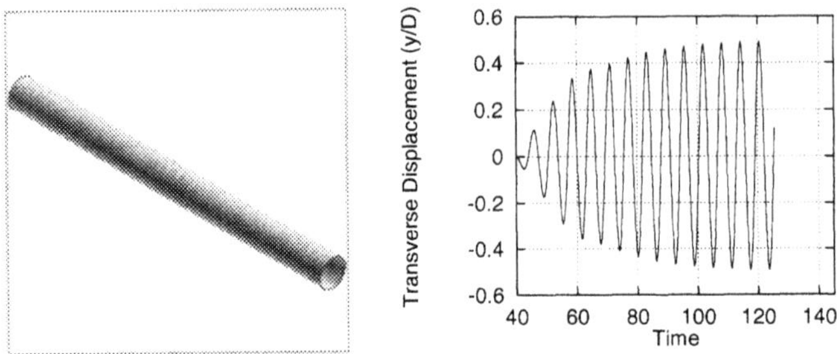

(a) Bare Cylinder

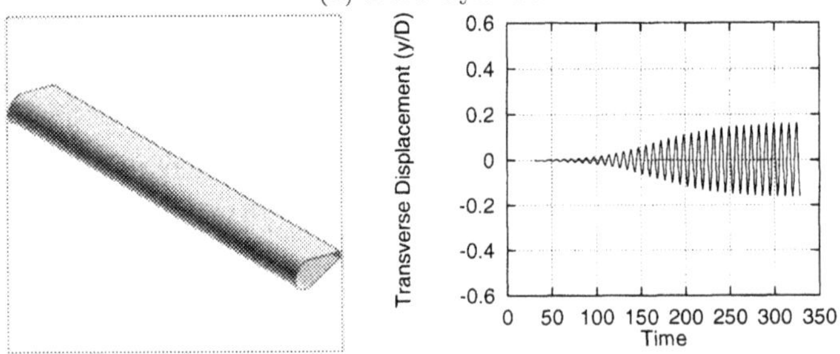

eamlined Fairing

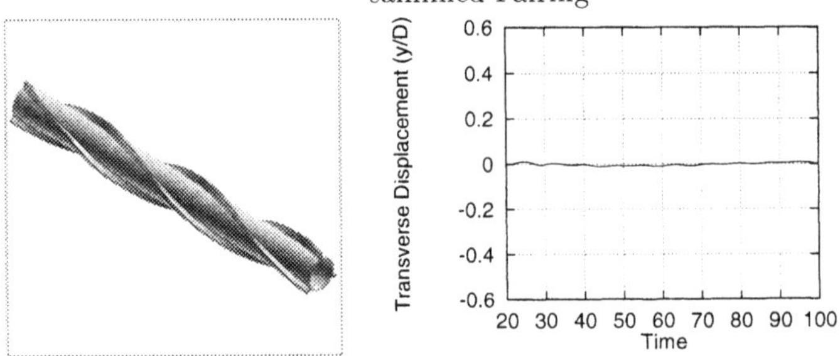

(c) Helical Strake

Figure 11: Geometries and VIV displacement histories for bare cylinder, streamlined fairing, and helical strakes with $L/D = 12$, $U_{red} = 6.5$, $Re = 1.27 \times 10^4$.

Acknowledgements

This work was supported in part by a Joint Industry Project in cooperation with BP-Amoco, Chevron, Deep Oil Technology, Exxon-Mobil, Global Marine, MARIN Netherlands, Shell, Statoil, and UNOCAL. Additional funding was obtained from the Offshore Technology Research Center (OTRC) and the Texas Advanced Technology Program (1999).

References

[1] C. Larsen and K. Halse. "Comparison of Models for Vortex Induced Vibrations of Marine Risers and Cables". In *Final Report of the Workshop on Vortex-Induced Vibrations of Marine Risers and Cables*, Trondheim, Norway, 1994.

[2] K. W. Schulz and Y. Kallinderis. "Unsteady Flow Structure Interaction for Incompressible Flows Using Deformable Hybrid Grids". *Journal of Computational Physics*, 143:569–597, 1998.

[3] T. S. Meling. "Numerical Prediction of the Response of a Vortex-Excited Cylinder at High Reynolds Numbers". In *Proceedings of the International OMAE Symposium*, Lisbon, Portugal, 1998.

[4] J. Dalheim. "An ALE Finite Element Method for Interaction of a Fluid and a 2D Flexible cylinder". In *ECCOMAS*, 1996.

[5] R. W. Yeung and M. Vaidhyanathan. "Flow Past Oscillating Cylinders". *Journal of Offshore Mechanics and Arctic Engineering*, 115(4):197–205, 1993.

[6] K. Herfjord, C. Larsen, G. Furnes, T. Holmås, and K. Randa. Fsi-simulations of vortex induced vibration of offshore structures. In *International Symposium Computational Methods for Fluid-Structure Interaction, '99*, 1999.

[7] Y. Kallinderis, A. Khawaja, and H. McMorris. "Hybrid Prismatic/Tetrahedral Grid Generation for Viscous Flows Around Complex Geometries". *AIAA Journal*, 34(2), 1996.

[8] P. R. Spalart and S. R. Allmaras. "A One-Equation Turbulence Model for Aerodynamic Flows". *AIAA Paper 92-0439-CP*, 1992.

[9] R. Craig. *Structural Dynamics*. Wiley, New York, 1981.

[10] J. Vandiver. "Dimensionless Parameters Important to the Prediction of Vortex-Induced Vibration of Long, Flexible Cylinders in Ocean Currents". *Journal of Fluids and Structures*, 7:423–455, 1993.

[11] E. Achenback and E. Heinecke. "On Vortex Shedding from Smooth and Rough Cylinders in the Range of Reynolds Numbers 6×10^3 to 5×10^6". *Journal of Fluid Mechanics*, 109:239–251, 1981.

[12] M. M. Zdravkovich. Flow induced oscillations of two interfering circular cylinders. In *International Conference on Flow Induced Vibrations in Fluid Engineering*, number D2, 1982.

[13] Shoi-Sheng Chen. *Flow-Induced Vibration of Circular Cylindrical Structures.* Hemisphere Publishing Company, 1987.

[14] H.J. Kim and P.A. Durbin. Investigation of the flow between a pair of circular cylinders in the flopping regime. *Journal of Fluid Mechanics*, 196:431–448, 1988.

[15] K. Chang and C. Song. Interactive vortex shedding from a pair of circular cylinders in a transverse arrangement. *International Journal for Numerical Methods in Fluids*, 11:317–329, 1990.

[16] P. Anagnostopoulos and P. W. Bearman. "Response Characteristics of a Vortex-Excited Cylinder at Low Reynolds Numbers". *Journal of Fluids and Structures*, 6:39–50, 1992.

[17] C. C. Feng. "The Measurement of Vortex-Induced Effects in Flow Past Stationary and Oscillating Circular and D-Section Cylinders". Master's thesis, University of British Columbia, Vancouver, 1968.

[18] O. M. Griffin and S. E. Ramberg. "Some Recent Studies of Vortex Shedding With Application to Marine Tubulars and Risers". *ASME Journal of Energy Resources Technology*, 204:2–13, 1982.

[19] G. Moe, K. Holden, and P. Yttervoll. "Motion of Spring Supported Cylinders in Subcritical and Critical Water Flows". In *Proceedings of the Fourth International Offshore and Polar Engineering Conference*, pages 468–475, 1994.

[20] T. Sarpkaya and M. Isaacson. *Mechanics of Wave Forces on Offshore Structures.* Van Nostrand Reinhold Company, 1981.

[21] D. W. Allen and D. L. Henning. "Vortex-Induced Vibration Tests of a Flexible Smooth Cylinder at Supercritical Reynolds Numbers". In *Proceedings of the 1997 ISOPE Conference*, volume III, pages 680–685, Honolulu, 1997.

[22] R. Blevins. *Flow-Induced Vibration.* Krieger Publishing Company, 1994.

Parallel Computational Fluid Dynamics – Trends and Applications
C.B. Jenssen et al. (Editors)

Dynamics controlled by magnetic fields: parallel astrophysical computations

R. Keppens

keppens@rijnh.nl, FOM-Institute for Plasma Physics 'Rijnhuizen',
P.O. Box 1207, 3430 BE Nieuwegein, The Netherlands*

I present the development history of and the obtained flexibility within the Versatile Advection Code [VAC, see http://www.phys.uu.nl/~toth], a software package for solving sets of – near – conservation laws in any dimensionality. Using a finite volume discretization on a static structured grid, the versatility resides in the choice of applications (Euler, Navier-Stokes, Magneto-Hydro-Dynamics), geometry, computer platform, shock-capturing spatial and temporal discretizations. For distributed memory execution, the VAC source code is automatically preprocessed to High Performance Fortran which results in fully scalable applications.

In its most challenging configuration, the code is used for simulating magnetized plasma dynamics in astrophysically relevant, three-dimensional geometries. The magnetic field can severely complicate plasma flow behaviour both in shock-dominated and otherwise (un)steady regimes. We present examples of (i) 'kinking' magnetic flux tubes, related to emerging sunspots on the solar surface; (ii) Kelvin-Helmholtz unstable magnetized jet flows, present at all scales in our universe; and (iii) coronal mass ejections in the continuously expanding solar magnetic environment. All these simulations would benefit greatly from dynamically controlled grid adaptivity, a research topic we are currently investigating.

1. VERSATILE ADVECTION CODE

1.1. Motivation: Astro-Plasma-Physics

Highly ionized plasma dynamics is encountered in virtually all astrophysical phenomena. In many of these, magnetic fields play a crucial role. Perhaps the best known examples are found throughout the solar atmosphere, where magnetic activity is at play in sunspots on the solar surface, in the intricate loop-like structuring of the solar corona, and throughout the entire solar heliosphere where the solar wind interacts with planetary magnetospheres. Particularly violent events are the coronal mass ejections, when

*Work done within the association agreement of Euratom and the 'Stichting voor Fundamenteel Onderzoek der Materie' (FOM) with financial support from the 'Nederlandse Organisatie voor Wetenschappelijk Onderzoek' (NWO) and Euratom. It is part of a project on 'Parallel Computational Magneto-Fluid Dynamics', an NWO Priority Program on Massively Parallel Computing. Use of computing facilities is funded by 'Nationale Computer Faciliteiten' (NCF). I thank G. Tóth, J.P. Goedbloed, A.J.C. Beliën, R.B. Dahlburg, P.A. Zegeling, M. Nool for fruitful collaborations.

destabilized magnetic field complexes erupt and lead to global disturbances of the solar environment. From a plasma physical viewpoint, these processes involve a wealth of linear and nonlinear wave dynamics, with bow shocks, current sheets, flux tubes, jet-like flows, as reoccuring basic concepts. Computational Magneto-Fluid Dynamics (CMFD) is rapidly evolving into the prime means to advance our understanding of such complex magnetized plasma behavior. The Versatile Advection Code was developed in a collaborative effort on parallel CMFD.

1.2. Philosophy and Discretizations

The Versatile Advection Code (VAC) was designed from the start [1] to solve systems of conservation laws in 1, 2, or 3-dimensional settings. A modular structuring of the code groups code into segments which are specific to I/O, to e.g. hydrodynamic (HD) or magnetohydrodynamic (MHD) equations, or to purely grid or discretization related matters. The code employs a finite volume discretization on a structured grid, and the dimensionality is unspecified in the source code through the use of the Loop Annotation SYntax (LASY) [2]. Using Perl scripts, the LASY-augmented Fortran 90 code can be preprocessed to Fortran 77, Fortran 90, or High Performance Fortran, making VAC easily portable between all UNIX-based platforms [3]. The code is available via registration on `http://www.phys.uu.nl/~toth`, where regularly updated publication lists document the progress. An extensive user manual, complete with examples of VAC usage, can also be found there.

Shock-capturing discretizations implemented [4] include variants of the Flux Corrected Transport type [5] and several Total Variation Diminishing (TVD) schemes [6]. The MHD simulations reported below make use of the one step TVD scheme with a Roe-type approximate Riemann solver [7], or of the robust and simple predictor-corrector TVD-Lax Friedrich (TVDLF) method [8] which just uses the fastest wave speed in the numerical flux definition. All problems advance the MHD equations explicitly in time, while there are options provided in VAC for problem-specific choices of semi-implicit and fully implicit time integration strategies [9,10]. The finite volume discretization with cell-centered variables lends itself naturally to the specification of boundary conditions through the use of ghost cells.

1.3. The Magnetohydrodynamic Module

For simulating magnetically controlled plasma flow, one selects the set of partial differential equations corresponding to one-fluid MHD:

$$\frac{\partial \rho}{\partial t} + \nabla \cdot (\mathbf{v}\rho) = S_\rho \tag{1}$$

$$\frac{\partial \rho \mathbf{v}}{\partial t} + \nabla \cdot (\mathbf{v}\rho\mathbf{v} - \mathbf{B}\mathbf{B}) + \nabla(p + \mathbf{B}^2/2) = \vec{S}_{\rho v} \tag{2}$$

$$\frac{\partial e}{\partial t} + \nabla \cdot \left(\mathbf{v}e + \mathbf{v}(p + \mathbf{B}^2/2) - \mathbf{B}\mathbf{B}\cdot\mathbf{v}\right) = S_e \tag{3}$$

$$\frac{\partial \mathbf{B}}{\partial t} + \nabla \cdot (\mathbf{v}\mathbf{B} - \mathbf{B}\mathbf{v}) = \vec{S}_B, \tag{4}$$

so that in 3D problems, the eight unknowns represent the density ρ, the components of the momentum vector $\rho\mathbf{v}$, the total energy density e, and the vector magnetic field $\mathbf{B}$.

Sources and sinks are collected in the right-hand-side terms, representing e.g. Ohmic losses through resistivity. These terms vanish for ideal MHD problems. The primitive pressure variable p is related to the energy density through $p = (\gamma - 1)(e - \rho \mathbf{v}^2/2 - \mathbf{B}^2/2)$ with adiabatic index γ. In the examples that follow, we take $\gamma = 5/3$ if not stated otherwise. Units of $\mathbf{B}$ are chosen such that the magnetic permeability is unity and this field must further obey the law $\nabla \cdot \mathbf{B} = 0$. This latter property poses a non-trivial complication for modern shock-capturing schemes, and a recent, thorough evaluation of seven different ways of ensuring it in multi-dimensional MHD simulations with VAC is found in [11]. We typically took a projection scheme approach [12] or followed Powell's eight-wave formulation [13].

2. COMPUTATIONAL MAGNETO-FLUID DYNAMICS

At first sight, the ideal MHD equations provide a straightforward extension of the Euler system: a Lorentz force appears as an additional means to transfer momentum, and the extra vector variable $\mathbf{B}$ obeys yet another conservation law. However, the dynamical influence of the magnetic field $\mathbf{B}$, its interwoven character with the flow field $\mathbf{v}$ through the induction equation, and the topological constraint arising from $\nabla \cdot \mathbf{B} = 0$, can lead to significant alterations of simplified flow problems. Physically, these complications are due to the presence of 7 wave speeds, instead of the 3 speeds $[v - c_s, v, v + c_s]$ dealt with in compressible hydrodynamic cases with sound speed c_s. In MHD, entropy disturbances still travel with the flow speed v, but there are now three pairs of co- and counter travelling slow magnetosonic, magnetic Alfvén, and fast magnetosonic wave signals. These waves are anisotropic: localized perturbations influence directions along and across the magnetic field differently.

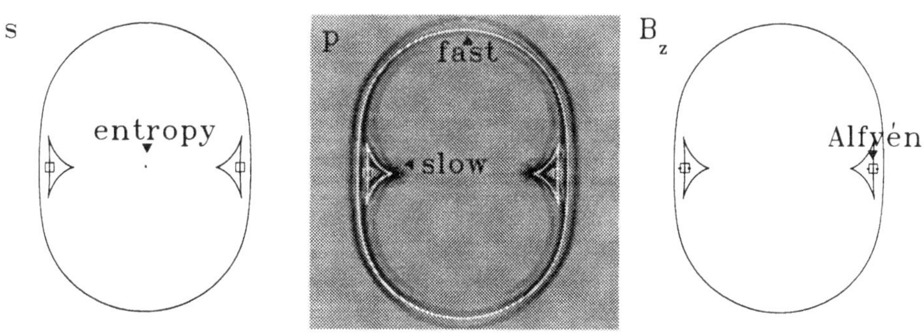

Figure 1. A 2.5D MHD simulation of a localized p, ρ, and v_z perturbation in a static, homogeneous plasma generates numerically the – overplotted – theoretical Friedrichs diagram. The background state has a horizontal magnetic field and is characterized by the ratio $c_s/v_A = 1.11$. Plotted are the entropy $s = p\rho^{-\gamma}$ [left], the thermal pressure p [middle], and the transverse component of the magnetic field B_z [right], at a finite time after the imposed perturbation. The different wave signals are labeled according to type. Note the extreme anisotropy between field-aligned and cross-field directions.

Figure 1 shows a VAC simulation result of impulsively generated MHD waves in a static, homogeneous, magnetized plasma. Taking $\rho = 1$, $p = 0.6$ and $\mathbf{B} = 0.9\hat{\mathbf{e}}_x$, a localized 10 % perturbation in density and pressure, together with a co-spatial velocity pulse $\mathbf{v} = 0.01\hat{\mathbf{e}}_z$ perpendicular to the simulated (x, y)-domain, triggers all MHD wave modes. In particular, the almost pointwise velocity pulse results in two transverse, similarly pointwise Alfvén waves that are only seen in the transverse B_z and v_z components. They are purely magnetically driven and travel only along the (horizontal) fieldlines at the characteristic Alfvén speed $v_A = B/\sqrt{\rho}$. For the time shown, these Alfvén signals are at the location of the squares in the diagram. The pressure field contains slow and fast wave signals, with anisotropic wavefronts. For a pure pointlike perturbation, the wavefronts are given by the theoretical Friedrichs diagram, which is overplotted in each frame. Note how the polarization of the individual wave signals are fully obeyed: there is no entropy or transverse B_z variation associated with the latter two waves. An entropy discontinuity is only seen at the origin, where the pulse was imposed in the static medium.

2.1. Steady and unsteady shock-dominated simulations

Ultimately, the different linear MHD wave modes can lead to complicated shock patterns in the ensuing nonlinear dynamics. Different shock types are distinguished when upstream and downstream states connect *super-* to *sub-* fast, Alfvén, and slow flow regimes. A *fast magnetosonic shock* keeps the flow super-Alfvénic but goes from super-fast upstream to sub-fast downstream. An *intermediate shock* occurs when the Alfvén speed is crossed, while a *slow shock* involves a jump across the slow magnetosonic speed. Depending on the relative importance of magnetic, thermal and kinetic energies, as e.g. characterized by the plasma beta $\beta = 2p/B^2$ and the Alfvén Mach number $M_A = v/v_A$, one distinguishes between magnetically dominated, pressure-dominated, or flow-dominated scenarios.

An intrinsic magnetically dominated effect was recently investigated by De Sterck *et al.* [14]. They analysed up-down symmetric bow shocks occuring when 2D, field-aligned, super-fast flow impinges on a perfectly conducting cylindrical obstacle. For upstream states in the *switch-on* regime, with $\beta < 2/\gamma$ and

$$1 < M_A < \sqrt{\frac{\gamma(1-\beta)+1}{\gamma-1}},$$

the bow shock contains several shock segments of different type. The shock front becomes concave-outward about the symmetry stagnation line, in sharp contrast with the more familiar case from hydrodynamic simulations, where a single concave-inward bow shock forms. This can have implications for magnetospheric bow shocks and for shock structures induced by coronal mass ejections. Figure 2 shows the magnetic field lines, together with the density pattern in a greyscale contour plot, for magnetically dominated upstream conditions as in [14], namely $\beta = 0.4$ and $M_A = 1.5$. The calculation was done with VAC on an elliptical, stretched grid of size 120^2. The figure zooms in on the interacting shock segments in front of the obstacle. For an extensive analysis of the elliptic and hyperbolic flow regions present, we refer to [15]. Two-dimensional, pressure-dominated, single MHD bow shock flows, as well as the **B**-dominated, multiple-shock scenarios from Figure 2, served to demonstrate in [9,10] the obtainable computational efficiency when using a fully implicit time integration scheme, instead of explicit time marching towards steady-state. These implicit calculations used Courant numbers of 100 up to 1000.

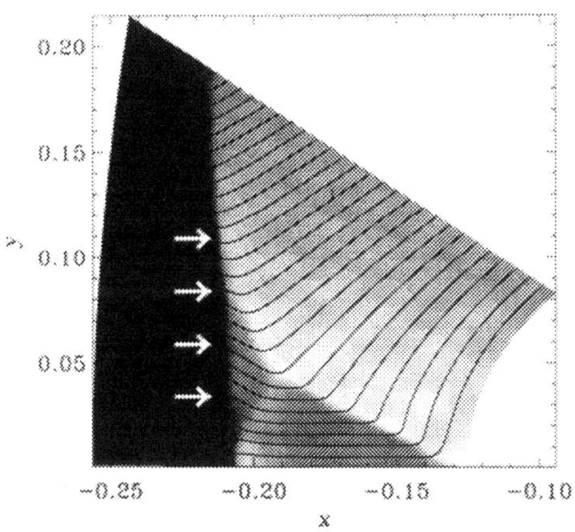

Figure 2. A magnetically dominated super-fast flow (arrows) upstream of a perfectly conducting cylindrical obstacle (bottom right) results in a steady bow shock consisting of multiple interacting shock segments. We show the density in greyscale and the magnetic fieldlines.

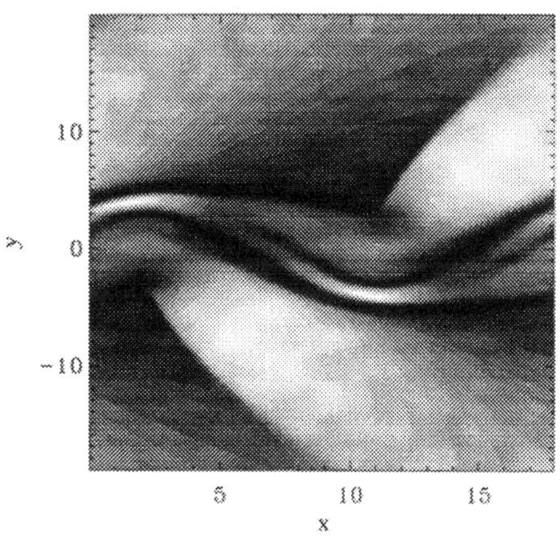

Figure 3. An unsteady magnetized wake can develop significant density variations due to sinuous instabilities. This snapshot of the density structure in the evolution of a supersonic, super-Alfvénic wake clearly reveals the presence of fast magnetosonic shocks above and below the wake.

Figure 3 contrasts the steady, magnetically dominated case from Figure 2 with a snapshot of an unsteady, flow-dominated simulation. Shown is the density structure in a section of an evolving magnetized wake [16], which must be periodically repeated along the streamwise x-direction. Originally, the wake configuration was given by a planar velocity profile $\mathbf{v} = [1 - \mathrm{sech}(y)]\hat{\mathbf{e}}_x$ which models a low speed 'wake' sandwiched between high speed flow regimes. A 3D, sheared magnetic field of constant magnitude pervades the wake, with components given by $B_x = A\,\tanh(y),\;\; B_z = A\,\mathrm{sech}(y)$. The field rotates about the y-axis from parallel alignment with the flow in the upper $y \gg 0$ regions, becomes perpendicular to the flow plane at $y = 0$, and is antiparallel to $\mathbf{v}$ in $y \ll 0$. We start with uniform density and pressure throughout – taking $\rho = 1$ and $p = 1/\gamma M^2$ – and consider a supersonic $M = 3$, super-Alfvénic $M_A = 1/A = 5$ case. A linear stability analysis – as in [17] – of this shear-flow, sheared-**B** system reveals that it is unstable to kink-like, transverse (y-direction) displacements. At the time shown, this sinuous instability has, after a period of exponential growth, led to the formation of rightwardly traveling, fast magnetosonic shocks both above and below the transversely swaying wake. The grid employed is a 200×200 vertically stretched Cartesian mesh. With a plasma beta $\beta = 3.33$, this configuration could serve to mimick conditions far out in the heliospheric current sheet, where the slow solar wind is found in between fast wind regions. It will be of interest to consider 3D scenarios where the background 'wake' flow accelerates from sub- to super-Alfvénic wind. In what follows, we give examples of 3D MHD studies which highlight the role of the magnetic field in fully nonlinear plasma dynamics.

2.2. Knotting a flux tube

This first example illustrates that the magnetic Lorentz force contains two potentially opposing contributions: it can be decomposed in an isotropic magnetic pressure $-\nabla\frac{B^2}{2}$ and a tension part $(\mathbf{B}\cdot\nabla)\mathbf{B}$ along the field lines. As a result, magnetized plasma gets accelerated from high to low magnetic field strength regions and curved field lines have a tendency towards straigthening. When one then considers an isolated magnetic flux tube with *twisted* field lines, the axial **B**-component tries to prevent helical deformations through tensional forces, while the azimuthal component B_φ acts destabilizing through the magnetic pressure. Therefore, a flux tube is *kink-unstable* when a critical twist $q = B_\varphi/rB_z > q_{crit}$ is exceeded.

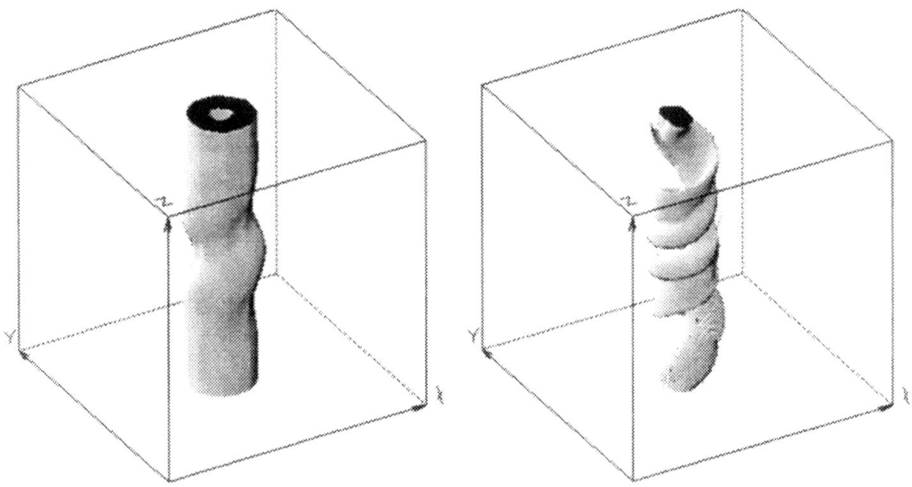

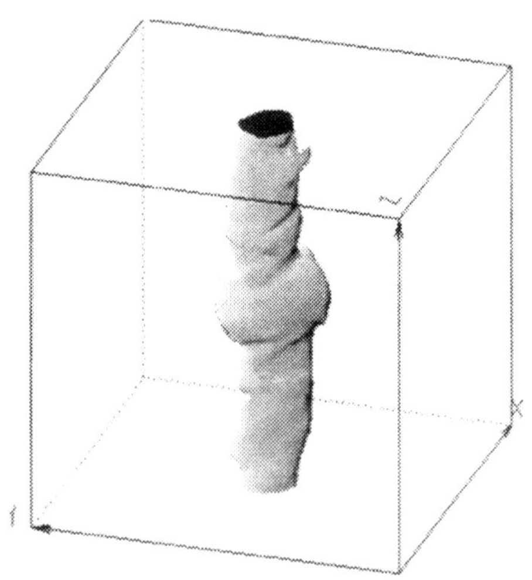

Figure 4. The evolution of a kink unstable flux tube as seen in isosurfaces of the magnetic field strength. Top row: a helical deformation self-amplifies. Kink instabilities of varying axial wavelength interact. Shown are isosurfaces of the instantaneous $\mid \mathbf{B} \mid_{max}/2$ at times $t = 2.5$ and $t = 3.5$. At right: constructive interference leads to the formation of a central 'knot', shown at time $t = 10$ as the isosurface $\mid \mathbf{B} \mid_{max}/3$.

Linton *et al.* [18] performed detailed numerical studies of such kink-unstable flux tubes using a 3D spectral code. The aim was to assess whether the complex magnetic field structure observed in so-called δ-sunspots could be related to this fundamentally magnetic instability. With VAC, we could recover their most striking result, namely that the nonlinear, constructive interference of several unstable, linear kink modes can lead to the formation of a concentrated 'knot'. The simulation considers a flux tube with an aspect ratio given by $R/L = 1/8$ and of constant normalized twist $\bar{q} \equiv qR = 5.98$ embedded in unmagnetized, uniform surroundings. Setting the axial field component to $B_z = (1 - r^2/R^2)^{1/4}$, the density to $\rho = 1$, and the pressure profile within the tube according to

$$p(r < R) = \frac{1}{2}\left(\beta_o + 1 - \frac{2\bar{q}^2}{3}\right) - \sqrt{1 - \frac{r^2}{R^2}}\left(\frac{1}{2} - \frac{\bar{q}^2}{3} + \frac{5\bar{q}^2}{6}\frac{r^2}{R^2}\right),$$

one can impose the value for the axial plasma beta β_o. Taking $\beta_o = 600$, it was argued that the physical conditions for flux tubes stored at the base of the solar convection zone were closely met. As the twist significantly exceeds the critical twist of $q_{crit}R = 0.5$, the tube is unstable to kink instabilities of varying axial wavelength. Instead of perturbing the tube with the exact linear kink eigenfunctions as done in [18], our simulation simply induced multi-mode kinks by applying a purely radial velocity perturbation given by

$$v_r = \delta v_r e^{-\left(\frac{r/R-1}{\sigma}\right)^2} \sum_{n=1}^{n=7} \frac{L/R}{2\pi n} \cos\left(\varphi - 2\pi n \frac{z}{L}\right).$$

We used a 100^3 Cartesian grid and took $\sigma = 0.2$ and $\delta v_r = 0.01$. Figure 4 shows isosurfaces of magnetic field strength at 2.5, 3.5, and 10 units of the Alfvén crossing time R/v_A. The first two frames clearly show the initial helical deformation and the appearance and interaction of several 'kinks'. The final frame depicts the 'knotted' structure that develops. When analyzing the magnetic field structure throughout this concentrated kink, Linton *et al.* found a striking similarity with δ-sunspot fields. Using the flexibility of VAC for adding gravitational source terms and switching from triple-periodic (as done for the spectral calculations) to boundary conditions consistent with a background stratification, it is the intention to simulate the development of these kinking flux tubes *as they rise through the solar convection zone.* This will make an explicit connection between kink-unstable flux tubes in the solar interior and complex δ-spot emergence observed on the solar surface.

2.3. Disrupting jets

Our second example serves to show that even in those circumstances where the magnetic field is weak initially and can thus safely be ignored for predicting the *linear stability* of stationary plasma equilibria, the *nonlinear evolution* can become magnetically controlled. This was demonstrated in [19], where Kelvin-Helmholtz unstable jets were investigated. The Kelvin-Helmholtz instability – well-known from wind-induced waves on the surface of a pond – is in essence a hydrodynamical instability occuring at the interface between fluids that move at different speeds. Considering a cylindrical jet flow with

$$\mathbf{v} = V_o \tanh \frac{r - R_{jet}}{0.1 R_{jet}} \hat{\mathbf{e}}_x,$$

in a uniform magnetic field $\mathbf{B} = B_o \hat{\mathbf{e}}_x$, 3D MHD simulations allow to address the role of compressibility and magnetic forces when the shear flow about $r = R_{jet}$ is Kelvin-Helmholtz destabilized. Keppens & Tóth [19] analyzed cases where the characterizing dimensionless parameters typified a pressure and flow-dominated situation with $\beta = 120$, $M_A = 5$, for a subsonic $M = 0.5$ jet with aspect ratio $R_{jet}/L = 1/2$. An analytic treatment of the quasi-linear regime neglected the influence of magnetic fields altogether and could correctly predict the first-excited couplings between linear (m, n) modes with variations $\cos(m\varphi)\sin(n2\pi z/L)$. However, in the further evolution, it was demonstrated that the magnetic field becomes locally amplified and influences the jet deformation dynamically.

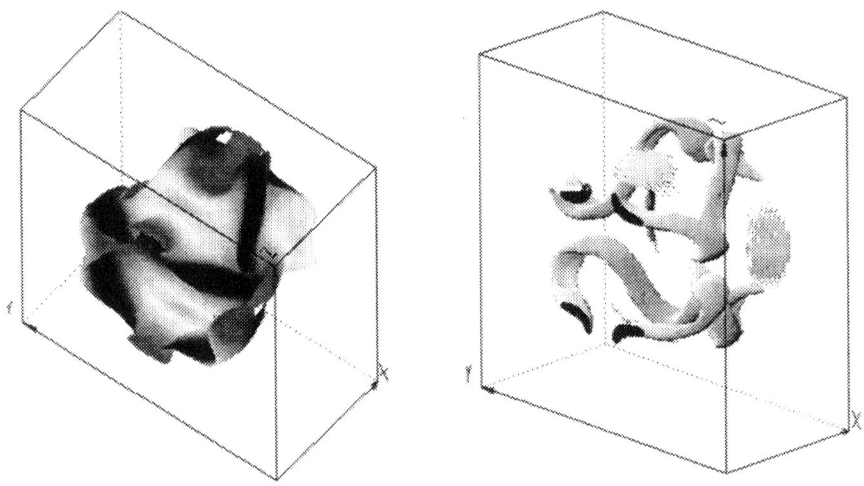

Figure 5. A Kelvin-Helmholtz unstable cylindrical jet deforms in a manner which is additionally regulated by locally amplified magnetic fields. Well within the nonlinear evolution, we show the jet surface shaded by thermal pressure at left. Note the fibril strands of dark, low pressure zones. These are cospatial with concentrated magnetic fields: the right panel shows isosurfaces where $\log_{10}(B_r^2 + B_\varphi^2) = -1.36$.

Figure 5 is generated by a 3D $50 \times 100 \times 100$ simulation of an unstable, magnetized jet. It shows the jet surface where v_x vanishes, shaded by the thermal pressure variation. Instead of an induced $(m, n) = (1, 1)$ excitation as was used for Figure 3 in [19], we perturbed the jet with a $(2, 1)$ radial velocity. The time depicted corresponds to four times the isothermal sound crossing time $L/\sqrt{p/\rho}$ within which the pressure has changed from uniform throughout to become $m = 4$ modulated about the jet circumference. Four low pressure, dark strands can be distinguished and they coincide with loci of amplified magnetic fields. This is demonstrated in the right panel showing an isosurface of strong poloidal fields. Besides four fibril-like concentrations of the magnetic field, one can distinguish four sheet-type structures which intersect high-pressure, high-density compression zones. These latter structures are reminiscent of what had previously been studied in initially uniformly magnetized, 2D shear flows [20]. Locally, Lorentz forces can no longer be ignored and the **B** field plays a prominent role in the further shaping and breakup of the jet flow. Future work can address the role of magnetic shear cospatial with the jet surface, as was done in two-dimensional resistive MHD simulations where reconnection events were triggered [20].

2.4. Solar mass ejections

Our final example of magnetized plasma dynamics relates to the global magnetic reconfigurations of the solar heliosphere associated with coronal mass ejections. These almost daily, violent mass loss events can potentially influence the earth's magnetosphere, making their study of central importance for *space weather* predictions. They occur within the continuously expanding solar corona, combining the intricacies of stationary transonic solar wind modeling [21,22] with highly time-varying dynamics.

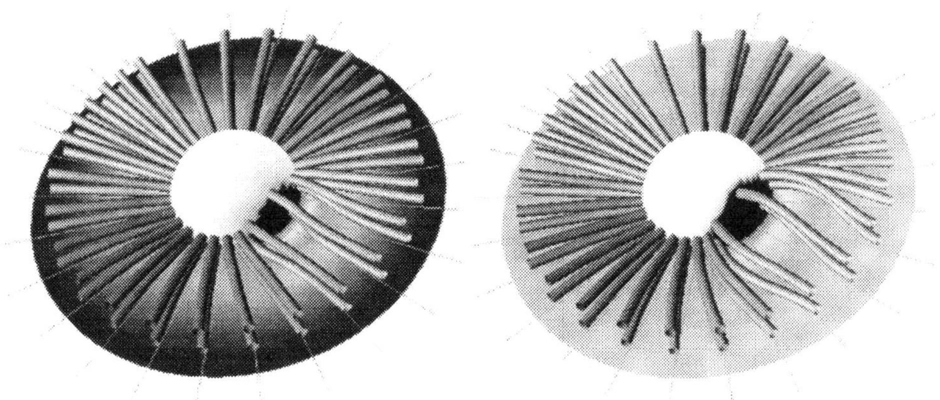

Figure 6. An analytic, 3D time-dependent MHD solution for a self-similarly expanding coronal mass ejection in the magnetized solar wind is recovered numerically. Looking down on the pole, the solar surface is shaded by the magnitude of the magnetic field: the black area marks the protrusion of the expanding bubble. Several field lines are drawn as tubular structures. The equatorial plane is shaded according to the density stratification. Two consecutive times show the CME expansion.

As a stepping stone towards ambitious realistic modeling, we first set forth to recover numerically an analytical, 3D time-dependent MHD solution by Gibson & Low [23] mimicking a coronal mass ejection. The formidable task of getting a closed form solution to the full set of MHD equations was achieved in [23] through a sequence of mathematical transformations. The first one involved getting rid of the explicit time dependence through a self-similar transformation to the variable $\zeta \propto r/t$ and assuming a purely radial outflow $\mathbf{v} = (r/t)\hat{\mathbf{e}}_r$. Starting from spherical coordinates (r, θ, φ), this reduces the set of MHD equations to a 3D 'magnetostatic' problem in (ζ, θ, φ) variables. The second realization from Gibson & Low was that the geometric stretching procedure $r \rightarrow r - a$ can relate different magnetostatic solutions in a straightforward manner. It also contracts an off-origin sphere σ into a tear-shaped 'bubble'. Hence, the combined transformations would result in an analytic solution for a self-similarly expanding bubble-shaped coronal mass ejection in the solar wind, when one can solve for a magnetostatic atmosphere both inside and outside an off-center sphere. The published model combined a potential $\mathbf{B}$ solution (with $\nabla \times \mathbf{B} = 0$) outside the sphere σ with an axisymmetric 'spheromak' $\mathbf{B}$ field configuration inside σ. The geometrical stretching breaks the axisymmetry of the internal field, resulting in a truly 3D time-dependent solution. The only extra complication is to perform a matching procedure across the deformed bubble boundary that ensures the pressure balance there.

Figure 6 shows two snapshots of a fully 3D time-dependent MHD simulation on a 50^3 spherical grid which recovers the self-similar expansion known from the explicit analytic solution. Since the initial condition considered a hydrostatic solution interior to the contracted sphere σ, with the magnetized exterior constructed as in [23], this particular evolution is axisymmetric about the axis connecting the origin and the σ-centre. This preserved symmetry serves as an extra check on the obtained numerical solution. Note how the hydrostatic bubble expands as it rises through the solar surface, pushing aside the magnetic field lines in the stratified surroundings. It is our intention to recover the 3D magnetized bubble evolution as well and explore meaningful variations that are no longer tractable analytically. These could be the inclusion of solar rotation, changes in the polytropic index ($\gamma = 4/3$ for the analytic solution), or simulating the CME detachment from the coronal base. All these effects would make the expansion no longer self-similar in r/t and bring the model closer to realistic scenarios.

3. OUTLOOK

The continued development and application of the Versatile Advection Code has already brought several new insights in algorithmic as well as specific plasma physical and astrophysical issues. Computational science achievements include: (1) the possibility for doing fair comparisons between different spatial and temporal discretizations adopted for solving the same multi-dimensional HD or MHD problems [4,9–11]; (2) the invention of the original Loop Annotation SYntax for writing a CFD code independent of the dimensionality [2]; (3) the development and succesful demonstration of new numerical algorithms [11]; (4) the obtained linear speedup on shared and distributed memory parallel platforms [3]. Besides the applications already discussed, physics driven achievements comprise among others: a new assesment of the efficiency of footpoint driven, thermally stratified solar coronal loops [24]; the simulation of bi-directional jet flows precursed and bounded by different MHD shock fronts comparing favorably with several aspects known from solar microflare observations [25]; getting accurate stationary transonic stellar wind solutions for varying magnetic field strength and topology, rotation rate, etc. [22].

In the examples above, we indicated several directions for future research. A lot can be learned from simplified, but at the same time sophisticated, model problems. Ultimately, realistic models of e.g. the solar heliosphere must combine all knowledge collected from

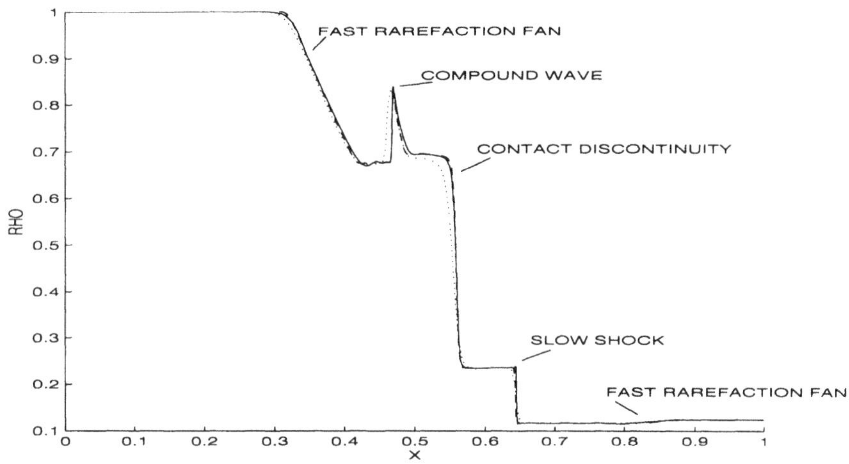

Figure 7. The density at time $t = 0.1$ for the Brio-Wu shock tube problem. Out of the initial two-state discontinuity, five nonlinear flow features develop which are named according to type. We compare a moving grid solution exploiting 250 grid points (solid line) with a solution from VAC on a static 250 grid (dotted) and on a high resolution 1000 point grid (dashed line).

idealized studies as described above: coronal loops act as waveguides for linear MHD waves, current and/or shear flow driven instabilities may reconfigure the magnetic topology locally, complex MHD shocks can be induced impulsively by coronal mass ejections or they may form as the natural outcome of instabilities associated with the heliospheric current sheet. For achieving a fair amount of realism in the numerical simulations while keeping the computational costs within reasonable limits, it is highly desirable to combine the flexibility of VAC with some means of dynamic grid adaptivity. We are presently exploring two routes [26]: (i) the use of adaptive mesh refinement [27] where hierarchically nested, subsequently finer grid patches are created and removed when needed; and (ii) the use of dynamic-regridding methods where the partial differential equations of the MHD system are coupled to adaptive grid PDEs describing the mesh movements. As an example, we show in Figure 7 a comparison of two static grid VAC solutions exploiting the one step TVD method, with one dynamically-regridded solution of a 1D magnetic shock tube problem. The description is taken from [28] and shown is the resulting density structure which spontaneously develops out of the initial Riemann data. Note how almost all nonlinear flow features appear in this very simple model problem. The adaptive method-of-line solution is clearly more accurate than the static solution employing an equal amount of grid points and achieves a resolution corresponding to a four times finer static grid. High resolution, parallel CMFD simulations, possibly exploiting solution-adaptive meshes, will certainly provide many more exciting insights in magnetically controlled plasma evolution.

REFERENCES

1. G. Tóth, Astrophys. Lett. & Comm. 34 (1996) 245.
2. G. Tóth, J. Comput. Phys. 138 (1997) 981.
3. R. Keppens and G. Tóth, Parallel Computing 26 (2000) 705.
4. G. Tóth and D. Odstrčil, J. Comput. Phys. 128 (1996) 82.
5. J.P. Boris and D.L. Book, J. Comput. Phys. 11 (1973) 38.
6. A. Harten, J. Comput. Phys. 49 (1983) 357.
7. P.L. Roe and D.S. Balsara, SIAM J. Appl. Math. 56 (1996) 57.
8. H.C. Yee, NASA TM-101088 (1989).
9. R. Keppens, G. Tóth, M.A. Botchev, and A. van der Ploeg, Int. J. for Numer. Meth. in Fluids 30 (1999) 335.
10. G. Tóth, R. Keppens, and M.A. Botchev, Astron. & Astrophys. 332 (1998) 1159.
11. G. Tóth, J. Comput. Phys. 161 (2000) 605.
12. J.U. Brackbill and D.C. Barnes, J. Comput. Phys. 35 (1980) 426.
13. K.G. Powell, ICASE Report No 94-24, Langley, VA (1994).
14. H. De Sterck, B.C. Low, and S. Poedts, Phys. of Plasmas 5 (1998) 4015.
15. H. De Sterck, B.C. Low, and S. Poedts, Phys. of Plasmas 6 (1999) 954.
16. R. Keppens, R.B. Dahlburg, and G. Einaudi, Proc. of '27th EPS conference on controlled fusion and plasma physics', Budapest, 12-16 June (2000).
17. R.B. Dahlburg and G. Einaudi, Phys. of Plasmas 7 (2000) 1356.
18. M.G. Linton, G.H. Fisher, R.B. Dahlburg, and Y. Fan, Astrophys. J. 522 (1999) 1190.
19. R. Keppens and G. Tóth, Phys. of Plasmas 6 (1999) 1461.
20. R. Keppens, G. Tóth, R.H.J. Westermann, and J.P. Goedbloed, J. Plasma Phys. 61 (1999) 1.
21. R. Keppens and J.P. Goedbloed, Astron. & Astrophys. 343 (1999) 251.
22. R. Keppens and J.P. Goedbloed, Astrophys. J. 530 (2000) 1036.
23. S.E. Gibson and B.C. Low, Astrophys. J. 493 (1998) 460.
24. A.J.C. Beliën, P.C.H. Martens, and R. Keppens, Astrophys. J. 526 (1999) 478.
25. D.E. Innes and G. Tóth, Solar Phys. 185 (1999) 127.
26. R. Keppens, M. Nool, P.A. Zegeling, and J.P. Goedbloed, Lect. Notes in Comp. Science 1823 (2000) 61.
27. M.J. Berger, SIAM J. Sci. Stat. Comput. 7 (1986) 904.
28. M. Brio and C.C. Wu, J. Comput. Phys. 75 (1988) 400.

Parallel Computational Fluid Dynamics – Trends and Applications
C.B. Jenssen et al. (Editors)

A Software Framework for Easy Parallelization of PDE Solvers

Hans Petter Langtangen and Xing Cai

Department of Informatics, University of Oslo,
P.O. Box 1080, Blindern, N-0316 Oslo, Norway
{hpl,xingca}@ifi.uio.no

We present three approaches to parallelizing partial differential equation solvers. The approaches are based on the philosophy of domain decomposition methods and are implemented in a generic software framework utilizing object-oriented programming techniques. In this framework, an application programmer can parallelize sequential code with minor effort. The efficiency of the proposed framework is evaluated in several case studies. In most cases a speed-up close to the optimal value is achieved, i.e., the overhead of our generic framework is usually negligible.

1. INTRODUCTION

Developing parallel partial differential equation (PDE) solvers is usually a much more time-consuming, frustrating, and challenging task than developing the corresponding sequential solvers. It might be even more difficult to migrate existing PDE software to parallel computing environments. Early domain decomposition (DD) workers had the idea of using the DD methodology to parallelize PDE solvers by writing a simple controlling program on top of the sequential software. However, as pointed out in [8], this strategy is in general unrealistic because most existing PDE solvers are too rigid and inflexible. The present paper describes a successful application of the simple DD idea for parallelization of existing sequential solvers and outlines the corresponding demands to the design of the sequential software and the parallelization framework.

The DD methods consist in dividing the computational domain into a set of subdomains [4,8,9]. We have used a particular type of DD, known as the *overlapping Schwarz method*, due to its simple algorithmic structure, which avoids solution of special interface problems that arise in non-overlapping DD methods. Basically, the overlapping Schwarz method involves solution of the original PDE problem restricted to each subdomain. Since the boundary values on the internal boundaries between the subdomains are unknown, an iterative procedure is invoked where the solution from the previous iteration is used as boundary values. All the subproblems in an iteration can then be solved in parallel.

Whether this type of the DD method converges is an open question. Moreover, the convergence is often slow. To speed up the convergence, one can apply a *coarse grid correction*, resulting in a kind of two-level multigrid method involving fine grids over the subdomains and a coarse grid over the global domain (see [8]). The overlapping Schwarz method can be used either as a stand-alone solver or as a preconditioner (e.g. in combination with a Conjugate Gradient-like method).

The overlapping Schwarz method has a couple of advantages besides being well suited for parallelizing sequential PDE software. Since different numerical methods, and even different mathematical models, can be used in the subdomains, it is easy to introduce special treatment of singularities, complicated geometries and so on. Furthermore, the overlapping Schwarz method is a very efficient method even on sequential computers, often allowing a problem with n degrees of freedom to be solved in $\mathcal{O}(n)$ operations.

The present paper extends the ideas in [1] to three different parallelization approaches implemented in a common software framework within the Diffpack programming environment.

2. AN OBJECT-ORIENTED PARALLELIZATION FRAMEWORK

Our objective is a software framework for parallelization, where an existing sequential PDE solver can be easily extended and modified to run on a parallel platform. The extra programming effort that the user is required to perform should be small, and the resulting parallel PDE solver should have both good numerical efficiency and high parallel efficiency. The briefly described overlapping Schwarz method, which can be used as a stand-alone solver or as a preconditioner, offers in theory an efficient numerical method, at least if the convergence is fast and the subdomain solvers (i.e. the sequential solvers) are efficiently implemented. High parallel efficiency requires that the overhead, e.g. in communication, introduced by a generic software framework, is negligible.

Although the original sequential simulators, which are to be used for the subdomain solves, are the most important component of our DD strategy, a parallel overlapping Schwarz method needs two other components to function as a whole. The two components are (i) global administration, which coordinates the subdomain solves and administers the information exchange between the subdomains, and (ii) inter-processor communication, which has specific routines for sending and receiving information between processors. Figure 1 shows a modular design involving the three components. Notice that we allow multiple subdomain solvers to reside on one processor.

The present software framework has been implemented on top of Diffpack [5,7], which is a package, written in C++, for solving PDEs in general. The sequential simulators are assumed to be available as standard Diffpack classes.

2.1. Subdomain simulators

A class hierarchy with base class `SubdomainSimulator` represents a general sequential subdomain simulator, as viewed from the generic `Administrator` object that administers the steps in the DD algorithm. Most of the member functions in class `SubdomainSimulator`

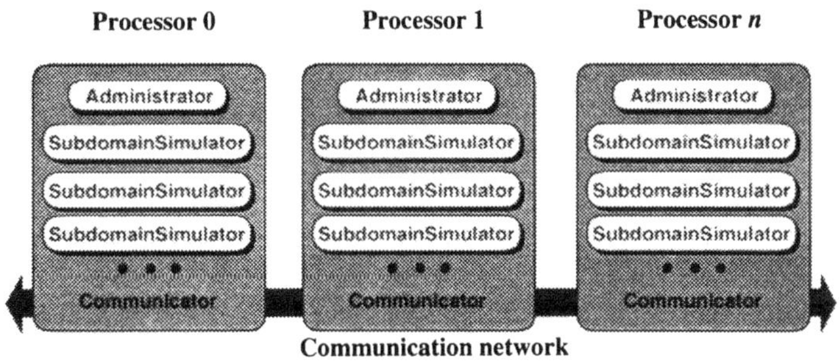

Figure 1. Three software modules on each processor: the administration of the DD algorithm, one or more subdomain solvers, and a communication tool for exchanging internal boundary values between subdomains.

are pure virtual, which need to be overridden in a derived subclass. These member functions constitute a generic standard interface shared by all the subdomain simulators. It is through this standard interface that the communication part and the global administrator of the implementation framework operate. Adapting an existing sequential simulator is easy, because most of the work consists merely of binding the pure virtual member functions in a subclass of `SubdomainSimulator` to existing member functions in the sequential simulator. We here assume that the sequential simulator is realized as a Diffpack class according to the standard in [7]; other types of sequential solvers can also be utilized, provided that they are wrapped in a Diffpack class.

Different classes in the `SubdomainSimulator` hierarchy are designed to be used in different situations. An example of such a subclass is `SubdomainFEMSolver` for simulators that solve a scalar/vector PDE discretized by finite element methods. For example, the coarse grid correction operations needed by two-level overlapping DD methods are implemented as a generic portion of class `SubdomainFEMSolver`.

Suppose an existing finite element-based simulator is available as a class `MySim`. The parallel version of this simulator is then a subclass (say) `MySimP` of *both* class `MySim` *and* class `SubdomainFEMSolver`. The pure virtual functions defined in `SubdomainSimulator` are (mostly) implemented in `MySimP` as simple calls to common functionality in class `SubdomainFEMSolver` and application-specific functionality in class `MySim`. The code of class `MySimP` often fits within a page.

2.2. Inter-processor communication

Between each DD iteration, neighboring subdomains have to exchange information from overlapping regions. This can be done in form of sending and receiving messages between processors. To relieve the user of the painstaking task of writing low-level message passing codes (e.g. in MPI), we have incorporated into the parallelization framework a toolbox that offers, among other things, high-level communication routines designed especially for parallel DD methods. For example, one call to the routine `updateGlobalValues` carries out the above described inter-processor communication. The user does not have to worry about issues such as collecting values from mesh nodes inside the overlapping regions or organizing the messages. Although MPI is used as the underlying message passing protocol due to the consideration of software portability and parallel efficiency, this is completely transparent to the user. Another message passing protocol can therefore easily replace MPI.

2.3. The global administrator

During the parallel solution of a PDE problem, one global administrator is to reside on each processor. The functionality of the global administrator is to coordinate the subdomain solves, invoke the necessary inter-processor communication, and combine the fine-grid subdomain solves with a coarse grid correction. The administrator also determines at run-time whether the DD method is to be used as a stand-alone solver or a preconditioner of a Conjugate Gradient-like method.

2.4. The simulator-parallel approach

The common parallelization approach using DD methods normally deals directly with sub-matrices and sub-vectors of the discrete PDE problem. Our approach consists in reusing the original sequential solver, in the way that the linear system generated by the sequential solver over a subdomain automatically provides sub-matrices and sub-vectors in a natural fashion. As a consequence, we term this parallelization technique as the *simulator-parallel* approach. This approach works at a higher level than the common parallel DD methods, because any object-oriented sequential PDE solver, which encapsulates data structures for its matrices/vectors and has a numerical discretization scheme plus a linear algebra toolbox, is capable of carrying out all the operations needed in the subdomain solves. In this parallelization approach we only work with local PDE problems, the data distribution is implied by an overlapping partition, so there is no explicit need for global representation of data. The local administration of each subdomain simulator allows flexible choice of the local solution method, preconditioner, stopping criterion etc. Most importantly, parallelization at the level of subdomain simulators opens up the possibility of reusing existing reliable and optimized sequential simulators. New improvements of the original sequential solver will immediately after a re-compilation be available in the parallel version of the solver, because the parallel version inherits the sequential solver.

To summarize, the simulator-parallel approach works with any sequential simulator that handles arbitrary grid and Dirichlet boundary conditions and is capable of assembling and solving a system of linear equations.

2.5. Parallel solution of an elliptic PDE

We have developed a parallel Poisson equation solver, which inherits a standard Diffpack finite element Poisson equation solver [7, ch. 3.2] and the `SubdomainFEMSolver` class. Table 1 contains the CPU-measurements of the parallel solver when applied to a 2D rectangular domain discretized using a global 481×481 mesh. The CPU-measurements in this table and all the subsequent tables of the paper are given in seconds and are obtained on an SGI Cray Origin 2000 machine with R10000 195MHz processors. The number of the subdomains is fixed at $M = 32$, while the number of processors P varies. That is, when only one processor is in use, the simulator has to solve all the 32 sub-problems consecutively in each DD iteration. A direct method based on fast Fourier transform is used as the subdomain solver, and coarse grid correction is employed to ensure a fast convergence.

Table 1
The simulator-parallel approach; solution of a Poisson equation using a parallel DD solver on different numbers of processors P (but with a fixed partition $M = 32$)

P	CPU-time	Speed-up	Efficiency
1	53.08	N/A	N/A
2	27.23	1.95	0.97
4	14.12	3.76	0.94
8	7.01	7.57	0.95
16	3.26	16.28	1.02
32	1.63	32.56	1.02

3. LOW-LEVEL PARALLELIZATION

The outlined simulator-parallel approach has its restrictions in that the numerical convergence properties of the underlying overlapping Schwarz method determines whether the resulting parallel PDE solver will work well. Besides, the user is required to do some implementation work, though quite straightforward. It is therefore desirable to also offer users another *effortless* parallelization approach. We have achieved this by devising an add-on parallelization software toolbox, which is already in extensive use in the communication part of the simulator-parallel approach. The standard sequential Diffpack libraries also underwent a few adjustments to allow a seamless coupling with the toolbox. The result, available in Diffpack v3.5 and its add-on parallelization toolbox, facilitates an automatic parallelization of any Diffpack PDE solver, provided that the solver applies iterative solution methods for solving systems of linear equations arising from the discretization.

The core of this low-level approach is to let the toolbox automatically substitute all the encountered global linear algebra operations with local linear algebra operations that are restricted to the subdomains plus inter-processor communication. We therefore denote this parallelization approach as the *linear-algebra* level approach. The only coding effort required of the user is to insert into the original sequential Diffpack PDE solver around 10 new code lines, which invoke the necessary user-friendly preparation and communication routines available from the add-on parallelization toolbox.

3.1. Parallel solution of an elliptic PDE

Having a standard Diffpack simulator [7] for the Poisson equation

$$-\nabla \cdot (K\nabla u) = f,$$

a user should be able, within half an hour, to insert the few code lines that will transform the standard Diffpack simulator into a parallel version using the "automatic" linear-algebra level parallelization approach. Table 2 lists some CPU-measurements of the parallel simulator when applied to a 2D domain discretized using a highly unstructured mesh of linear triangular elements, with a total of 130,561 degrees of freedom. The global Conjugate Gradient iterations are preconditioned by a blockwise ILU preconditioner obtained by adding up the results from local ILU operations restricted to the overlapping subdomains. The preconditioning effect of the blockwise ILU preconditioner varies with P, as Table 2 shows (between 481 and 691 iterations are needed to achieve the same convergence criterion). This also explains the seemingly bad speed-up results. However, if the speed-up figures are measured in terms of CPU-time spent per CG iteration, it can be seen that super linear speed-up is actually obtained for $P = 12$ and $P = 16$, which is due to the cache effect.

Table 2
Some CPU-measurements of an elliptic PDE solver parallelized by the linear-algebra level approach

P	# iter.	CPU	Speed-up	Efficiency
1	480	420.09	N/A	N/A
3	660	200.17	2.10	0.70
4	691	156.36	2.69	0.67
6	552	83.87	5.01	0.84
8	541	60.30	6.97	0.87
12	586	38.23	10.99	0.92
16	564	28.32	14.83	0.93

3.2. Parallel solution of Navier-Stokes equations

Table 3 lists some CPU-measurements of a parallel Navier-Stokes simulator, based on a velocity-correction strategy and a Poisson equation solve at each time step. Essentially, this is a matter of parallelizing Poisson equation solves and a series of explicit updates. The parallelization is carried out by the linear-algebra level approach, see [3] for the details. We remark that the relatively bad speed-up results are due to the small size of the problem.

Table 3
Some CPU-measurements of a fast FE Navier-Stokes simulator parallelized by the linear-algebra level approach

P	CPU	Speed-up	Efficiency
1	1418.67	N/A	N/A
2	709.79	2.00	1.00
3	503.50	2.82	0.94
4	373.54	3.80	0.95
6	268.38	5.29	0.88
8	216.73	6.55	0.82

4. A COMBINED PARALLELIZATION APPROACH

A new parallelization approach arises from combining the simulator-parallel and linear-algebra level approaches. More specifically, the add-on toolbox is used to parallelize automatically the global linear algebra operations required by any Conjugate Gradient-like method, such as matrix-vector multiplication and inner-product between two vectors. The programmer of the parallel application is responsible for developing a parallel DD iteration as the preconditioner. This is done in the way as in section 2, i.e., the programmer derives a new subclass that handles the subproblem solves. The following test cases are examples of using the flexible combined parallelization approach.

4.1. Parallel solution of a linear elasticity problem

The following vector PDE,

$$-\mu\Delta\vec{u} - (\mu + \lambda)\nabla\nabla \cdot \vec{u} = \vec{f},$$

models elastic deformations (displacements $\vec{u}$) of a homogeneous body and is solved on a quarter of a hollow disk. One parallel DD iteration is used as the preconditioner of the BiCGStab method. One multigrid ([6]) V-cycle works as the subdomain solver. The exceptionally good speed-up results in Table 4 are due to the fact that different numbers of BiCGStab iterations I are needed for achieving the convergence for different P.

Table 4
Combined parallelization approach; solution of a 2D linear elasticity problem

P	CPU	Speed-up	I	Subdomain grid
1	66.01	N/A	19	241×241
2	24.64	2.68	12	129×241
4	14.97	4.41	14	129×129
8	5.96	11.08	11	69×129
16	3.58	18.44	13	69×69

4.2. Parallel simulation of nonlinear water waves

We consider the following system of PDEs modeling fully nonlinear 3D water waves with an inviscid fluid model:

$$
\begin{aligned}
-\nabla^2\varphi &= 0 \quad \text{in the water volume,} \\
\eta_t + \varphi_x\eta_x + \varphi_y\eta_y - \varphi_z &= 0 \quad \text{on the free surface,} \\
\varphi_t + \frac{1}{2}(\varphi_x^2 + \varphi_y^2 + \varphi_z^2) + g\eta &= 0 \quad \text{on the free surface,} \\
\frac{\partial\varphi}{\partial n} &= 0 \quad \text{on solid boundaries.}
\end{aligned}
$$

Here, the primary unknowns are the velocity potential φ and the the free surface elevation η. The global 3D mesh is of size $41 \times 41 \times 41$ and one parallel DD iteration is used as the preconditioner of the CG method. One multigrid V-cycle works as the subdomain solver. See [2] for more details.

Table 5
Combined parallelization approach; simulation of 3D nonlinear water waves. CPU-times (in seconds) for the whole simulation for different choices of P, where the partition is fixed with $M = 16$

P	CPU-time	Speed-up	Efficiency
1	1404.40	N/A	N/A
2	715.32	1.96	0.98
4	372.79	3.77	0.94
8	183.99	7.63	0.95
16	90.89	15.45	0.97

4.3. Parallel simulation of porous media flow

The following system of PDEs describes incompressible two-phase flow in a porous medium:

$$s_t + \vec{v} \cdot \nabla(f(s)) = 0 \quad \text{in } \Omega \times (0, T] \tag{1}$$

$$-\nabla \cdot (\lambda(s)\nabla p) = q \quad \text{in } \Omega \times (0, T] \tag{2}$$

$$\vec{v} = -\lambda(s)\nabla p.$$

In the above equations, s and p are primary unknowns, which represent the saturation of water and the pressure distribution, respectively. The global 2D domain is discretized by a 241×241 mesh and one parallel DD iteration is used as the preconditioner of the BiCGStab iterations. For subdomain solves we use one multigrid V-cycle. In Table 6, I denotes the average number of BiCGStab iterations needed for achieving the converge per time step.

Table 6
Combined parallelization approach; simulation of 2D porous media flow

P	Total CPU	Speed-up	Subgrid	CPU P_{eq}	I	CPU S_{eq}
1	4053.33	N/A	241 × 241	3586.98	3.10	440.58
2	2497.43	1.62	129 × 241	2241.78	3.48	241.08
4	1244.29	3.26	129 × 129	1101.58	2.97	134.28
8	804.47	5.04	129 × 69	725.58	3.93	72.76
16	490.47	8.26	69 × 69	447.27	4.13	39.64

5. CONCLUDING REMARKS

Object-oriented programming techniques have been used at different levels when we built the presented parallelization framework. The result is a user-friendly, structural and extensible framework that easily transforms any sequential (Diffpack) PDE solver into a parallel PDE solver, either at the simulator-parallel level or at the linear algebra level, or at a level in between. The framework is not application-specific and consists of generic library modules that can be easily extended to treat new situations. Flexibility and portability arise from those generic library modules, which e.g. hide the details of message passing programming and offer instead high-level and user-friendly inter-processor communication routines. In this way, the programmer is relieved of the hard and error-prone task of writing a parallel PDE solver at the usual low abstraction level. In other words, the proposed framework allows the developer of parallel applications to concentrate more on the physical, mathematical and numerical issues of solving PDEs.

ACKNOWLEDGMENT

We acknowledge support from the Research Council of Norway through a grant of computing time (*Programme for Supercomputing*).

REFERENCES

1. A. M. Bruaset, X. Cai, H. P. Langtangen, A. Tveito, *Numerical solution of PDEs on parallel computers utilizing sequential simulators.* In Y. Ishikawa et al. (eds): Scientific Computing in Object-Oriented Parallel Environment, Springer-Verlag Lecture Notes in Computer Science 1343 (1997) pp. 161–168.
2. X. Cai, *Numerical Simulation of 3D Fully Nonlinear Water Waves on Parallel Computers.* In B. Kågström et al. (eds): Applied Parallel Computing, PARA'98, Springer-Verlag Lecture Notes in Computer Science 1541 (1998) pp. 48–55.
3. X. Cai, H.P. Langtangen, and O. Munthe, An Object-Oriented Software Framework for Building Parallel Navier-Stokes Solvers. In the proceedings of *Parallel CFD'99* (1999).
4. T. F. Chan and T. P. Mathew, *Domain decomposition algorithms.* Acta Numerica (1994) pp. 61–143.
5. Diffpack Home Page, *http://www.diffpack.com.*
6. W. Hackbusch, *Multi-Grid Methods and Applications.* Springer (1985).
7. H. P. Langtangen, *Computational Partial Differential Equations – Numerical Methods and Diffpack Programming.* Springer (1999).
8. B. F. Smith, P. E. Bjørstad, W. D. Gropp, *Domain Decomposition, Parallel Multilevel Methods for Elliptic Partial Differential Equations.* Cambridge University Press (1996).
9. J. Xu, *Iterative methods by space decomposition and subspace correction.* SIAM Review 34 (1992) pp. 581–613.

Parallel Computational Fluid Dynamics – Trends and Applications
C.B. Jenssen et al. (Editors)

Parallel Computing of Non-equilibrium Hypersonic Rarefied Gas Flows

Yoichiro MATSUMOTO[a], Hiroki YAMAGUCHI[a] and Nobuyuki TSUBOI[b]

[a]Department of Mechanical Engineering, The University of Tokyo, 7-3-1, Hongo, Bunkyo-ku, Tokyo 113-8656, Japan

[b]Research Division for Space Transportation, Institute of Space and Astronautical Science, Yoshinodai, Sagamihara, Kanagawa, 229-8510, Japan

Multi-scale analysis of non-equilibrium hypersonic rarefied diatomic gas flow by using parallel computers is presented in this paper. DSMC(direct simulation Monte Carlo) simulations are employed with the Dynamic Molecular Collision (DMC) model and the Multi-Stage (MS) model based on Molecular Dynamics (MD) simulation of nitrogen molecules. Those models do not need any empirical parameters such as inelastic collision parameter and can predict non-equilibrium between translational and rotational temperature rationality. The DSMC results over a flat plate in hypersonic rarefied gas flow show that non-equilibrium between translational and rotational temperature are obtained behind the leading edge over the plate and that the leading edge angle and the gas-surface interaction model has considerable effects on the flow structure. However, the three-dimensional effects exist near the span edge but would be small near the symmetric line of the plate in this flow conditions. The parallel implementation of the DSMC code shows to have linear scalability using the dynamic load balancing technique.

1. INTRODUCTION

Many researchers have been studying hypersonic rarefied gas flows around space vehicles with respect to the non-equilibrium characteristics. In rarefied regime, mean free pass of molecule is so large that non-equilibrium between translational and rotational temperature appears in the shock wave and in the interaction between shock wave and boundary layer around configurations. However, the non-equilibrium between these temperature did not studied. Therefore, such the non-equilibrium phenomena in the rarefied regime should be researched in detail.

In numerical simulation of the hypersonic flow in rarefied regime, DSMC (direct simulation Monte Carlo) is valid due to the treatment of particles directly [1], [2]. The important procedures in the DSMC method for diatomic molecular flow are the molecular collision and the gas-surface interaction. The DSMC simulations require significant amounts of computational time in near-continuum conditions and three-dimensional simulations because of the increase in the number of molecules and the collision frequency with increasing density. Therefore parallel implementation for the DSMC code is attractive method to reduce CPU time per processor.

The purpose of the present paper is to show the simulation of the hypersonic rarefied gas flow over a flat plate using a parallel DSMC method with the DMC model for nitrogen molecular collision and the MS model for the nitrogen/graphite gas-surface interaction.

2. NUMERICAL METHOD

2.1. DSMC Method

In numerical simulations in rarefied regime, DSMC method is valid because the continuum approach may break down in a low density flow. Gas-gas molecular collision model and gas-surface interaction model have significant roles on the accuracy of the simulation. The multi-scale models are recently developed by molecular dynamics simulation in order to contain the information about the micro-scale phenomena. The schema of the multi-scale model is shown in Fig. 1. A bow shock wave with non-equilibrium in internal degrees of freedom is generated around a re-entry vehicle. In the shock wave, large amount of molecular collision relaxations between diatomic molecules exists to exchange between the translational and rotational energy. With respect to two molecules, interaction between atomic nucleus and electron appears. Because it needs complicated procedure and computational cost to solve the Schrödinger equation to obtain the interaction between them, it is useful to assume the potential such as Lennard-Jones potential. Next, the molecular dynamics simulations with various translational and rotational energies are conducted to obtain the probability density function of translational and rotational energy exchange before and after the molecular collisions and to construct gas-gas collision model and

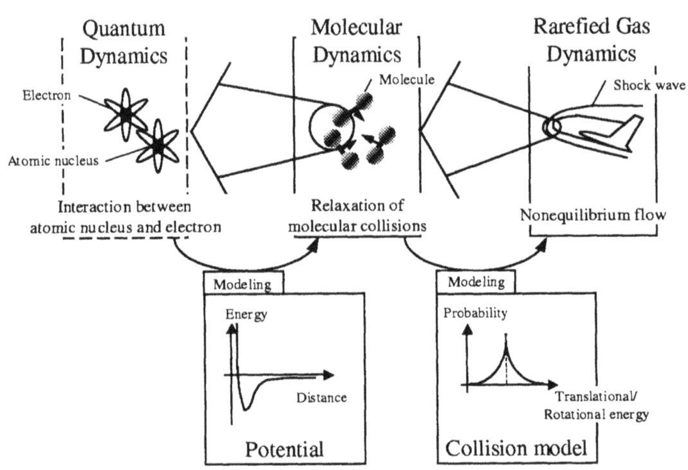

Figure 1. Schematic diagram of the present multi-scale analysis model.

gas-surface collision model to estimate aerodynamic characteristics around the re-entry vehicle. Tokumasu and Matsumoto have constructed the DMC model [3] for nitrogen molecules which is able to capture the non-equilibrium characteristics in the rarefied gas flow below 2,000 K. DMC model is based on the cross sections and energy distributions after the collisions obtained by molecular dynamics simulation for diatomic molecules. Yamanishi and Matsumoto have presented the MS model for gas-surface interaction which is based on the molecular dynamics simulations [4]. The MS model is also applicable for nitrogen or oxygen gas molecules interacting with graphite surface.

2.2. Parallel Implimentation

In order to reduce CPU time per processor for three-dimensional simulation, the parallel implementation of the computational code is efficient method. Vectorization is one way to reduce CPU time and is effective for a continuum simulation such as the Navier-Stokes simulation. However, the DSMC code has many DO loops for the number of particles which changes at each iteration and the vectorization would be not effective method for the DSMC code.

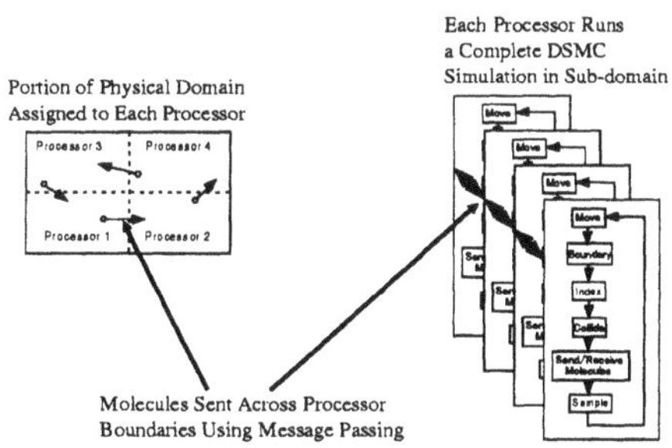

Figure 2. Schematic of the parallel DSMC method.

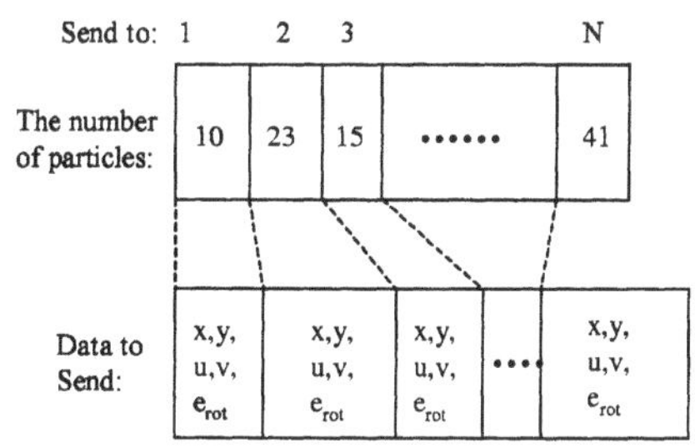

Figure 3. Procedure of send data in each processor.

In the present research, we adopted the domain decomposition method to allocate subdomain in each PE(Processor Element) where the calculation were done for the allocated particles [6], [7]. When the particles pass the belonged subdomain, the information such as the position, velocity and rotational energy are communicated with the other PEs allocated to the other subdomain as shown in Fig. 2.

The problem to conduct the parallel DSMC simulation is that the large unbalance for CPU time and the number of particles in each PE arises because particle distributions between initial and final situation in the simulation are quite different. To avoid the problem, an active load balancing method for the allocated subdomain was applied during the parallel simulation by controlling the amount of the number of particles in each subdomain.

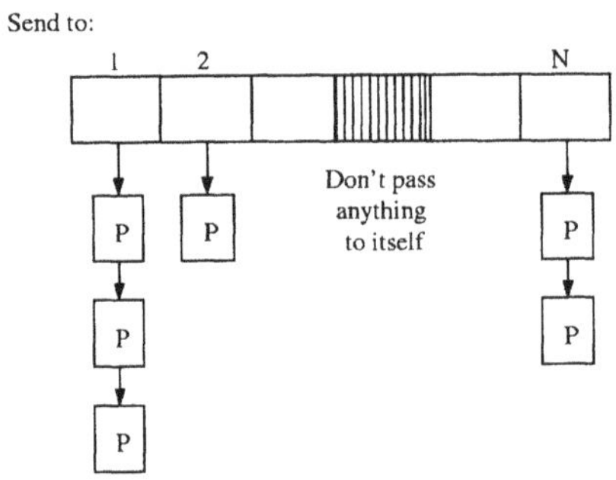

Figure 4. Schematic of the process to send a present processor to other processors. P denotes particle to be sent to other processors.

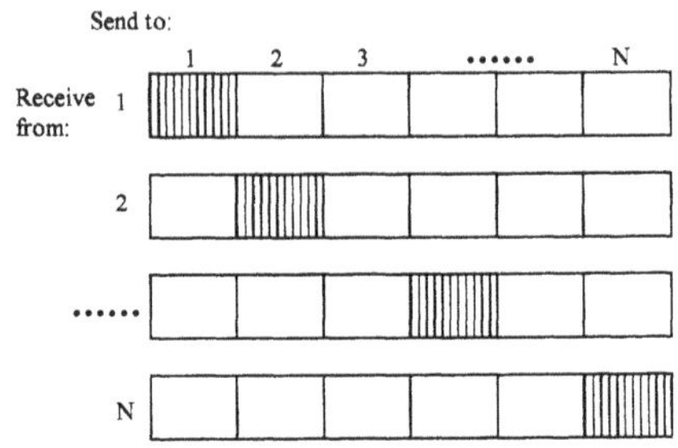

Figure 5. Grobal point of view for send/receive data.

The information which each PE requires are particle's velocity, position and its rotational energy. After the number of particles to cross the subdomain boundary is checked, the information is stored in a temporal array as shown in Fig. 3. The reason to need this process is that the number of called communication subroutines has to be reduced because those subroutines need a few times before starting to communicate data.

Then the data in the temporal array such as velocity, et al. are communicated with the other PEs. The data such as the total number of particles in each PE, velocity, position and rotational energy received from the other PEs are stored in the actual arrays.

There are two kinds of the parallel languages: one is the language which is constructed for a special compiler to add some compiler indicator statements in an existed language; another is the parallel language library to afford the parallel processes by Message Passing Library (MPL). The former is mainly afforded as a special parallel language for a specific special parallel computer. The latter can use for many parallel computers because the language is a message passing library. Parallel Virtual Machine (PVM) and Message Passing Interface (MPI) belong to the message passing library. Because MPI is one of the common languages for parallel computing and is not dependent on the parallel computer architecture, MPI was used to parallelize the present DSMC code.

In the above communication between PEs, the point-to-point non-blocking communication routines were used. If the point-to-point blocking communication routines are used, waiting status appears and much communication time arises until a send buffer is able to use again or a receive buffer receives all data.

The above process conducts the send process from the specified PE as shown in Fig.4. But the PE has to receive from other PEs so that a matrix communication in Fig.5 is carried out in whole PEs. The simulations are synchronized between the movement and the collision routines at each time step.

Table 1
Computational conditions in the present simulations.

No.	1	2	3	4	5	6
2D/3D simulation	2D				2D	3D
Leading Edge angle[deg.]	–				20	20
Mach number M_∞	20.2					
Working Gas	Nitrogen					
Stagnation pressure p_0[Pa]	3.5×10^5					
Stagnation temperatureT_0[K]	1,100					
Freestream temperature T_∞[K]	13.32					
Freestream pressure p_∞[Pa]	0.06831					
Freestream velocity V_∞[m/s]	1,503					
Reynolds number Re_∞(based on L=0.05[m],2L=plate length)	566					
Wall temperature T_w[K]	290					
Knudsen number Kn_∞ (based on L)	0.047					
Wall boundary condition	Diffuse	MS		CLL	Diffuse	
Accomm. coeff. $\alpha n/\alpha t/\alpha r$	–	–		1/0.986/1	–	–
Note		with thermal equilibrium for MS	without thermal equilibrium for MS			

3. RESULTS AND DISCUSSIONS

Simulation conditions are shown in Fig. 1 [5]. In the simulations, the plate has a leading edge angle of 20 degrees. Plate length is 100 mm, plate width is 100 mm, plate thickness is 5 mm and leading edge angle is 20 degrees, respectively. Flow field near the leading edge in the present simulation conditions is merged layer interacted between shock wave and boundary layer.

3.1. Parallel Performance

Figure 6(a)(b) show the computational grid and density contours after steady state in the two-dimensional simulation(no.1). Figure 6(c) $\sim$ (d) show the subdomain boundary distributions for initial and steady state with dynamic load balancing for 64 PEs. Though the average number of computational grid cells for all PEs is allocated in each PE in the initial state, the subdomain area near the wall becomes narrower than the other subdomain area. But each subdomain area has an average area after steady state.

The parallel computing performance of the two-dimensional DSMC code was measured in the case of no.1 in table 1 on IBM SP2 and Hitachi SR2201. Peak performances in each processor are 266.4 MFLOPS for SP2 and 300 MFLOPS for SR2201. Interprocessor data payload communication rates are 40 Mbyte/sec. for SP2 and 266 Mbyte/sec. for SR2201. The CPU measurement for each case was conducted for a fixed number of time step, after the steady state was reached, and did not include the initialization or I/O costs. Figure 7 shows the results of the speed-up of the DSMC code with and without dynamic load balancing. The maximum speed-up without dynamic load balancing is 15.13

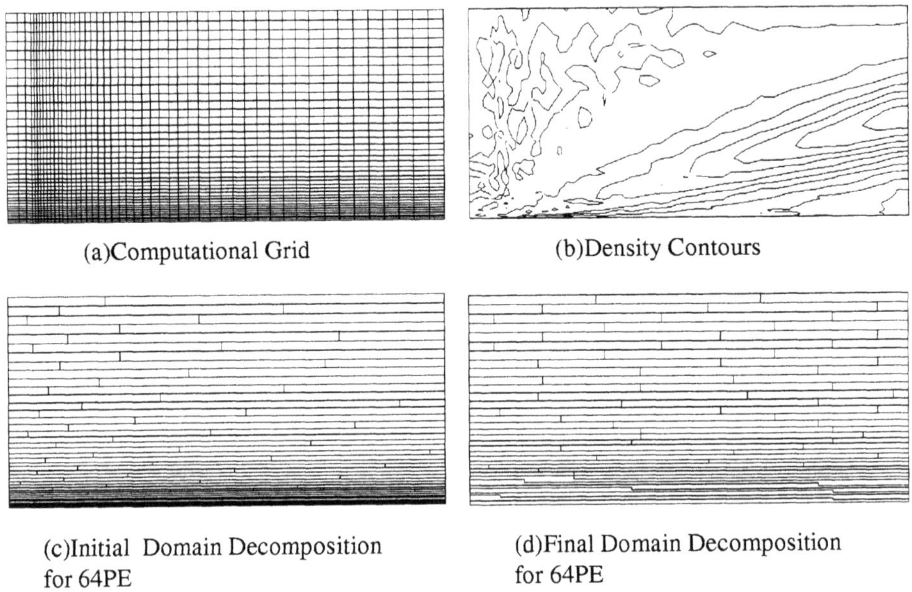

Figure 6. Domain decomposition for the two-dimensional simulation. Simulation condition is no.1.

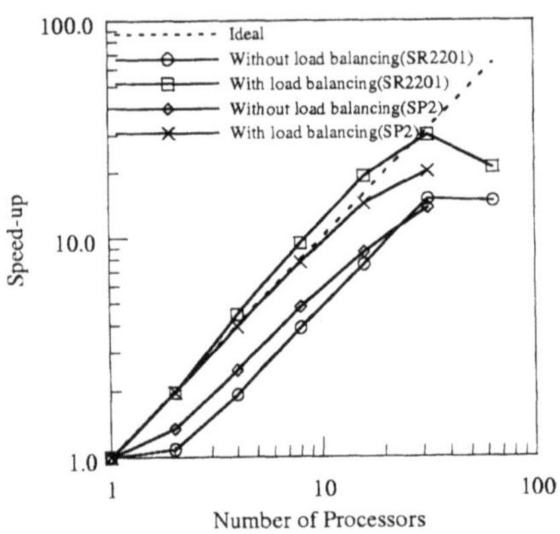

Figure 7. Speed-up for the two-dimensional code. Simulation condition is no.1

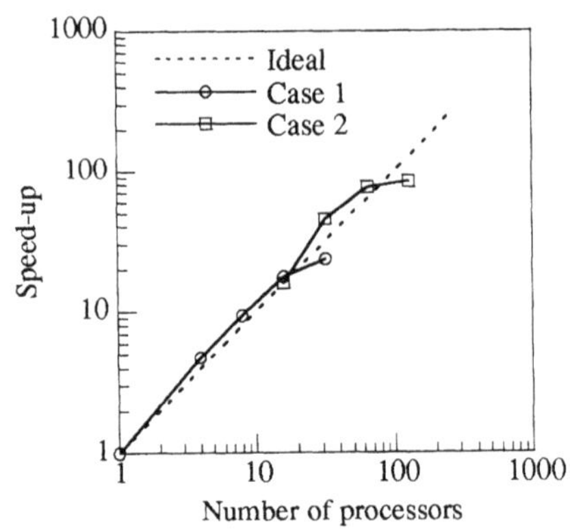

Figure 8. Speed-up for the three-dimensional code. Simulation condition is no.6.

on the SR2201(32CPUs) and 13.78 on SP2(32CPUs). However, that with dynamic load balancing is 30.0 on SR2201(32CPUs) and 20.32 on SP2(32CPUs), and linear speed-up is attained at the various numbers of processors except 64 processors. Figure 8 shows the parallel speed-up in the three-dimensional simulation. The parallel performance of the code was measured on the HITACHI SR2201 for no.6. The number of particles used in case 1 was about 250,000, and that in case 2 was about 50,000. Speed up in case 1 is based on 1 CPU, but that in case 2 is based on 16 CPUs because memory requirement with 1 CPU in case 2 exceeds the SR2201's 1 CPU memory 256 MB. The speed-up performance with 32 CPUs in case 1 is 23.9, and that with 64 CPUs in case 2 is 84.68. The parallel implementation of the present DSMC code shows to have linear scalability by using the dynamic load balancing technique.

3.2. Two-Dimensional Simulation Results with and without Leading Edge

Figure 9 shows effects of leading edge on density profiles at X/L=1.5. It is shown that the flowfield over the plate is affected by the thick shock wave generated by the lower leading edge surface. Maximum density location for LE=20(Leading Edge angle of 20 degrees) moves toward the upper direction than that for LE=0. However, the DSMC results are slightly different from the experimental results. Figure 10 shows heat transfer distributions on the plate. Results for LE=20 shifts toward the leading edge and maximum value for LE=20 increase more than that for LE=0. But the DSMC results are larger than the experimental results. From the above discussions, leading edge effects exist, however, other effects would exist then the effects of the gas-surface interaction were estimated in the next section.

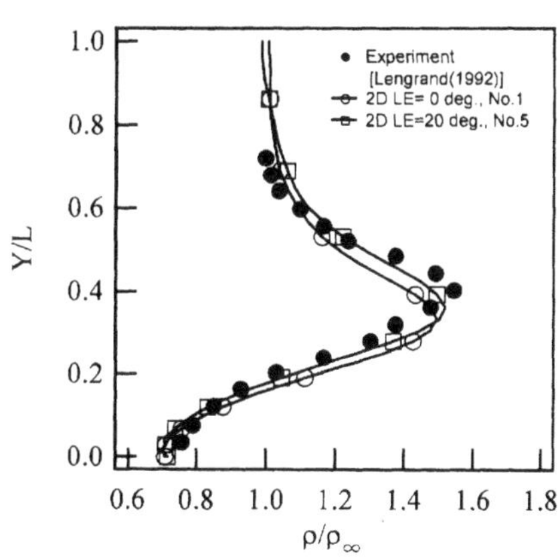

Figure 9. Effects of leading edge on density profiles at X/L=1.5.

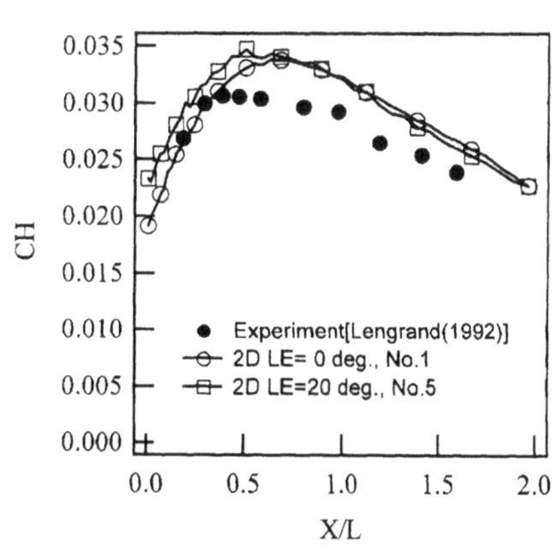

Figure 10. Effects of leading edge on heat transfer rate distributions on the plate.

3.3. Two-Dimensional Simulation Results with Gas-surface Interaction Model

In Fig.11, it is shown that the maximum density with the diffuse reflection model and thermal equilibrium case for the MS model are higher than other cases. The density distributions near the plate in no.3 shows the least values of others. However, all simulation results disagree with the experimental results. The reasons for the discrepancy are that there would be significant effects of leading edge and three-dimensional flow. Especially, the results of MS(no.3) in Fig.12 is close to the experimental results near the trailing edge where there are less effects of the shape of the leading edge. The results reveal that there are no region on the plate where the diffuse reflection and thermal equilibrium dominate.

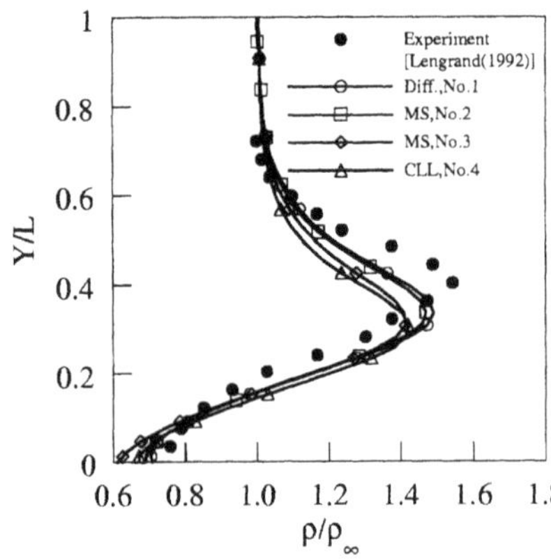

Figure 11. Effects of gas-surface interaction model on density profiles at X/L=1.5.

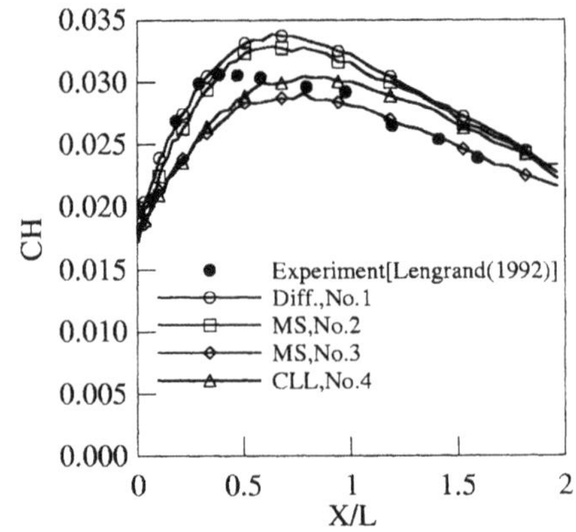

Figure 12. Effects of gas-surface interaction model on heat transfer rate distributions on the plate.

3.4. Three-Dimensional Simulation Results

Density distributions of three-dimensional numerical simulation with finite leading edge angle are shown in Fig. 13. From the results, a formation of a shock wave is captured clearly and the size of the leading edge effect and the finite span effect are appears. The finite span effects appears $Z/L > 0.5$ at $X/L = 1.5$ over the plate. However, flowfield at $0 < Z/L < 0.5$ has quasi two-dimensional flow at X/L=1.5. The finite span effects are due to three-dimensional viscous effects near the plate tip.

Comparison between two- and three-dimensional results is shown in Fig.14 and 15. It is shown that there are small discrepancy between two- and three-dimensional simulation because finite spanwise effects are limited near the span edge as shown in Fig. 13. It is concluded that the three-dimensional effects can be negligible near the symmetry axis on

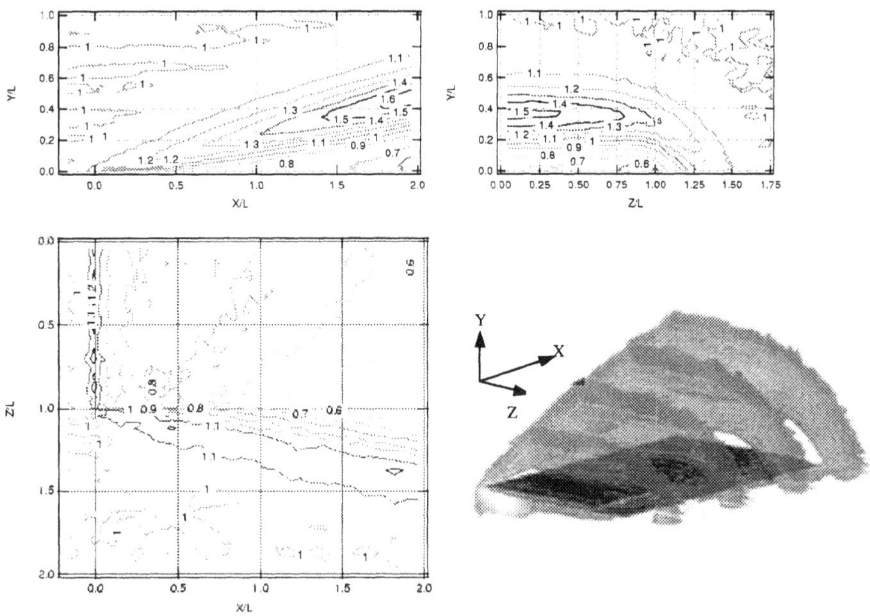

Figure 13. Density contours over the flat plate (upper left: cross section at $Z/L = 0$, upper right: cross section at $X/L = 1.5$, lower left: cross section at $Y/L = 0$, lower right: density contours at each cross section over the half width of the plate).

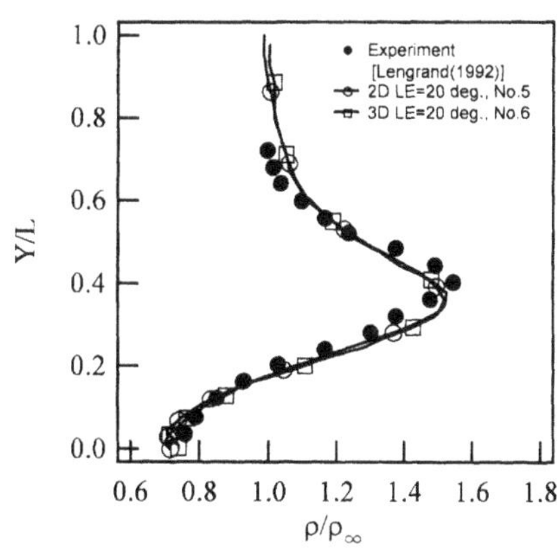

Figure 14. Comparison between two- and three-dimensional density profiles at X/L=1.5.

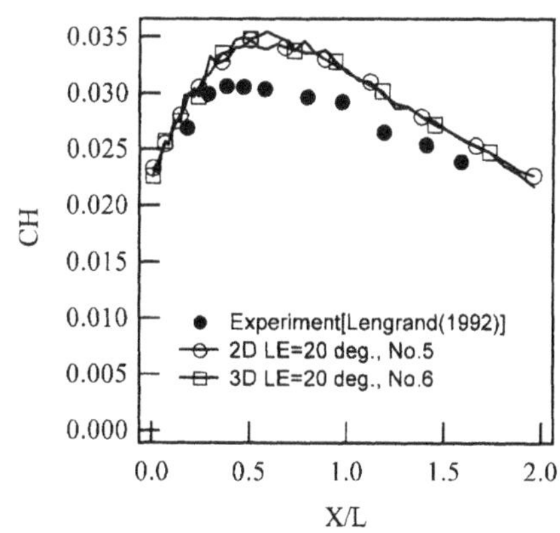

Figure 15. Comparison between two- and three-dimensional heat transfer rate distributions on the plate.

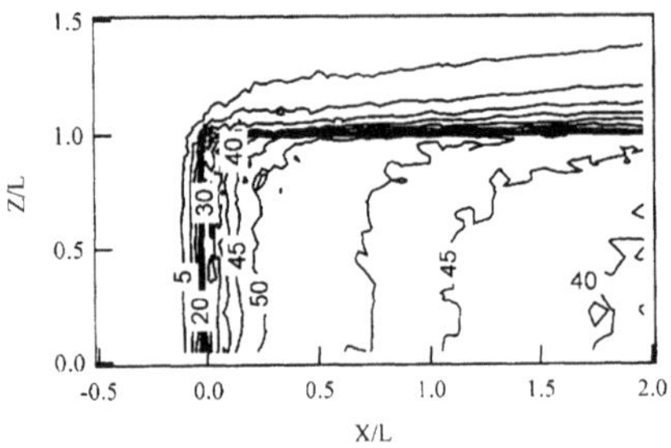

Figure 16. Normalized translational temperature contours on the plate in the three-dimensional simulation.

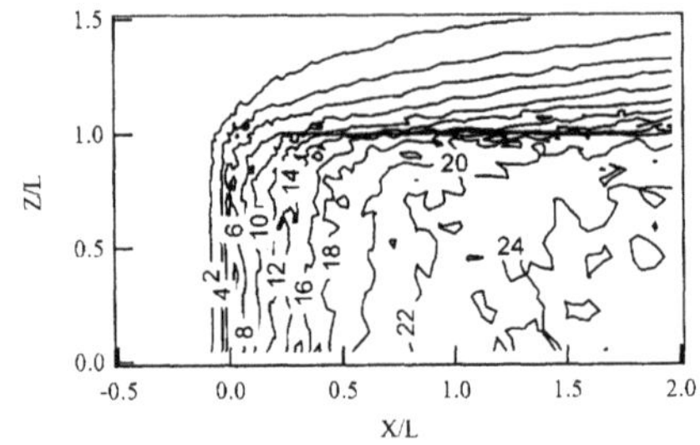

Figure 17. Normalized rotational temperature contours on the plate in the three-dimensional simulation.

the plate and the flow can be treated as the approximation of the two-dimensional flow.

Figures 16, 17 show the normalized translational and rotational temperature contours on the plate. Both temperatures are normalized by the freestream temperature of 13.32 K and the normalized wall temperature is equal to 21.77. Translational temperature increases rapidly near the leading edge whereas rotational temperature slowly increases. The difference between their temperatures is about 300 K at X/L=1.0 on the symmetry line and large non-equilibrium is appeared on the whole domain of the plate.

The above comparisons show that the DSMC results did not coincide with the experimental results. The factors for the discrepancy in the experiment side would be considered: (i) non-uniformity flow at the nozzle exit, (ii) rotational temperature freezing in the nozzle, (iii) vibrational excitation. For non-uniformity flow at the nozzle exit, Allégre [8] measured density distributions downstream the nozzle exit. The results shows that the density gradient exists at the nozzle exit due to the thick boundary layer developed in the nozzle and the use of conical nozzle. For the rotational temperature freezing in the nozzle, the influence would be considered to be significant, however, the value was not estimated. Finally, the vibrational excitation for $T_0 = 1,100[K]$ is considered to be small, but the excitation rate for $T_0 = 1,100[K]$ would be about 10%. Furthermore, degree of vibrational temperature freezing is larger than that of rotational temperature and is thought to be freezing completely. Therefore, the above effects in the experiment should be estimated in order that the experimental data are utilized for the validation of the simulation.

However, we have constructed the efficient parallel two- and three-dimensional DSMC code and revealed the three-dimensional effects.

4. CONCLUSIONS

Multi-scale analysis of non-equilibrium hypersonic rarefied diatomic gas flow was presented by using a parallel DSMC method with the DMC model for a diatomic gas molecular collision and with the MS model for a gas-surface interaction model. The parallel implementation of the DSMC code shows to have linear scalability using the dynamic load balancing technique. The DSMC simulations revealed that the leading edge angle, gas-surface interaction effects affected on the flow over the plate, however, the three-dimensional effects would be small near the symmetric line of the plate in this flow conditions. From the three-dimensional simulations, the three-dimensional flow structure exists due to the viscous effects near the span edge.

REFERENCES

1. Bird,G.A., Molecular Gas Dynamics, Calrendon Press, Oxford, 1976.
2. Nanbu, K., "Stochastic Solution Method of the Model Kinetic Equation for Diatomic Gas," J. Phys. Soc. Jpn. , Vol.49, p.2042-2049, 1988.
3. Tokumasu, T. and Matsumoto, Y., "Dynamic Molecular Collision (DMC) Model for Rarefied Gas Flow Simulations by the DSMC Method," Physics Fluids Vol.11, No.7, p.1907-1920, 1999.
4. Yamanishi, N. and Matsumoto, Y., "Multistage Gas-Surface Interaction Model for the direct simulation Monte Carlo Method," Physics Fluids Vol.11, No.11, p.3540-3551, 1999.
5. Lengrand,J., Allégre,J., Chpoun,A., and Raffin,M., *18th Int. Symp. on Rarefied Gas Dynamics*, **160**, 276, 1992.
6. Dietrich S. and Boyd I.D., "Parallel Implementation on the IBM SP-2 of the Direct Simulation Monte Carlp Method," AIAA paper 95-2029,1995.
7. Richard, G.W., "Application of a Parallel Direct Simulation Monte Carlo Method to Hypersonic Rarefied Flows," AIAA Journal, Vol.30, p.2447-2452, 1992.
8. Allégre,J., Bisch,D. and Lengrand,J., *Journal of Spacecraft and Rockets*, 714-718, **34**, 6 (1997).

Parallel Computational Fluid Dynamics – Trends and Applications
C.B. Jenssen et al. (Editors)

Large-Eddy Simulations of turbulence: towards complex flow geometries

O. Métais [a]

[a]Laboratoire des Ecoulements Géophysiques et Industriels, BP 53, 38041 Grenoble Cédex 9, France

1. INTRODUCTION

Direct-numerical simulations of turbulence (DNS) consist in solving explicitly all the scales of motion, from the largest l_I to the Kolmogorov dissipative scale l_D. It is well-known from the statistical theory of turbulence that l_I/l_D scales like $R_l^{3/4}$, where R_l is the large-scale Reynolds number $u'l_I/\nu$ based upon the rms velocity fluctuation u'. Therefore, the total number of degrees of freedom necessary to represent the whole span of scales of a three-dimensional turbulent flow is of the order of $R_l^{9/4}$ in three dimensions. In the presence of obstacles, around a wing or a fuselage for instance, and if one wants to simulate three-dimensionally all motions ranging from the viscous thickness $\delta_v = \nu/v_* \approx 10^{-6}$ m up to 10 m, it would be necessary to put 10^{21} modes on the computer. At the present, the calculations performed in reasonable computing time on the biggest machines take about 2. 10^7 grid points, which is a long way from the above estimation. Even with the unprecedented improvement of scientific computers, it may take several decades (if it ever becomes possible) before DNS permit to simulate situations at Reynolds numbers comparable to those encountered in natural conditions.

Statistical modelling based on Reynolds Averaged Navier-Stokes (RANS) equations are particularly designed to deal with statistically steady flows or with flows whose statistical properties vary "slowly" with time, that is to say of characteristic time scale much larger than a characteristic turbulent time scale. The application of phase averaging constitutes another alternative which allows for the modelling of time periodic flows. With the RANS approach all the turbulent scales are modelled. First order as well as second order RANS models involve many adjustable constants and it is therefore impossible to design models which are "universal" enough to be applicable to various flow configurations submitted to diverse external forces (rotation, thermal stratification, etc ...). However, since RANS models compute statistical quantities, they do not require temporal or spatial discretizations as fine as the ones necessary for DNS or even LES. They are therefore applicable to flows in complex geometries.

Large-Eddy Simulations (LES) techniques constitute intermediate techniques between DNS and RANS in the sense that the large scales of the flow are deterministically simulated and only the small scales are modelled but statistically influence the large-scale motion. LES then explicitly resolve the large-scales inhomogeneity and anistropy as well as the large-scales unsteadiness. This is important from an engineering point of view since the large scales are responsible for the major part of turbulent transfers of momentum or heat for example. Most subgrid-scale models which parameterized the action of the small-scales are based upon "universal" properties of small-scales turbulence: those can therefore be applied to various flows submitted to various external effects without being modified. In this respect, they constitute "universal" models directly applicable to various flow configurations. However, they require much finer spatial and temporal discretizations than RANS and lie inbetween DNS and RANS as far as CPU time consumption is concerned. Once confined to very simple flow configurations such as isotropic turbulence or periodic flows, the field is evolving to include spatially growing shear flows, separated flows, pipe flows, riblet walls, and bluff bodies, among others. This is due to the tremendous progress in scientific computing and in particular of parallel computing. As will be seen in the few examples presented below, LES are extremely useful in particular towards the understanding of the dynamics of coherent vortices and structures in turbulence. We will show below that this is of special importance for flow control problems, for detached flows and their aeroacoustics predictions and for flows submitted to compressibility effects and density differences.

2. LARGE-EDDY SIMULATION (LES) FORMALISM

LES have been the subject of many review articles. Details concerning the LES formalism and new developments in LES can be found, for instance, in [11], [12], [14]. LES consist in considering a spatial filter G of width Δx, which filters out the *subgrid-scales* of wavelength $< \Delta x$. The filtered field is defined as

$$\bar{u}_i(\vec{x},t) = \int u_i(\vec{y},t)G(\vec{x}-\vec{y})d\vec{y} = \int u_i(\vec{x}-\vec{y},t)G(\vec{y})d\vec{y} \quad , \tag{1}$$

and the subgridscale field is the departure of the actual flow with respect to the filtered field:

$$u_i = \bar{u}_i + u_i' \quad . \tag{2}$$

The application of the filter to the Navier-Stokes equations leads to the classical closure problem because of the non-linear nature of the equations. Unknown tensors related with the subgrid-scale quantities appear which have to be modelled: a subgrid-scale model has then to be introduced. Many subgrid-scale models make eddy-viscosity and eddy-diffusivity assumptions (Boussinesq's hypothesis) in order to model the unknown subgrid-scale tensors. The reader is referred to [12] for further details. All the computations presented below are LES based on the structure-function subgrid-scale model developed in our Grenoble group.

3. LES: A TOOL FOR FLOW CONTROL

Our goal is to demonstrate the ability of the LES to control turbulent flows by manipulation of inflow conditions. We here concentrate on the turbulent jet. The control of the turbulent jets find numerous industrial applications in thermohydraulics, aeronautics, industrial processes or even the dispersion of pollutants. For these applications , it is particularly interesting to control certain flow characteristics such as the mixing efficiency, the acoustic generation, etc.. We will show below that an efficient control requires a precise knowledge of the spatial and temporal flow organization to manipulate the three-dimensional coherent vortices. The detailed results are presented in Urbin (1997) [22], Urbin and Métais [23] and Urbin *et al.* [24], we here just recall the main results. The use of large-eddy simulations (LES) techniques allow us to reach high values of the Reynolds number: here, Re is 25000. The LES filtered Navier-Stokes equations are solved using the TRIO-VF code. This is an industrial software developed for thermal-hydraulics applications at the Commissariat a l'Energie Atomique de Grenoble. It has been thoroughly validated in many LES of various flows such as the backward facing step. It uses the finite volume element method on a structured mesh.

We here consider a computational domain starting at the nozzle and extending up to 16 jet diameters downstream. We succesively consider two jets configurations: the "natural" jet which is forced upstream by the top-hat profile to which is superposed a weak 3D white noise; the "excited" jet development is controlled with the aid of a given deterministic inflow forcing (plus a white noise) designed to trigger a specific type of three-dimensional coherent structures.

3.1. The natural jet

We have thoroughly validated our numerical approach by comparing the computed statistics with experimental results for the mean and for the r.m.s. fluctuating quantities. The frequency spectra have furthermore revealed the emergence of a predominant vortex-shedding Strouhal number, $Str_D = 0.35$ in good correspondance with the experimental value. A usual way to characterize large scale coherent vortices consists in considering vorticity or pressure isosurfaces. Another way is to use the so-called Q-criterion proposed by Hunt *et al.* [8]. This method is particularly attractive since it consists in isolating the regions where the strain rate is lower than the vorticity magnitude. Hunt *et al.*[8] define a criterion based on the second invariant of the velocity gradient Q with $Q = (\Omega_{ij}\Omega_{ij} - S_{ij}S_{ij})/2$ where Ω_{ij} is the antisymmetrical part of $\partial u_i/\partial x_j$ and S_{ij} the symmetrical part. $Q > 0$ will define zones where rotation is predominant (vortex cores). These different methods of visualization will be used in the present paper. The experimental studies by Michalke and Hermann [15] have clearly shown that the detailed shape of the mean velocity profile strongly influences the nature of the coherent vortices appearing near the nozzle: either axisymmetric structures (vortex rings) or helical structure can indeed develop. The temporal linear stability analysis performed on the inlet jet profile we have used predicts a slightly higher amplification rate for the axisymmetric (varicose) mode than for the helical mode (see Michalke and Hermann [15]). The 3D visualization (figure 1) indeed shows that the Kelvin-Helmholtz instability along the border of the jet yields, further downstream, vortex structures mainly consisting in axisymmetric toroidal shape. However, the jet exhibits an original vortex arrangement subsequent to the varicose mode growth: the "alternate pairing". Such a structure was previously observed by Fouillet [6] in a direct simulation of a temporally evolving round jet at low Reynolds number ($Re = 2000$). The direction normal to the toroidal vortices symmetry plane, during their advection downstream, tends to differ from the jet axis. The inclination angle of two consecutive vortices appears to be of opposite sign eventually leading to a local pairing with an alternate arrangement.

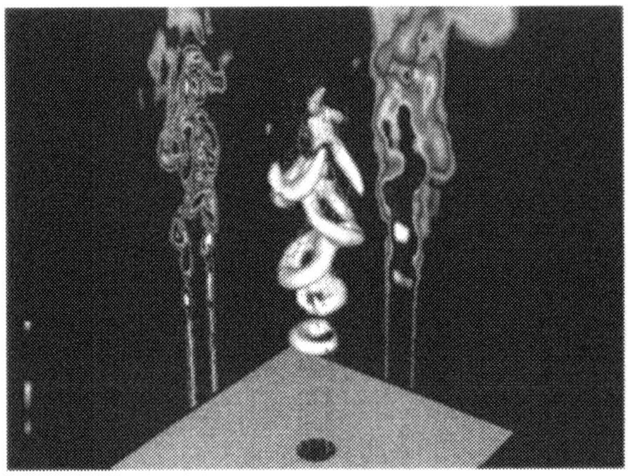

Figure 1. Natural jet: instantaneous visualization. Light gray: low pressure isosurface; wired isosurface of the axial velocity $W = W_o/2$; YZ cross-section (through the jet axis) of the vorticity modulus; XZ cross-section of the velocity modulus (courtesy G. Urbin).

3.2. The forced jet

We here show how a deterministic inflow perturbation can trigger one particular flow organization. We apply a periodic fluctuation associated with a frequency corresponding to $Str_D = 0.35$ for which the jet response is known to be maximal. The inflow excitation is here chosen such that alternate-pairing mode previously described is preferentially amplified. The resulting structures are analogous to figure 1 except that the alternatively inclined vortex rings now appear from the nozzle (see Figure 2). These inclined rings exhibit localized pairing and persist far downstream till $Z/D = 10$. One of the striking features is the very different spreading rates in different directions: the streamlines originally concentrated close to the nozzle tend to clearly separate for $Z/D > 4$. Furthermore, the alternatively inclined vortex-rings seem to separate and move away from the jet centerline to form a Y-shaped pattern. Note that the present jet exhibits strong similarities with the "bifurcating" jet of Lee and Reynolds [9]. One of the important technological application of this peculiar excitation resides in the ability to polarize the jet in a preferential direction.

3.3. Coaxial jets

Coaxial jets are present in numerous industrial applications such as combustion chambers, jet engine, etc ... The figure 3 shows the three-dimensional coherent structures obtained through a highly resolved DNS, at Reynolds 3000, of a coaxial jet with the inte-

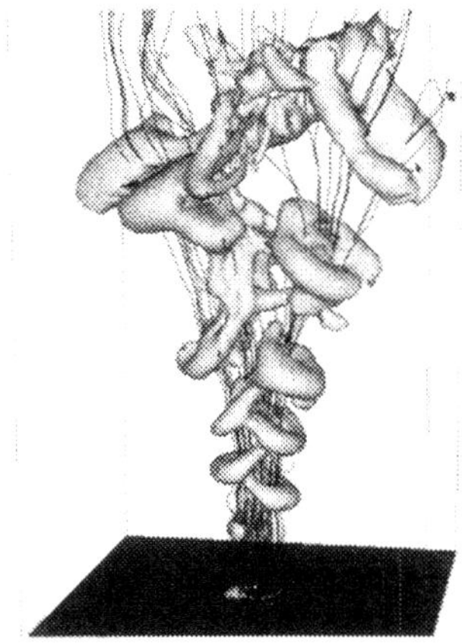

Figure 2. Bifurcation of the jet with alternate-pairing excitation. Instantaneous vizualisation of streamlines emerging from the nozzle. Low pressure isosurface in grey ($P = 25\% P_{min}$) (courtesy G. Urbin).

rior of the jet faster than the outer. One sees vortex rings which, like in a plane miximg layer, pair, while stretching intense alternate longitudinal vortices. By the depression they cause, these vortices are responsible for important sources of noise during take-off of transport planes, and are in particular a major concern for future supersonic commercial aircrafts. The control of this flow is therefore of vital importance for problems related to noise generation. One may notice that the large vortices violently breakdown into very intense developed turbulenec at smale scales. Details of this computation are described in [16].

4. SEPARATED FLOWS

The effect of a spanwise groove (whose dimensions are typically of the order of the boundary layer thickness) on the vortical structure of a turbulent boundary layer flow has recently regained interest in the field of turbulence control (Choi & Fujisawa [1]). The groove belongs to the category of passive devices able of manipulating skin friction in turbulent boundary layer flow. Depending on the dimensions of the cavity, the drag downstream of the groove can be increased or decreased.

In order to investigate the effects of a groove on the near-wall structure of turbulent boundary layer flows, Dubief and Comte [5], [4] have performed a spatial numerical simulation of the flow over a flat plate with a spanwise square cavity embbeded in it. The goal here is to show the ability for the LES to handle geometrical singularities.

The width d of the groove is of the order of the boundary layer thickness, $d/\delta_0 = 1$. The computational domain is sketched in figure 4. We here recall some of Dubief and Comte's results. The simulation is slightly compressible: the Mach number is 0.5. The reader is referred to [10] for the LES formalism of compressible flows. Computations are

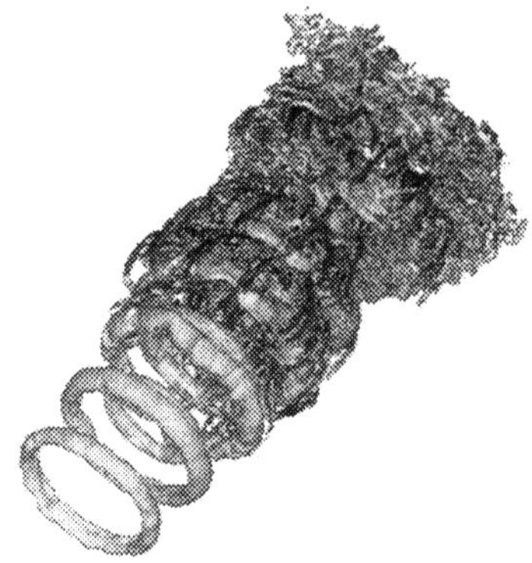

Figure 3. Three-dimensional vortex structures in the numerical simulation of an incompressible coaxial jet (courtesy C. Silva, LEGI, Grenoble).

performed with the *COMPRESS* code developed in Grenoble. The numerical method is a Mac Cormack-type finite differences (see [3], [2]). The numerical scheme is second order accurate in time and fourth order accurate in space. Periodicity is assumed in the spanwise direction. Non reflective boundary conditions (based on the Thompson characteristic method, Thompson, [21])are prescribed at the outlet and the upper boundaries. The computational domain is here decomposed into three blocks. The computational domain is sketched in figure 4. The large dimension of the upstream domain is required by the inlet condition. The coordinate system is located at the upstream edge of the groove. The resolution for the inlet, the groove and the downstream flat plate blocks are respectively $101 \times 51 \times 40$, $41 \times 101 \times 40$ and $121 \times 51 \times 40$. The minimal grid spacing at the wall in the vertical direction corresponds to $\Delta y^+ = 1$. The streamwise grid spacing goes from $\Delta x^+ = 3.2$ near the groove edges to 20 at the outlet. The spanwise resolution is $\Delta z^+ = 16$. The Reynolds number of the flow is 5100, similar to the intermediate simulation of Spalart [20] at $R_\theta = 670$.

One of the difficulty, for this spatially developing flows, is to generate a realistic turbulent flow at the entry of the computational domain. An economical way to generate the inflow is to use the method proposed by Lund *et al.* [13]. This method is based on the similarity properties of canonical turbulent boundary layers. At each time step, the fluctuating velocities, temperatures and pressures are extracted from a plane, called the recycling plane and rescaled at the appropriate inlet scaling. The statistics are found in good agreement with Spalart's data.

Figure 5 shows an instantaneous visualisation of the isosurface of the fluctuation of the streamwise velocity component u. We recognize the well known streaky structures of the boundary layer which are elongated in the flow direction (see [11] for details): these are

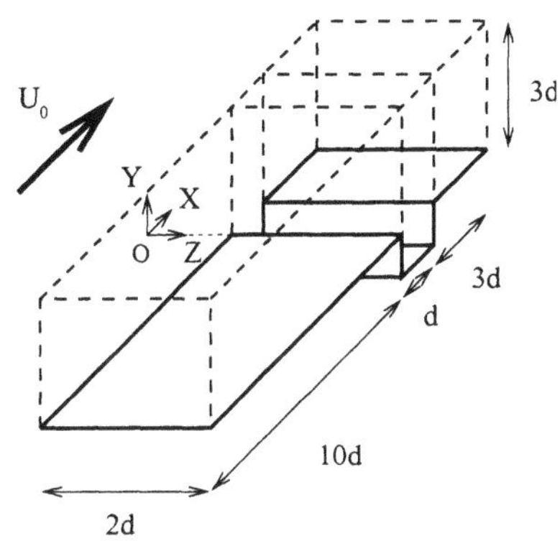

Figure 4. Sketch of the computational domain (courtesy Y. Dubief).

constituted of the well known low- and high-speed streaks. The vertical extent of low-speed streaks is increased as they pass over the groove. The vorticity field is plotted using isosurfaces of the norm of the vorticity, conditioned by positive $Q = (\Omega_{ij}\Omega_{ij} - S_{ij}S_{ij})/2$. The structures downstream of the groove are smaller and less elongated in the streamwise direction (figure 6). It was checked that the statistics show a return towards a more isotropic state downstream of the groove. It was checked that the flow inside the groove is also highly unsteady and there is obviously a high level of communication between the recirculating vortex and the turbulent boundary layer.

5. HEATED FLOWS

The understanding of the dynamics of turbulent flows submitted to strong temperature gradients is still an open challenge for numerical and experimental research. It is of vital importance due to the numerous industrial applications such as the heat exchangers, the cooling of turbine blades, the cooling of rocket engines, etc ... The goal of the present study is to show the ability for LES to adequately reproduce the effects of an asymetric heat flux in a square duct flow. The details of the computations are reported in [17] and [18]. We solve the three-dimensional compressible Navier-Stokes equations with the *COMPRESS* code previously described. We have successively considered the isothermal duct, at a Reynolds number $Re_b = 6000$ (based on the bulk velocity), with the four wall at the same temperature and the heated duct for which the temperature of one of the walls is imposed to be higher than the temperature of the three other walls ($Re_b = 6000$). It is important to note that moderate resolutions are used: the grid consists of $32 \times 50 \times 50$ nodes in the isothermal case and of $64 \times 50 \times 50$ nodes in the heated case along x (streamwise), y and z (transverse) directions. This moderate resolution renders the computation very economical compared with a DNS. One crucial issue in LES is to have a fine description of the boundary layers. In order to correctly simulate the near-wall regions, a nonuniform

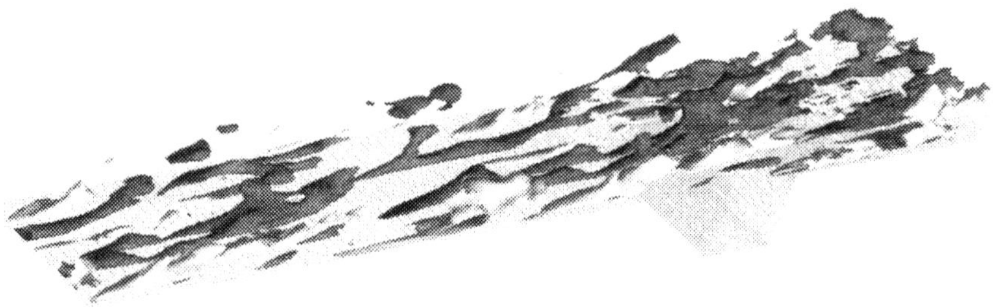

Figure 5. Isosurfaces of streamwise velocity fluctuations. Black $u' = -0.17$, white $u' = 0.17$ (courtesy Y. Dubief).

(orthogonal) grid with a hyperbolic-tangent stretching is used in the y and z directions: the minimal spacing near the walls is here 1.8 wall units. The Mach number is M=0.5 based upon the bulk velocity and the wall temperature.

We have first validated our numerical procedure by comparing our results, for the isothermal duct, with previous incompressible DNS results [7]: a very good agreement was obtained at a drastically reduced computer cost. The flow inside a duct of square cross section is characterized by the existence of secondary flows (Prandtl's flow of second kind) which are driven by the turbulent motion. The secondary flow is a mean flow perpendicular to the main flow direction. It is relatively weak (2-3% of the mean streamwise velocity), but its effect on the transport of heat and momentum is quite significant. If a statistical modelling approach is employed, elaborate second-order models have to been employe to be able to accurately reproduce this weak secondary flow. Figure 7 a) shows the contours of the streamwise vorticity in a quarter of a cross section. The secondary flow vectors reveal the existence of two streamwise counter-rotating vortices in each corner of the duct. The velocity maximum associated with this flow is 1.169% of the bulk velocity: this agrees very well with experimental measurements. It shows the ability for LES to accuratly reproduced statistical quantities. Figure 7 b) shows the instantaneous flow field for the entire duct cross-section. As compared figure 7 a), it clearly indicates a very pronounced flow variability with an instantaneous field very distinct from the mean field. The maximum for the transverse fluctuating velocity field is of the order of ten times the maximum for the corresponding mean velocity field. As far as the vorticity is concerned, the transverse motions are associated with streamwise vorticity generation, whose maximum is about one third of the transverse vorticity maximum.

In the heated case, Salinas and Métais ([19]) have investigated the effect of the heating intensity by varying the temperature ratio between the hot wall and the other walls.

Figure 6. Isosurfaces of the norm the vorticity filtred by positive Q. $\omega = 0.3\omega_i$ (courtesy Y. Dubief).

When the heating is increased, an amplification of the mechanism of ejection of hot fluid from the heated wall is observed. Figure 8 shows temperature structures near the heated wall of the duct. Only one portion of the duct is here represented. As shown on figure 8, these ejections are concentrated near the middle plane of the heated wall. This yields a strong intensification of the secondary flow. It is also shown that the turbulent intensity is reduced near the heated wall with strong heating due to an increase of the viscous effect in that region.

6. CONCLUSION

Turbulence plays a major role in the aerodynamics of cars, trains and planes, combustion in engines, acoustics, cooling of nuclear reactors, dispersion of pollution in the atmosphere and the oceans, or magnetic-field generation in planets and stars. Applications of turbulence, industrial in particular, are thus immense. Since the development of computers in the sixties, so-called industrial numerical models have been created. These models solve Reynolds ensemble-averaged equations of motions (RANS), and they require numerous empirical closure hypotheses which need to be adjusted on given particular experimentally-documented cases. RANS are widely used in the industry. However, it has become clear than RANS models suffer from a lack of universality and require specific adjustments when dealing with a flow submitted to such effects as separation, rotation, curvature, compressibility, or strong heat release.

Classical turbulence modelling, based on one-point closures and a statistical approach allow computation of mean quantities. In many cases, it is however necessary to have access to the fluctuating part of the turbulent fields such as the pollutant concentration or temperature: LES is then compulsory. Large-eddy simulations (LES) of turbulent flows are extremely powerful techniques consisting in the elimination of small scales by a

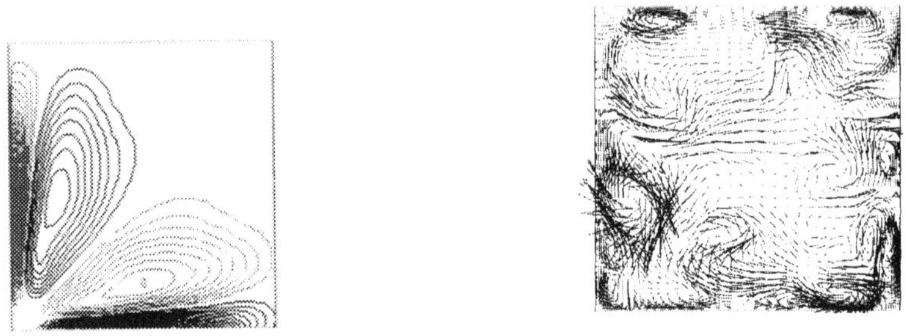

a b

Figure 7. *(a) Ensemble averaged streamwise vorticity contours; (b) Vectors of the instantaneous velocity field (courtesy M. Salinas-Vasquez).*

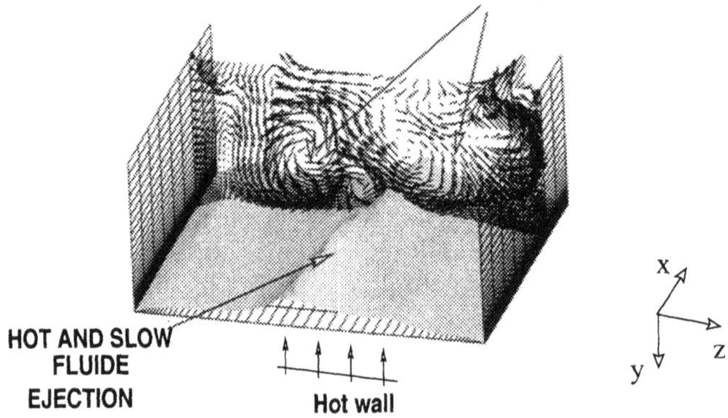

Figure 8. *Large scale motion over the hot wall in a heated duct ($T_h/T_w = 2.5$). Instantaneous transversal vector field and a isosurface of temperature ($T/T_w = 2.1$) (courtesy M. Salinas-Vasquez).*

proper low-pass filtering, and the formulation of evolution equations for the large scales. The latter have still an intense spatio-temporal variability. History of large-eddy simulations (LES) started also at the beginning of the sixties with the introduction of the famous Smagorinsky's (1963) eddy viscosity. Due to the tremendous progress in scientific computing and in particular of parallel computing, LES, which were first confined to very simple flow configurations, are able to deal with more and more complex flows. We have here shown several examples of applications showing that LES are an invaluable tool to decipher the vortical structure of turbulence. Together with DNS, LES is then able to perform deterministic predictions (of flows containing coherent vortices, for instance) and to provide statistical information. The last is very important for assessing and improving one-point closure models, in particular for turbulent flows submitted to external forces (stratification, rotation, ...) or compressibility effects. The ability to deterministically capture the formation and ulterior evolution of coherent vortices and structures is very important for the fundamental understanding of turbulence and for designing efficient turbulent flow control.

The complexity of problems tackled by LES is continuously increasing, and this has nowadays a decisive impact on industrial modelling and flow control. Among the current challenges for LES in dealing with very complex geometries (like the flow around an entire car) are the development of efficient wall functions, the use of unstructured meshes and the use of adaptative meshes. Furthermore, the design of efficient industrial turbulence models will necessarily require an efficient coupling of LES and RANS techniques.

Acknowledgments The results presented have greatly benefitted from the contributions of P. Comte, Y. Dubief, M. Lesieur, M. Salinas-Vasquez, C. Silva, G. Urbin. We are indebted to P. Begou for the computational support. Some of the computations were carried out at the IDRIS (Institut du Développement et des Ressources en Informatique Scientifique, Paris).

REFERENCES

1. Choi, K.S. and Fujisawa, N., 1993, Possibility of Drag Reduction using a *d*-type Roughness, *Appl. Sci. Res.*, **50**, 315–324.
2. Comte, P., 1996, Numerical Methods for Compressible Flows, in *Computational Fluid Dynamics, Les Houches 1993*, Lesieur et al. (eds), Elsesevier Science B.V., 165–219.
3. Comte, P., Silvestrini, J.H. and Lamballais, E., 1995, in *77th. AGARD Fluid Dynamic Panel Symposium "Progress and Challenges in CFD Methods and Algorithms"*, Seville, Spain, 2–5.
4. Dubief, Y., 2000. "Simulation des grandes échelles de la turbulence de la région de proche paroi et des écoulements décollés", PhD thesis. National Polytechnic Institute, Grenoble.
5. Dubief, Y. and P. Comte, 1997, Large-Eddy simulation of a boundary layer flow passing over a groove, in *Turbulent Shear Flows 11*, Grenoble, France, 1-1/1-6.
6. Fouillet, Y., 1992, *Contribution à l'étude par expérimentation numérique des écoulements cisaillés libres. Effets de compressibilité.* PhD thesis. National Polytechnic Institute, Grenoble.

7. Gavrilakis, S., 1992, "Numerical simulation of low Reynolds number turbulent flow through a straight square duct" *J. of Fluis Mech.* **244**, 101.
8. Hunt, J.C.R., Wray, A.A. and Moin, P., 1998, Eddies, stream, and convergence zones in turbulent flows. Center for Turbulence Research Rep., **CTR-S88**, 193.
9. Lee, M., Reynolds, W.C., 1985, Bifurcating and blooming jets at high Reynolds number.*Fifth Symp. on Turbulent Shear Flows*, Ithaca, New York 1.7-1.12.
10. Lesieur, M. and Comte, P., 1997. "Large-eddy simulations of compressible turbulent flows". dans *Turbulence in Compressible flows, AGARD/VKI course, AGARD report 819, ISBN 92-836-1057-1.*
11. Lesieur, M., 1997, *Turbulence in Fluids, Third Revised and Enlarged Edition*, Kluwer Academic Publishers, Dordrecht.
12. Lesieur, M., and Métais, O. (1996) New trends in large-eddy simulations of turbulence", *Annu. Rev. Fluid Mech.* **28**, 45–82.
13. Lund, T.S., Wu, X. and Squires, K. D., 1996, On the Generation of Turbulent Inflow Conditions for Boundary Layer Simulations, *Ann. Res. Briefs*, Stanford, 287–295.
14. Métais, O., Lesieur, M. & Comte, P., 1999, "Large-eddy simulations of incompressible and compressible turbulence", in *Transition, Turbulence and Combustion Modelling*, A. Hanifi *et al.* (eds), ERCOFTAC Series, Kluwer Academic Publishers, 349–419.
15. Michalke, A. and Hermann, G., 1982, On the inviscid instability of a circular jet with external flow. *J.Fluid Mech*, **114**, 343–359.
16. da Silva, C.B. and Métais, O., 2000, "Control of round and coaxial jets", in *Advances in Turbulence VIII*, proceedings of *Eight European Turbulence Conference*, C. Dopazo *et al.* (Eds), CIMNE, pp. 93–96.
17. Salinas-Vazquez, M., 1999. Simulations des grandes échelles des écoulements turbulents dans les canaux de refroidissement des moteurs fusée, PhD thesis. National Polytechnic Institute, Grenoble.
18. Salinas-Vazquez, M., and Métais, O., 1999, Large-eddy simulation of the turbulent flow in a heated square duct, in *Direct and Large Simulation III*, P.R. Voke *et al.* Eds, Kluwer Academic Publishers, 13–24.
19. Salinas-Vazquez, M., and O. Métais, 2000, Large-eddy Simulation of a turbulent flow in a heated duct, in *Advances in Turbulence VIII*, proceedings of *Eight European Turbulence Conference*, C. Dopazo *et al.* (Eds), CIMNE, p. 975.
20. Spalart, P.R., 1988, Direct Simulation of a Turbulent Boundary Layer up to $R_\theta =$ 1410, *J. Fluid Mech.*, **187**, 61–98.
21. Thompson, K.W., 1987, Time Dependent Boundary Conditions for Hyperbolic Systems, *J. Comp. Phys.*, **68**, 506–517.
22. Urbin, G., 1998, Etude numérique par simulation des grandes échelles de la transition à la turbulence dans les jets. PhD thesis. National Polytechnic Institute, Grenoble.
23. Urbin, G. and Métais, O., 1997, Large-eddy simulation of three-dimensional spatially-developing round jets, in *Direct and Large-Eddy Simulation II*, J.P. Chollet, L. Kleiser and P.R. Voke eds., Kluwer Academic Publishers, 35–46.
24. Urbin, G., Brun, C. and Métais, O., 1997, Large-eddy simulations of three-dimensional spatially evolving roud jets, 11th symposium on Turbulent Shear Flows, Grenoble, September 8-11, 25-23/25-28.

Parallel Computational Fluid Dynamics – Trends and Applications
C.B. Jenssen et al. (Editors)

Direct Numerical Simulations of Multiphase Flows*

G. Tryggvason[a] and B. Bunner[b]

[a]Department of Mechanical Engineering, Worcester Polytechnic Institute, 100 Institute Rd., Worcester 01609, USA

[b]Department of Mechanical Engineering, University of Michigan, Ann Arbor, MI 48109, USA

Direct numerical simulations of flows containing many bubbles are discussed. The Navier-Stokes equations are solved by a finite difference/front tracking technique that allows the inclusion of fully deformable interfaces and surface tension, in addition to inertial and viscous effects. A parallel version of the method makes it possible to use large grids and resolve flows containing O(100) three-dimensional finite Reynolds number buoyant bubbles.

1. INTRODUCTION

Multiphase and multifluid flows are common in many natural and technologically important processes. Rain, spray combustion, spray painting, and boiling heat transfer are just a few examples. While it is the overall, integral characteristics of such flow that are of most interest, these processes are determined to a large degree by the evolution of the smallest scales in the flow. The combustion of sprays, for example, depends on the size and the number density of the drops. Generally, these small-scale processes take place on a short spatial scale and fast temporal scale, and in most cases visual access to the interior of the flow is limited. Experimentally, it is therefore very difficult to determine the exact nature of the small-scale processes. Direct numerical simulations, where the governing equations are solved exactly, offer the potential to gain a detailed understanding of the flow. Such direct simulations, where it is necessary to account for inertial, viscous and surface tension forces in addition to a deformable interface between the different phases, still remains one of the most difficult problems in computational fluid dynamics.

Here, a numerical method that has been found to be particularly suitable for direct simulations of flows containing moving and deforming phase boundary is briefly described. Applications of the method to the study of bubbly flows are reviewed in some detail.

2. NUMERICAL METHOD

We consider the three-dimensional motion of a triply periodic monodisperse array of buoyant bubbles with equivalent diameter d, density ρ_b, viscosity μ_b, and uniform surface

*Support by NSF and NASA

tension σ in a fluid with density ρ_f and viscosity μ_f. The array of bubbles is repeated periodically in the three spatial directions with periods equal to L. In addition to the acceleration of gravity, $\mathbf{g}$, a uniform acceleration is imposed on the fluid inside and outside the bubbles to compensate for the hydrostatic head, so that the net momentum flux through the boundaries of the computational domain is zero. The initial condition for the velocity field is zero.

The fluids inside and outside the bubbles are taken to be Newtonian and the flow is taken to be incompressible and isothermal, so that densities and viscosities are constant within each phase. The velocity field is solenoidal:

$$\nabla \cdot \mathbf{u} = 0. \tag{1}$$

A single Navier-Stokes equation with variable density ρ and viscosity μ is solved for the entire computational domain. The momentum equation in conservative form is

$$\frac{\partial \rho \mathbf{u}}{\partial t} + \nabla \cdot \rho \mathbf{u}\mathbf{u} = -\nabla P + (\rho - \rho_0)\mathbf{g} + \nabla \cdot \mu(\nabla \mathbf{u} + \nabla^T \mathbf{u}) + \int \sigma \kappa' \mathbf{n}' \delta^\beta(\mathbf{x} - \mathbf{x}') dA'. \tag{2}$$

Here, $\mathbf{u}$ is the velocity, P is the pressure, $\mathbf{g}$ is the acceleration of gravity, σ is the constant surface tension coefficient, ρ_o is the mean density, κ' is twice the mean local curvature of the front, $\mathbf{n}'$ is the unit vector normal to the front, and dA' is the area element on the front. $\delta^\beta(\mathbf{x} - \mathbf{x}')$ is a three-dimensional δ-function constructed by repeated multiplication of one-dimensional δ-functions. $\mathbf{x}$ is the point at which the equation is evaluated and $\mathbf{x}'$ is a point on the front. This delta function represents the discontinuity of the stresses across the interface, while the integral over the front expresses the smoothness of the surface tension along the interface. By integrating equations 1 and 2 over a small volume enclosing the interface and making this volume shrink, it is possible to show that the velocities and tangential stresses are continuous across the interface and that the usual statement of normal stress discontinuity at the interface is recovered:

$$\left[-P + \mu(\nabla \mathbf{u} + \nabla^T \mathbf{u})\right] \mathbf{n} = \sigma \kappa \mathbf{n}. \tag{3}$$

Here the brackets denote the jump across the interface.

The two major challenges of simulating interfaces between different fluids are to maintain a sharp front and to compute the surface tension accurately. A front tracking method originally developed by Unverdi & Tryggvason [1] and improved by Esmaeeli & Tryggvason [2] is used here. A complete description is available in Tryggvason *et al.* [3]. In addition to the three-dimensional fixed grid on which the Navier-Stokes equation is solved, a moving, deformable, two-dimensional mesh is used to track the boundary between the bubble and the ambient fluid. This mesh consists of marker points connected by triangular elements. The surface tension is represented by a distribution of singularities (delta-functions) located on the moving grid. The gradient of the density and viscosity also becomes a delta function when the change is abrupt across the boundary. To transfer the front singularities to the fixed grid, the delta functions are approximated by smoother functions with a compact support on the fixed grid. At each time step, after the front has been advected, the density and the viscosity fields are reconstructed by integration of the smooth grid-delta function. The surface tension is then added to the nodal values of the discrete Navier-Stokes equations. The front points are advected by the flow velocity, interpolated

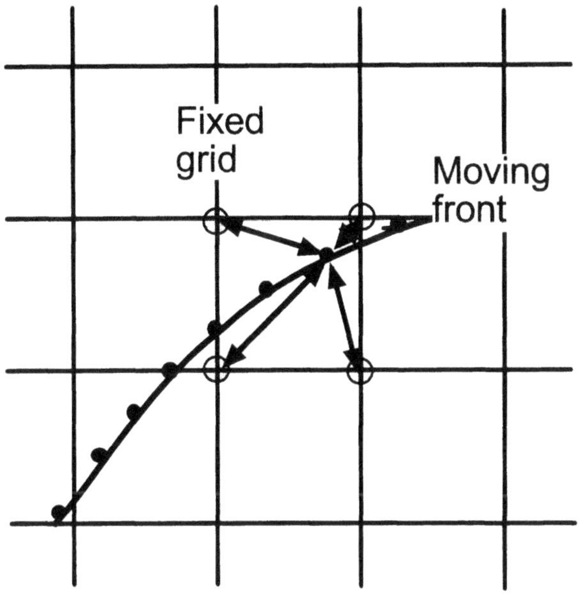

Figure 1. A sketch of the fixed grid and the moving front. The front singularity is approximated by a smoothed function on the fixed grid and the front velocities are interpolated from the fixed grid.

from the fixed grid. See figure 1. Equation 2 is discretized in space by second order, centered finite differences on a uniform staggered grid and a projection method with a second order, predictor-corrector method is used for the time integration.

Because it is necessary to simulate the motion of the bubbles over long periods of time in order to obtain statistical steady state results, an accurate and robust technique for the calculation of the surface tension is critical. This is achieved by converting the surface integral of the curvature over the area of a triangular element ΔS into a contour integral over the edges $\partial\Delta S$ of this element. The local surface tension $\Delta \mathbf{F}_e$ on this element is then:

$$\Delta \mathbf{F}_e = \sigma \int_{\Delta S} \kappa \mathbf{n} dA = \sigma \int_{\partial \Delta S} \mathbf{t} \times \mathbf{n} dl. \tag{4}$$

The tangent and normal vectors $\mathbf{t}$ and $\mathbf{n}$ are found by fitting a paraboloid surface through the three vertices of the triangle ΔS and the three other vertices of the three adjacent elements. To ensure that the two tangent and normal vectors on the common edge of two neighboring elements are identical, they are replaced by their averages. As a consequence, the integral of the surface tension over each bubble remains zero throughout its motion. As a bubble moves, front points and elements accumulate at the rear of the bubble, while depletion occurs at the top of the bubble. It is therefore necessary to add and delete points and elements on the fronts in order to maintain adequate local resolution on the front. The criteria for adding and deleting points and elements are based on the length of the edges of the elements and on the magnitude of the angles of the elements (Tryggvason *et al.*, [3]).

A single bubble of light fluid rising in an unbounded flow is usually described by the Eötvös number (sometimes also called Bond number), $Eo = \rho_f g d^2/\sigma$ and the Morton number, $M = g\mu_f{}^4/\rho_f\sigma^3$ (see [4]). For given fluids, the Eötvös number is a characteristic of the bubble size and the Morton number is a constant. At low Eötvös number, a bubble is spherical. At a higher Eo, it is ellipsoidal and possibly wobbly if the Morton number is low, which is usually the case in low viscosity liquids like water. At a still higher Eo, the bubble adopts a spherical-cap shape, with trailing skirts if the Morton number is high. As they rise, the bubbles move into the other periodic cells in the vertical direction through buoyancy and in the horizontal direction through dispersion. The bubbles are not allowed to coalesce, so that N_b is constant. A fifth dimensionless parameter for this problem is the void fraction, or volume fraction of the bubbly phase, $\alpha = N_b\pi d^3/6L^3$. Since both fluids are assumed to be incompressible, α is constant throughout a simulation. Values of α ranging from 2% to 24% have been considered.

The number of bubbles in the periodic cell, N_b, is an additional parameter, and its effect has been studied by looking at systems with N_b ranging from 1 ro 216 bubbles. It is found that the rise velocity depends only weakly on N_b when N_b is larger than about ten, but the velocity fluctuations and dispersion characteristics of the bubbles are significantly affected by N_b.

Accurate and fast simulations of large, well-resolved, three-dimensional bubble systems can only be obtained on parallel computers. The finite difference/front tracking method was therefore reimplemented for distributed-memory parallel computers using the Message Passing Interface (MPI) protocol (see [5]. Different strategies are employed for the fixed grid and the front due to the different data structures used for these grids. While the fixed grid data, such as velocity, density, viscosity, and pressure, is stored in static arrays, the information describing the front points and elements is stored in several linked lists. The Navier-Stokes solver is parallelized by Cartesian domain decomposition. The computational domain is partitioned into equisized subdomains, where each subdomain is computed by a different processor, and boundary data is exchanged between adjacent subdomains. The front is parallelized by a master-slave technique which takes advantage of the nature of the physical problem to limit programming complexity and provide good performance. When a bubble is entirely within the subdomain of one processor, this subdomain or processor is designated as the 'master' for this bubble. When a bubble is spread over more than one subdomain, the subdomain which contains the largest part of the bubble is master for the bubble, while the other subdomains are the 'slaves'. The master gathers the data for each bubble, performs front restructuring and curvature calculation, and sends the data to the slaves. At each instant, each processor is typically a master for some bubbles and a slave for other bubbles. The main advantage of this approach is to preserve the linked list data structure of each bubble. Therefore, the algorithms developed in the serial code for the front restructuring and curvature can be used in the parallel code with no modification. The only overhead due to parallelization (in addition to the communication time required to exchange the front data between processors) is the additional memory needed to duplicate the front data on several processors.

This memory overhead is aproximately 10% of the entire memory needed for a typical simulation and does not represent a serious penalty on the IBM SP2 parallel computers used here. An alternative approach is to break up the linked list across processors so that each processor supports only the front points which are inside its subdomain, plus a few additional 'ghost' points needed for restructuring and curvature calculation. This approach is computationally more complex because it requires matching of the points and elements at the interprocessor boundaries in order to maintain data coherency.

The solution of the non-separable elliptic equation for the pressure, is by far the most expensive computational operations in our method. The MUDPACK multigrid package [6] was used in the serial code. In the parallel code, we developed a parallel multigrid solver for a staggered mesh. The grid arrangement is vertex-centered, V cycling is used, and the relaxation method at each grid level is red-and-black Gauss-Seidel iteration. The convergence parameters are chosen so that the dimensionless divergence, $(d/g)^{1/2}\nabla \cdot \mathbf{u}$, is about 10^{-8}. Even with the acceleration provided by the multigrid method, 60% to 90% of the total CPU time is spent in the solution of the pressure equation, depending on problem size and void fraction. About half of the remainder is spent on front calculations. The grid and front communications represent between 5 and 10% of the total CPU time.

Since the bubbles are distributed uniformly throughout the flow field, on average, the parallel code is naturally load balanced. However, the parallelization efficiency is degraded by the multigrid solver. Multigrid methods achieve their efficiency gain by coarsening the original grid, and since boundary information must be exchanged among neighboring subdomains at all grid levels, they incur large communication overheads compared to more traditional iteration techniques like SOR. It is important to note that the computational cost of the method depends only moderately on the number of bubbles.

3. RESULT

To examine the behavior of complex multiphase flows, we have done a large number of simulations of the motion of several bubbles in periodic domains. Esmaeeli and Tryggvason [2] examined a case where the average rise Reynolds number of the bubbles remained relatively small, 1-2, and Esmaeeli and Tryggvason [8] looked at another case where the Reynolds number is 20-30. In both cases the deformation of the bubbles were small. The results showed that while freely evolving bubbles at low Reynolds numbers rise faster than a regular array (in agreement with Stokes flow results), at higher Reynolds numbers the trend is reversed and the freely moving bubbles rise slower. Preliminary results for even higher Reynolds numbers indicate that once the bubbles start to wobble, the rise velocity is reduced even further, compared to the steady rise of a regular array at the same parameters. We also observed that there is an increased tendency for the bubbles to line up side-by-side as the rise Reynolds number increases, suggesting a monotonic trend from the nearly no preference found by Ladd [9] for Stokes flow, toward the strong layer formation seen in the potential flow simulations of Sangani and Didwania [10] and Smereka [11]. In addition to the stronger interactions between the bubbles, simulations with a few hundred two-dimensional bubbles at O(1) Reynolds number by Esmaeeli and Tryggvason [7] showed that the bubble motion leads to an inverse energy cascade where the flow structures continuously increase in size. This is similar to the evolution of stirred two-dimensional turbulence, and although the same interaction is not expected in three dimensions, the simulations demonstrated the importance of examining large systems with many bubbles. To examine the usefulness of simplified models, the results were compared with analytical expressions for simple cell models in the Stokes flow and the potential flow limits. The simulations were also compared to a two-dimensional Stokes flow simulation. The results show that the rise velocity at low Reynolds number is reasonably well predicted by Stokes flow based models. The bubble interaction mechanism is, however, quite different. At both Reynolds numbers, two-bubble interactions take place by the "drafting, kissing, and tumbling" mechanism of Joseph and collaborators [12]. This is, of course, very different from either a Stokes flow where two bubbles do not change their relative orientation unless acted on by a third bubble, or the predictions of potential flow where a bubble is repelled from the wake of another one, not drawn into it. For moderate Reynolds numbers (about 20), we find that the Reynolds stresses for a freely evolving two-dimensional bubble array are comparable to Stokes flow while in three-dimensional flow the results are comparable to predictions of potential flow cell models. Most of these computations were limited to relatively small systems, and while Esmaeeli and Tryggvason [7] presented simulations of a few hundred two-dimensional bubbles at a low Reynolds number, the three-dimensional simulations in Esmaeeli and Tryggvason [2] [8] were limited to eight bubbles. For moderate Reynolds numbers the simulations had reached an approximately steady state after the bubbles had risen over fifty diameters, but for the low Reynolds numbers the three-dimensional results had not reached a well defined steady state. The two-dimensional time averages were, on the other hand, well converged but exhibited a dependency on the size of the system. This dependency was stronger for the low Reynolds number case than the moderate Reynolds number one. The vast majority of the simulations done by Esmaeeli and Tryggvason assumed two-dimensional flow. Although many of the qualitative aspects of a few bubble interactions are captured by two-dimensional simulations, the much stronger interactions between two-dimensional bubbles can lead to quantitative differences.

Using a fully parallelized version of the method we have recently simulated several three-dimensional systems with up to 216 three-dimensional buoyant bubbles in periodic domains, Bunner and Tryggvason ([13], [14], [15], [16]). The governing parameters are selected such that the average rise Reynolds number is about 20-30, depending on the void fraction, and deformations of the bubbles are small. Although the motion of the individual bubbles is unsteady, the simulations are carried out for a long enough time so the average behavior of the system is well defined. Simulations with different number of bubbles have been used to explore the dependency of various average quantities on the size of the system. The average rise Reynolds number and the Reynolds stresses are essentially fully converged for systems with 27 bubbles, but the average fluctuation of the bubble velocities requires larger systems. Examination of the pair distribution function for the bubbles shows a preference for horizontal alignment of bubble pairs, independent of system size, but the distribution of bubbles remains nearly uniform. The energy spectrum for the largest simulation quickly reaches a steady state, showing no growth of modes much longer than the bubble dimensions.

To examine the effect of bubble deformation, we have done two sets of simulations using 27 bubbles per periodic domain. In one the bubbles are spherical, in the other the

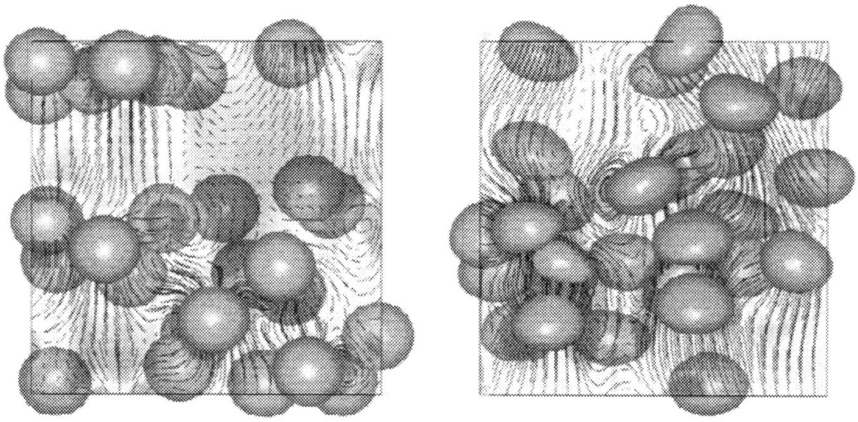

Figure 2. Two frames from simulations of 27 bubbles. In the left frame, the bubbles remain nearly spherical, but in the right frame, the bubble deformations are much larger.

bubbles deform into ellipsoids of an aspect ratio of approximately 0.8. The nearly spherical bubbles quickly reach a well-defined average rise velocity and remain nearly uniformly distributed across the computational domain. The deformable bubbles generally exhibit considerably larger fluctuations than the spherical bubbles and bubble/bubble collisions are more common. Figures 2 shows the bubble distribution along with the streamlines and vorticity for one time from a simulation of 27 bubbles in a periodic domain. Here, N= 900, the void fraction is 12%, and Eo=1 in the left frame and Eo=5 in the right frame. The streamlines in a plane through the domain and the vorticity in the same plane are also shown. In a few cases, usually for small void fractions, and after the bubbles have risen for a considerable distance, the bubbles transition to a completely different state where they accumulate in vertical streams, rising much faster than when they are uniformly distributed. This behavior can be explained by the dependency of the lift force that the bubbles experience on the deformation of the bubbles. For nearly spherical bubbles, the lift force will push bubbles out of a stream, but the lift force on deformable bubbles will draw the bubbles into the stream. Although we have not seen streaming in all the simulations that we have done of deformable bubbles, we believe that the potential for streaming is there, but since the system require fairly large perturbations to reach the streaming state, it may take a long time for streaming to appear. Simulations starting with the bubbles in a streaming state shows that deformable bubbles say in the stream but spherical bubbles disperse.

4. CONCLUSION

The results presented here show the feasibility of using direct numerical simulations to examine the dynamics of finite Reynolds number multiphase flows. Large-scale simulations of systems of many bubbles have been used to gain insight into the dynamics of such flows and to obtain quantitative data that is useful for engineering modeling. The methodology has also been extended to systems with more complex physics, such as surface effects and phase changes.

REFERENCES

1. S. O. Unverdi and G. Tryggvason, "A Front-Tracking Method for Viscous, Incompressible, Multi-Fluid Flows," *J. Comput Phys.* **100** (1992), 25-37.
2. A. Esmaeeli and G. Tryggvason, "Direct Numerical Simulations of Bubbly Flows. Part I—Low Reynolds Number Arrays," *J. Fluid Mech.*, **377** (1998), 313-345.
3. G. Tryggvason, B. Bunner, O. Ebrat, and W. Tauber. "Computations of Multiphase Flows by a Finite Difference/Front Tracking Method. I Multi-Fluid Flows." In: *29th Computational Fluid Dynamics.* Lecture Series 1998-03. Von Karman Institute for Fluid Dynamics.
4. R. Clift, J.R. Grace, and M.E. Weber, *Bubbles, Drops, and Particles.* Academic Press, 1978.
5. W. Gropp, E. Lusk, & A. Skjellum, A. *Portable parallel programming with the message-passing interface.* The MIT Press, 1995.
6. J. Adams, "MUDPACK: Multigrid FORTRAN Software for the Efficient Solution of Linear Elliptic Partial Differential Equations," *Applied Math. and Comput.* **34**, p. 113, (1989).
7. A. Esmaeeli and G. Tryggvason, "An Inverse Energy Cascade in Two-Dimensional, Low Reynolds Number Bubbly Flows," *J. Fluid Mech.*, **314** (1996), 315-330.
8. A. Esmaeeli and G. Tryggvason, "Direct Numerical Simulations of Bubbly Flows. Part II—Moderate Reynolds Number Arrays," *J. Fluid Mech.*, **385** (1999), 325-358.
9. A.J.C. Ladd, "Dynamical simulations of sedimenting spheres," *Phys. Fluids* A, **5** (1993), 299-310.
10. A.S. Sangani and A.K. Didwania, "Dynamic simulations of flows of bubbly liquids at large Reynolds numbers." *J. Fluid Mech.*, **250** (1993), 307-337.
11. P. Smereka, "On the motion of bubbles in a periodic box." *J. Fluid Mech.*, **254** (1993), 79-112.
12. A. Fortes, D.D. Joseph, and T. Lundgren, "Nonlinear mechanics of fluidization of beds of spherical particles." *J. Fluid Mech.* **177** (1987), 467-483.
13. B. Bunner and G. Tryggvason "Direct Numerical Simulations of Three-Dimensional Bubbly Flows." *Phys. Fluids*, **11** (1999), 1967-1969.
14. B. Bunner and G. Tryggvason, "An Examination of the Flow Induced by Buoyant Bubbles." *Journal of Visualization*, 2 (1999), 153-158.
15. B. Bunner and G. Tryggvason, "Dynamics of Homogeneous Bubbly Flows: Part 1. Motion of the Bubbles." Submitted to *J. Fluid Mech.*
16. B. Bunner and G. Tryggvason, "Effect of Bubble Deformation on the Stability and Properties of Bubbly Flows." Submitted to *J. Fluid Mech.*

Parallel Computational Fluid Dynamics – Trends and Applications
C.B. Jenssen et al. (Editors)

Aerodynamic Shape Optimization and Parallel Computing Applied to Industrial Problems

Per Weinerfelt[a] and Oskar Enoksson[b]*

[a]SAAB Aerospace, Linköping, Sweden

[b]Department of Mathematics, Linköping University, Linköping, Sweden

The present paper describes how aerodynamic shape optimization can be applied to industrial aeronautical problems. The optimization algorithm is based on steady flow solutions of the Euler and its adjoint equations from which gradients are computed. Since these calculations are computational intensive parallel computers have to be used. The parallel performance as well as optimization results for some typical industrial problems are discussed.

1. INTRODUCTION

Optimization has become increasingly important for many industries today. By using optimization technique the cost can be reduced and the performance of a product improved. For the aircraft industry multi disciplinary optimization, taking both structure, aerodynamic and electromagnetic aspects into account, have to be performed when designing a complete aircraft.

Concerning aerodynamic shape optimization, which is the topic of the present paper, several issues have to be considered. During take off and landing the flow around an aircraft is subsonic and strongly viscous and hence has to be modelled by the Navier-Stokes equations. A relevant optimization problem is then to design the highlift system of the aircraft so that the ratio L/D (lift over drag) is maximized under both physical and geometrical constraints.

Under transonic cruising condition the Euler or Potential equations are often suitable models for describing the flow. In order to reduced the fuel consumption, and hence the cost, the drag has to be minimized at constant lift and pitching moment as well as geometrical constraints. If we finally consider supersonic flows, the drag from the fore body of an aircraft or a missile can be reduced by controlling the aera distribution of the body. Another way to reduce drag, for an aircraft with delta wings, is to supress the vortex separation at the leading edge of the wing by drooping the wing.

We will in the remaining part of the paper focus on the transonic flow optimization problem.

*The work has been supported by the Swedish national network in applied mathematics (NTM) and the Swedish national graduate school in scientific computing (NGSSC).

Many methods used today in aerodynamic optimization are based on gradient computations. Instead of using finite difference methods for obtaining approximative gradients, gradient methods developed during the last decade by Jameson [1] and others [2]-[6] are preferrable. These methods compute the gradient from the solutions to the flow equations and its adjoint equations. The computational cost is almost independent of the number of design variables which means that this approach is superior to finite difference approximations.

In [3] and [4] a new efficient method for computing the gradient was presented. The main result showed that the gradient can be expressed as a simple surface integral over the design surface. The formulation of the optimization problem as well as the gradient expression are described in the next session.

During the optimization process several steady Euler flow equations have to be computed. The time consuming part, which is the flow and the adjoint computations, are however well suited to parallel computing. As will be shown in section 4.1 these computations scale well on distributed memory machines.

Results from some typical industrial optimization problems are presented in section 5 together with the final conclusions in section 6.

2. MATHEMATICAL FORMULATION OF THE OPTIMIZATION PROBLEM

We will in this section consider a transonic flow optimization problem.

The objective is to minimize the drag on an aircraft under the following constraints

- The Euler flow equations have to be fulfilled
- Prescribed constant lift
- Prescribed constant pitching moment
- Geometrical constraints, such as constant volume or requirements on the shape of the surface

The Euler equations for a 3D steady inviscid fluid read

$$\frac{\partial f_i(w)}{\partial x_i} = 0 \text{ where } w = \begin{pmatrix} \rho \\ \rho \overline{u} \\ \rho E \end{pmatrix} \text{ and } f_i = \begin{pmatrix} \rho \\ \rho \overline{u} \\ \rho H \end{pmatrix} u_i + pI_i \tag{1}$$

Here $\rho, \overline{u}, p$ and H denote the density, velocity, pressure and enthalpy. For future purpose we will split the flux f_i into two parts $f_i = f_{Ni} + f_{Fi}$ where $f_{Ni} = w_H u_i$ and $f_{Fi} = pI_i$ (cf. (1) above). On a solid wall we have the boundary condition $f_{Ni} dS_i = 0$ where $d\overline{S}$ is the surface vector.

The objective function and the physical constraints on the lift and the pitching moment can all be formulated as surface integrals over the solid surface of the aircraft. The pressure force in the direction $\overline{n}$ on the surface $B_W(a)$ reads

$$F_n = \int_{B_W(a)} p\, n_i\, dS_i, \tag{2}$$

and the total moment around an axis $\overline{n}$ at x_0

$$M_n = \int\limits_{B_W(a)} p\,\varepsilon_{kji}(x_k - x_{0k})n_j\,dS_i, \tag{3}$$

The computation of the gradients of (2) and (3), with respect to a design variable a, will be discussed in the next section.

2.1. The gradient formulation

Since our optimization technique is based on both function and gradient evaluations derivatives of (2) and (3), with respect to a design variable a, have to computed. The expressions in (2) and (3) lead us to consider the following general surface integral

$$I(a) = \int\limits_{B_W(a)} \varphi_i(x, p(w(x,a)))dS_i \tag{4}$$

By using the main result from reference [3] and [4] we can express the derivative of the integral (4) as

$$\frac{d}{da}\int\limits_{B_W(a)} \varphi_i dS_i = \int\limits_{B_W(a)} \frac{\partial \varphi_i}{\partial a}dS_i + \int\limits_{B_W(a)} \frac{\partial \varphi_i}{\partial x_i}\frac{\partial x_k}{\partial a}dS_k \tag{5}$$

Let us introduce the fields ψ and η and the Lagrangian $\mathcal{L}$

$$\mathcal{L}(a) = \int\limits_{B_W(a)} \left(\varphi_i - \eta^t f_{Ni}\right) dS_i - \int\limits_{D(a)} \psi^t \frac{\partial f_i}{\partial x_i}dV \tag{6}$$

where $D(a)$ is the flow domain. Observe that $\mathcal{L}(a) = I(a)$ due to the Euler equations and boundary conditions. Differentiating $\mathcal{L}$ with respect to a and applying (5) to the first integral in (6) we get

$$\frac{d}{da}\int\limits_{B_W(a)} \left(\varphi_i - \eta^t f_{Ni}\right) dS_i =$$

$$\int\limits_{B_W(a)} \left(\frac{\partial \varphi_i}{\partial w} - \eta^t \frac{\partial f_{Ni}}{\partial w}\right)\frac{\partial w}{\partial a}dS_i + \int\limits_{B_W(a)} \frac{\partial}{\partial x_i}\left(\varphi_i - \eta^t f_{Ni}\right)\frac{\partial x_k}{\partial a}dS_k \tag{7}$$

For the second integral in (6) we have

$$\frac{d}{da}\int\limits_{D(a)} \psi^t \frac{\partial f_i}{\partial x_i}dV = \int\limits_{D(a)} \psi^t \frac{\partial}{\partial a}\frac{\partial f_i}{\partial x_i}dV = \int\limits_{\partial D(a)} \psi^t \frac{\partial f_i}{\partial w}\frac{\partial w}{\partial a}dS_i - \int\limits_{D(a)} \frac{\partial \psi^t}{\partial x_i}\frac{\partial f_i}{\partial w}\frac{\partial w}{\partial a}dV \tag{8}$$

Summing up (7) and (8) leads to

$$\frac{d\mathcal{L}}{da} = \frac{d}{da}\int\limits_{B_W(a)} \varphi_i dS_i = \int\limits_{B_W(a)} \left(\frac{\partial \varphi_i}{\partial w} - \eta^t \frac{\partial f_{Ni}}{\partial w}\right)\frac{\partial w}{\partial a}dS_i$$

$$+ \int\limits_{B_W(a)} \frac{\partial}{\partial x_i}\left(\varphi_i - \eta^t f_{Ni}\right)\frac{\partial x_k}{\partial a}dS_k$$

$$- \int\limits_{\partial D(a)} \psi^t \frac{\partial f_i}{\partial w}\frac{\partial w}{\partial a}dS_i + \int\limits_{D(a)} \frac{\partial \psi^t}{\partial x_i}\frac{\partial f_i}{\partial w}\frac{\partial w}{\partial a}dV \tag{9}$$

The derivative $\partial w/\partial a$ can be eliminated by letting ψ be the solution to the adjoint equation below and by putting $\eta = -\psi$ on the boundary.

$$\frac{\partial \psi^t}{\partial x_i}\frac{\partial f_i}{\partial w} = 0 \quad in \quad D$$
$$\left(\frac{\partial \varphi_i}{\partial w} - \psi^t \frac{\partial f_{Fi}}{\partial w}\right) dS_i = 0 \quad on \quad B_W$$
$$\psi^t = 0 \quad on \quad \partial D - B_W$$

The only remaining terms in (9) are

$$\frac{d}{da}\int\limits_{B_W(a)} \varphi_i dS_i = \int\limits_{B_W(a)} \frac{\partial}{\partial x_i}\left(\varphi_i + \psi^t w_H u_i\right)\frac{\partial x_k}{\partial a} dS_k \tag{10}$$

Equation (10) is the final expression for the gradient. As can be seen from the formula only integration over the solid surface has to be considered. We will end this session by applying equation (10) to the aerodynamic force and moment described earlier. For the force in the direction $\overline{n}$ in equation (2) we have

$$\frac{dF_n}{da} = \int\limits_{B_W(a)} \frac{\partial}{\partial x_i}(p n_i + \psi^t w_H u_i)\frac{\partial x_k}{\partial a} dS_k$$
$$(\psi^t I_i - n_i) dS_i = 0 \tag{11}$$

where (11) is the adjoint solid wall boundary condition.

For the pitching moment around an axis $\overline{n}$ at x_0 in equation (3), we have a similar expression as in (11)

$$\frac{dM_n}{da} = \int\limits_{B_W(a)} \frac{\partial}{\partial x_i}(p\,\varepsilon_{kji}(x_k - x_{0k})n_j + \psi^t w_H u_i)\frac{\partial x_k}{\partial a} dS_k$$
$$(\psi^t I_i - \varepsilon_{kji}(x_k - x_{0k})n_j)\, dS_i = 0$$

3. OPTIMIZATION METHOD

From equation (4) and (10) follows the approximation

$$\delta I \approx \int\limits_{B_W(a)} G\delta x_k n_k \, dS \tag{12}$$

where $G = \dfrac{\partial}{\partial x_i}(\varphi_i + \psi^t w_H u_i)$. Equation (12) can be considered as a scalar product, denoted by $< \cdot, \cdot >$, between the gradient and the projected surface correction $\delta x_k n_k$ where $\overline{n}$ is the surface unit normal vector. Assume that the surface correction is written

$$\delta x_k = \sum_j \sum_i c_{ij_k} b_{ij} \tag{13}$$

where c_{ij_k} are coefficients and b_{ij} arbitrary basis functions. Inserting (13) into (12) results in

$$\delta I \approx \sum_j \sum_i c_{ij_k} < G, n_k b_{ij} >$$

Observing that the last sum is a tensor inner product, here denoted by $(\cdot, \cdot)$, we finally obtain the following expression for the variation δI

$$\delta I \approx (c, g) \tag{14}$$

where c and g are the tensors defined by $(c)_{ijk} = c_{ij_k}$, $(g)_{ijk} =< G, n_k b_{ij} >$.

The original optimization problem is nonlinear and thus has to be solved iteratively. In each iteration step the linear approximation below is obtained by linearization

$$\begin{cases} \min\limits_c \left(c, g^0\right) \\ (c, g^m) = \Delta^m, \quad m = 1, ..., M \\ (c, h^n) = \Delta^n, \quad n = 1, ..., N \end{cases} \tag{15}$$

where g^0 is the gradient of the objective function, g^m the gradients of M physical constraints, h^n the gradients of N geometrical constraints and $\Delta^{m,n}$ deviations from the target values of the constraints. We also need to impose upper and lower bounds on the coeffiecients c in order to assure a bounded solution.

Our experience is that the solution to (15) might result in too large values on the coefficents c which in turn leads to an unphysical design. We have instead replaced the minimization formulation above by the following problem

$$\begin{cases} \min\limits_c ||c||^2, \\ \left(c, g^0\right) = \Delta^0 \\ (c, g^m) = \Delta^m, \quad m = 1, ..., M \\ (c, h^n) = \Delta^n, \quad n = 1, ..., N \end{cases} \tag{16}$$

which is reasonable from engineering point of view. Δ^0 is a user defined parameter determining the decrease of the objective function in each design step.

The method above can be considered as a constraint steepest descent method similar to the one described in [7].

3.1. Surface modification and parametrization

When the solution c to (16) is determined, a new surface grid is created by adding the corrections, obtained from (13), to the existing surface grid. A number of different basis functions, describing the surface modification, has been implemented. The following options are avaiblable at present

- Smoothed gradients
- Set of wing profiles
- Sinusoidal bumpfunction
- B-splines functions

The last three functions above are one dimensional but the extension to a surface is obtained by simply taking the tensor product of the basis functions in each surface coordinate direction.

4. DESCRIPTION OF THE OPTIMIZATION CODE/SYSTEM

When working in an industrial environment emphasis has to be put on robustness, efficiency and flexibility of computer programs. To meet these requirements the well known Jameson scheme, for structured multiblock grids, has been employed to both the flow and the adjoint solver. The equations have thus been discretized in space by a cell centered finite volume method. Second and fourth order artificial viscosity is used to capture shocks and damp spurious oscillations. A Runge-Kutta scheme is applied as the basic time stepping method, and multigrid and local time stepping are used to accelerate convergence to steady state. In order to fulfill a prescribed lift constraint the angle of attack α is adjusted until the constraint is satisfied. The Euler and adjoint solver have also been parallelized using MPI.

The solver consist to a large extend of modules written in an object oriented language (C++). A few time consuming subroutines were written in FORTRAN in order to ensure high efficiency on vector and parallel computers. The main reason for using an object oriented approach is that different cost functions and constraints, on both the flow solution and the design variables, are (and will be) implemented and hence the modularity of the program has high priority. We have also taken into account future extension of the program to new applications such as coupled structure/fluid optimization.

4.1. Parallelization

The Euler and adjoint solver are parallelized using MPI. The multiblock structure makes the parallelization straightforward. A load balancing of the original problem is first computed. Block splitting can be performed by using a graphical user interface. The blocks are then distributed, according to the result from the load balancing, over the number of processors. The flow in each block is updated by the time stepping scheme and the new boundary data, computed at each time step, is exchanged between the processors by message passing.

The program has been tested and validated on workstations such as SGI, Digital, Sun and PC-linux as well as the super/parallel computer IBM SP2 and SGI Power Challenge.

4.2. The optimization system cadsos

The optimization code has been integrated into an optimization system called **cadsos** (Constraint Aero Dynamic Shape Optimzation System). An overview of the system is shown in figure 1 below. The Euler and Adjoint solver compute solutions from which gradients are calculated. In order to obtain the gradients of the objective function and the physical constraints an adjoint solution has to be computed for each of them. If the optimality criteria is not fulfilled then the function values and gradients are passed to the surface updating module which is written in MATLAB. A number of different basis functions, describing the surface modifications, have been implemented as we have seen in section (3.1). After modifying the surface grid, according to the method in section 3, a volume grid is computed. This can either by done by means of a mesh generator, for single wings, or by a volume perturbation technique. The surface modifications are in the last case propagated from the surface into the volume and added to the existing grid. The new volume grid is finally fed into the flow solver and the optimization loop is then completed.

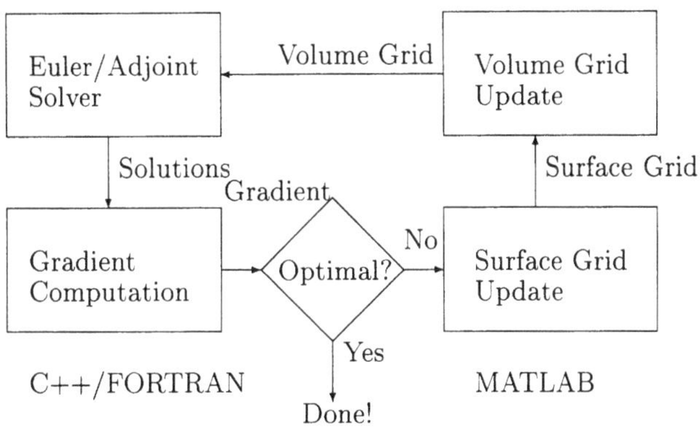

Figure 1. Overview of the optimization system cadsos

5. RESULTS

The cadsos system has been applied to several 2D and 3D problems. We will in this section discuss three typical problems of industrial interest.

5.1. Optimization of a 2D wing profile

In the first example a 2D wing profile optimization is considered. The flow is assumed to be inviscid and modelled by the Euler equations. The objective is to design a drag free airfoil, (this is only possible in 2D inviscid flows) with prescribed lift and pitching moment as well as thickness constraints on the airfoil. As starting geometry the ONERA M6 wing profile was chosen. The flow at the free stream condition $M = 0.84$ and $\alpha = 3.0^o$ was first computed around the original geometry in order to get constraint values on the lift and pitching moment. Optimization was then performed for three types of surface modifications

i) a set of 12 wingprofiles
ii) a set of 24 wingprofiles
iii) a set of 20 sinusoidal bump functions

The drag converence histories are displayed in the figures 2-4 below. For all cases convergence was achived within less than 20 design cycles. The lowest drag is obtain by using the sinusoidal bump functions.

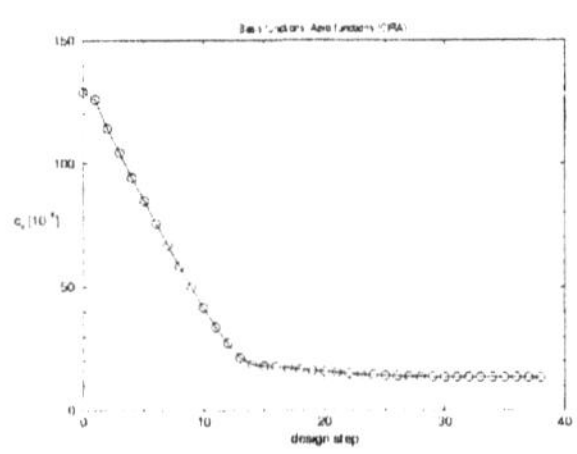

Figure 2. Drag convergence history using surface modification i) in section 5.1.

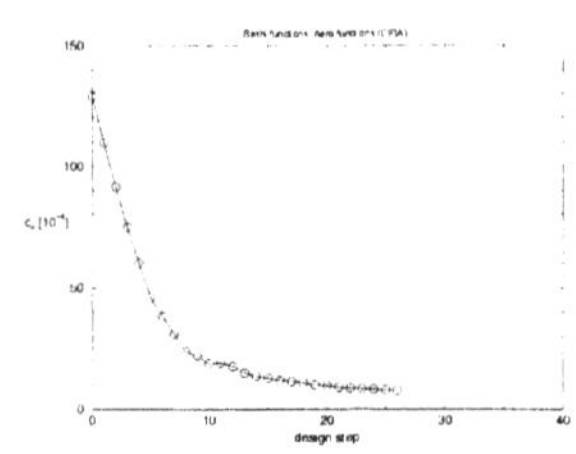

Figure 3. Drag convergence history using surface modification ii) in section 5.1.

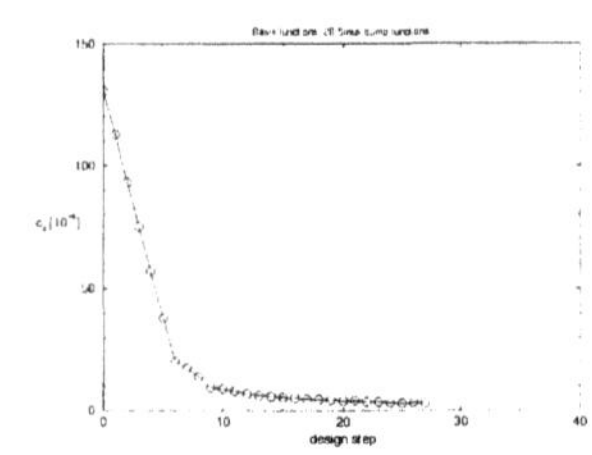

Figure 4. Drag convergence history using surface modification iii) in section 5.1.

The original and optimized wing profiles are displayed in the figures 5-7. Notices the similarity of the optimized profiles. In figure 8-10 finally the C_p distribution is plotted. The strong shock wave, which is present in the original pressure distribution, has been completely removed. Since the only drag contribution comes from the shock wave, a drag close to zero is achieved after optimization (ses figures 2-4).

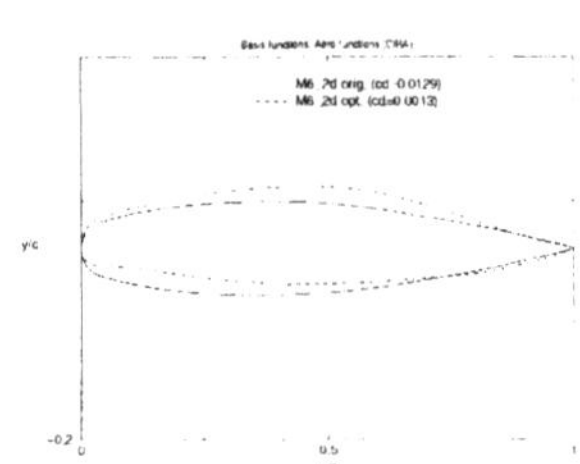

Figure 5. Original and optimized wing profiles using surface modification i) in section 5.1.

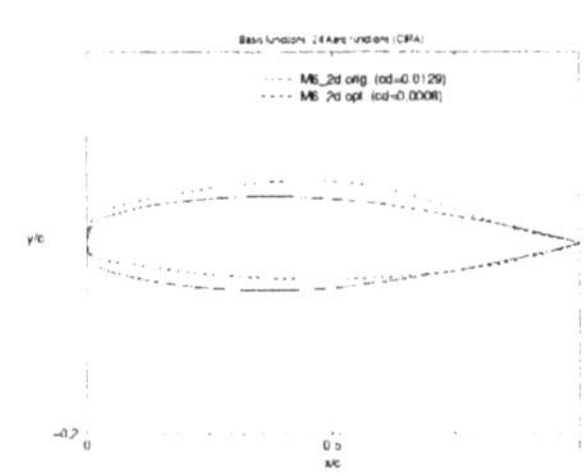

Figure 6. Original and optimized wing profiles using surface modification ii) in section 5.1.

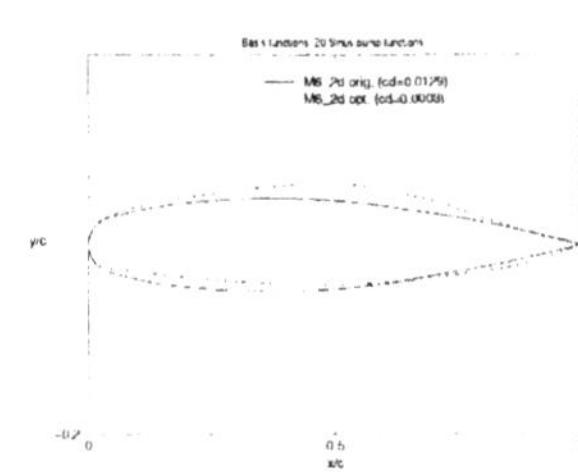

Figure 7. Original and optimized wing profiles using surface modification iii) in section 5.1.

5.2. Optimization of a 3D wing

In the second example minimization of the drag over the ONERA M6 wing was studied. The same free stream condition as in the previous example was chosen. A grid consisting

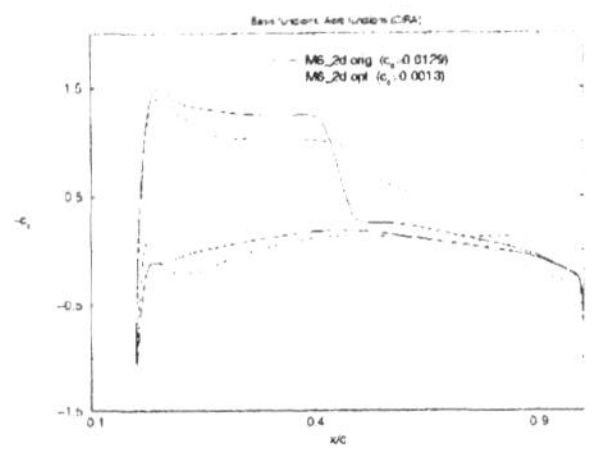

Figure 8. The c_p distribution over the original and optimized wing profile using surface modification i) in section 5.1.

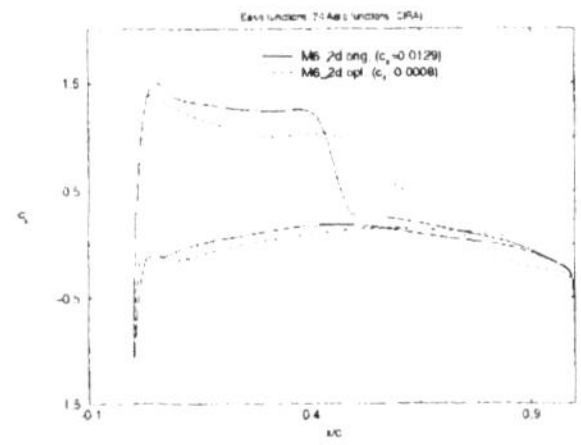

Figure 9. The c_p distribution over the original and optimized wing profile using surface modification ii) in section 5.1.

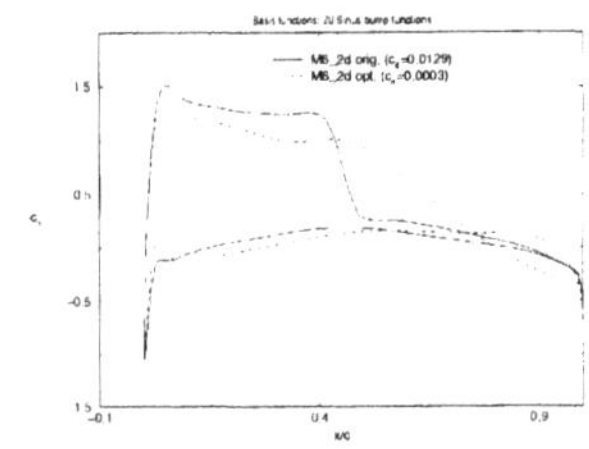

Figure 10. The c_p distribution over the original and optimized wing profile using surface modification iii) in section 5.1.

of totally 295 000 cells was generated around the wing. For parallel computations up to 8 block was used. The optimization was performed at fixed lift and pitching moment using the basis functions i) in the previous section. The pressure distribution over the original and optimized wing are diplayed in figure 11 and 12. We can clearly see that the lambda shock pattern on the original wing has disappeared after optimization. This can also be seen in the plots 13-15 below. The strength of the first shock is slightly reduced whereas the second one is almost gone. The drag has decreased from 152 to 114 drag counts[2] in 10-15 design steps (see figure 19) resulting in a drag reduction of 25% . Figure 16-18 show the original and the optimized wing at three span stations 15%, 50% and 95%.

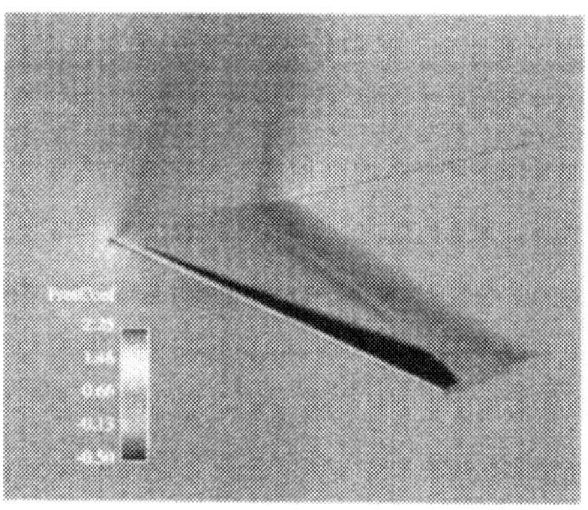

Figure 11. C_p distribution over the original ONERA M6 wing.

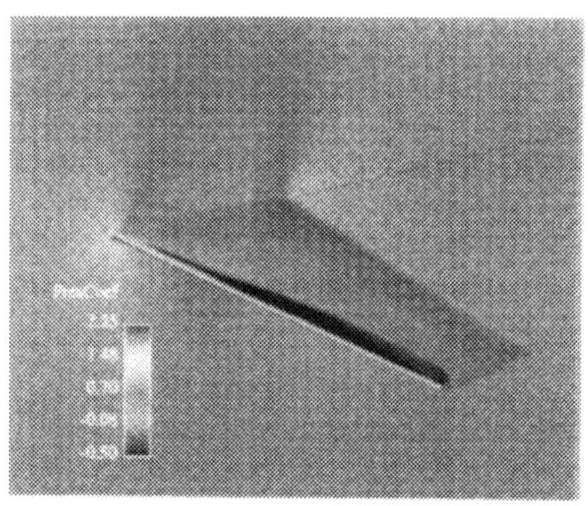

Figure 12. C_p distribution over the optimized wing.

[2](1 drag count=$1 \cdot 10^{-4}$)

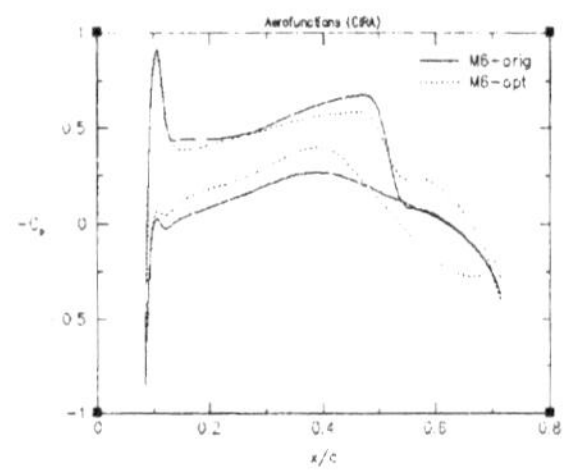

Figure 13. C_p distribution at 15% span station of the original and optimized ONERA M6 wing.

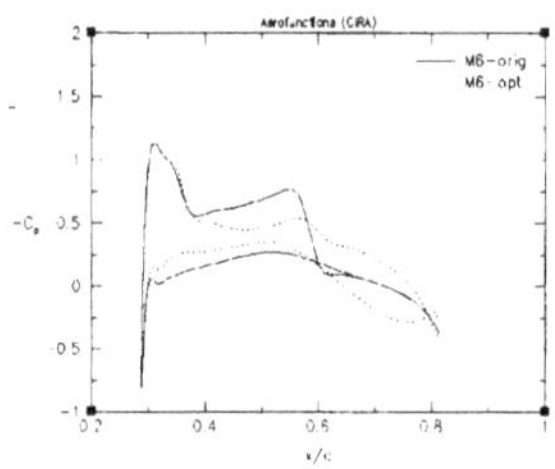

Figure 14. C_p distribution at 50% span station of the original and optimized ONERA M6 wing.

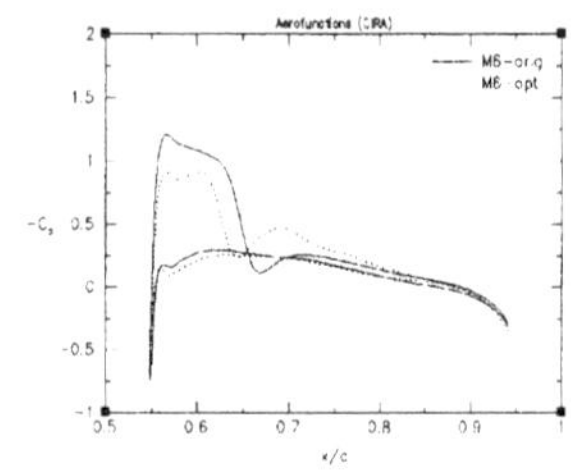

Figure 15. C_p distribution at 95% span station of the original and optimized ONERA M6 wing.

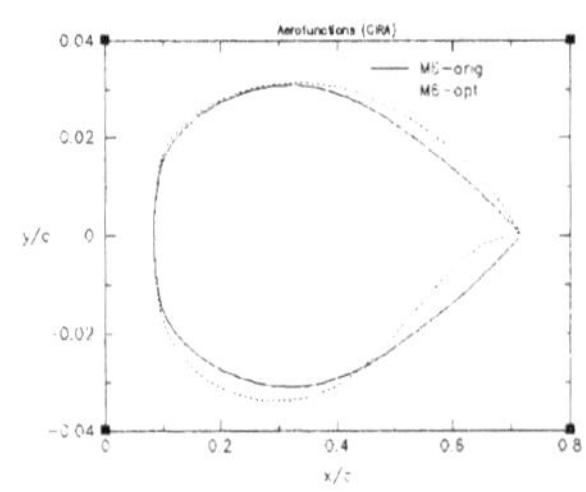

Figure 16. Wing profiles at 15% span station of the original and optimized ONERA M6 wing.

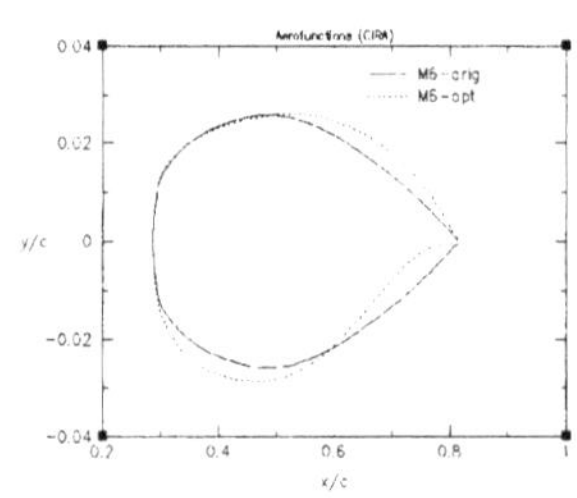

Figure 17. Wing profiles at 50% span station of the original and optimized ONERA M6 wing.

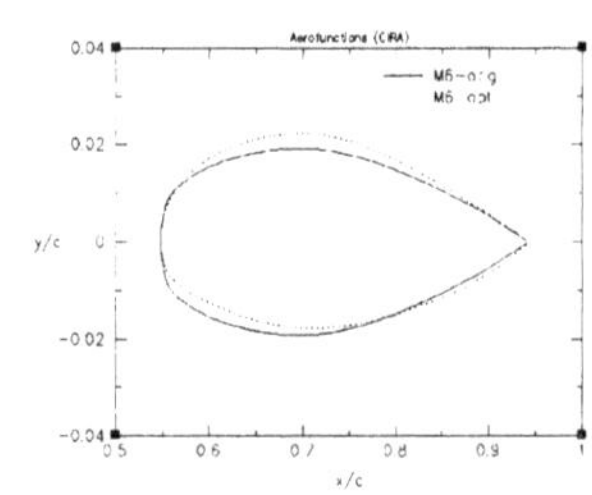

Figure 18. Wing profiles at 95% span station of the original and optimized ONERA M6 wing.

In order to measure the parallel performance of the code the flow calculations were done on an SGI Power Challenge system. An almost linear speed up curve was obtained (see figure 20) for both the Euler and the adjoint calculations.

5.3. Optimization of an aircraft

The last example shows how aero dynamic shape optimization can be used within an industrial project. The optimization aim was to reduce the drag and the pressure load at the wing tip of an UAV (unmanned aerial vehicle). Euler calulations were perfomed on a multi block grid consisting of 18 blocks and 792 000 cells. The free stream condition was $M = 0.8$ and $\alpha = 3.0^o$. The lift coefficient was fixed during the optimization. The optimization was done in two steps. First an optimal twist distribution was computed (figure 21). Secondly the wing profile form was improved (figure 22).

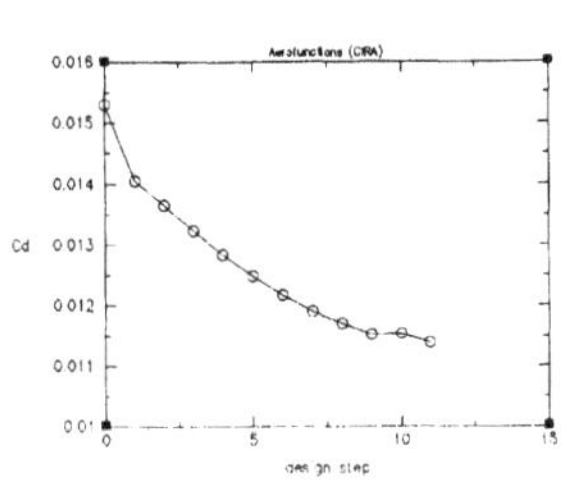

Figure 19. Drag convergence history ONERA M6 wing optimization.

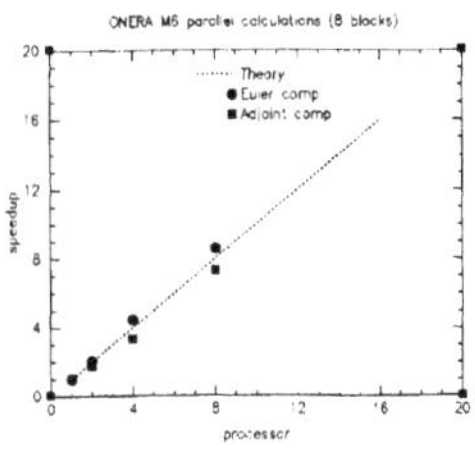

Figure 20. Speed up results, parallel flow computations for the ONERA M6 wing.

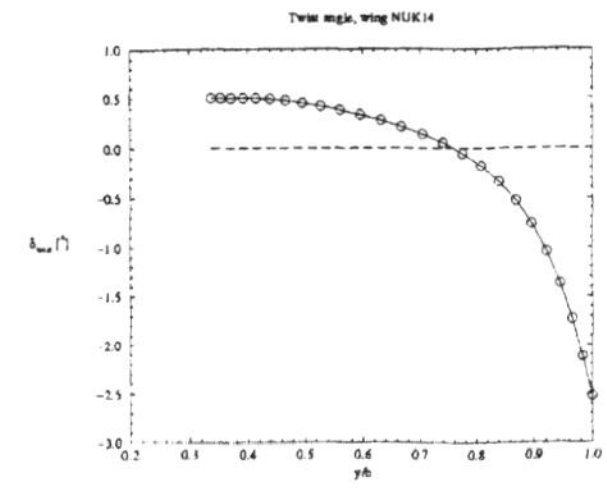

Figure 21. Twist distribution of the UAV wing.

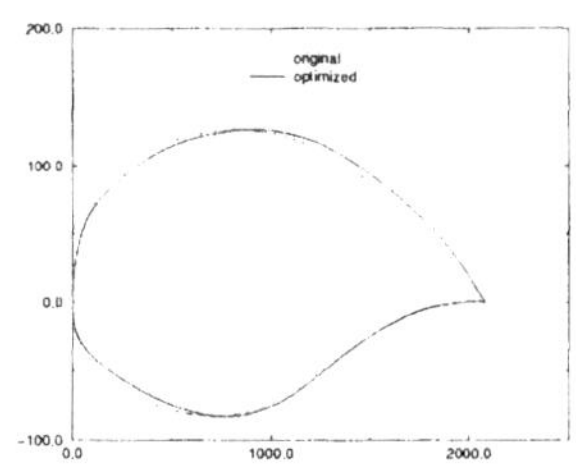

Figure 22. UAV wing profile at the 56% span station.

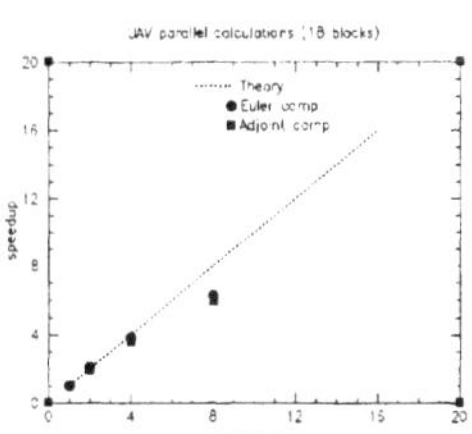

Figure 23. Speed up results 18 blocks, parallel flow computations for the UAV.

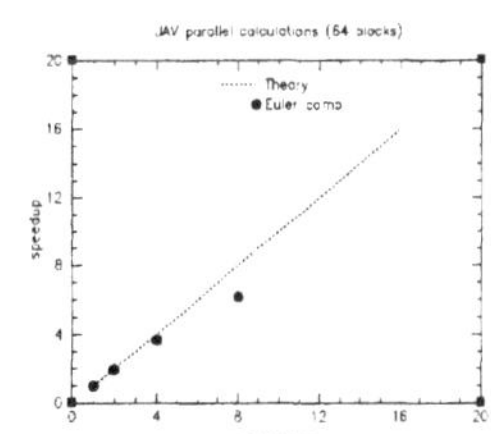

Figure 24. Speed up results 64 blocks, parallel flow computations for the UAV.

This resulted totally in a drag reduction of 7%. We can see in the figure 25 and 26 that the pressure load at the wing tip has been decreased after optimization. This is due to the fact that the modyfied twist distribution leads to a better flow attachment at the leading edge.

Figure 23 and 24 finally show that good speedup results can be obtained also for realistic 3D flow calcultions and optimization.

6. CONCLUSION

We have in the present paper demonstrated the capability and applicability of a gradient based optimization method to 2D and 3D industrial flow problems. We have discussed efficient methods for computing the gradients by using the Euler and its adjoint equations. Our optimzation system, **cadsos**, fulfills criteria such as generality, modularity and robustness. We have finally demonstrated that the optimization process can be efficienly parallelized using MPI on distributed memory computers.

Figure 25. C_p distribution over the original UAV.

Figure 26. C_p distribution over the optimized UAV.

REFERENCES

1. A. Jameson, Optimum Aerodynamic Design Using, Control Theory, CFD Review, Wiley,1995, pp.495-528
2. J. Reuther et. al., Constrained Multipoint Aerodynamic Shape Optimization, Adjoint Formulation and Parallel Computers, AIAA paper no. AIAA 97-0103
3. P. Weinerfelt & O. Enoksson, Numerical Methods for Aerodynamic Optimization, Accepted for publication in CFD Journal 2000
4. O. Enoksson, Shape Optimization in Compressible Inviscid Flow, LiU-TEK-LIC-2000:31, ISBN 91-7219-780-3, Department of Mathematics, Linkping University, Sweden
5. P. Weinerfelt & O. Enoksson, Aerodynamic Optimization at SAAB, Proceedings to the 10th Conference of the European Consortium for Mathematics in Industry (ECMI 98), June 22-27 1998 in Gothenburg, Sweden
6. B.I. Soemarwoto, Airfoil optimization using the Navier-Stokes Equations by Means of the Variational Method, AIAA paper no. AIAA 98-2401
7. J. Elliot & J. Peraire, Constrained, Multipoint Shape optimization for Complex 3D Configurations, The Aero - nautical Journal, August/Septeber 1998, Paper no. 2375, pp.365-376

2. Affordable Parallel Computing

Parallel Computational Fluid Dynamics – Trends and Applications
C.B. Jenssen et al. (Editors)

Accurate Implicit Solution of 3-D Navier-Stokes Equations on Cluster of Work Stations

Ü.Gülçat[a] and V.Ü.Ünal[b]
[a]Faculty of Aeronautics and Astronautics, Istanbul Technical University, 80626, Maslak, İstanbul, Turkey
[b]Faculty of Science, Yeditepe University

Parallel implicit solution of Navier-Stokes equations based on two fractional steps in time and Finite Element discretization in space is presented. The accuracy of the scheme is second order in both time and space domains. Large time step sizes, with CFL numbers much larger than unity, are used. The Domain Decomposition Technique is implemented for parallel solution of the problem with matching and non-overlapping sub domains. As a test case, lid driven cubic cavity problem with 2 and 4 sub domains are studied.

1. INTRODUCTION

Stability requirements for explicit schemes impose severe restrictions on the time step size for analyzing complex viscous flow fields which are, naturally, to be resolved with fine grids. In order to remedy this, implicit flow solvers are used in analyzing such flows. The time and space accuracy of a numerical scheme is an important issue in the numerical study of complex flows. The higher order accurate schemes allow one to resolve a flow field with less number of grid points while taking large time steps. Resolving a flow field with less number of points gives a great advantage to implicit schemes, since the size of the matrix to be inverted becomes small.

In this study a second order accurate scheme, both in time and space, is developed and implemented for parallel solution of N-S equations. A modified version of the two step fractional method, [1], is used in time discretization of the momentum equation which is implicitly solved for the intermediate velocity field at each time step. The space is discretized with brick elements. The pressure at each time level is obtained via an auxiliary scalar potential which satisfies the Poisson's equation. The Domain Decomposition Technique, [2,3,4], is implemented saperately for parallel solution of the momentum and pressure equations using non-overlapping matching grids.

Lid-driven flow in a cubic cavity with a Reynolds number of 1000 is selected as the test case to demonstrate the accuracy and the robustness of the method used. The mesh employed here has 2x(25x13x13) for 2 domain and 4x(25x13x7) grid points for 4 domain solutions. The speed up is 1.71 as opposed to ideal value of 2., and overall parallel efficiency is 85 %.

* This work is supported by TUBITAK: Project No. COST-F1

A cluster of DEC Alpha XL266 work stations running Linux operating sytem, interconnected with a 100 Mbps TCP/IP network is used for computations. Public version of the Parallel Virtuel Machine, PVM 3.3, is used as the communication library.

2. FORMULATION

2.1 Navier-Stokes equations

The flow of unsteady incompressible viscous fluid is governed with the continuity equation

$$\nabla.\mathbf{u} = 0 \tag{1}$$

and the momentum (Navier-Stokes) equation

$$\frac{D\mathbf{u}}{Dt} = -\nabla p + \frac{1}{Re}\nabla^2 \mathbf{u} \tag{2}$$

The equations are written in vector form(here on, boldface type symbols denote vector or matrix quantities). The velocity vector, pressure and time are denoted by **u**, p and t, respectively. The variables are non-dimensionalized using a reference velocity and a characteristic length. Re is the Reynolds number, $Re = Ul/\nu$ where U is the reference velocity, l is the characteristic length and ν is the kinematic viscosity of the fluid.

2.2 FEM formulation

The integral form of Eqn. (2) over the space-time domain reads as

$$\int_{\Omega}\int_{t} \frac{\partial \mathbf{u}}{\partial t}\mathbf{N}\,d\Omega\,dt = \int_{\Omega}\int_{t}(-\mathbf{u}.\nabla\mathbf{u} - \nabla p + \frac{1}{Re}\nabla^2\mathbf{u})\mathbf{N}\,d\Omega\,dt \tag{3}$$

where **N** is an arbitrary weighting function. The time integration of both sides of Eqn. (3) for half a time step, $\Delta t/2$, from time step n to n +1/2 gives

$$\int_{\Omega}(\mathbf{u}^{n+1/2} - \mathbf{u}^{n})\mathbf{N}d\Omega = \frac{\Delta t}{2}\int_{\Omega}(-\mathbf{u}.\nabla\mathbf{u}^{n+1/2} - \nabla p^{n} + \frac{1}{Re}\nabla^2\mathbf{u}^{n+1/2})\mathbf{N}d\Omega. \tag{4}$$

At the intermediate time step the time integration of Eqn. (3), where the convective and viscous terms are taken at n $+\frac{1}{2}$ and pressure term at time level n, yields

$$\int_{\Omega}(\mathbf{u}^{*} - \mathbf{u}^{n})\mathbf{N}d\Omega = \Delta t\int_{\Omega}(-\mathbf{u}.\nabla\mathbf{u}^{n+1/2} - \nabla p^{n} + \frac{1}{Re}\nabla^2\mathbf{u}^{n+1/2})\mathbf{N}d\Omega. \tag{5}$$

For the full time step, the averaged value of pressure at time levels n and n+1 is used to give

$$\int_{\Omega}(\mathbf{u}^{n+1} - \mathbf{u}^{n})\mathbf{N}d\Omega = \Delta t\int_{\Omega}(-\mathbf{u}.\nabla\mathbf{u}^{n+1/2} + \frac{1}{Re}\nabla^2\mathbf{u}^{n+1/2} - \nabla\frac{p^{n} + p^{n+1}}{2})\mathbf{N}d\Omega. \tag{6}$$

Subtracting (5) from (6) results in

$$\int_{\Omega}(\mathbf{u}^{n+1}-\mathbf{u}^{*})\mathbf{N}d\Omega=\frac{\Delta t}{2}\int_{\Omega}-\nabla(p^{n+1}-p^{n})\mathbf{N}d\Omega. \quad (7)$$

If one takes the divergence of Eqn. (7), the following is obtained;

$$\int_{\Omega}\nabla.\mathbf{u}^{*}\mathbf{N}d\Omega=-\frac{\Delta t}{2}\int_{\Omega}\nabla^{2}(p^{n+1}-p^{n})\mathbf{N}d\Omega. \quad (8)$$

Subtracting (4) from (5) yields

$$\mathbf{u}^{*}=2\mathbf{u}^{n+1/2}-\mathbf{u}^{n}. \quad (9)$$

2.3 Numerical Formulation

Defining the auxiliary potential function $\phi=-\Delta t(p^{n+1}-p^{n})$ and choosing N as trilinear shape functions, discretization of Eqn. (4) gives

$$\left(\frac{2\mathbf{M}}{\Delta t}+\mathbf{D}+\frac{\mathbf{A}}{Re}\right)\mathbf{u}_{\alpha}^{n+1/2}=\mathbf{B}_{\alpha}+p_{e}\,\mathbf{C}_{\alpha}+\frac{2\mathbf{M}}{\Delta t}\mathbf{u}_{\alpha}^{n}, \quad (10)$$

where α indicates the Cartesian coordinate components x, y and z, M is the lumped element mass matrix, D is the advection matrix, A is the stiffness matrix, C is the coefficient matrix for pressure, B is the vector due to boundary conditions and E is the matrix which arises due to incompressibility.

The discretized form of Eqn. (8) reads as

$$\frac{1}{2}\mathbf{A}\phi=-\frac{1}{2}\mathbf{A}\left(p^{n+1}-p^{n}\right)\Delta t=2\mathbf{E}_{\alpha}\mathbf{u}_{\alpha}^{n+1/2}. \quad (11)$$

Subtracting Eqn. (5) from Eqn. (6) and introducing the auxiliary potential function ϕ, one obtains the following;

$$\mathbf{u}_{\alpha}^{n+1}=\mathbf{u}_{\alpha}^{*}-\frac{1}{2}\mathbf{E}_{\alpha}\phi\Delta t=2\mathbf{u}_{\alpha}^{n+1/2}-\mathbf{u}_{\alpha}^{n}-\frac{1}{2}\mathbf{E}_{\alpha}\phi\Delta t. \quad (12)$$

The element auxiliary potential ϕ_e is defined as

$$\phi_{e}=\frac{1}{vol(\Omega_{e})}\int_{\Omega_{e}}N_{i}\,\phi_{i}d\Omega_{e}, \qquad i=1,\ldots\ldots\ldots,8,$$

where Ω is the flow domain and N_i are the shape functions.

The following steps are performed to advance the solution one time step.

i) Eqn. (10) is solved to find the velocity field at time level n+1/2 with domain decomposition,

ii) Knowing the half step velocity field, Eqn. (11) is solved with domain decomposition to obtain the auxiliary potential ϕ.

iii) With this ϕ, the new time level velocity field $\mathbf{u}^{n+1}$ is calculated via Eqn.(12).

iv) The associated pressure field p^{n+1} is determined from the old time level pressure field p^n and ϕ obtained in step ii).

The above procedure is repeated until the desired time level. In all computations lumped form of the mass matrix is used.

3. DOMAIN DECOMPOSITION

The domain decomposition technique, [7,8,9], is applied for the efficient parallel solution of the momentum, Eqn. (10) and the Poisson's Equation for the auxiliary potential function, Eqn. (11). This method consists of the following steps, [8].

Initialization: Eqn. (10) is solved in each domain Ω_i with boundary of $\partial\Omega_i$ and interface S_j, with vanishing Neumann boundary condition on the domain interfaces.

$$\overline{\mathbf{A}}y_i = f_i \quad \text{in } \Omega_i \qquad\qquad g^o = \mu^o - (y_2 - y_1)S_j$$

$$y_i = g_i \quad \text{on } \partial\Omega_i \qquad\qquad w^o = g^o$$

$$\frac{\partial y_i}{\partial n_i} y_i = 0 \quad \text{on } S_j \qquad\qquad \mu^o\text{: arbitrarily chosen}$$

where $\overline{\mathbf{A}} = \dfrac{2\mathbf{M}}{\Delta t} + \mathbf{D} + \dfrac{\mathbf{A}}{\text{Re}}$ in Eqn. (10) and $y_i = \left\{\mathbf{u}_\alpha^{n+1/2}\right\}$.

Unit Problem: A unit problem is then defined as

$$\overline{\mathbf{A}}x_i^n = 0 \quad \text{in } \Omega_i$$

$$x_i^n = 0 \quad \text{on } \partial\Omega_i$$

$$\frac{\partial x_i^n}{\partial n_i} = (-1)^{i-1} w^n \quad \text{on } S_j$$

Steepest Descent

$$aw^n = \left(x_1^n - x_2^n\right)S_j$$

$$g^o = \left(y_1^o - y_2^o\right)S_j \qquad\qquad g^{n+1} = g^n - \beta^n aw^n$$

$$\beta^n = \frac{\sum_j \int_{S_j} \left| g^n \right|^2 ds}{\sum_j \int_{S_j} (aw^n) w^n ds} \qquad s^n = \frac{\sum_j \int_{S_j} \left| g^{n+1} \right|^2 ds}{\sum_j \int_{S_j} (g^n)^2 ds}$$

$$w^{n+1} = g^{n+1} + s^n w^n \qquad \mu^{n+1} = \mu^n - \beta^n w^n$$

Convergence check: $\left| \mu^{n+1} - \mu^n \right| < \varepsilon$

Finalization: Having obtained the correct Neumann boundary condition for each interface, the original problem is solved for each domain.

$$\overline{\mathbf{A}} y_i = f_i \qquad \text{in } \Omega_i$$

$$y_i = g_i \qquad \text{on } \partial\Omega_i$$

$$\frac{\partial y_i}{\partial n_i} = (-1)^{i-1} \mu^{n+1} \qquad \text{on } S_j$$

For the pressure equation:

After the velocity field at half time level is obtained, the Eqn. [11] is solved in each domain Ω_i with boundary of $\partial\Omega_i$ and interface S_j, with vanishing Neumann boundary condition on the domain interfaces. The steps indicated above for the momentum equation is repeated, but now $\overline{\mathbf{A}} = \mathbf{A}$ in Eqn.[11] and $y_i = \{\phi : \text{auxiliary potential function}\}$. In this chapter, subscripts *i* and *j* indicate the domain and the interface respectively, superscript *n* denotes iteration level.

4. PARALLEL IMPLEMENTATION

During parallel implementation, in order to advance the solution single time step, the momentum equation is solved implicitly with domain decomposition. Solving Eqn. (10) gives the velocity field at half time level which is used at the right hand sides of Poisson's Eqn. (12), in obtaining the auxiliary potential. The solution of the auxiliary potential is obtained with again domain decomposition where an iterative solution is also necessary at the interface. Therefore, the computations involving an inner iterative cycle and outer time step advancements have to be performed in a parallel manner on each processor communicating with the neighbouring one. Part of a flow chart concerning the parent (master) and the child (slave) processes are shown in Figure 1.

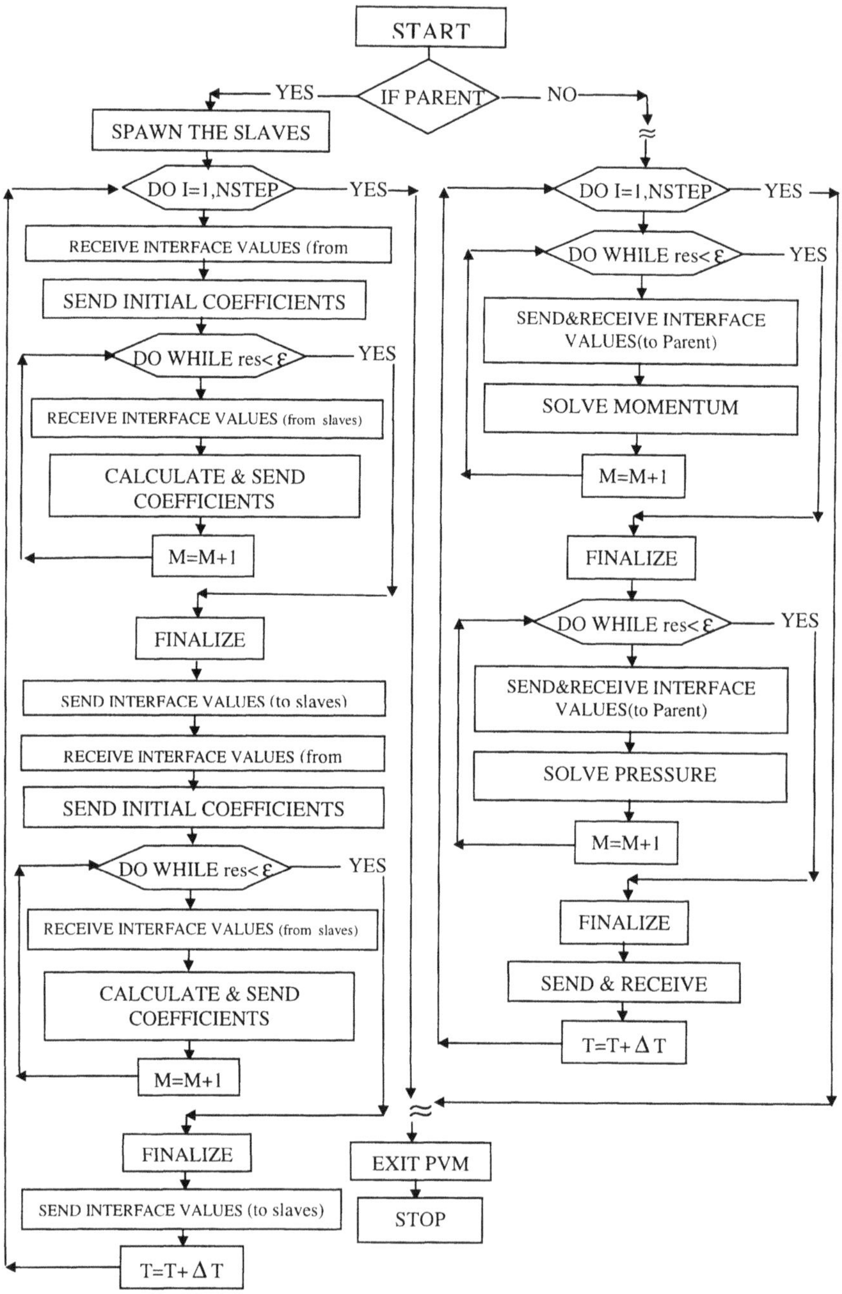

Figure 1. Flow chart of the parallel implementation concerning the parent and child processes

5. RESULTS AND DISCUSSION

Lid-driven flow in a cubic cavity with a Reynolds number of 1000 is selected as a test case to demonstrate the efficiency and accuracy of the method presented above. Results with 2 domain and 4 domain partitionings are presented in this study. For two domain partitioning 2x(25x13x13), and for 4 domain partitioning 4x(25x13x7) grid points are used as shown in Figure 2. In Figure 3., the 4 domain velocity vector field and the pressure iso-lines at the symmetry plane of the cavity are shown at dimensionless time t = 25 where the steady state is reached with a large time step size of 0.1. Based on the 2 domain solution, the overall parallel efficiency of the 4 domain solution is 85% and the speed-up is 1.71 as opposed to its ideal value, 2. The efficiency and the speed-up values can be improved in future work.

REFERENCES

1. Ü.Gülçat and A.R.Aslan, International Journal for Numerical Methods in Fluids, 25, 985-1001, 1997.
2. R. Glowinski and J. Periaux, "Domain Decomposition Methods for Nonlinear Problems in Fluid Dynamics", Research Report 147, INRIA, France, 1982.
3. Q.V.Dinh, A.Ecer, Ü.Gülçat, R.Glowinski and J.Periaux, "Concurrent Solutions of Elliptic
Problems via Domain Decomposition, Applications to Fluid Dynamics", Parallel CFD 92, May 18-20, Rutgers University, 1992.
4. A.R.Aslan, F.O.Edis and Ü.Gülçat, Parallel CFD 98, Edited by C.A.Lin et al, Elsevier, Amsterdam, 1999.

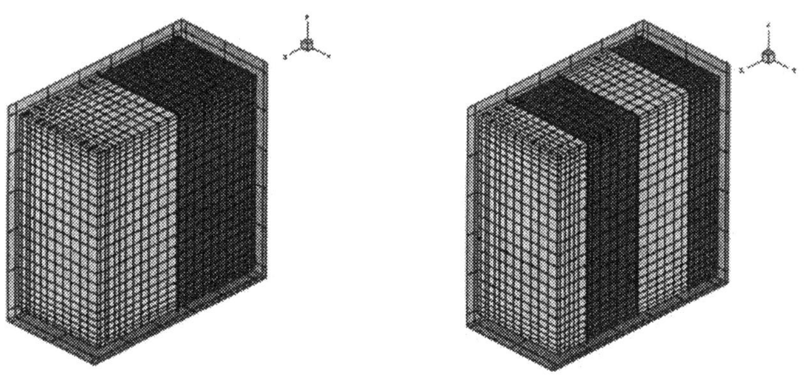

Grid for 2 domains
2 x (25x13x13) points

Grid for 4 domains
4 x (25x13x7) points

Figure 2. Grids for 2 and 4 domain partitionings

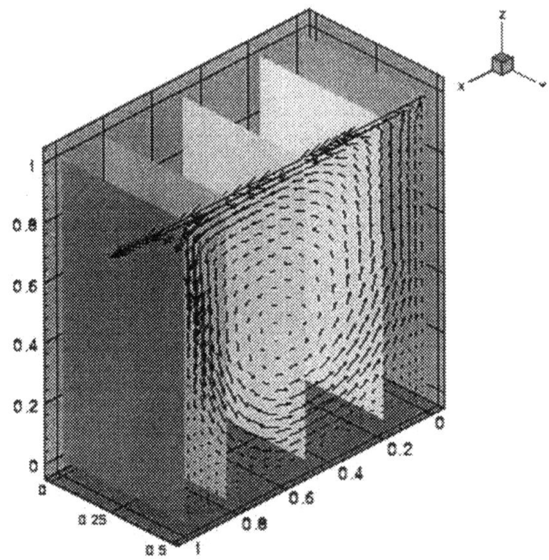

THE VELOCITY FIELD, Re=1000
Dimensionless time=25, Δt=0.1

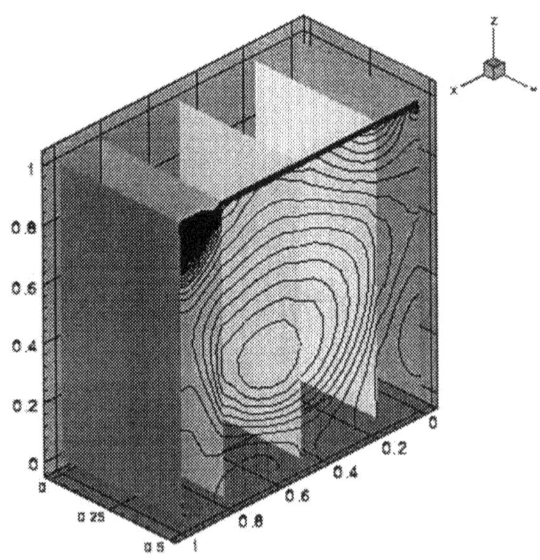

PRESSURE CONTOURS, Re=1000
Dimensionless time=25, Δt=0.1

Figure 3. The velocity vector field and the pressure iso-lines at the symmetry plane

Parallel Computational Fluid Dynamics – Trends and Applications
C.B. Jenssen et al. (Editors)

Performance of a Parallel CFD-Code on a Linux Cluster

Petri Kaurinkoski[a], Patrik Rautaheimo[a],Timo Siikonen[a] and Kimmo Koski[b]

[a]CFD-group, Helsinki University of Technology, P.O. Box 4400, 02015 HUT, Finland

[b]Center for Scientific Computing, P.O. Box 405, 02101 Espoo, Finland

So far, commercial UNIX workstations have been superior in comparison with personal computers as far as computational speed is concerned. However, the gap between the speeds has been decreasing. Today's personal computers are fast and relatively cheap with respect to traditional workstations, and, therefore, a comparison between PCs and workstations is made in this work. It will be shown that, with ideal cases, the speed and parallel performance figures of a cluster of personal computers are competitive with those of a workstation. As an example of practical CFD simulations, the flow past a delta wing and the flow past the Onera-M6 wing is simulated on a cluster of Linux workstations and several other hardware environments.

1. Introduction

The recent reduction in the price of computational power has made CFD an affordable option for small-sized companies and research groups, because a supercomputer is not necessarily needed, even in challenging applications. In addition, the idea of employing idle desktop computers for parallel work at night time has been discussed for several years. However, only a few such solutions have emerged [1, 2] for productive work.

The concept of a Linux-based cluster of PCs is very interesting, since it offers the alternatives of a standard DOS/Windows-based desktop tool for daytime, and a UNIX-type operating system for the night time at a relatively low investment in the hardware. In this work, dual-boot ability is not investigated, but the work is concentrated on the computational capacity of a cluster of Linux workstations.

Parallelization has been used to enhance the efficiency of flow solvers for over a decade. A common approach is to divide the computational domain into equally sized sub-domains, i.e. blocks, and to apply message passing between the blocks. In this paper, the parallelization of a multi-block Navier–Stokes software program, called FINFLO [3] is briefly described. A more detailed description of the code can be found in Refs. 4–7. The parallelization is based on the Message Passing Interface (MPI) standard [8].

To complete the comparison between different hardware environments, the performance of different parallelization strategies of the code are compared on a shared memory system. Some of the test cases are studied using an MPI-parallelized and a directive-parallelized version of the code on an SGI Origin 2000 system. Up to 32 CPUs are employed in that comparison.

2. Flow Solver

2.1. Governing Equation

The flow simulation is based on the solution of the Reynolds–averaged Navier–Stokes equations:

$$\frac{\partial U}{\partial t} + \frac{\partial(F-F_v)}{\partial x} + \frac{\partial(G-G_v)}{\partial y} + \frac{\partial(H-H_v)}{\partial z} = Q \tag{1}$$

where $U = (\rho \ \rho u \ \rho v \ \rho w \ \rho E \ \rho k \ \rho\epsilon)^T$ is the vector of the dependent variables, F, G, H and F_v, G_v, H_v represent the inviscid and viscous parts of the fluxes, and Q is a possible source term vector. This solver employs the finite volume method. For the finite volume method, Eq.(1) is cast in an integral form, and for an arbitrary fixed region V with a boundary S the equations are

$$\frac{\mathrm{d}}{\mathrm{d}t}\int\limits_V U \,\mathrm{d}V + \int\limits_S \vec{F}(U)\cdot \mathrm{d}\vec{S} = \int\limits_V Q \,\mathrm{d}V \tag{2}$$

where U is the vector of conservative variables, and $\vec{F}(U)$ is the flux vector. Performing the integrations for a computational cell i yields

$$V_i\frac{\mathrm{d}U_i}{\mathrm{d}t} = \sum_{\text{faces}} -S\hat{F} + V_iQ_i \tag{3}$$

where S is the area of the cell face, and the sum is taken over the faces of the computational cell.

2.2. Solution Algorithm

The flow solver utilizes a structured multiblock grid. The solution algorithm is implicit pseudo-time integration for which Eq.(3) is linearized around time-level n to obtain the following equation

$$V\frac{U^{n+1}-U^n}{\Delta t} = V\frac{\Delta U}{\Delta t} = R^n + \frac{\partial R}{\partial U}\Delta U \tag{4}$$

where the subscript i has been dropped and R is used as shorthand for the right-hand side of Eq. (3). Rearranging terms, we obtain a matrix-equation for ΔU which is not suitable for direct inversion in any practical 3-D case. Instead, the Diagonally Dominant ADI-factorization (DDADI) [9] is employed. This is based on approximate factorization and on the splitting of the Jacobians of the flux terms. The resulting implicit stage consists of a backward and forward bidiagonal sweep in every computational coordinate direction plus a matrix multiplication. In addition, the linearization of the source term is factored out of the spatial sweeps. After factorization, the implicit stage can be written as

$$\begin{aligned}&\left[I + \frac{\Delta t_i}{V_i}\left(\partial_i^- S_{i+1/2}A_i^+ - \partial_i^+ S_{i-1/2}A_i^-\right)\right] \times \left[I + \frac{\Delta t_j}{V_j}\left(\partial_j^- S_{j+1/2}B_j^+ - \partial_j^+ S_{j-1/2}B_j^-\right)\right] \times \\ &\left[I + \frac{\Delta t_k}{V_k}\left(\partial_k^- S_{k+1/2}C_k^+ - \partial_k^+ S_{k-1/2}C_k^-\right)\right] \times \left[I - \Delta t_i D_i\right]\Delta U_i = \frac{\Delta t_i}{V_i}R_i\end{aligned} \tag{5}$$

where I is an identity matrix, $\partial^-_{i,j,k}$ and $\partial^+_{i,j,k}$ are first-order spatial difference operators in the i, j and k-directions, A, B and C are the corresponding Jacobian matrices, $D = \partial Q/\partial U$, and R_i is the right-hand side of Eq. (3).

The solution proceeds blockwise employing explicitly defined boundary conditions. For the boundary conditions, the block boundaries are divided into 'patches', which can be connected with each other in an arbitrary way [4]. Thus, a block surface can contain several patches with different boundary conditions. The block-to-block connectivity boundary conditions are handled using two layers of ghost cells on the block boundaries to ensure continuity of the solution. The values of the dependent variables at the ghost cells are updated using message passing, if necessary.

2.3. Parallelization

The block structure forms an ideal base for the parallelization. All the essential procedures operate on one block at a time, including the updating of the boundary conditions. The parallelization is based on the Message Passing Interface (MPI) standard [8]. The updating of boundaries between different processes is done using the basic `MPI_SSEND` and `MPI_RECV` commands. Also `MPI_BCAST` and `MPI_GATHER` are used to give input parameters to the processes, and to gather convergence histories.

At the beginning of the calculation the connectivity information is analyzed, and an optimal order of communication is determined. The exchanged data is rearranged into vectors that are sent to other processes where the data is rearranged back into the desired order.

During an ordinary run the workload is distributed as follows: One processing element (PE0) is the master and the others are slaves. Parallelization is realized so that only the master process reads the input parameters, including files where boundary conditions are specified and the grid is defined, and sends the desired input parameters and the appropriate parts of the grid to the slaves. The master process also reads and writes the possible restart file. In a restart, the number of processors can be changed. After every iteration cycle, the slave processes send convergence parameters (global residuals etc.) to the master, which prints them on the screen and stores in a convergence monitoring file. Because the processes are highly independent of each other, the memory requirement per process comes from the size of the block(s) that a process simulates. Since the possibility to calculate a different number of differently sized blocks has been retained, an independent dynamic memory allocation is performed in each process.

One example of block decomposition can be seen in Fig. 1. In the figure, the original single-block grid is divided into 16 blocks, which are partially shown.

3. The Linux-Cluster Environment

The computer environment for this study was provided by CSC - Scientific Computing Ltd. in which a specific pilot project for PC clusters was carried out during 1999. In this project, a number of scientific software packages was tested and various application benchmarks were run. One of the main applications was the parallel CFD code described in this paper.

The employed LINUX cluster consisted of 16 dual-processor Dell Precision 410 workstations and a Dell PowerEdge 2300 frontend server. The systems were attached together with a standard Ethernet using a BayStack 450-24T switch. The processor was a Pentium II with 400 MHz frequency. Each computing node had 128 MB of central memory and a 4-GB local SCSI disk, except the frontend system in which the amount of memory was 512 MB with an 18-GB disk.

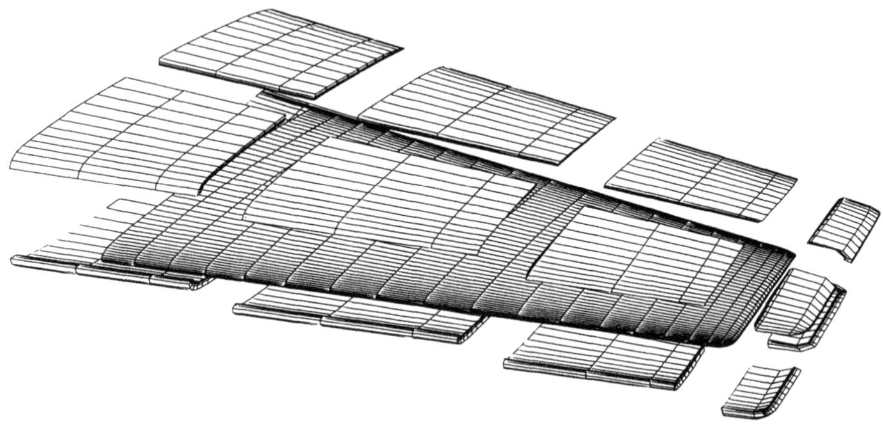

Figure 1. Block decomposition for the ONERA M6 wing. Only the original surface grid with every second gridline and the divided blocks adjacent to the surface are presented. For clarity, the divided blocks have been lifted off the surface, and only every fourth gridline is displayed.

The systems were running LINUX RedHat 5.2 with kernel level 2.2.5. Multiple compilers were used, such as public domain GNU (C, C++, Fortran 77) compilers and commercial Portland Group and Absoft compilers (Fortran77, Fortran90). For message passing, the MPI implementation from Argonne National Laboratories with a standard p4 (TCP/IP) communication layer was employed.

The communication network was based on switched 100 Mbit/s Ethernet. The switch had 24 ports of which 16 were used for the computing nodes and one for the server. The fast ethernet was based on a 3Com chip embedded in the mother board of each system. The largest parallel jobs were run with 32 processors using both processors from all the systems.

The cluster had a dedicated private network. The login server connected both the public local area network and the private cluster network. The design was such that users did not need to login to the nodes. The login server was used for starting the jobs with a script automatically starting the jobs in the selected nodes.

In all the simulations described in this paper, the Absoft Fortran 90 compiler was used.

4. Performance of the Code

4.1. Distributed Memory Systems

To start with, the single-CPU performance was first studied with a flow over a delta wing using a single-block grid. The grid contained 32^3 computational cells, and the simulation required roughly 20 MB of memory.

Table 1 depicts a comparison of the single-CPU performance on different platforms in the above case. The speed is expressed in MFLOPS (the number of flops per iteration cycle was obtained from the C94 performance monitoring tools), and also as a ratio to the speed of the

Cray T3E. The performance with the Linux PC is 73 MFLOPS, which is larger than that for one CPU on the Cray T3E, and also only about 50 % lower than a value of a somewhat old UNIX server (Origin 2000 with R10k). Note also that the performance on the C94 would have been better ($\approx$ 500 MFLOPS), if a larger grid had been used.

Table 1
Single-CPU performance on different platforms.

Platform	Speed (MFLOPS)	Speed (Cray T3E)
C94 (240MHz Vector processor)	367	6.92
SGI Indigo^2 (R10k 195 MHz IP28 / 1 MB SC)	90	1.70
SGI Origin 2000 (R10k 250 MHz IP27 / 4 MB SC)	140	2.64
COMPAQ AlphaServer 8400 (525 MHz EV6 / 4 MB SC)	244	4.60
T3D (150 MHz Digital alpha 21064 / 8 kB SC)	11	0.21
T3E (375 MHz EV5 / 96 kB SC)	53	1.00
T3E (with no streams)	48	0.91
Dell precision 410 (400 MHz Pentium II)	73	1.38

There are two commonly used test methods in the parallelization. One is to keep the size of the computational task of one processor constant, and the other is to keep the size of the global task constant dividing it to smaller ones. Here, the former one is called scaling and the latter one is called blocking.

In the scaling case, the grids were generated so that each block had a size of $32 \times 32 \times 32$. The computational domain size varied from 1 to 64 different blocks, and each block was calculated in a different processor. On the PC cluster, the largest case was 32 blocks. Thus the coarsest grid had 32 768 and the densest grid 1 048 576 computational cells on the PC cluster.

The speed-up was determined directly from the absolute time spent in the calculation, and the results are presented in Fig. 2 for both scaling (left) and blocking (right). It can be seen that a perfect scaling is achieved in these test runs with the T3E up to 64 CPUs. The PC cluster has good scaling up to 16 CPUs. With 32 CPUs, there is a sudden drop in efficiency and, therefore, more tests were conducted with 16 CPUs. The parallelization efficiency with 32 CPUs is 0.7 on the PC-cluster and consequently the maximum speed is $32 \cdot 0.7 \cdot 73$ MFLOPS $= 1.64$ GFLOPS.

Since the PCs in the cluster were dual-processor computers, the simulation with 16 blocks was repeated involving only 8 PCs. With this configuration the performance of the cluster was also poor, indicating difficulties for the Linux operating system to balance the load. The code version used on the cluster of SGI workstations was a bit older, and the communication was carried out through a low-speed ethernet (10 Mbit/s) without a switch. However, the performance is good up to 16 CPUs. It is seen that the speed-up is linear on the T3E, and almost linear up to 16 CPUs on the PC cluster.

With the blocking approach, the grid sizes were not the same on the T3E and the PC cluster. On the T3E, a grid with $64 \times 64 \times 32 = 131\,072$ cells was used, whereas on the PC cluster the grid had dimensions of $192 \times 80 \times 40 = 614\,400$. In both cases, the number of grid points was

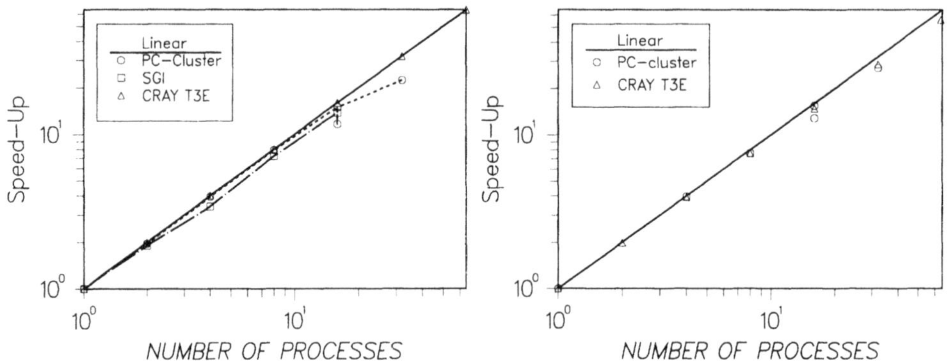

Figure 2. Speed-up of the parallelization in scaling (left) and blocking (right).

limited by the processor memory size. The corresponding block sizes varied from 131 072 to 2 048 on the T3E simulations, and from 614 400 to 19 200 on the PC cluster.

Fig. 2 shows some partly unexpected results. With 16 CPUs, the parallelization efficiency seems to be better for the PC cluster than for the T3E. There are primarily two reasons for this. Firstly, the blocks are larger in the PC cluster than in the T3E simulation. Secondly, the single-CPU performance increases in a PC, as the simulated case gets smaller, because the data fits the cache better. As evidence of this behaviour, further test runs were made on the PC cluster. The computing time per iteration per computational node was taken from a single-CPU run with two different grid sizes. The time spent in one computational cell was 143 μs with the smaller grid and 152 μs with the larger grid, and they contained 32 768 and 614 400 points, respectively. The improvement in the single-CPU performance is 5 % when the grid size is decreased.

It seems that the cache size of the T3E is far too small to store the data of even the smaller grid case. Therefore, no speed-up is seen when reducing the block size. On the other hand, in order to optimize the global performance, the surface-to-volume ratio of the grid should be kept as low as possible to minimize the relative time spent with boundary conditions. In effect, this means larger problem dimensions.

With these observations, it is clearly seen that the block size should be sufficiently large; in this case 8 000 cells seemed to be the lower limit. Also, in blocking with 32 CPUs the performance is not good on the PC cluster, and the simulation with 16 CPUs is repeated with only 8 PCs involved as in scaling. Again, the performance drops dramatically, when each PC is fully loaded, as is seen from Fig. 2. In blocking with 32 CPUs the parallelization efficiency was 0.848 and the speed in this case was $32 \cdot 0.848 \cdot 142.98/151.85 \cdot 73$ MFLOPS $= 1.87$ GFLOPS, i.e. more than in scaling.

An example of a calculation performed on the Origin 2000 is the simulation of the flow field around the BAe Hawk Mk 51 jet trainer. The computational domain consists of 3 800 000 cells, and it was split into 28 blocks. The case was calculated using 8 CPUs, and one iteration cycle required 35 s per CPU. The total wall-clock time consumption was 39 hours. Without parallelization this case would have taken almost two weeks.

4.2. Shared Memory Systems

As a comparison between shared memory and distributed memory systems, the flow past the Onera-M6 wing shown in Fig. 1 was simulated on the Origin 2000 system employing both the shared memory model and the distributed memory model for parallelization. The number of CPUs was varied between 1 and 32 with the blocking approach.

Up to 4 CPUs, the shared memory parallelized code performed as well or slightly better. The differences, however, were nominal. With 8 CPUs involved, the MPI-paralleization outperformed directives by 25 %. With 16 or more CPUs, the directive-approach was a waste of resources in comparison with message passing. The conclusion is that, as the size of the parallel region of the code decreases, the overhead associated with synchronizing different threads will dominate. Quite evidently, the MPI paralleization in this solver is less sensitive to this problem because the synchronization of the processes is realized implicitly. This way, the processes will wait for other processes only when it is really necessary.

4.3. Experiences from the Cluster

These tests confirm that the Linux-based PC cluster is a competitive alternative to the traditional UNIX-based computing servers. There are, however, some drawbacks in the Linux environment in comparison with a commercial UNIX system. As shown previously, the system was not able to efficiently use two CPUs on one host, which remarkably reduced the throughput capacity of the system. It became evident that all the necessary tools have not yet reached maturity on that platform. For instance, the system does not include a free Fortran compiler, and the commercially available ones seemed to have problems with standard Fortran-90 code. The compilers available in this work could not inline functions, and the flexibility of the compilers did not quite meet the criteria of a development environment.

In addition, one has to pay attention to the administrative difference between a single multiprocessor system and a cluster of tens of PCs connected with a network. In practice, the maintenance of the latter system requires a dedicated person, which also costs money.

5. Conclusions

A comparison was carried out between a PC cluster, a UNIX workstation cluster and a massively parallel computer. With the UNIX workstation cluster, the performance curve obtained is almost linear up to 16 processes, despite the fact that the workstations were connected to a standard, 10 Mbit/s Ethernet. With the Cray T3E, test runs indicate an excellent parallelization. With the PC cluster, the parallelization is excellent up to 16 CPUs. For both scaling and blocking the parallelization efficiency is over 90 % up to 16 CPUs. In the case of 32 CPUs the performance collapses because both of the CPUs in a single PC are in use simultaneously. It seems that the Linux operating system cannot use dual processors hardware efficiently.

A comparison between message passing and directive parallelization on a shared memory system shows that with 8 CPUs or more, message passing is the only reasonable way of computation. With up to 4 CPUs, the different methods were equally effective.

Since the price of a CPU on a PC is considerably lower than on a commercial UNIX platform, the Linux operating system is a tempting alternative for traditional commercial UNIX systems. With respect to computational speed, the cluster of 32 Linux nodes is roughly equivalent to 44 CPUs on the Cray T3E, 4 CPUs on the Cray C94, and 14 R12k CPUs on the Origin 2000. This comparison, however, is based on the present perfectly scalable test cases. In real-life cases, these figures would probably not be as good for the Linux cluster. Nevertheless, offices with lots of PCs have a potential supercomputer-level capacity available if harnessed for distributed computing.

One drawback of the Linux-based systems is the lack of traditional commercial support. This lifts the threshold for jumping into a Linux cluster higher, because it is evident that the labour costs involved with the necessary in-house support will be higher in comparison with a fully commercial UNIX system. On the other hand, the net community often provides solutions faster than a fully loaded commercial support organization.

All the cases studied in this paper are optimal for parallel work. In reality, the grids can very seldom be divided ideally, and therefore the performance is not as good. Although not tested in this study, it is expected that, for massively parallel works, for the use of over 100 CPUs, the performance of the PC cluster will be poor. The supercomputers are still needed for solving very large computing tasks.

REFERENCES

1. Bush, R., Jasper, D., Parker, S., Romer, W., and Willhite, P., "Computational and Experimental Investigation of F/A-18E Sting Support and Afterbody Distortion Effects," *Journal of Aircraft*, Vol. 33, Mar-Apr 1996, pp. 414–420.
2. McMillan, W., Woodgate, M., Richards, B. E., Gribben, B. J., Badcock, K. J., Masson, C. A., and Cantaiti, F., "Demonstration of cluster computing for three-dimensional CFD simulations," *The Aeronautical Journal*, Sep 1999, pp. 443–447. Paper No. 2467.
3. Siikonen, T., "An Application of Roe's Flux-Difference Splitting for the $k - \epsilon$ Turbulence Model," *International Journal for Numerical Methods in Fluids*, Vol. 21, 1995, pp. 1017–1039.
4. Rautaheimo, P., Salminen, E., and Siikonen, T., "Parallelization of a Multi-Block Navier–Stokes Solver," in *Proceedings of the Third ECCOMAS Conference*, (Paris), John Wiley & Sons, Ltd., Sept. 1996.
5. Rautaheimo, P., Salminen, E., and Siikonen, T., "Parallelization of a Multi-block Flow Solver," Helsinki University of Technology, Laboratory of Applied Thermodynamics, 1997. ISBN 951–22–3416–5.
6. Kaurinkoski, P. and Hellsten, A., "Numerical Simulation of a Supersonic Base Bleed Projectile with Improved Turbulence Modelling," *Journal of Spacecraft and Rockets*, Vol. 35, Sept-Oct 1998, pp. 606–611.
7. Kaurinkoski, P. and Hellsten, A., "FINFLO: the Parallel Multi-Block Flow Solver," Helsinki University of Technology, Laboratory of Aerodynamics, 1998. ISBN 951–22–3940–X.
8. Message Passing Interface Forum, "MPI: A Message-Passing Interface Standard.," Computer Science Dept., University of Tennessee, Knoxville,TN, 1994.
9. Lombard, C., Bardina, J., Venkatapathy, E., and Oliger, J., "Multi-Dimensional Formulation of CSCM — An Upwind Flux Difference Eigenvector Split Method for the Compressible Navier–Stokes Equations," in *6th AIAA Computational Fluid Dynamics Conference*, (Danvers, Massachusetts), pp. 649–664, Jul 1983. AIAA Paper 83-1895-CP.

Parallel Computational Fluid Dynamics – Trends and Applications
C.B. Jenssen et al. (Editors)

Utilising Existing Computational Resources to Create a Commodity PC Network suitable for Fast CFD Computation

R. A. Law and S. R. Turnock

Fluid-Structure Interaction Group,
School of Engineering Sciences, University of Southampton,
Highfield, Southampton, SO17 1BJ, UK

The use of computational simulations of fluid flow around complex bodies is increasingly popular as an alternative to experimental testing in research and industry. Such simulations often require the use of substantial computing facilities, which are often unavailable or restricted in many small research organisations and design companies. In this paper, the idea of harnessing existing PC computers that usually sit idle for a substantial part of their life span is explored as an alternative high performance computational resource suitable for CFD computation.

1 INTRODUCTION

The use of Computational Fluid Dynamics (CFD) in industry has previously been a rather elitist tool, demanding substantial computational resources significantly more powerful and expensive than the best desktop computers. Such facilities are usually only available to large and medium sized companies or research groups through time share schemes, which too often become expensive to maintain or frustrating to use when working within short cycle times.

Over the past decade, the cost of computing has reduced significantly and the convergence of the price and performance of high-end workstations with affordable Personal Computers (PC's) has been particularly rapid over the past few years. Through such advances in computing technology it is now feasible to consider the long-term role of PC's in the field of CFD.

Through the utilisation of network technology it is possible to connect clusters of PC's together. When used with parallel and distributed computing software library's such as MPI [1], it is possible to obtain Super-Computer level machines at the fraction of cost [2] of their stand-alone predecessor machines such as the CRAY TE4. These machines are burdened with enormously high setup and support costs, whereas a PC cluster offers a minimal cost environment with the potential for incremental improvements in performance.

This paper explores the development and performance evaluation of such a system, developed from an existing cluster of teaching PC's used for the Ship Science courses at the University of Southampton. The teaching cluster was purchased to service the need of students for dedicated ship design software, CAD packages, programming skills and general office software. For a small cost (£9,000), the network was adapted to allow dual use - providing a teaching resource during term time between 8 am and 9 pm and for the remainder acting as a networked cluster available for carrying out CFD computations.

2 CREATION OF A DUAL USE COMPUTATIONAL FACILITY

The original network of PC's consisted of 52 PC units of at least 350Mhz processing speed, and 64Mb+ RAM available, connected via 10MHz Ethernet directly to the NT4 server through 10MHz switches.

This network is then connected to the main UNIX computing facility via an Internet connection. Red Hat Linux 6 was selected as the target operating system for the high performance network for its increased stability over Windows, its ease of porting and similar environment to UNIX. A dual boot facility was targeted to accommodate both teaching support through NT and high performance processing using Linux.

Two high-end dual processing PC workstation were purchased to provide a base for building computationally intensive problems, which could be extended later in future purchasing plans. One of these workstations was chosen to serve as the Linux server, as well as a desktop workstation during the daytime. To accommodate a finite budget, a hardware trade off was made in favor of network requirements for high processor speed, 10/100 Ethernet and a fast 36Gb Ultra2 Wide SCSI hard drive. An affordable 512Mb of memory was added to both workstations such that small CFD problems could ultilise the dual 500MHz PIII processors. A file server was created using NFS, and user and password authentication through the existing UNIX server. File backup is currently carried out through the UNIX sever, but will later use the University global backup facility via the Intranet significantly reducing costs.

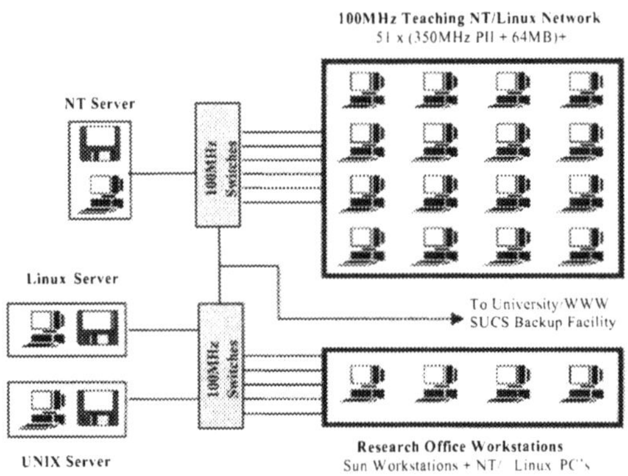

Figure 1: Schematic diagram of computing facility layout

The speedup scalability of the commodity cluster is mainly limited to network speed performance [3], originally 10MHz, and an upgrade to an affordable 100MHz was carried out as shown in Figure 1, so that high performance CFD computations could be realised. The original computing facility was based around two networks, the NT and UNIX, which remained as separate entities so that the performance of one would not degrade the other. Communication between networks was carried out through the external communication line that connected the UNIX network to the University Intranet and outside world. High speed cabling that could accommodate a faster 100MHz network already existed for the PC cluster, and its full network upgrade to 100MHz required only new network switches. To allow scalability of the commodity cluster beyond the 52 PS's housed in a single computer room, any new PC's entering the department for research use should also be accommodated. The research offices are networked through the UNIX 100MHz switches, and a simple upgrade of network cabling was needed to offer full scalability.

The Linux server carries out parallel process scheduling to avoid either processor or network overload. The MPI libraries are in current use for three in-house flow solvers as well as for a parallel processor Genetic Algorithm and would form the main software for use on the network.

3 EVALUATION OF THE COMPUTATIONAL RESOURCE

In order to evaluate the performance of the new computing resource, tests have been conducted using two different computational problems typical of the day to day computational work carried out within the department. An unstructured Euler solver [4] used on a variety of different computational platforms was ported into Linux using GNU gcc compiler, using the MPICH 1.2.2 library from Argonne National Laboratory for message passing between mesh partitions. A parallel Genetic Algorithm (GA) [5] coupled with XFOIL [6] is used for optimisation of airfoil sections at low Reynolds numbers. MPICH is used to facilitate a global Master-Slave distributed implementation of the GA. Porting into Linux was achieved using GNU g++, and f77 compilers for the GA and XFOIL respectively. Only minor porting problems were found with the Euler solver where the communication buffers created using MPI_Struct procedures, required some extra attention to ensure that the data was correctly aligned within the buffer, without overwriting itself.

3.1 Network Performance for Unstructured Euler Solver

3.1.1 Numerical Model

The Euler equations describing the conservation laws of mass, momentum, and energy, can be written in vector form as:

$$\frac{\partial U}{\partial t} + \vec{\nabla} \cdot \vec{F} = 0, \tag{1}$$

where $\vec{F} = (f, g, h)$, and

$$U = \begin{bmatrix} \rho \\ \rho u \\ \rho v \\ \rho w \\ \rho E \end{bmatrix} \quad f = \begin{bmatrix} \rho u \\ \rho u^2 + P \\ \rho uv \\ \rho uw \\ \rho uH \end{bmatrix} \quad g = \begin{bmatrix} \rho v \\ \rho vu \\ \rho v^2 + P \\ \rho vw \\ \rho vH \end{bmatrix} \quad h = \begin{bmatrix} \rho w \\ \rho wu \\ \rho wv \\ \rho w^2 + P \\ \rho wH \end{bmatrix}$$

E and H are the total energy and stagnation enthalpy per unit mass respectively. Equation (1) can be integrated over an arbitrary finite volume Ω to construct the integral form of the equation.

Using Gauss' Divergence Theorem this can be expressed as

$$\frac{\partial}{\partial t} \int_{\Omega} U d\Omega = -\oint_{\partial\Omega} \left(f n_x + g n_y + h n_z \right) dS \tag{2}$$

where $\partial\Omega$ represents the contour around the volume Ω and S represents the surface of the volume. An average change of the conserved variables, denoted by $\overline{U}$ can be expressed for a discrete finite volume as

$$\left(\frac{\partial \vec{U}}{\partial t}\right)_{\Omega} = \frac{-1}{V_{\Omega}} \sum_{k=1}^{n} \left(fS_x + gS_y + hS_z \right) dS \tag{3}$$

where the summation is over all the faces of the discrete volume, S_x, S_y, and S_z represent the projected areas of these faces, and V_{Ω} is the volume. Equation (3) can be used over any discrete volume thus the method can be applied on any grid topology.

A Cell Vertex scheme is used to simplify the boundary condition implementation. Additional volumes or 'Control Volumes' are constructed over which the equations are integrated. The flux values on the faces are calculated using Roe's upwind scheme [7] in which an approximation to the Riemann problem is sought on the control volume faces.

The numerical flux, equivalent to the terms inside the right hand integral of Equation (2), can be expressed as

$$f^{*}_{i+1/2} = \frac{1}{2}\left[f\left(U^{R}\right) + f\left(U^{L}\right) - \left|A_{Roe}\right|\left(U^{R} - U^{L}\right) \right. \tag{4}$$

Where $\left|A_{Roe}\right|$ is the flux Jacobean evaluated using Roe's fluid state [7]. A detailed description of the implementation of the solver is given by Roycroft [4].

3.1.2 Distributed Implementation

The parallel strategy for explicit field methods is straightforward in that the domain is partitioned according to the number of processors available. A satisfactory decomposition is achieved through the Jostle [8] graph-partitioning program. Using a knowledge of the nature in which the Cells are ordered, each cell is allocated to a process, and the remaining geometrical objects required to describe the Cells are compiled and allocated to the processors accordingly.

On partition interfaces, the Cells are duplicated in order to complete their descriptions on both partitions, and to provide enough information for the numerical flux calculation. To ensure that flux calculations are not duplicated on neighboring processes, a root processor is assigned to each interface Cell to carry out its flux calculations. The locations of interface cells that shadow the root Cell on neighboring partitions are assigned a 'shadow' flag, holding the location of the root Cell.

To implement the parallel solver algorithm, the partitioned grid is based around a Node to Node connectivity map, joined together by Edges. Control volumes are constructed around each of the Nodes forming the Cells with a dual grid associating each Cell face cutting between a Node-Node Edge on the original grid. The calculation of the face fluxes is carried out by an Edge based loop, up dating residuals on all nodes except those that are shadows. Contributing boundary faces to the node residuals are then included.

A semi-scheduled message-passing algorithm is implemented to update the shadow residuals residing on neighboring partitions. Essentially the sending of residuals from the host partition is implemented through a message-scheduling algorithm with non-blocking communication. This allows the send process to start before a matching receive is posted, minimising dead-time events where processes are held waiting for recipient partitions.

A node sweep is used to update all non-shadow nodes, with root partitioned nodes communicated to their adjacent shadows via the same semi-scheduled algorithm as before. Residual statistics are updated on the master process, which determines and implements convergence, and full field-data checkpointing requirements. Checkpointing is achieved through a full sweep of processors sending all flow variables to the master for storage.

3.1.3 Distribution performance on the Linux cluster

To determine the efficiency of the network for Euler solver computation, network performance timings were measured during the computation of a NACA0012 wing. The computational domain used, consisted of a three-dimensional unstructured grid made out of 48672 control volume Cells. The measurements were based on 1000 time step iterations for each partition topology. Performance tests were made on the original 10MHz network before upgrade as well as the new 100MHz network. Separate tests were conducted allowing for regular checkpointing at 20 iteration intervals on both networks.

Speedup performance results shown in Figures 2 and 3 show substantial gain in computation speed for up to 16 processors at 100MHz-network speed. Even with regular saving of flow values to disk, a reasonable performance is sustained. The results from the original 10MHz network highlight the need for good network speed for such solutions. Closer evaluation of iteration efficiency of the cluster reveals that further improvements to speedup performance can be sought if the size of the partition domains were maximised for the performance tests. By partitioning the domain to maximise processor memory use, iteration efficiency can be further recovered. The maximum partition size for an unstructured grid that can be adequately be used on each processor consists of approximately 30,000 nodes. Using just 16 of the 52 available processors a 0.5M cell solution can be expected with good speedup. The maximum domain size available through the entire cluster is approximately 1.5-2M cells.

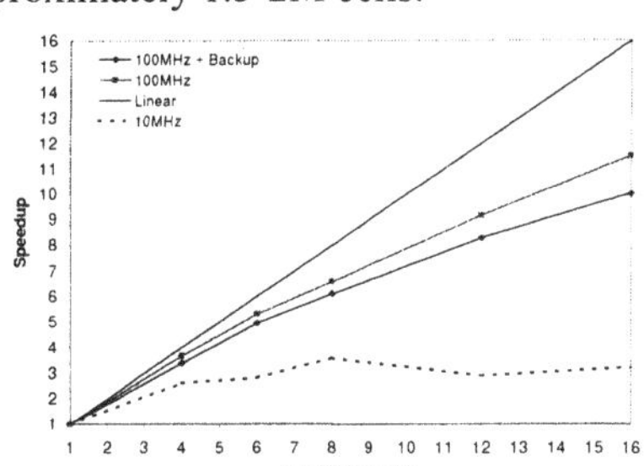

Figure 2: Speedup performance PC Cluster

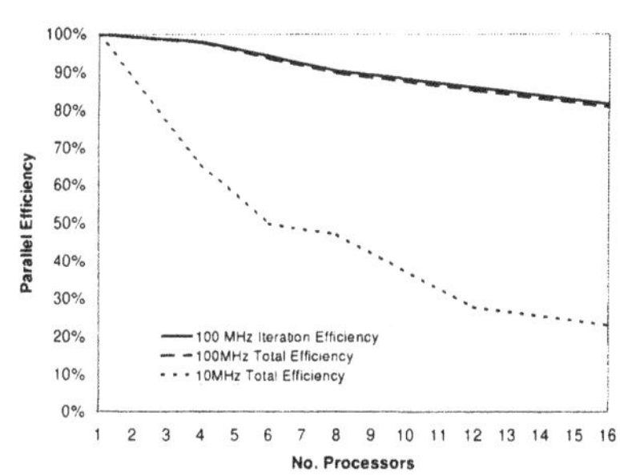

Figure 3: Parallel efficiency of PC Cluster

3.2 Distribution Performance for a Genetic Algorithm

Genetic Algorithm (GA) is a stochastic search algorithm based on the mechanics of evolution and natural selection. During the search process, a population of design points is sampled simultaneously. Members of this population then compete with one another to participate in the next generation of the process. New design points (offspring) are reproduced by the winning members through a sexual operator known as crossover where a portion of a member's chromosome is swapped with that of another Member. Mutation of the chromosome is often employed to help maintain diversity within the population.

In the implementation considered here, the low Reynolds viscous coupled panel solver

XFoil is to be used to evaluate the airfoil sections described by the individuals of the population. A Bezier [9] curve is used to construct the airfoil section from a set of control poles defined by the member's chromosome string. In this manner, the operators of crossover and mutation are directly altering the shape of the airfoil. The Bezier representation is divided into separate upper and lower section curves to reduce the order of the spline used with tangency imposed at the leading edge. The leading end trailing edge control points are fixed with the remaining nodes free to move in the vertical direction alone. With this parameterisation, a total of twenty design variables define the search space for airfoil design.

The objective of the search considered, is to minimise drag for a low speed airfoil section considered for in a deep-sea underwater thruster unit. The objective function is determined at a set lift coefficient, imposing constraints on the section shape in that it should remain real with a minimum thickness of 3% chord near the trailing edge. A pitching moment constraint is also used to ensure a usable result.

A global parallelisation strategy was originally used to distribute the evaluation phase of the population members only, based on the Master-Slave implementation given in [5]. This scheme was constructed using the MPICH library distributing an initial subset of the population to the available processors, and then passing the remainder on to each processor in turn on its completion with its current member. Due to the convergent based nature of XFoil however, the efficiency of this method is degraded on larger cluster sizes when all members have been passed for evaluation, and processors are left waiting for remainder slaves to complete before the next generation can be created and the distribution process continued. For a population size of 200, distributed using 50 slave processors, distribution inefficiencies of up to 35% were found on some generations.

To alleviate this problem, an asynchronous strategy was sought that decouples the evaluation processes from the synchronised selection phase of the GA. Aysnchronisation was achieved based on the island Genetic Algorithm (iGA) used by Doorly [9]. In the iGA implementation, subpopulations know as demes, are used to partition the population, each evolving independently with occasional exchange of genetic information between demes via a process termed migration. One feature of this strategy is that super linear speedup is often encountered, and is thought to be a result of the increased diversity formed by partitioning the parallel evolution processes, forming natural niches.

In this implementation, all evolving demes are contained on the master process, with members sent to slaves for evaluation using an asynchronous communication strategy. When a deme has passed all its members through for evaluation, but remains idle waiting for the last few evaluated members to return from their slave processors, members from the next deme are used to occupy idle slave processors.

Figure 4 shows the convergence for both distribution implementations based on a global population size of 200. For the iGA, ten demes were used with a sub-population size of 20. Migration was based on an interval of 5 generations with 10% migration between demes using a ring migration topology described by Doorly [9]. Population diversity was enhanced in both GA implementations using niching, based on deterministic crowding [10] and fitness sharing as described by Yin and Germay [11]. The use of niching has significantly reduced the difference in convergence observed by Doorley between the two implementations, although the iGA scheme continued to successfully find better solutions. Although the GA implementations converged close to on another when measured by objective fitness, the resulting sections shown in Figure 5 are remarkably different, indicating the complex modality of the fitness landscape searched.

The iGA implementation significantly reduced idle time previously found on slave processors between generations. Residual traces of idle time found on slave processors was significantly small in comparison to the evaluation time of the objective function resulting in a net iteration efficiency of over 99% for 52 processors.

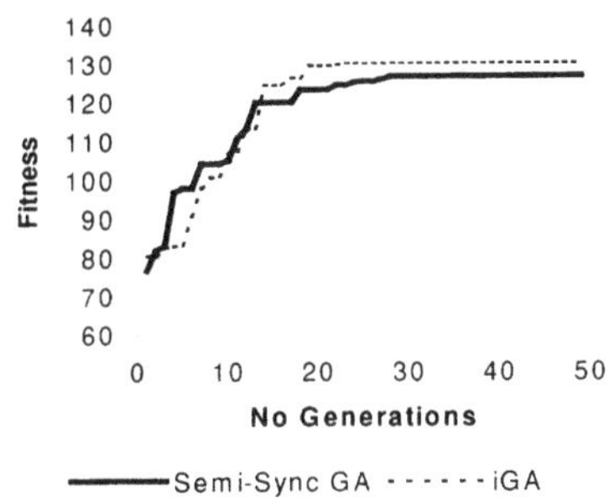

Figure 4: Convergence performance of Semi-synchronous and Asynchronous (iGA) implementations

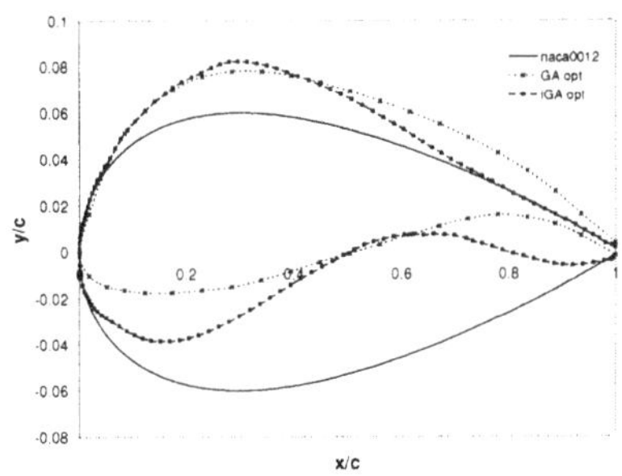

Figure 5: Resulting sections from different GA implementations

4 CLUSTER PERFORMANCE ISSUES CONCERNING DUAL BOOT IMPLICATION

The creation of a Linux computational facility from resources intended for a different daily purpose using Windows NT was easily achieved through the use of dual boot. Several issues were highlighted while implementing CFD problems using this strategy.

The dual boot feature poses an immediate barrier in gaining access to the computational resource when it is booted into the wrong operating system. Where the intention is to use the resource primarily for overnight computation, the individual processors can be configured to automatically reboot into the default Linux boot partition, at a designated time, which is the case used for this work. This strategy although very simple to implement, can cause problems both to NT users wishing to work during such times and computational resource users requiring access to Linux processors during the daytime. In this implementation, a small cluster of processors resides in a separate room to the rest of the processors. These processors reboot into Linux at a designated time, without causing significant disruption to users who are made aware in advance of this feature.

The second problem regarding the robust use of the facility for larger distributed processing using MPI, is when a processor is manually rebooted during computation. Without controlled management, such a reboot will cause the entire MPI process to crash on all associated processors losing all data. While this problem is unavoidable since the primary use of the cluster is for NT based activities. The devastation of such an activity has been limited through encouragement of regular checkpointing of program data with a small associated computational cost as illustrated in Figure 3. Additionally, rebooting through the action of pressing CTRL-ALT-DEL on the keyboard has been trapped on all processors, sending an appropriate signal to all residing programs. This signal is trapped in both the GA and Euler solver, and activates a final checkpointing procedure before a controlled exit of the program on all associated processors. A restart script residing on the master processor re-establishes a new set of available processors and attempts to continue the distributed computational job.

Some configuration time was required to establish the various processors required to increase the robustness and usability of the cluster when using a dual boot feature, however the process was straight forward and leads to substantial gains in the overall performance of the facility.

5 CONCLUSIONS

A commodity-processing network has been established from existing resources without conflicting on their original usage intent. Almost linear speedup has been achieved for the parallel computation of over 10,000 designs using a Panel Solver in a Genetic Algorithm design search. Good speedup was measured during the computation of an Euler solver based on an unstructured grid for up to 16 processors. The use of dual boot can significantly affect the robustness of the computation facility, and some progress was made in this implementation to reduce this effect. Overall, an excellent computational facility can be harnessed from existing PC resources at little additional cost.

REFERENCES

1. Gropp, W., Lusk, E., Doss, N., and Skjellum, A., "A high-performance, portable implementation of the MPI message passing interface standard", Parallel Computing, 22(6), pp 789-828, 1996.
2. Nicole, D.A., Takeda, K. and Wolton, I.C., "HPC on DEC Alpha and Windows NT", Proc. HPCI Conf. 98, Manchester 12-14, 1998.
3. Takeda, K., Tutty, O.R. and Nicole, D.A., "Parallel Discrete Vortex Methods on Commodity Supercomputers; An Investigation into Bluff Body Far Wake Behavior", Proc. 3rd International Workshop on Vortex Flow and Related Numerical Methods, Toulouse, August 1998.
4. Rycroft, N.C., "An Adaptive, Three-Dimensional, Finite Volume, Euler Solver for Distributed Architectures Using Arbitary Polyhedral Cells", PhD. Thesis, University of Southampton, 1998.
5. Goldberg, D.E., "Genetic Algorithms in Search, Optrimization and Machine Learning", Addison-Wesley, 1989.
6. Drela, M., "XFOIL: An Analysis and Design System for Low Reynolds Number Airfoil Aerodynamics", in Low Reynolds Number Aerodynamics, ed. Mueller, T.J., Lecture Notes in Engineering 54, Springer-Verlag, 1989.
7. Roe, P.L., "Approximate Riemann Solvers, parameter Vectors and Difference Schemes", Journal of Computational Physics, 43, pp. 357-372, 1981.
8. Walshaw, C., Cross, M., and Everett, M., "Mesh partitioning and load-balancing for distributed memory parallel systems", In B. H. V. Topping, editor, Advances in Computational Mechanics for Parallel & Distributed Processing, pages 97-104. Saxe-Coburg Publications, Edinburgh, 1997.
9. Doorly, D.J., "Parallel Genetic Algorithms", Ch. 13 of Genetic Algorithms in Engineering and Computer Science, ed. Winter, G., et al, Wiley, 1995.
10. Mahfond, S.W., "Niching Methods for Genetic Algorithms", IlliGAL Report No. 99007, Illinois Genetic Algorithms Laboratory, University of Illinois, Urbana, May 1995.
11. Yin, X., and Germay, N., "A Fast Genetic Algorithm with Sharing Scheme Using Cluster Methods in Multimodal Function Optimization", in Proceedings of the International Conference on Artificial Neural Nets and Genetic Algorithms, Springer-Verlag, Innsbruck, 1993.

Parallel Computational Fluid Dynamics – Trends and Applications
C.B. Jenssen et al. (Editors)
2001 Elsevier Science B.V.

Use of Commodity based cluster for solving Aeropropulsion applications

Isaac Lopez[a], Thaddeus J. Kollar[b] and Richard A. Mulac[c]

[a]ARL, U.S. Army Vehicle Technology Directorate, 21000 Brookpark Rd. NASA/Glenn Research Center, Cleveland, OH 44135 USA

[b]Integral System , Inc., 21000 Brookpark Rd. NASA/Glenn Research Center, Cleveland, OH 44135 USA

[c]AP Solutions, Inc., 21000 Brookpark Rd. NASA/Glenn Research Center, Cleveland, OH 44135 USA

1. Background

The rising cost of developing air-breathing propulsion systems is causing a strong demand for minimizing cost while still meeting the challenging goals for improved product performance, efficiency, emissions and reliability. Computational simulations represent a great opportunity for reducing design and development costs by replacing some of the large scale testing currently required for product development. NASA personnel are interested in computational simulations because of the potential for fielding improved air propulsion systems with lower development cost, greater fuel efficiency and greater performance and reliability. A greater use of simulations would not only save some of the costs directly associated with testing, but also enable design trade-offs to be studied in detail early in the design process before a commitment to a design is made. A detailed computational simulation of an engine could potentially save up to 50% in development time and cost.

Within NASA's High Performance Computing and Communication (HPCC) program, the NASA Glenn Research Center (GRC) in collaboration with the ARMY Research Lab (ARL) is developing an environment for the analysis and design of propulsion engines called the Numerical Propulsion System Simulation (NPSS) [1]. One of the goals for NPSS is to create a "numerical test cell" enabling full engine simulations overnight on cost-effective computing platforms. In order to achieve this goal NASA and ARL personnel at the GRC have been involved since the early 1990s in applying cluster computing technology in solving aeropropulsion applications. In 1993, researchers at the Glenn Research Center successfully demonstrated the use of an IBM 6000 cluster to solve Computational Fluid Dynamic applications. Later on, in a partnership with the US aerospace industry, a special project was started to demonstrate that a distributed network of workstations is a cost-effective, reliable high-performance computing platform. This project demonstrated that cluster computing was indeed a cost effective high-performance computing platform and showed performance and reliability levels equivalent to 1994 Vector Supercomputers at 8% of the capital cost. The next step for ARL/GRC personnel was to apply the experience gained in cluster computing to the emerging commodity hardware for executing aeropropulsion applications. From this foundation, GRC has proposed to research in commodity based clusters.

This paper presents; (1) a detailed description of the commodity based cluster at GRC, (2) a detailed description of a turbomachinery application ported to the cluster, and (3) a cost/performance comparison of the commodity based cluster to a SGI Origin 2000 high performance computer.

2. Commodity based cluster – Aeroshark

ARL/GRC researchers utilize the Aeroshark cluster as the primary test bed for developing NPSS parallel application codes and system software. The Aeroshark cluster provides sixty-four Intel Pentium II 400MHz processors, housed in thirty-two nodes. The networking configuration resembles a star, with a gigabit switch connected to eight routers, and each router connecting to four compute nodes, with one full duplex 100BT wire per node. There is one system server providing Network File System (NFS), Network Information Service (NIS), and other management services, and one interactive system through which users gain access to the entire cluster.

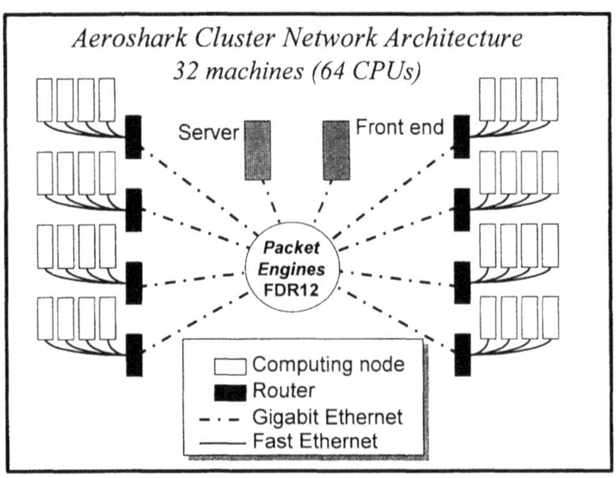

Figure 1. Aeroshark Cluster Network Architecture

Uptime on Aeroshark is in excess of 99%; when the absolute uptimes on all the individual nodes are added together it totals nearly a century. Despite this record, there are a few problems that have been encountered with the configuration. One is that the routers cannot provide the full bandwidth to the compute nodes under a maximal load. This 'star' network configuration was conceived when fast Ethernet switches were still very expensive. Its intention was to allow the incremental addition of ports by inserting another quad port fast Ethernet card into the routers. The cost of this was low compared to a new switch.

It has been found that the gigabit card, a first-generation model, used in the routers cannot perform at more than 200Mb/s. The result is that only two of its compute nodes could operate at full speed; there is a throughput drop-off with traffic to each additional node. Figure 2 shows the results of several Transmission Control Protocol (TCP) performance tests via ttcp (TCP for Transactions). This benchmark was used to time the transmission of data between two compute nodes using the TCP protocol. In these tests, the concurrent throughput capabilities of one to four compute nodes behind a single 'target router' (see Figure 1 for router configuration) was measured by directing traffic at them generated from other parts of the cluster. Each point on the graph in Figure 2 is an average of several tests.

The routers represent single points of failure, although in practice they rarely crash. The failure of a router can cause the temporary loss of eight processors. Since fast Ethernet switches have become very inexpensive, the star network configuration has become less desirable, although it has only had a slight negative impact on codes. The next Aeroshark upgrade will replace the routers with a Cisco Catalyst 6509 switch. There will be two fast Ethernet networks, one

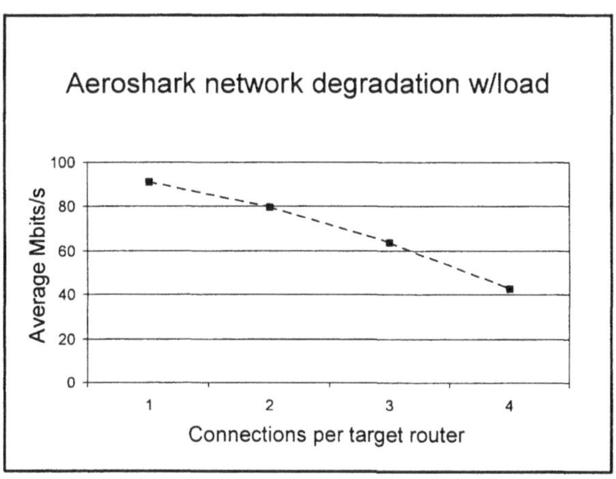

Figure 2. TCP performance tests

for system I/O and one for message passing, so the system traffic will not impact network-bound parallel codes at all.

Infrequently, the NFS server becomes unstable when too many requests are received at once. The result is that the server remounts the file system read-only and occasionally needs to be rebooted. There will be twice as many nodes after the upgrade, so this problem will be addressed by splitting the load across several machines. To do this, NFS will be provided by four systems; they will share a Global File System (GFS), which will allow them to access a large shared disk as if it were local. In addition, the other services (NIS, NTP, Domain Name Server, etc), which are provided by lord, will be moved to separate systems.

The last point of failure that will be addressed in the upgrade is the single login node. When this system goes down (not that it does very often), the entire cluster is inaccessible to users. This node will be replaced by a load-sharing sub-cluster running Linux Virtual Server (LVS), which will consist of three login nodes and one control node. Users will connect to a virtual Internet Protocol address, the control node will decide which of the other nodes is least loaded, and transparently redirect their request. Users will also be able to connect to a specific login node if they need to.

3. Aeropropulsion application – APNASA

The APNASA code was developed by a government / industry team for the design and analysis of turbomachinery systems. The code has been used to evaluate

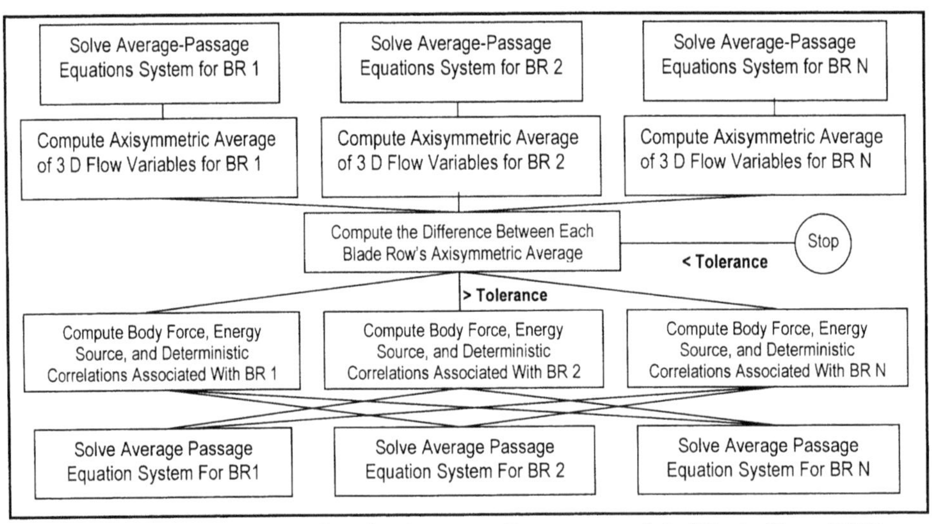

Figure 3. Solution algorithm for the Average-Passage model. {Blade Row (BR)}

new turbomachinery design concepts, from small compressors to large commercial aircraft engines. When integrated into a design system, the code can quickly provide a high fidelity analysis of a turbomachinery component prior to fabrication. This results in a reduction in the number of test rigs required and a lower total development cost. APNASA or the methodology on which it is based has been incorporated into the design systems of six major gas turbine manufacturers.

The code itself is based on the Average-Passage flow model [2] which describes the three-dimensional, deterministic, time-averaged flow field within a typical blade row passage of a multiple blade row turbomachinery configuration. The equations governing such a flow are referred to as the Average-Passage equation system. For multiple blade row configurations, the model describes the deterministic flow field within a blade passage as governed by the Reynolds-averaged form of the Navier-Stokes equations.

An APNASA simulation consists of running each blade row independently a number of iterations (typically 50) through a Runge-Kutta process until certain local convergence criteria are met. This part of the solution procedure has been termed a "flip". At the end of every flip, various information (body forces, correlations, ... etc.) is then communicated between the individual blade rows to update the effects of neighboring blade rows based on current information. The preceding two-step procedure is then repeated until certain overall convergence criteria are met based on each blade row's axisymmetric solution (typically 50+ flips). A flow diagram describing the entire solution algorithm for the Average-Passage model is shown in Figure 3. The solution procedure is very amenable to

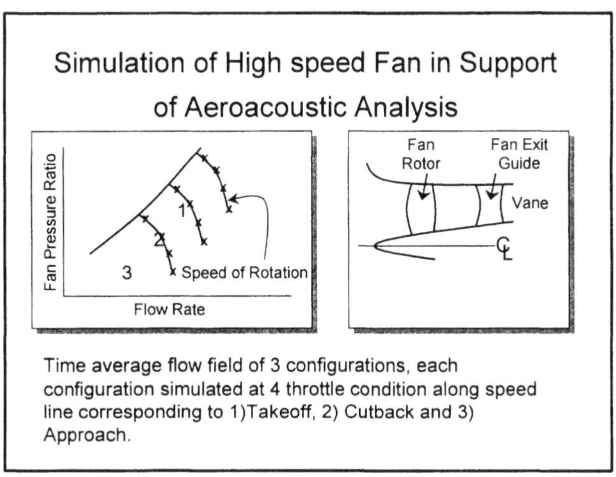

Figure 4. Solution algorithm for the Average-Passage model

parallel processing since communication between the blade rows is minimal once a flip is initiated.
The Average-Passage code APNASA has evolved over the last 15 years from a series of codes written for execution on high-speed multiprocessor computing platforms. SSTAGE, the original code developed in 1985, was written specifically for the CRAY-1. SSTAGE simulated the flow through a single turbomachinery stage by running each of the stage's two blade rows alternately on the CRAY-1's single CPU. The multiprocessor CRAY-XMP and CRAY-YMP systems arrived in the mid to late 80's, respectively. Access to these types of systems led to the development of the MSTAGE code, which allowed for the practical simulation of multistage turbomachinery by running each of the blade rows concurrently in parallel. This dramatically reduced the wall-clock time required for a multistage simulation. An N blade row simulation run across an N CPU system could be completed in the wall-clock time required for the simulation of a single blade row. By the mid-90's, supercomputers such as the CRAY C90 were starting to receive competition from high-end compute servers such as those manufactured by SGI. Average-Passage simulations of upwards of 10 stages were becoming economically viable due to the decreasing cost of compute cycles. Now with the advent of relatively low cost LINUX-based PC clusters, the high-end UNIX compute server market is being challenged as the platform of choice for APNASA.

A recent application simulated on the Glenn Research Center cluster Aeroshark using APNASA consisted of a single stage fan which was being analyzed to determine noise levels associated with various designs. The design matrix consisted of three different rotor geometries based on takeoff (100%), cutback (87.5%), and approach (61.7%) engine wheel speeds, each paired with three different vane geometries. This resulted in nine (3x3) different configurations to be simulated, and each of these specific configurations would be run at four different flow rates to map out a speedline as shown in Figure 4. For this project, access was granted to run on twelve Aeroshark nodes (24 CPUs). The cases to be simulated were then grouped based on the three different rotor wheel speeds. The first week the rotor geometry at takeoff would be simulated paired with each of the three vane designs at four different flow rates:

1 rotor x 3 vanes x 4 flow rates = 12 cases x 2 blade rows (rotor, vane) -> 24 CPUs

The same type of grouping would also be performed for the other two rotor wheel speeds. In total, three weeks were required to simulate all 36 cases utilizing 12 Aeroshark nodes (24 CPUs).

3. Cost/Performance comparison

Compilation of the code itself was very straightforward on the cluster using The Portland Group's Fortran 90 compiler, pgf90. There is even a compiler option "-byteswapio" which forces the code to perform file reads and writes in the IEEE

format compatable with most UNIX platforms. This allowed for easy porting of mesh and restart files between the cluster and various SGI systems.
For each single stage fan case (with a mesh size of 407 x 51 x 51 for each blade row), a single "flip" took approximately 6500 seconds of wall-clock time to run the fan's two blade rows in parallel on a 2 CPU node of the Aeroshark cluster. This compares to 2750 seconds of wall-clock time to run the same case on an SGI Origin 2000 system composed of 250 Mhz R10000 MIPS processors. This equates to roughly a factor of 2.36 when comparing the processor-to-processor speed of the Intel based Aeroshark cluster to the MIPS based Origin system for this application. The cost of a 24 processor SGI Origin 2000 is 22.3X greater than the cost of a 24 processor segment of the Aeroshark cluster. A cost/performance ratio of 9.4 in favor of the Aeroshark cluster is obtained.

Conclusion

Clearly the use of commodity based cluster has a tremendous potential of providing a computing platform on which detailed aeropropulsion simulations can be executed in a time compatible with the engine design cycle. In addition the cost/performance ratio shown by the cluster was impressive considering the cost differential between commodity based clusters and traditional UNIX workstation clusters.

As a result of this work the aeroshark cluster will be upgraded to address all the performance issues reported in this paper.

---

[1] A. L. Evans, J. Lytle, J., G. Follen, and I. Lopez, An Integrated Computing and Interdisciplinary Systems Approach to Aeropropulsion Simulation, ASME IGTI, June 2, 1997, Orlando, FL.

[2] Adamczyk, J.J., "Model Equation for Simulating Flows in Multistage Turbomachinery," NASA TM-86869, ASME Paper No. 85-GT-226, Nov. 1984

Parallel Computational Fluid Dynamics - Trends and Applications
C.B. Jenssen et al. (Editors)

Using a Cluster of PC's to Solve Convection Diffusion Problems

R. S. Silva[a] and M. F. P. Rivello[b]

[a]Computational Mechanics Dept., Laboratório Nacional de Computação Científica, Av. Getúlio Vargas 333, Petrópolis, RJ, Brazil, 25651-070, rssr@lncc.br

[b]Computer Science Dept., Universidade Católica de Petrópolis, Brazil

In this work we present our earlier experience in porting a convection diffusion code, which is designed to run in a general purpose network of workstations, to a Cluster of PC's. We present the effort to improve the algorithm scalability by changing the local solvers in the Krylov-Schwarz method and a identification of some bottlenecks in the code as consequence of the improvement of the communication network, which will lead to improvements in the code in the future

1. Introduction

In the last years Computational Fluid Dynamics (CFD) simulations are becoming an important, and in certain cases, dominant part of design process in the industry. When used correctly and implemented *efficiently*, they lead to great reductions in development costs. Cost effective designs require an equilibrium among modelling complexity and execution time. The modeling complexity comes from the necessity of modelling some physical phenomena like shocks, separation, boundary layers and/or turbulence. This requires reliable numerical formulations, more sophisticated numerical time schemes, adaptive methods and so on, possibly implying in an increase in grid points, small time steps and large data structures. The solution of these type of discrete problems requires the resolution of large, sparse and unsymmetric system of algebraic equations, better solved using iterative methods. With the development of parallel and distributed machines domain decomposition methods have been rediscovered and improved to deal with a large class of problems. Among them the overlapping Additive-Krylov-Schwarz method (KSM) has become a powerful tool because it combines high degree of parallelism with simplicity.

However the access to supercomputers sometimes is limited or very expensive to research groups, medium and small companies. One of the solutions to avoid this is to use clusters of machines. A common type of machine to be used in a cluster is the workstation (COW), but the price to keep them dedicated for a long time is still a limiting factor. The accelerated growth of the computational performance of microprocessors, in special the Intel Pentium family, and the increasing number of new network technologies turned the prices very accessible, creating the opportunity of increasing the productivity by using a cluster of dedicated PCs as a distributed system, at low cost. An important point of this type of machine is related to educational and research institutions where it can be used to teach parallel programming, leaving the massive parallel machines to the *production* codes. In this work we present our earlier experience in porting a convection diffusion code, which is designed to run in a general purpose network of workstations, to a Cluster of PC's. We present the effort to improve the algorithm scalability by changing the local solvers in the Krylov-Schwarz method and a identification of some bottlenecks in the code as consequence of the improvement of the communication network, which will lead to improvements in the code in the future

This work is organized as follows. In Section 1 a scalar convection dominated convection-diffusion problem. In Section 2 a distributed Krylov-Schwarz solver and the local solvers are presented.

In section 3 we present the PC cluster used to solve this kind of problem. In Section 4 we present the numerical results used to evaluate the performance for two different topologies. In Section 5 the conclusions are drawn.

2. Convection Diffusion Problems

In this work we are interested in solving the stationary, linear, convection-dominated, convection-diffusion problem of the form

$$\mathbf{u} \cdot \nabla\varphi + \nabla \cdot (-\mathbf{K}\nabla\varphi) = f(\mathbf{x}) \qquad \text{in} \quad \Omega \quad , \tag{1}$$

with boundary conditions

$$\begin{array}{rll} \varphi(\mathbf{x}) = g(\mathbf{x}) \ ; & & \mathbf{x} \in \Gamma_g; \\ -\mathbf{K}\nabla\varphi \cdot \mathbf{n} = q(\mathbf{x}) \ ; & & \mathbf{x} \in \Gamma_q, \end{array} \tag{2}$$

where the bounded domain $\Omega \in \Re^n$ has a smooth boundary $\Gamma = \Gamma_g \cup \Gamma_q$, $\Gamma_g \cap \Gamma_q = \emptyset$, with an outward unit normal $\mathbf{n}$. The unknown field $\varphi = \varphi(\mathbf{x})$ is a physical quantity to be transported by a flow characterized by the velocity field $\mathbf{u} = (u_1, \ldots, u_n)$, the (small) diffusion tensor $\mathbf{K} = \mathbf{K}(\mathbf{x})$, subject to the source term $f(\mathbf{x})$. The functions $g(\mathbf{x})$ and $q(\mathbf{x})$ are given data. To solve this problem a Finite Element Method with the SUPG formulation [5] is used

3. Additive Krylov Schwarz Method

Domain decomposition algorithms have been subjected to active research in the last few years [7] due to the intrinsic divide-and-conquer nature of the method as well as the diffusion of parallel and distributed machines. In this work we focus on the Overlapping Schwarz Methods (OSM), with emphasis on the additive version (ASM). The Additive version consists in dividing the original domain Ω into a number of smaller overlapping subdomains Ω_i, solving the original problem in each subdomain using the solution of the last iteration as the boundary conditions on the artificial interfaces created by the partition of the original domain.

The ASM can be viewed as the Preconditioned Richardson method with a damped factor equal 1, for NP subdomains, where the preconditioner matrix is:

$$M^{-1} = \sum_{i=1}^{NP} R_i^t A_i^{-1} R_i \quad . \tag{3}$$

A_i are the local matrices and R_i and R_i^t are the restriction and extension matrices defined in [7]. It is well known that the convergence of the Richardson method is very slow. Thus, in order to accelerate the convergence we used a Flexible variation of the restarted GMRES called FGMRES(k) introduced by Saad [9], because it allows the use of an iterative method for solving the preconditioner. The Additive Krylov-Schwarz algorithm is the following:

1. **Start:** Choose x_0 and a dimension k of the Krylov subspaces.
2. **Arnoldi process:**
   (a) Compute $r_0 = b - Ax_0$, $\beta = \|r_0\|$ and $v_1 = r_0/\beta$.
   (b) For $j = 1, \ldots, k$ do
      - Compute $z_j := \sum_{i=1}^{P} R_i^t A_i^{-1} R_i v_j$
      - Compute $w := Az_j$
      - For $i = 1, \ldots, j$, do
        $h_{i,j} := (w, v_i)$
        $w := w - h_{i,j} v_i$
      - Compute $h_{j+1,j} = \|w\|$ and $v_{j+1} = w/h_{j+1,j}$.

   (c) Define $Z_k := [z_1, \ldots, z_k]$ and $\bar{H}_k = \{h_{i,j}\}_{1 \le i \le j+1; 1 \le j \le k}$.
3. **Form the approximate solution:**
   Compute $x_k = x_0 + Z_k\, y_k$,
   where $y_k = argmin_y \left\| \beta e_1 - \bar{H}_k\, y \right\|$
   and $e_1 = [1, 0, \ldots, 0]^t$.
4. **Restart:** If satisfied stop,
   else set $x_0 \longleftarrow x_k$ and go to 2.

The Additive Krylov-Schwarz algorithm with right Preconditioning

3.1. Local Solvers

The solution of the preconditioner phase in the FGMRES(k) algorithm depends on the solution of the local problems A_i^{-1}. Those accuracy strongly affect not only the convergence, but also the computational cost of the Krylov-Schwarz Method [11].

Usually a direct solver is used, but only focusing the accuracy, in this work in order to improve the computational scalability we present the results using three iterative methods DILU, ILU(0), MILU(0).

The Incomplete LU(0), ILU(0), method is a traditional preconditioner. This methods consist in evaluate the LU factorization with no *fill-in*, meaning that if the zero pattern of the LU factorization will be equal to the original zero pattern of A_i [10]. The modified ILU(0), MILU(0), is a variation of ILU(0) where the elements of the matrix A_i that were dropped out are subtracted from the diagonal to guarantee that the row sums of A_i are equal to the factorization [10]. The Diagonal ILU factorization [4] , DILU, is a simplified version of the ILU(0) where only the diagonal terms are factorized.

3.2. Parallel Krylov-Schwarz Method

The first concern of the parallel version is to determine how to split the data among the processors. Thus, two factors have to be taken into account. The first one is that, as we are using the Additive Krylov Schwarz method, the preconditioner step is reduced to a series of local problems that can be solved in parallel. The second one is that, for the type of the topology usually used, hubs, allows all processors to logically communicate directly with any other one but physically supports only one 'send' at a time. For these reasons we chose one-dimensional partition of the domain reducing the overhead on 'startups' in the network and the number of neighbors, so that each subdomain should be assigned to a processor.

Independent of the type of partition, each processor has the matrix A_i and parts of the initial residual vector r_0 and the initial solution x_0. To evaluate the matrix-vector products, each processor has to execute a series of local communications with its neighbors, usually called nearest-neighbor communications, to obtain the contribution of the interface degrees of freedom. At the end, each processor has v_i^m with $m = 0, \ldots, NP$, where NP is the number of processors. Thus, the orthogonalization step is performed in parallel, where the partial results ($\|r_0\|$,$\|v_i\|$, $\beta_{i+1,j}$) are exchanged among the processors. Obviously this part of the algorithm requires a great number of communications, precisely $(k^2 + k)$. To overcome this, it will be necessary to change the FGMRES algorithm or to use another accelerator to the Schwarz method. The nearest-neighbor communications (preconditioner and matrix-vector products) use point-to-point send-receives pairs between neighbors and, for the global communications ($\|r_0\|$,$\|v_i\|$, $\beta_{i+1,j}$), two strategies can be used depending on the topology of the machine, a bi-directional ring (Array) [8] or a recursive doubling global collect algorithms (RDGC) [3].

4. Carcará Machine

Carcará[1] is a Beowulf class machine assembled at LNCC in July 1998, with the primary goal of obtaining the know-how of building and operating a cluster of PC's and to investigate the possibility of using it as a platform to solve computational mechanics problems. It is composed by 9 Intel Pentium II 266 MHz with 128 MBytes of RAM, 512 KBytes of secondary cache and 3.2 GBytes disk, yielding a total 25.6 GBytes of disk space and around 1 GBytes of RAM. They are connected by 2 Fast Ethernet networks. Following the Beowulf philosophy, the operational system is the LINUX, kernel 2.0.36, using the message passing libraries MPICH version 1.1 and PVM 3.4.

These two networks permit to reduce the contention in the network by dividing the communication traffic in two parts: administration and the application. Besides, this configuration allows using the channel bonding procedure from Beowulf Project [1], which increases the bandwidth by routing packages over multiple networks.

To fulfill its goals Carcará is in constant testing phase in which new software or hardware is installed and tested to verify the performance improvement. To this end we selected the CG benchmark from the NAS benchmark version 2.3 [2]. The CG is a CFD benchmark and it has almost the same structure of operations and communications of our solver.

The Table 1 shows one example of the evolution of the performance Carcará with

[1] A small bird of pray found in South America (Polyborus plancus brasiliensis).

different topologies and using the channel bonding. The average times (T_{avg}) and the average millions of float point operations per second ($Mfop/s$) of three runs are shown, comparing with the results obtained running the NPB2.3 in the IBM SP2 at LNCC, used as reference.

5. Numerical Results

The selected convection diffusion problem is a simple CFD problem, but has the advantages of being linear and having the same solver of our compressible Navier-stokes code. In this example the velocity field $\mathbf{u}$ is constant ($|\mathbf{u}| = 1$), the medium is assumed homogeneous with a constant physical diffusivity tensor $K = 10^{-6}$ and no reaction term in a square domain [0x1;0x1]. All computations were performed using piecewise linear elements.

It is well known that the convergence of the KSM depends on both the overlap size and the type of the partition of the domain [12,7,6]. Thus, to reduce the number of variables, the overlap size is fixed to one element and one-dimensional partitions in the x-direction are used. The Krylov dimension (k) was kept constant, equal to five, and the FGMRES tolerance is 10^{-6}. A hybrid master-slave model is adopted and the PVM version 3.4 is used as message passing library.

In Figures 1 and 2 the average time of each iteration is presented for different sizes of problems (ndof) for different local solvers. For the LU solver, the curves for different number of processors present the same exponential behavior, which compromise the scalability of algorithm. However the other solvers present a linear behavior, as shown in Figures 1(b), 2(a) and 2(b)

Due to the Beowulf flexibility the topology of Carcará was improved by using two switches. We expect that the performance of the code will increase significantly. Our surprise is that, even with a better communication algorithm (RGDC) the difference is not so significative as in the CG example. In Figure 3 is present the average time of each iteration with the switch topology for the two types of algorithm used for global communications. With these curves and the results of Table 1, we can concluded that the communication involved with the preconditioner and the matrix-vector products is the cause Because the CG and the KSM are very similar and the only difference is the partition of the domains, (unidimensional for KSM and in the CG is a cyclic type [3]).

6. Conclusions

This work presents local solvers that increase the computational scalability of the KSM. However future work has to be done to recovery the convergence rate of the KSM. This can be done for example to find new solvers, like ILUTP [10] or increase the degree of *fill-ins* of the ILU or MILU.

Another point is that this work highlights the importance of the combination algorithm - topology and this study can only be done due to the flexibility of a Beowulf-class machine that permits to change the topology easily.

The improvement in the topology usually improves the performance. However for applications that have algorithm restrictions, like the preconditioner step in our solver, the improvement in the communication on the most expensive communication part of the

	Class B							
	Carcará						SP2	
	Hubs		Switches					
	Channel Bonding		Fast Ethernet		Channel Bonding		Thin node	
NP	$T_{avg}(s)$	$Mfop/s$	$T_{avg}(s)$	$Mfop/s$	$T_{avg}(s)$	$Mfop/s$	$T_{avg}(s)$	$Mfop/s$
1	*	*	*	*	*	*	6530.60	8.38
2	1988.85	27.51	2004.51	27.29	1992.58	27.46	2709.52	20.19
4	1153.90	47.41	1109.93	49.29	1082.07	50.56	1483.43	36.88
8	891.99	61.34	595.02	91.94	574.52	95.23	655.78	89.95

Table 1
NAS CG results for Class B. T_{avg} in seconds. Entries marked with $*$ did not fit in memory

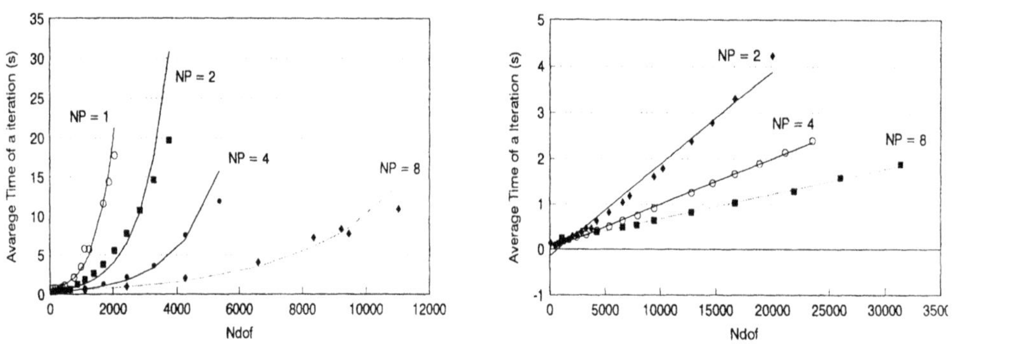

(a) LU Local solver

(b) DILU Local solver

Figure 1. Average time oa a iteration with ILU(0) and MILU(0) local solvers

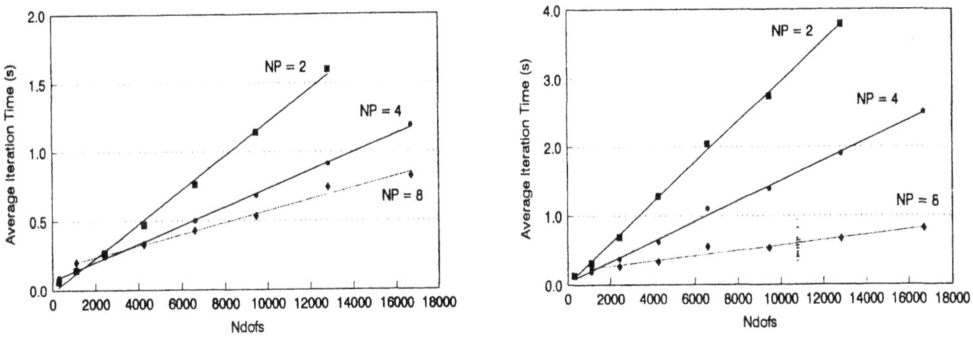

(a) ILU(0) Local solver

(b) MILU(0) Local solver

Figure 2. Average time oa a iteration with ILU(0) and MILU(0) local solvers

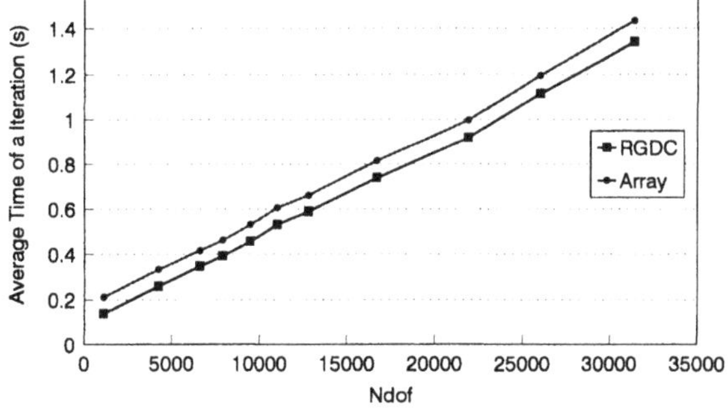

Figure 3. Average time of iteration using a switch topology with different algorithms

algorithm is not enough to get improvements. Besides, the high degree of parallelism of the additive version of the Schwarz method, is not enough to guarantee performance and some effort has to be done in the future on the partitioning of the domain to achieve the same performance level of the NAS-CG.

Acknowledgments

The authors would like to thank the PRONEX LNCC-FINEP project number 76.97.0999.00, PIBIC/CNPq - LNCC, that support part of this work and the DCE / LNCC which provided the switches.

REFERENCES

1. http://www.beowulf.org/.
2. D. Bailey, T. Harrtis, W. Saphir, R. van der Wijngaart, A. Woo, and M. Yarrow. The NAS parallel benchmarks 2.0. Technical Report NAS-95-020, NASA, December 1995.
3. M. Barnett, R. Littlefield, D. G. Payne, and R. A. van de Geijn. On the efficiency of global combine algorithms for 2-D meshes with wormhole routing. Technical Report 05, The University of Texas, Departament of Computer Science, 1993.
4. R. Barret, M. Berry, T. Chan, and J. Demmel et al. *Templates for the Linear Systems: Building Blocks for Iterative Methods.* Society for Industrial and Applied Mathematics, Philadelphia, PA, 1994.
5. A. N. Brooks and T. J. R. Hughes. Streamline upwind Petrov-Galerkin formulations for convection dominated flows with particular emphasis on the incompressible Navier-Stokes equations. *Computer Methods in Applied Mechanics and Engineering*, 32:199–259, 1982.
6. X. C. Cai, W. D. Gropp, and D. Keyes. A Comparisom of Some Domain Decomposition Algorithms for Nonsymmetric Elliptic Problems. In S. for Industrial and A. Mathematics, editors, *Fifth International Symposium on Domain Decomposition Methods for Partial Differential Equations*, pages 224–235, 1992.
7. T. F. Chan and T. Mathew. *Domain Decomposition Algorithms.* Acta Numerica, 1994.
8. J. J. Dongarra, R. A. van de Geijn, and R. C. Whaley. A users'guide to BLACS. http://www.netlib.org.
9. Y. Saad. A Flexible Inner-Outer Preconditioner GMRES Algorithm. Supercomputer Institute Research Report 279, The University of Minnesota, 1991.
10. Y. SAAD. *Iterative Methods for Sparse Linear Systems.* PWS, 1995.
11. R. S. Silva and R. C. Almeida. Iterative Local Solvers for Distributed Krylov-Schwarz Method applied to convection-diffusion problems. *Computer Methods in Applied Mechanics and Engineering*, pages 353–362, 1997.
12. P. L. Tallec. Domain Decomposition Methods in Computational Mechanics. *Computational Mechanics Advances*, 1(2), February 1994.

Parallel Computational Fluid Dynamics – Trends and Applications
C.B. Jenssen et al. (Editors)

Building PC clusters: An object-oriented approach

A. Soulaïmani[a], T. Wong[b] and Y. Azami[c]

[a]Mechanical engineering department, École de technologie supérieure, 1100 rue Notre-Dame Ouest, Montréal, H3C 1K3, Canada

[b,c]Automated production engineering department, École de technologie supérieure

This paper addresses a methodology for designing PC clusters using object-oriented technology. The proposed methodology simplifies the design process and manages specification differences in the academic and the industrial environments. This is to respond to the need to implement an existing CFD code on PC clusters for research purposes and future industrial usage. The object-oriented technology is well suited to solve this kind of design problems. The design procedure and computational results are given in this paper.

1. INTRODUCTION

Computational fluid dynamics (CFD) methods are becoming popular in research and industrial environments due to the increase use of numerical simulation in solving practical problems. The industrial use of CFD often requires computation of flows in or around large and complex geometry such as aircraft, automotive vehicles, engines, or hydraulic structures. Many of the geometric details have to be reproduced to obtain sufficient accuracy for engineering purposes. Realistic three-dimensional flow computations require fast execution speeds, one to three-orders of Giga-flops, and a large memory pool (1 to 100 GB of RAM). Presently, such processing power can only be achieved by distributing the computational task onto a number of processors.

Large commercial multiprocessors are costly to operate and to maintain. University research laboratories generally cannot afford large investment on a regular basis. Since such equipment has to be upgraded, if not completely overhauled, periodically in order to satisfy the ever-increasing computational demands. The Beowulf project at NASA's CESDIS [1] proved that a collection of desktop PCs connected by a fast internal network and running free software can deliver a competitive performance at a very low initial implementation cost. The results from the Beowulf project enable new possibilities for low-cost super-computing that are beneficial to the academic research environment. Also, high-performance PC clusters are attractive for many industrial applications because of their low operational and maintenance costs. Even though desktop clusters are low-cost and easy to operate, they represent a new challenge in terms of machine design and equipment maintenance. This is due to the extremely fast-paced hardware and software evolutions in the component-based personal computer market. Because of this component-based paradigm and the rather fragmented PC market, system designers have the daunting task of deciding what and which components to integrate into their computing machinery. Moreover, it is very laborious to keep track of system components because of their regular updates and upgrades. An object-oriented approach can alleviate the burden by organizing and representing the design and maintenance processes in a well-defined manner.

2. PROPOSED METHODOLOGY

This section presents a design procedure for a PC cluster using an object-oriented approach. This cluster contains 16 desktop PCs interconnected locally through a 100 Mbit/sec Ethernet network interface. The system will host a multi-OS execution environment and should be capable of running message-centered or memory-centered communication packages such as MPI and PVM libraries. Furthermore, each PC can also be mono or multiprocessor. In the proposed methodology, the application of object-oriented analysis and design techniques provides a flexible environment where hardware and software components can be viewed as interacting objects. By defining the proper hierarchical relationships among the objects, it is then possible to study the behavior of individual hardware and software components and their impact on the overall performance of the computing system.

2.1 UML Modeling

The Unified Modeling Language (UML) describes object relationships and their static and dynamical behavior in a standard manner [2]. Since this language is internally consistent, there exists a set of tools for automated design error checking and documentation. As well, object components represent a natural partitioning of the complete computing system into manageable units and each of these objects can be replaced by another object of the same class. Thus, the mapping between objects and actual system components can be done directly using UML.

2.2 Design steps

The proposed object-oriented approach is an iterative and incremental process. The approach is iterative since there is a continuous looping of the following steps: *i*) requirement analysis; *ii*) design process; *iii*) architectural review; *iv*) testing [3, 4]. This approach is also incremental because improvements are made at each iteration. The end result is a fully documented PC cluster satisfying the original system specification [5]. In order to identify system requirements correctly, the use case model is used to derive domain objects. These domain objects represent the hardware and software components of our PC cluster. In the design process, all domain objects derived from the use case model are mapped into classes by the use of UML. All properties, relationships and events pertaining to the domain objects are also identified in this step.

In the architectural review, all system related information are gathered and inspected. The goal of this step is to make sure that the system is concordant with the results obtained from the requirement analysis and the design process. This is actually a crucial step since misunderstanding can arise when project information is passed back and forth between the requirement steps and the design steps. Without an adequate architectural review, results from the testing of the system may be useless.

2.3 System requirement

The initial task in this step is the so-called *requirement elicitation*. Requirement elicitation usually involves the design team, the user groups and other decision-making bodies. Its goal is to define the desired system in plain language. The system requirement analysis begins after the desired PC cluster characteristics are defined. Use case models are then derived from the system definitions. These models represent the desired functionality of the system from the user's point of view [6]. As shown in figure 1, three actors and two general use cases are derived for our PC cluster. The actors represent different user groups and the use cases represent possible usage of the PC cluster. Inside each use case is a detailed description explaining how the PC cluster is being used in plain language. This defines the role of each user group, the proper usage of the system by users and the access control policy for the system.

The next task in the requirement step is to define classes and objects, which are entities manipulated by the users and the system. This user-centric object model is shown in figure 2a. The ordinal beneath the classes denotes the cardinality of objects that can be instantiated. Since our PC cluster is to host a number of operating systems, different classes representing different PC clusters are aggregated inside a common base class. In order to formalize the use cases models, it is necessary to transform its plain language description into UML diagrams, which identifies user-to-system and system-to-system interactions. Hence, each use case is supported by a set of diagrams similar to the sequence diagram shown in figure 2b. This diagram shows that the maintainer user group can change or modify system hardware and software without any restriction applied.

2.4 System design

In this step, several models are produced. These models are generated by performing the following tasks: *i*) subsystem decomposition; *ii*) hardware/software mapping; *iii*) interconnect management; *iv*) user access control. For our PC cluster, subsystem decomposition consists of abstracting all relevant system components by a set of proxy classes. These proxy classes represent a particular view of the system by the users. For example, figure 3 shows the physical view of the PC cluster. In hardware/software mapping, the goal is to identify completely the basic hardware and software requirement of our PC cluster. Part of these models is shown by the set of hardware and software proxy classes given in figure 4 and figure 5. Note that the details can be very fine-grained depending on the needs and the design goal. The interconnect management task defines the type of network filesystems and transport protocols needed by the PC cluster. Normally, filesystem specifications are dictated by the operating system and transport protocols are established by examining the filesystem and the parallel library's communication requirement.

3 FEM MULTIPURPOSE CODE (PFES)

PFES is an in-house research finite element code designed for solving some multi-physics problems such as aeroelasticity, magnetohydrodynamics and thermo-hydrodynamics. A parallel version has been developed using the domain decomposition approach [7, 8], MPI library for communication and PSPARSLIB library for parallel data organization and solution algorithms (such as ILUT-preconditioned GMRES). Promising performance has been achieved on multi-processor machines (over 80% speed-up for Euler and RANS computations). This code is used in the system-testing step of our design procedure. In the following, a brief description of the parallel implementation of the FE code on distributed machines is presented.

3.1 Parallel implementation

Domain decomposition has emerged as a quite general and convenient paradigm for solving partial differential equations on parallel computers. Typically, a domain is partitioned into several sub-domains and some technique is used to recover the global solution by a succession of solutions of independent subproblems associated with the entire domain. Each processor handles one or several subdomains in the partition and then the partial solutions are combined, typically over several iterations, to deliver an approximation to the global system. All domain decomposition methods (d.d.m.) rely on the fact that each processor can do a big part of the work independently. In this work, a decomposition-based approach is employed using an Additive Schwarz algorithm with one layer of overlapping elements. The general solution algorithm used is based on a time marching procedure combined with the quasi-Newton and the matrix-free version of GMRES algorithms. The MPI library is used for communication among processors and PSPARSLIB is used for pre-processing the parallel data structures.

3.2 Data structure

In order to implement a domain decomposition approach we need a number of numerical and non-numerical tools for performing the pre-processing tasks required to decompose a domain and map it into processors, as well as to set up the various data structures, and solving the resulting distributed linear system. PSPARSLIB [7], a portable library of parallel sparse iterative solvers, is used for this purpose. The first task is to partition the domain using a partitioner such as METIS. PSPARSLIB assumes a vertex-based partitioning (a given row and the corresponding unknowns are assigned to a certain domain). However, it is more natural and convenient for FEM codes to partition according to elements. The conversion is easy to do by setting up a dual graph which show the coupling between elements. Assume that each subdomain is assigned to a different processor. We then need to set up a local data structure in each processor that makes it possible to perform the basic operations such as computing local matrices and vectors, the assembly of interface coefficients, and preconditioning operations. The first step in setting up the local data-structure mentioned above is to have each processor determine the set of all other processors with which it must exchange information when performing matrix-vector products, computing global residual vector or assembly of matrix components related to interface nodes.

When performing a matrix-by-vector product or computing a residual global vector (as actually done in the present FEM code), neighboring processors must exchange values of their adjacent interface nodes. In order to perform this exchange operation efficiently, it is important to determine the list of nodes that are coupled with nodes in other processors. These local interface nodes are grouped processor by processor and are listed at the end of the local node list.

Once the boundary exchange information is determined, the local representations of the distributed linear system must be built in each processor. If it is needed to compute the global residual vector or the global preconditioning matrix, we need to compute first their local representation to a given processor and move the interface components from remote processors for the operation to complete. The assembly of interface components for the preconditioning matrix is a non-trivial task. A special data structure for the interface local matrix is built to facilitate the assembly operation, in particular when using the Additive Schwarz algorithm with *geometrically non-overlapping* subdomains. The boundary exchange information contains the following items:

3.3 Algorithmic aspects

The general solution algorithm employs a time marching procedure with local time steeping for steady state solutions. At each time step, a non-linear system is solved using a quasi-Newton method and the matrix-free GMRES algorithm.

The preconditioner used is the block-Jacobian matrix computed, and factorized using ILUT algorithm, at each 10 time steps. Interface coefficients of the preconditoner are computed by assembling contributions from all adjacent elements and subdomains. Another aspect, which is worth mentioning, is the fact that the FEM formulation requires a continuous solution vector in order to compute a residual vector. However, when applying the preconditoner (i.e. multiplication of the factorized preconditioner by a vector) or at the end of Krylov-iterations, a discontinuous solution at the subdomains interfaces is obtained. To circumvent this inconsistency, a simple averaging operation is applied to the solution interface coefficients.

3.4 Comparison tests

We consider here performance tests of PFES on the PC cluster. The classical Euler flow (Mach= 0.8447 and an angle of attack of 5.06 degrees) over the Onera-M6 wing is used as the benchmark tests. The mesh used has 15460 nodes and 80424 elements and generate 80000 d. of f. Computations are carried out using a SUN-Enterprise-6000 multiprocessor machine with 165 MHz processors and a 550 MHz PC cluster. Note that for the PC cluster, a stackable 100Mb/s Ethernet switch was used to link the cluster computers together.

4 CONCLUSION

In this paper we described an on-going effort to use an object-oriented approach to solve the problem of PC cluster design and maintenance. The application of object-oriented analysis and design techniques enabled the modeling of hardware and software components as interacting objects and permitted the definition of a set of hierarchical relationships among proxy objects using UML notations. We also performed system tests that showed the performance of a PC cluster compared to that of a commercial multiprocessor system.

Table 1. Speedup versus number of processors.

	# processors			
	1	2	3	4
Compute time (min)				
PC cluster	428 (+7h)	242 (+5h)	210 (+3h)	143 (+2h)
Multiprocessor	1145 (+19h)	619 (+10h)	474 (+7h)	313 (+5h)
GMRES Iterations	1553	1632	1966	1651
Speedup				
PC cluster	1	1.77	2	2.99
Multiprocessor	1	1.85	2.4	3.7

REFERENCES

[1] The Beowulf Project. http://beowulf.gsfc.nasa.gov/

[2] UML documentation at Rational Rose. http://www.rational.com/uml

[3] P.B. Kruchten, "The 4+1 View Model of architecture," *IEEE Software*, Vol. 12, Issue 6, pp. 42 –50,1995.

[4] W.M. Ho, F. Pennaneac'h; N. Plouzeau, "UMLAUT: a framework for weaving UML-based aspect-oriented designs," *Proc. 33rd Int. Conf. on Technology of Object-Oriented Languages*, pp. 324 –334, 2000.

[5] B. Selic, "A Generic Framework for Modeling Resources with UML," *IEEE Computer*, pp. 64-69, June, 2000.

[6] J. Lee, N-L. Xue, "Analyzing user requirements by use cases: a goal-driven approach," *IEEE Software*, Vol. 16, Issue 4 , pp. 92 –101, 1999.

[7] A. Soulaïmani, Y. Saad and A. Rebaine, ''Parallelization of the Edge Based Stabilized Finite Element Method Using PSPARSLIB,'' in *Parallel Computational Fluid Dynamics, Towards Teraflops, Optimization and Novel Formulations.* D. Keyes, A. Ecer, N. Satofuka, P. Fox and J. Periaux editors, pp. 397-406, North-Holland, 2000.

[8] A. Soulaïmani, N. Ben Salah and Y. Saad, ''Enhanced GMRES acceleration techniques for some CFD problems,'' *Computer Methods in Applied Mechanics* (submitted).

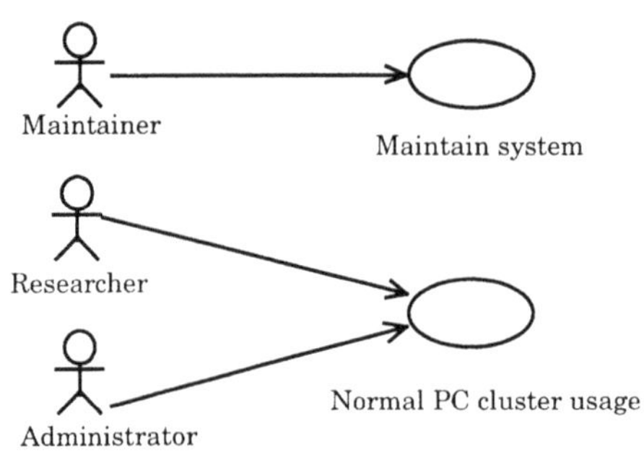

Figure 1. User-centric functional model.

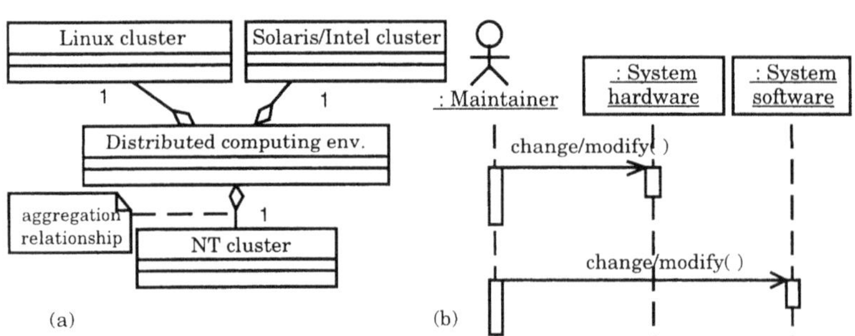

Figure 2. (a) User-centric class/object model. (b) Sequence diagram.

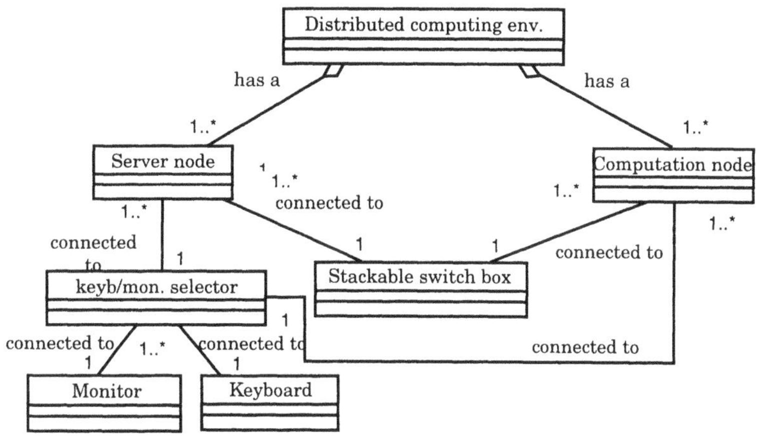

Figure 3. Physical view of the PC cluster.

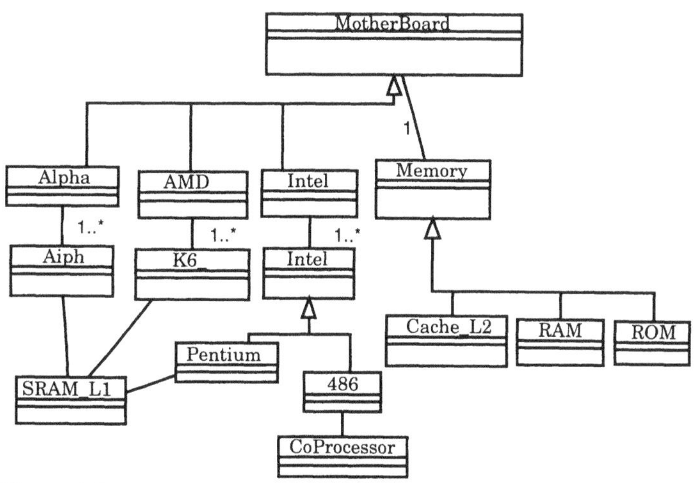

Figure 4. Hardware proxy classes.

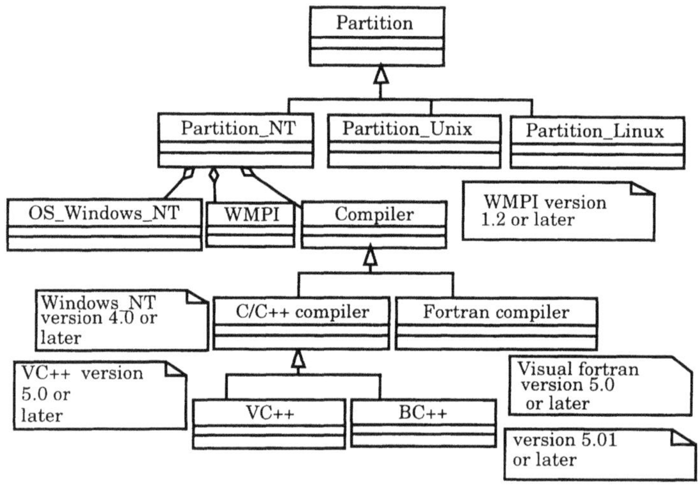

Figure 5. Software proxy classes.

Parallel Computational Fluid Dynamics – Trends and Applications
C.B. Jenssen et al. (Editors)

The Solution of Pitching and Rolling Delta Wings on a Beowulf Cluster

M. A. Woodgate, K. J. Badcock, B.E. Richards [a]

[a]Department of Aerospace Engineering, University of Glasgow, Glasgow, G12 8QQ, United Kingdom

A parallel, unfactored and implicit time-marching method for the solution of the unsteady three dimensional Euler equations is presented. For robustness the convective terms are discretised using an upwind TVD scheme while the unsteady equations are discretised using the implicit pseudo time approach. The pseudo steady state problem is solved using an unfactored implicit method. The linear system arising from each pseudo time step is solved using a Krylov subspace method with preconditioning based on a block incomplete lower-upper factorisation. The method running in parallel yields high efficiency on a cluster of PC's. Results are shown for the WEAG TA-15 wing in forced pitching and the WEAG TA-15 wing, body and sting in forced rolling. The rolling motion produces asymmetric leading edge vortices, which can lead to roll instabilities as was observed during the flight testing of modern high performance aircraft.

1. INTRODUCTION

Fully implicit methods have a stability advantage over explicit and approximately factored methods and this provides the potential to calculate steady or unsteady fluid flow problems in a small number of time steps. By grouping a number of PC's together into a cluster and pooling the memory and local storage, solution of these large systems of equations becomes practical on modest hardware.

The current paper describes the ongoing development of an implicit parallel method for solving the three dimensional fluid flow equations [1]. The features of the method are the iterative solution [2] of an unfactored linear system for the flow variables, approximate Jacobian matrices [3], an effective preconditioning strategy for good parallel performance [4,5], and the use of a Beowulf cluster to demonstrate high performance computing on commodity machines [6]. The method has been used on a wide range of applications which include unsteady aerofoil flows and multielement aerofoils [7], an unsteady delta wing common exercise [8] and the unsteady NLR F-5 CFD validation exercise [9].

The following work will focus on two aspects. First a low cost PC cluster has been used to calculate and analyse vortical flow fields about delta wings. Secondly if the periodic nature of the forced pitching/rolling wing is taken into account the amount of I/O increases by $O(10^2)$ meaning that it becomes an important factor in the wall clock time and parallel efficiency.

2. GOVERNING EQUATIONS

The three-dimensional Cartesian Euler equations can be written in conservative form as

$$\frac{\partial \mathbf{W}}{\partial t} + \frac{\partial \mathbf{F}}{\partial x} + \frac{\partial \mathbf{G}}{\partial y} + \frac{\partial \mathbf{H}}{\partial z} = 0, \tag{1}$$

where $\mathbf{W} = (\rho, \rho u, \rho v, \rho w, \rho E)^T$ denotes the vector of conservative variables. The inviscid flux vectors $\mathbf{F}$, $\mathbf{G}$, $\mathbf{H}$ are the direct 3D extension to the fluxes given in [7].

3. NUMERICAL METHOD

The unsteady Euler equations are discretised on a curvilinear multi-block body conforming mesh using a cell-centred finite volume method that converts the partial differential equation into a set of ordinary differential equations, which can be written as

$$\frac{\mathrm{d}}{\mathrm{d}t}(V_{i,j,k}^{n+1}\mathbf{W}_{i,j,k}^{n+1}) + \mathbf{R}_{i,j,k}(\mathbf{W}_{i,j,k}^{n+1}) = 0. \tag{2}$$

The convective terms are discretised using either Osher's [10] or Roe's [11] upwind methods. MUSCL variable extrapolation is used to provide second-order accuracy with the Van Albada limiter to prevent spurious oscillations around shock waves.

Following Jameson [12], the time derivative is approximated by a second-order backward difference and equation (2) becomes

$$\mathbf{R}^{\star}{}_{i,j,k}(\mathbf{W}_{i,j,k}^{n+1}) \equiv \frac{3V_{i,j,k}^{n+1}\mathbf{W}_{i,j,k}^{n+1} - 4V_{i,j,k}^{n}\mathbf{W}_{i,j,k}^{n} + V_{i,j,k}^{n-1}\mathbf{W}_{i,j,k}^{n-1}}{2\Delta t} + \mathbf{R}_{i,j,k}(\mathbf{W}_{i,j,k}^{n+1}) = 0. \tag{3}$$

To get an expression in terms of known quantities the term $\mathbf{R}^{\star}{}_{i,j,k}(\mathbf{W}_{i,j,k}^{n+1})$ is linearized with respect to the pseudo-time variable $t^{\star}$. Hence one implicit time step takes the form

$$\left[\left(\frac{V}{\Delta t^{\star}} + \frac{3V}{2\Delta t}\right)\mathbf{I} + \frac{\partial \mathbf{R}}{\partial \mathbf{W}}\right]\left(\mathbf{W}^{m+1} - \mathbf{W}^{m}\right) = -\mathbf{R}^{\star}(\mathbf{W}^{m}) \tag{4}$$

where the superscript m denotes a time level $m\Delta t^{\star}$ in pseudo-time. Equation (4) is solved in the following primitive form

$$\left[\left(\frac{V}{\Delta t^{\star}} + \frac{3V}{2\Delta t}\right)\frac{\partial \mathbf{W}}{\partial \mathbf{P}} + \frac{\partial \mathbf{R}}{\partial \mathbf{W}}\frac{\partial \mathbf{W}}{\partial \mathbf{P}}\right]\left(\mathbf{P}^{m+1} - \mathbf{P}^{m}\right) = -\mathbf{R}^{\star}(\mathbf{W}^{m}). \tag{5}$$

4. IMPLEMENTATION

The implementation of the code follows the same approach as [5]. Improvement in the serial performance of the code is possible by using the periodic nature of the solution. This is done simply by using the initial guess at time level $n+1$ to be the solution at time level $n+1-c$, where c is the number of timesteps per cycle. The use of extrapolation as an initial guess for the next time level has been found when strong moving shocks are present to casue robustness problems.

This approach means that full solutions must be saved to disk at every iteration. This amounts to tens of Megabytes per iteration even for modest problem sizes. The sequential method of storing all the data in one file and having each processor read the whole file can have a large impact on run times as the number of processors increases. Table 1 shows the effect of reading in the data located on a single file server. In the single file case each processor reads in the whole file while in the multiple file case each processor just reads the information that its needs. Files can be written in either ASCII or binary format. In can be seen for the ASCII single file the read times remain constant until 8 nodes where it takes $\approx 10\%$ longer. This is due either to the saturation of the network bandwidth or the accessing of data from the hard disk. For binary reads from a single file this saturation starts much earlier. It is always faster to read a binary file but the advantage is reduced from a ratio $8:1$ for a single node to less than $3:1$ for 8 nodes. If multiple files are used the ASCII mode shows some favourable parallel speedup where the binary file times remain constant. It should be noted however that it is still always faster to read the binary files even though they scale like 0(1).

Table 1
IO Performance with no caching

Procs	ASCII		Binary	
	Single	Multiple	Single	Multiple
1	86s	86s	10.6s	10.6s
2	86s	42.5s	13.2s	11.2s
4	88s	21.9s	18.2s	12.9s
8	98s	12.5s	38.3s	8.0s

It is possible to remove the effects of the disk access time and data transfer rate in the file server by caching all the data into memory. This means all the times in table 2 are just limited by network saturation and/or congestion. It can be seen that all the conclusions still hold true with the only marked difference being that the multiple binary reads for more than one processor are approximately halved.

Table 2
IO Performance with file server caching

Procs	ASCII		Binary	
	Single	Multiple	Single	Multiple
1	86s	86s	7.0s	7.0s
2	86s	42.5s	10.2s	5.1s
4	88s	21.7s	16.8s	4.3s
8	95s	11.9s	37.3s	4.6s

In table 3 the best solution is illustrated, which is to use any local disk which is available. Here the results are as expected with the multiple file reads scaling perfectly while the single reads remaining constant.

Table 3
IO Performance with local disk usage

Procs	ASCII		Binary	
	Single	Multiple	Single	Multiple
1	74s	74s	1.1s	1.1s
2	74s	37.2s	1.1s	0.56s
4	74s	18.5s	1.1s	0.28s
8	74s	9.2s	1.1s	0.14s

In the above nothing was mentioned about writing the data to disk but it should be noted that writing multiple files is also much cheaper as you do not have to communicate the data back to a master node for writing. However, having a process only write data to the local disk means that if a process is migrated to another node either a large amount of local disk information must also be migrated, the local disks must all be mountable on any node or the last solution is used.

5. RESULTS

5.1. Force Pitched WEAG TA-15 Wing

The first test case considered is flow around the pitching WEAG TA-15 wing with freestream Mach number of 0.4, a mean angle of attack of 21° an amplitude of oscillation of 6° and a reduced frequency of 0.56. This test case was part of the WEAG common exercise IV [8]. Figure 5.1 shows the pressure loss though the vortex at 80% chord. It can be seen that the vortex separates from the wing in the downstoke as well as becoming weaker. By the time $\alpha = 21^o$ on the upstroke the vortex has reattached and is growing in size. In fact the vortex bursts in this case and this has been shown in [8]. The calculation at a mean incidence of 21° requires around 24 hours for 3 complete cycles using 40 real time steps per cycle on 4 Pentium Pro 200 processors. The first cycle takes 14 hours to complete while the third cycle only takes 3. By using the method of multiple binary files stored on the local disk it is possible to cut down the number of linear solves in the third cycle from about 30 per real timestep to around 5.

5.2. Rolling WEAG TA-15 Wing

The second test case considered is flow around the rolling WEAG TA-15 wing with sharp leading edge and symmetric body, see Figure 2. The body begins at the apex of the wing and has circular cross section. The grid is a single block containing 73 points in the streamwise direction, 29 points normal to the surface and 193 points in the spanwise direction. This grid was then partitioned into 16 blocks ranging from 6,720 to 52,416 cells. This case has an artificially high ω_r since the values used in the experiments of Common Execise V did not show asymmetric leading edge vortices at lower values of ω_r. Figure 3 shows the unsteady pressure loss contours at the 60% span. The symmetry seen at lower roll rates in the 60° and 120° angles is totally lost and at 90° small vortices exist on both the upper and lower surfaces unlike that at slower roll rate.

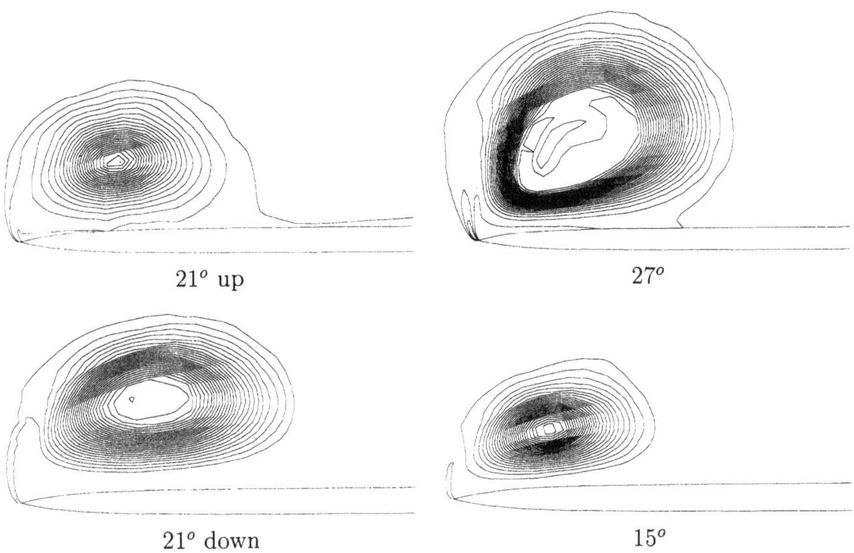

Figure 1. Pressure Loss Contours at $X/c = 0.8$, $\alpha = 21^o$

6. CONCLUSIONS

An unfactored implicit time-marching method for solving the three dimensional unsteady Euler equations in parallel has been presented. The present code shows that it is possible to obtain Euler solutions on a pitching and rolling delta wing in a less than 50 hours, for 3 full cycles, on a cluster of PPro 200 PC's. Asymmetric leading edge vortices and other significant unsteady dynamics is captured by the code.

The use of local disk space, if avaiable, can greatly aid the parallel efficently of any IO that is not required for post processing. This allows the re-use of the solutions from the whole previous cycle to be used reduce the cost of periodic calculations by 50% and more.

REFERENCES

1. Badcock, K., Woodgate, M., Cantariti, F., and Richards, B., "Solution of the unsteady 3-D Euler equations using a fully-unfactored method" *38th AIAA Aerospace Sciences, Reno, Nevada, 10–13 January* (2000).
2. Badcock, K.J., Xu, X., Dubuc, L. and Richards, B.E., "Preconditioners for high speed flows in aerospace engineering", *Numerical Methods for Fluid Dynamics*, V. Institute for Computational Fluid Dynamics, Oxford, pp 287-294, 1996
3. Cantariti, F., Dubuc, L., Gribben, B., Woodgate, M., Badcock, K. and Richards, B., "Approximate Jacobians for the Solution of the Euler and Navier-Stokes Equations", Department of Aerospace Engineering, Technical Report 97-05, 1997.

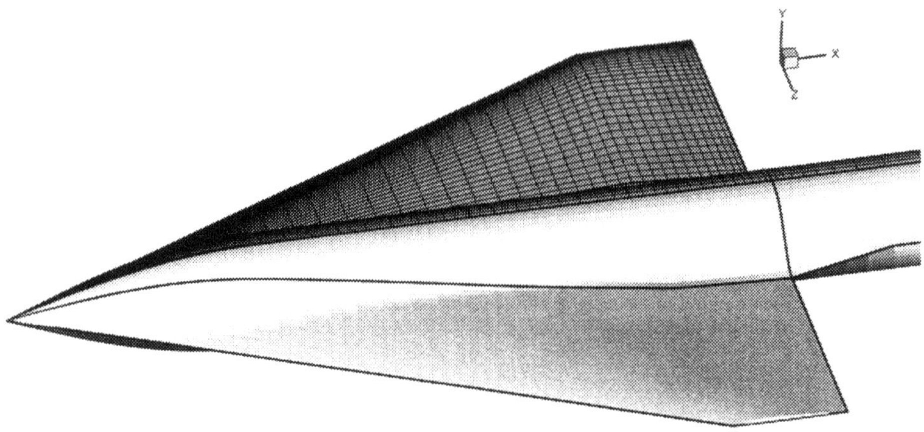

Figure 2. The mesh and surface grid of the WEAG TA-15 delta wing

4. Badcock, K.J., McMillan, W.S., Woodgate, M.A., Gribben, B., Porter, S., and Richards, B.E., "Integration of an impilicit multiblock code into a workstation cluster environment", in *Parallel CFD 96*, pp 408. Capri, Italy, 1996
5. Woodgate, M.A., Badcock, K.J., Richards, B.E., and Gatiganti, R., "A Parallel 3D Fully Implicit Unsteady Multiblock CFD Code Implemented on a Beowulf Cluster", in *Parallel CFD 99*, Williamsburg, U.S.A, 1999
6. McMillan, W.,Woodgate, M., Richards, B., Gribben, B., Badcock, K., Masson, C. and Cantariti, F., "Demonstration of Cluster Computing for Three-dimensional CFD Simulations", *The Aeronautical Journal* Sep 1999, pp. 443-447 Paper No. 2467
7. Dubuc, L., Cantariti, F., Woodgate, M., Gribben, B., Badcock, K. and Richards, B.E., "Solution of the Euler Equations Using an Implicit Dual-Time Method", *AIAA Journal*, Vol. 36, No. 8, pp 1417–1424, 1998.
8. Arthur, M.T., Brandsma F., Ceresola N., Kordulla W., "Time accurate Euler calculations of vortical flow on a delta wing in pitching motion", *AIAA Applied Aerodynamics Conference, 17th, Norfolk, VA, June 28-July 1* pp 89–99 (1999).
9. Henshaw, M., Bennet, R., Guillemot, S., Geurts, E., Pagano, A., Ruiz-Calavera, L. and Woodgate, M., "CFD Calculations for the NLR F-5 Data - Validation of CFD Technologies for Aeroelastic Applications Using One AVT WG-003 Data Set", presented at CEAS/AIAA/ICASE/NASA Langley International forum on aeroelasticty and structural dynamics, Williamsburg, Virginia USA, June 22-25, 1999
10. Osher, S. and Chakravarthy, S., "Upwind Schemes and Boundary Conditions with Applications to Euler Equations in General Geometries", *Journal of Computational Physics*, Vol. 50, pp 447-481, 1983.
11. Roe, P.L., "Approximate Riemann Solvers, Parameter Vectors and Difference

Schemes", *Journal of Computational Physics*, vol. 43, 1981.
12. Jameson, A. "Time dependent calculations using multigrid, with applications to unsteady flows past airfoils and wings", AIAA Paper 91-1596, 1991

3. Performance Issues

Parallel Computational Fluid Dynamics – Trends and Applications
C.B. Jenssen et al. (Editors)

Serial and parallel performance using a spectral code

G. Amati[a] and P. Gualtieri[b]

[a] CASPUR, p.le A. Moro 5, I-00185 Roma, Italy; e-mail: g.amati@caspur.it

[b] Dipartimento di Meccanica e Aeronautica, Universitá "La Sapienza", Via Eudossiana 18, I-00184 Roma, Italy; e-mail: paolo@casciola.ing.uniroma1.it.

Serial optimization and parallelization, using OpenMP directives, of a spectral code are described. Performances are verified on different machines in order to evaluate computer architectures.

1 INTRODUCTION

The aim of this work is to investigate possible issues that may improve performance of a spectral code both in serial and parallel computations.
We are interested in studying turbulent features like intermittency and long-term statistics of various flows such as 3D homogeneous and isotropic turbulence, passive scalar turbulence and shear flow turbulence [1], [2], [3]. A sketch of the complexity of these problems can be seen in fig.1 where low and high speed streaks for a homogeneous shear flow are presented [4].
The computational kernel of all these simulations is an in-house developed spectral code. Our analysis needs to perform many simulations with various Re numbers, different aspect ratios, different shear parameters S and so on. So we need to maximize performance to reduce total time, saving as much as possible portability in order to reduce also turnaround time using the architectures available at the moment. A single simulation for one Reynolds number Re, using a 128^3 modes resolution[1], asks for about 10^5 time steps to ensure enough statistics. On a workstation this calls for 100-200 days of serial uninterrupted computation and produce many GB of dumped configurations. We need to reduce in a drastic way total time to make the statistical analysis of these flows feasible.
Here we present results about serial and parallel performance, and in particular:

- Use of two different FFT libraries;
- Dependence from the problem size and efficient use of memory;
- Speed-up on different SMP architectures using OpenMP.

[1]Here and in the following we always consider anti-aliasing in all the three directions.

Table 1: Average time in seconds for a single Runge-Kutta step, computed on a Compaq EV56 533 MHz workstation using both VECFFT (up) and FFTPACK (down), together with total Mflops, FFT Mflops and local operation Mflops and the ratio of time needing to copy data from memory to memory.

Dimension	T (s)	Total Mflops	FFT Mflops	Local Mflops	Copy to memory
32^3	0.23	122	246	57	34 %
64^3	4.7	59	90	35	25 %
128^3	62	42	52	36	18 %

Dimension	T (s)	Total Mflops	FFT Mflops	Local Mflops	Copy to memory
32^3	0.28	110	198	64	22 %
64^3	3.40	85	248	36	50 %
128^3	31.7	84	285	35	56 %

2 SERIAL OPTIMIZATION

2.1 Use of different FFT libraries

Here we study two FFT libraries: VECFFT [6], and FFTPACK [5]. The first has been developed by Henningson et al., starting from NCAR's FFTPACK library, explicitly tuned for CRAY vector machines. The Henningson's code, a turbulent channel flow, has been the starting point for our developments. Chebychev polynomial expansion in the normal to wall direction was replaced by a Fourier expansion. Time integration was a low-storage third order Runge-Kutta method. It must be noticed that the different interface convention of FFT libraries forced us to write two distinct codes: VECFFT uses two real vector and stride can be different from one while FFTPACK use one real or complex vector and stride must be unitary: this yields a different impact in memory subsystem between the two developed codes.

The main problem of VECFFT is that computational performance decreases remarkably as the size of the problem increase, as shown in tab.1, because of cache effects[2]. Starting from a total performance of 122 Mflops for a 32^3 resolution[3] we only attain 42 Mflops for a 128^3 resolution. If we count only floating point operation due to FFT subroutines, performance decrease from 246 Mflops to 52 Mflops. Local operation (e.g. calculation of vorticity in Fourier space) are quite costly in time using a four-dimensional matrix where the last index is the direction (x, y or z). The remaining time is spent in memory-to-memory copies need to cope with the calling interface convention of FFT library. Local operations are extremely inefficient and FFT performance are much lower than theoretical peak performance (1066 Mflops).

Those issues induced us to look for a different FFT library. In tab.1 the performances

[2]Mflops and events like cache missing are computed using an hardware counter available on Compaq Alpha EV56 processor.

[3]The memory occupation of a 32^3 simulation is bigger than 4MB L2/L3 cache.

Table 2: Average time in seconds for a single Runge-Kutta step, computed on various machines, using different FFT libraries, for a 128^3 resolution.

Processor	VECFFT	FFTPACK
Compaq EV56 533 MHz	62	32
Compaq EV6 500 MHz	35	28
Sun Ultra2 336 MHz	40	41
IBM Power3 200 MHz	65	36

of the new code using public domain FFTPACK library are reported. The FFT are now much more efficient, we reach 285 Mflops for 128^3. On the contrary, much more time is spent in memory-to-memory copies, up to 50% of total time for a 128^3 resolution: nevertheless the total elapsed time is reduced by a factor 2.
In tab.2 average time for various architectures are presented. Time for single time step was decreased on all architectures except Sun, in which the VECFFT code and FFTPACK code present the same performances. Because FFTPACK is not able to deal with non-unitary stride we were forced to make extra memory copies to cope FFT interface. Our Sun machine suffers from those extra memory-to-memory copies and new FFT doesn't present advantages in time.
FFT operation on EV6 reach something more than 400 Mflops, 40% of peak performances, but more than 50% of time is spent in moving data, stressing memory bandwidth.

2.2 Memory padding

A typical bottleneck for cache-based architectures is inefficient use of memory hierarchy, producing serious decrease of performances because of cache misses or trashing. This problem is much more relevant for power of 2 memory access: they are the ideal candidates for cache trashing problems. This can be avoided, or at least cured, using memory padding, namely adding unused memory locations to induce a better data addressing. In tab.3 average time of for single Runge-Kutta step are presented with and without using padding in velocity ($\vec{u}$) and vorticity ($\vec{\omega}$) fields. The use of explicit padding is extremely favourable for IBM and Compaq EV6. Padding produces a reduction of time devoted to move data from memory to memory and memory access for local operation, while FFT doesn't present an increase in performance. (tab.4).
For EV6 most of the time is spent in moving data, while for Power3 the most expensive section is still FFT. These results show the different "philosophy" of these vendors. For Compaq CPU velocity is the main concern, while for IBM is memory bandwidth.
Another technique to save time is to trade memory vs. computation: we abandoned low-storage technique using an extra 3-D array in favour of sparing a 3-D forward FFT for the computation of the nonlinear term of Navier-Stokes equation. This increases the memory request by a factor 1.5: we must use now 3 3-D arrays instead of 2. A 128^3 simulation requires now for about 500 MB instead of 350 MB. This may lead to limitations for bigger simulations: a 256^3 simulation would call for about 4 GB instead of 2.6

Table 3: Average time in seconds for a single Runge-Kutta step, computed on various machines with and without padding, for a 128^3 resolution.

Processor	no-padding	padding
Compaq EV6 500 MHz	28	15
Sun Ultra2 336 MHz	41	37
IBM Power3 200 MHz	36	18

Table 4: Total Mflops, FFT Mflops and local operation Mflops and the ratio of time needing to move data memory to memory for various architectures using padding

Processor	Total FFT	FFT	Local Op.	Copy to memory
Compaq EV6 500 MHz	178	435	157	54%
Sun Ultra2 336 MHz	72	162	36	45%
IBM Power3 200 MHz	147	217	98	26%

GB, exceeding available physical memory for some SMP systems. In scalar cache-based architecture using more memory could be faster than making extra FFTs: in this case a great gain in time due to this little modification. A saving of about 30 % in time, is exhibited by Sun.

Moreover using vendor's FFT libraries we have achieved a further gain using Sun's Sunperf (from 28 to 24 seconds/RK step). This happens because Sunperf are the vendor's implementation of the FFTPACK library. Other vendor's libraries (ESSL for IBM, DXML for Compaq) doesn't present any gain because have a different calling interface convention. So we are obliged to introduce extra memory to memory overheads to cope with these different interfaces, if we want to preserve the main structure of the developed code.

The use of profiling driven compilation doesn't produce gain in time, due probably to the relative simplicity of the code. In tab.5 the best timing obtained for single Runge-Kutta step for the fully optimized serial code on every platform are presented.

Table 5: Average time in seconds for a single Runge-Kutta step, computed on various machines, for the original FFTPACK code and the fully optimized code for a 128^3 resolution.

Processor	FFTPACK	FFTPACK fully optimized
Compaq EV56 533 MHz	28	19.9
Compaq EV6 500 MHz	15	11.3
Sun Ultra2 336 MHz	37	24.3
IBM Power3 200 MHz	18	14.3

Table 6: Total time in second with speed-up, between parenthesis, for Sun E4500 at 366 MHz, Compaq ES40 at 500 MHz and IBM SP 8-way node at 222 MHz, using 128^3 resolution.

# proc.	Sun (S_p)	Compaq (S_p)	IBM (S_p)
serial	2187	932	1271
1	2166 (1.0)	930 (1.0)	1265 (1.0)
2	1096 (2.0)	610 (1.5)	634 (2.0)
3	746 (2.9)	446 (2.1)	444 (2.8)
4	571 (3.8)	370 (2.5)	324 (3.9)
5	484 (4.5)		266 (4.7)
6	406 (5.3)		225 (5.6)
7	374 (6.0)		199 (6.4)
8	321 (6.7)		189 (6.7)
9	297 (7.3)		
10	274 (7.9)		
11	263 (8.2)		
12	236 (9.2)		
13	224 (9.6)		
14	222 (9.7)		

3 PARALLEL SPEED-UP

The OpenMP directives [7] were used to make a simple "do-loop" work distribution scheme on the serial code without spending time in loop modification. This was based on profiling information previously collected from the serial optimization phase, after a check of the thread-safeness of FFTPACK library. We have inserted about 30 "parallel do" directives on a total of more then 4000 Fortran lines. For all architectures we have used KAP/Pro toolset by KAI [8]. Results on various SMP servers are shown in tab.6 where the total time of serial and multi-threaded code are reported [9]. Regarding parallel performances we can see that the Sun machine, who presents the worst serial time also exhibits the better speed-ups ($\simeq$ 10 with 14 processors). On the other hand Compaq machine gives a poor speed-up (2.5 using 4 processor): this probably happens because the memory bandwidth is stressed by using very fast processors. Many other factors could explain this behaviour, as different performance of runtime library of the OpenMP tool used on different architectures.

We also like to underline that the OpenMP version using only one thread spent the same time of the serial version.

4 CONCLUSIONS

One "trivial" conclusion of this work is to underline how serial optimization is crucial for parallelization.
Working on serial optimization we have gained a factor 2 or more (see EV6 processor from the original 35 to 11.3 seconds), except on Sun machine where the gain is only 40 % for the 128^3 simulation. Moreover now the performances are almost the same on various architectures allowing the use of different systems to reduce total turnaround time, a fundamental aspect in this work.
An other important aspect of this work is to show OpenMP programming as a straightforward extension of standard serial programming that allows for an incremental parallelization and can yield good speed-ups even with limited efforts, maintaining only one

Table 7: Estimated time for a 128^3 simulation, in hour

	Parallel	Serial (1)	Serial (2)
4 proc. EV56 at 400 Mhz	1000	2150	5670
4 proc. EV6 at 500 Mhz	343	941	2920
8 proc. Pwr3 at 222 Mhz	180	1190	5420
8 proc. Ultra2 at 336 Mhz	298	2070	3333
14 proc. Ultra2 at 336 Mhz	205	2070	3333

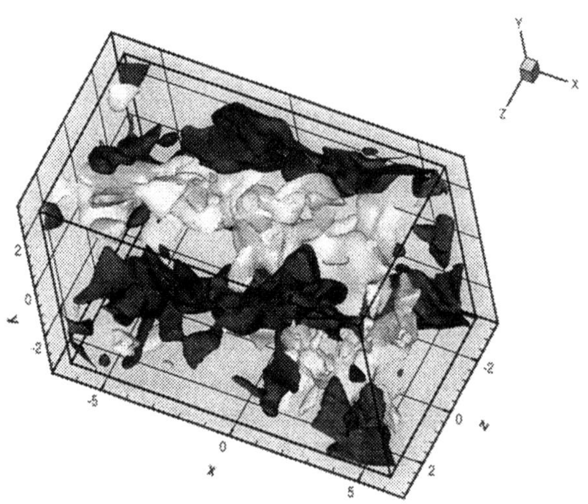

Figure 1: Low and high speed streaks for a homogeneous shear flow.

version of the code, the same both for serial or parallel computations on different architectures.
OpenMP approach doesn't allow the use of a great number of processors but it could be sufficient for our purpose: our long-term simulation (10^5 time steps) now will ask from 1 to 2 weeks on a single SMP machine, as can be seen in tab. 7. This makes our study now feasible and allows farming using various architectures.

ACKNOWLEDGEMENT

We like to acknowledge F. Massaioli for useful discussion and support, D. Henningson who furnished the original channel flow code and L. Bognetti for his help in the first step of this work.

References

[1] Kolmogorov, A.N., *A refinement of previous hypothesis concerning the local structure of turbulence in a viscous incompressible fluid at high Reynolds number*, J. Fluid Mech., 1962, 82 - 85, 6.

[2] Amati, G., Succi, S., Piva, R., and Toschi, F., *Scaling exponents in turbulent channel flow*, Proceeding of: European Turbulence Conference-7, 159 - 162, 1998, Nizza.

[3] Piva, R., Casciola, C.M., Amati, G., Gualtieri, P. *Vorticity structures and intermittency in near wall turbulence* Proceeding of: Iutam99 symposium on geometry and statistics of turbulence, 1999, Tokyo.

[4] Gualtieri, P., Casciola, C.M., Amati, G., and Piva, R., *Scaling laws and vortical structures in homogeneous shear flows* Proceeding of: European Turbulence Conference-8, 779 - 782, 2000, Barcellona.

[5] FFTPACK FFT library, see: *http://www.netlib.org*

[6] Lumbdladh, A., Henningson, D.S., Johansson, A.J. *An efficient spectral integration method of the Navier-Stokes equations* Technical Report; FFA-TN 1992-28,1992

[7] OpenMP application program interface, see: *http://www.openmp.org*

[8] KAP/Pro Toolset, see: *http://www.kai.com*

[9] Bailey, D.H. *Twelve Ways to Fool the Masses When Giving Performance Results on Parallel Computers* Supercomputing Review,Aug. 1991,54-55

Parallel Computational Fluid Dynamics – Trends and Applications
C.B. Jenssen et al. (Editors)

On the Design of Robust and Efficient Algorithms that combine Schwartz method and Multilevel Grids

A Ecer**, M. Garbey* and M. Hervin*-**,
CDCSP* ISTIL/University Lyon 1 Claude Bernard
** IUPUI /Department of Mechanical Engineering,
Purdue School of Engineering and Technology at Indianapolis

1. Introduction:

In the development of a parallel code for the industry, the simplicity of the code and robustness of the solution method is extremely important. The engineering time spent to develop and run a parallel code is the leading part of the cost, in comparison with the arithmetic efficiency of the solver. Classical additive Schwarz methods or similar iterative relaxation procedures with Dirichlet Neuman Boundary Conditions [8] are extremely attractive from this point of view because they provide a practical tool to get a parallel version of an existing CFD code with minimum effort in terms of human power. The procedure is standard. One need, first, a good partitioning of the grid of computation which produces overlapping subdomains; each subdomain is then solved by a given processor and one can re-use existing sequential classical block solvers for each processor. Second, the main modification of the code consists to organise exchange of information between subdomains by adding few MPI calls in order to exchange artificial boundary of subdomains between processors. This can be done easily with tools such as GPAR [3]. Robust load balancing is also available to use efficiently heterogeneous network of computers with tools such as DLB [2]. We know that the additive Schwarz algorithm is rather slow and inefficient numerically [8] except when applied to operators associated with singular perturbation problems [4,5].

There have been several methodologies designed to accelerate this basic procedure. Most often, there are based on a coarse grid preconditioner. Here, we present an attempt, to make the additive Schwarz procedure more efficient. While keeping as much as possible the simplicity of the original method, a cascade algorithm with three levels of grids for acceleration purpose is used for verification.

Most of the concepts describe thereafter, can be applied in a straightforward way to FE or FV discretization with an unstructured grid. But for simplicity, we will report on the second order FD solution of the well-known Bratu problem in a square.

2. Description of the method:

The problem can be written as,

$N[u] = \Delta u + \lambda\ e^{u} = 0 in \qquad \Omega = (0,1)^2, \qquad u_{|\partial\Omega} = 0$

The discretized problem can then be written as,

$$N^h[U]=0 \equiv \begin{cases} \dfrac{u_{i+1,j} - 2u_{i,j} + u_{i-1,j}}{h_x^2} + \dfrac{u_{i,j+1} - 2u_{i,j} + u_{i,j-1}}{h_y^2} + \lambda\ e^{u_{i,j}} = 0, \quad i = 1...N_x - 1, \quad j = 1...N_y - \\ u_{1,\bullet} = u_{N_x,\bullet} = u_{\bullet,1} = u_{\bullet,N_y} = 0 \end{cases},$$

where $h_x (resp\ h_y)$ denotes the space step in x direction (resp y direction).

We consider a basic decomposition of the domain into nd overlapping strips, $(x_A(k), x_B(k)) \times (0,1), k = 1...nd$ with arbitrary overlap of q meshes, i.e.,

$x_B(k) - x_A(k+1) = q \cdot h_x, \quad k = 1...nd - 1$, q being a positive integer.

At the continuous level, the algorithm can be written as follows:

$$\left| \begin{array}{l} for\ \ k = 1,...,nd \\ N\left[U^{k,n+1}\right] = 0 \\ U^{k,n+1}(x_a(k),\bullet) = U^{k-1,n}(x_A(k),\bullet) \\ U^{k,n+1}(x_B(k),\bullet) = U^{k+1,n}(x_B(k),\bullet) \\ end \end{array} \right.,$$

or in our notation formally $U^{0,n} \equiv U^{nd+1,n} \equiv 0$

This process constructs a multivalued piecewise solution because of the overlap and we define the global solution as follows:

$$\text{For}: X \in (x_A(k), x_B(k)): u^{n+1}(x,y) = \sum_{k=1}^{nd} \aleph^k\ u^{k,n+1}(x,\bullet) + (1 - \aleph^k) \begin{cases} u^{k-1,n+1}(x,\bullet) if\ x > \dfrac{x_{A^1} x_B}{2} \\ u^{k+1,n+1}(x,\bullet) if\ x \le \dfrac{x_{A^1} x_B}{2} \end{cases}$$

with $\aleph^k$ a smooth partition of unity that is 1 in $(x_B(k-1), x_A(k+1))$ and 0 outside $(x_A(k), x_B(k))$.

We know that the convergence of this algorithm is linear and very slow. We refer to [7] for a detailed description of the method. First, there is obviously no need to solve exactly each non-linear problem in each sub-block, since the domain decomposition procedure is an iterative process.

Let us denote U to be the exact solution and U^h to be the exact solution of the discretized problem N^h. Let R^h denotes the truncation error $N^h U$. Also, let us denote E^h the numerical error $U - U^h$; from a non-linear maximum principle. With small enough h, we can write the maximum norm,

$$\|E^h\|_\infty < Ct \quad \min(1, 1/\lambda) \quad \|R^h\|_\infty$$, with Ct independent of h.

E^h can then be decomposed into the numerical error of the approximation method : $E_O^h = U - U^h$, and the error due to the unperfected convergence of the iterative solver $E_I^{h,n} = U^h - U^{h,n}$, where n denotes the nth Schwartz iterate: $E^h = E_O^h + E_I^{h,n}$.

In general, the stopping criterion of the iterative procedure is only required to produce an error $E_I^{h,n}$ less than the truncation error E_O^h.

Let us denote $R_B^{h,j}$ the residual in subdomain (or block) j produced by the iterative block solver and $E_B^{h,j}$ the corresponding error. We can stop the iterative block solver procedure when $E_B^{h,j}$ is less or equal to $E^{h,}$.

If $[\![U^{n,j}]\!]$ denotes the jump of the multivalued piecewise solution at artificial interface j with minimum overlap, it can be shown that the stop criterion for the block solver is $R_B^{h,j} < Ct/h_x [\![U^{n,j}]\!]$, with Ct independent of h_x.

Let us denote tol^j this maximum tolerance for the residual. If the overlap is more than one mesh, one can obtain similar estimates with Ct actually depending on the size of the overlap.

Since the measure of the jump $[\![U^{n,j}]\!]$ is not available until the blocks are completely solved, we use a priori the fact that the additive Schwarz has a linear rate of convergence. We proceed to an estimate of the final error at the nth iteration of Schwarz procedure by a linear least square extrapolation of the error based on previous iterates, i<n. The block solver is actually a non-linear solver based on the Newton algorithm as follows:

$$+\Delta V + \lambda(\exp U^k)V = -\Delta U^k - \lambda \exp U^k$$
$$U^{k+1} := U^k + V$$

Let V^h denotes the solution of the linear system corresponding to the discretized problem:

$$+\Delta V^h + \lambda(\exp U^{h,k})V^h = -\Delta U^{h,k} - \lambda \exp U^{h,k}$$

We used a pre-conjugate gradient algorithm with incomplete LU factorisation as the linear solver. Stopping criterion for the resolution of linear system satisfied by V^h is based at first order on the tolerance $tol^j / \min(1, \|V^h\|)$.

As we will report later on, in our numerical experiments, the efficiency of the overall method is modest compare to the method where one apply the iterative solver to the global problem with no domain decomposition.

Let us notice also that one can use many different class of iterative block solvers. A large choice of Krylov method in combination with appropriated preconditioning is available. Multigrid methods are still among the most efficient technique when the discretization grid can be decomposed properly into multilevel, [1,6].

In our basic strategy to enhance the additive Schwarz algorithm, we introduce three levels of grids G^m, m=1,2,3. For simplicity, we restrict ourselves to discretization ratio:
$hx^1 / hx^2 = hx^2 / hx^3 = hy^1 / hy^2 = hy^2 / hy^3 = 2$, although it will be seen later on that this simplification may not be necessary. The classical idea of cascade algorithm is to provide as an initial guess for the iterative solution process on the grid level m which is obtained from the discretized solution on the coarser grid m-1. This very old idea is called nested loop when Successive Over Relaxation (SOR) is the block solver.

It is known that in terms of arithmetic efficiency, the nested loop method should be limited to two levels of grids. As a matter of fact, the invention of Multigrid methods has been a breakthru to improve the basic nested loop idea and involve as many levels of grids as one may define [1,6]. In cascade algorithm, one use any kind of iterative block solver of his choice, rather than the relaxation method (SOR). It is also useful to provide a good initial guess for the next iterative solve on the finer grid by solving the rough grid problem first.

The implementation is very simple, since one always go from the coarse grid to the next finer grid level, as long as grid level can be defined properly. With FV or FE unstructured grid, it does not seem unrealistic to define three level of embedded grids.

The main purpose of using three levels of grids instead of two will be to provide sufficient information in order to proceed code verification. We will denote in the following $U^{(m,h)}$the discrete solution obtained by our iterative method on grid level m.

In summary, the proposed algorithm involves, first, the solution of the discretized problem on the grid G^1, with the additive Schwarz and an iterative block solver. Then, we project the solution on grid G^2 and use an interpolation procedure, to define the initial guess everywhere. Here, linear interpolation seems like a natural tool..

The same solution procedure is then reproduced on the grid G^2. If we basically project the solution obtained on G^2 into G^3, we miss a very important property of our approximation method. The discretized solution of Bratu problem should converge to the exact solution with two order of accuracy in space, as $h = (h_x, h_y)$ goes to zero, that is $U - U^h = 0(h_x^2 + h_y^2)$. Using the classical Richardson extrapolation principle reported as in Roache [13] for code verification, we may therefore produce an initial guess for the iterative solution procedure on grid G^3 that is much better than a basic projection of U^2 onto G^3.

We define the initial guess for the iterative solution on G^3 to be
$U_I^{(3)} = 4/3U^{(2,h)} - 1/3\,U^{(1,h)}$
We encounter two difficulties. First of all $U_I^{(3)}$ is defined only on grid $G^{(1)}$. We do need therefore to interpolate $U_I^{(3)}$ with an interpolation procedure that keeps the Richardson extrapolation procedure effective on $G^{(3)}$. If $U^{(2,h)}$ and $U^{(1,h)}$ are known at second order on $G^{(1)}$, we can expect $U_I^{(3)}$ to be at least a third order approximation of $U^{(3,h)}$ on $G^{(1)}$.

Actually with regular grids of constant space steps, one obtains a fourth order approximation. In order to preserve as much as possible the quality of this information, we should use third order interpolation method to extend our grid solution $U_I^{(3)}$ from $G^{(1)}$ to $G^{(3)}$. In our case, we have applied cubic bilinear interpolation as well as spline interpolation. The second difficulty is due to the fact, that $U^{(1,h)}$ and $U^{(2,h)}$ are computed with an iterative procedure. The Richardson extrapolation principle applies to an exact discrete solution. Then, the error is given by a truncation error formula, based on Taylor expansion of the discrete operator with respect to discrete parameter h. The leading coefficients of this error are a priori independent of h. As opposed to the iterative solution procedure on a single grid, one therefore needs to compute an approximation of $U^{(j,h)}$, j=1,2 on grids $G^{(j)}$, j=1,2 with a residual of the order the expected rate of convergence of Richardson extrapolation method, that is at least h^3.

This procedure combining Richardson extrapolation and high order interpolation allows then the production of a good initial guess for the iterative solution on grid $G^{(3)}$, and the code can be verified by simply proceeding with the computation of $U^{(3,\ )}$. In principle, the number of Schwarz iterates should be small if the construction of $U^{(3)}$ and the resolution of $U^{(j,\ )}$, j=1,2 are correct. The solution process is however robust because $U^{(3)}$ is only used as an initial guess.
In the mean time the order B(x) of the method can be checked with formula:
(**) $B = \log_{10}\left|U^{(1)} - U^{(2)}\right| / \left|U^{(2)} - U^{(3)}\right| / \log_{10}(2)$ at each point x.

This formula can be used in combination to the computation of the residual in order to verify the stopping criterion of the iterative computation of the solution on the finest grid $G^{(3)}$ that in practice dominates the overall cost of the method.

3. Numerical result :

We report in the following on numerical experiments realised with **matlab**. Our computation is rather modest and should motivate further validation of our solution procedure with more complex grid and model problems. From our point of view arithmetic efficiency of the method is not the key issue.

However, in order to provide an indication of the efficiency of this method, we compare the number of floating point operations realised with our algorithm to the flops performance of Pre-Conjugate Gradient (PCG). Here, we use incomplete LU factorisation, without domain decomposition on a single fine grid, and the trivial initial guess on $G^{(3)}$.

Parallel efficiency is not an issue for the proposed method as well, since it is known that the additive Schwarz scales very well even on MIMD parallel systems with a slow network, as long as the load on each processor is sufficiently high [2,8].

In order to illustrate the method, we report thereafter on the numerical solution of the Bratu problem with $\lambda = 6$.

The Schwarz algorithm is applied with minimum overlap on grid $G^{(j)}$, j=1,2. The size of the overlap is chosen on $G^{(3)}$ in order to minimise the overall flops performance.

The evolution of the residual and the difference in maximum norm with the exact discrete solution as a function of the Schwarz iterate with our cascade of three consecutive runs on grid $G^{(1)}$ to $G^{(3)}$ goes asymptotically to a straight line as expected. The order of convergence of the method B(x) computed on the coarse grid is 2 within 5%.

Table 1 shows the global efficiency of the iterative solver by listing: the total number of Mflops used on the cascade algorithm, the error in maximum norm at each grid level, as calculated against the fine grid solution on grid $G^{(4)}$ with a direct solver, and the number of Schwarz iterates for each grid level. The optimal overlap on the fine grid $G^{(3)}$ is 5 meshes.

It is interesting to notice that the cascade algorithm combined to Schwarz algorithm with PCG block solver can be more efficient than the basic PCG applied to solve the discrete problem on $G^{(3)}$ with no domain decomposition and the trivial initial guess.

The efficiency is depending very much on the size of the blocks, and the choice of the overlap on the fine grid. In all our experiments, we observed that the fine grid computation is dominant in terms of flops. In theory, as the interpolation method used to extend the Richardson extrapolation from $G^{(1)}$ to $G^{(3)}$ is improved, less Schwarz iterations are needed. It should be noticed that monitoring the order of the method with formula $(**)B(x)$ avoids premature iteration stops of the Schwarz iteration process. It is also interesting to notice that the overall number of flops, required to rich the correct solution, is relatively insensitive to the number of subdomains. However, the number of Schwarz iterations grows as expected with the number of subdomains. This has an impact on the parallel efficiency of the method which is usually limited with the network of communications, i.e. as the number of messages increase with the Schwarz iterations, the parallel efficiency decreases.

Finally, one can look artificially at a modified Bratu problem with space dependant coefficient $\lambda(x)$ that is discontinuous in space. The solution is then at most $C^2(\Omega)$, and the second order accuracy of the FD differences approximations breaks down in the neighbourhood of the jump of μ. This example shows that our solution method for varying order method may not be efficient and that the verifications of the code can be difficult [5].

4. Conclusions:

We have presented a practical way of combining efficiently the solution process and code verification by using standard additive Schwartz algorithm with cascade of 3 grid levels. For simplicity, we have restricted our study to an elementary non-linear problem with second order FD in a square. However, in the development of our solution, we restrict ourselves carefully to a method that should be easily generalised to non-structured grid and FV or FE discretization. The iterative block solvers are just standard and general, external high order interpolation tools on unstructured grids can be built independent of the numerical methods; defining 3 levels of grids might still be a modest task for a grid generator. Finally, the availability of a good approximate solution on three embedded grid is important. In our opinion, it is more valuable than a single "good" discrete solution, in order to better understand the validity of a numerical method [7] for which it is rare that all mathematical hypotheses are fullfilled correctly.

References:

[1] W.L. Briggs, A Multigrid tutorial, SIAM, Phildelphia, PA, 1987.
[2] Y.P. Chien, F. Carpenter, A Ecer and H.U. Akay, "Load Balancing for Parallel Computation of Fluid Dynamics Problems", Computer Methods in Applied Mechanics and Engineering, Vol. 120, pp. 119-130, 1995.
[3] D. Ercoskun, H. Wang and N. Gopalaswamy, "GPAR- A Grid Based Database System for Parallel Computing" CFD Laboratory Report, Department of Mechanical Engineering. IUPUI, Latest Revision January 1996. This report with file named "Gpar.ps" may be retrieved from the sub-directory pub by anonymous ftp to 134.68.81.171.
[4] M. Garbey, « A Schwarz alternating procedures for singular perturbation problems », SISC Vol17, No5, pp1175-1201, 1996.
[5] M. Garbey and H.G.Kaper, « Heterogeneous Domain Decomposition for Singular perturbed elliptic boundary value problems ». Siam J.Numer. Anal. Vol 34, No4, pp. 1513-1544, 1997.
[6] S.E. Mc Cormik, « Multilevel Adaptative Methods for PDES », Siam Philadelphia, 1989.
[7] P.J. Roache, « Verification and Validation in Computational Science and Engineering », Hermosa Publishers, Albuquerque, New Mexico, 1998.
[8] Series of Domain Decomposition Conferences, http//www.ddm.org

Performance of Cascade - Newton - Schwarz on 2D Bratu problem with 65 * 61 unknowns					
Number of subdomains	**2**	**3**	**4**	**5**	**6**
error on G3	**3.0 10-5**	**3.0 10-5**	**3.0 10-5**	**3.0 10-5**	**3.0 10-5**
error on G2	**1.8 10-4**	**1.8 10-4**	**1.8 10-4**	**1.8 10-4**	**1.8 10-4**
error on G1	**7.5 10-4**	**7.7 10-4**	**7.8 10-4**	**8.1 10-4**	**8.4 10-4**
speed up ratio versus no DD	**1.69**	**1.06**	**0.95**	**0.95**	**0.99**
Mflops of cascade	**89.5**	**142**	**120**	**120**	**115**
nb of Schwarz iterates on G1	**79**	**100**	**119**	**150**	**202**
nb of Schwarz iterates on G2	**88**	**131**	**149**	**168**	**196**
nb of Schwarz iterates on G3	**8**	**15**	**12**	**13**	**14**

Table 1

Parallel Computational Fluid Dynamics – Trends and Applications
C.B. Jenssen et al. (Editors)

2-D to 3-D conversion for Navier–Stokes codes: parallelization issues

J. M. McDonough[a] and Shao-Jing Dong[b]

[a]Department of Mechanical Engineering, University of Kentucky, Lexington, KY USA 40506-0108

[b]Computing Center, University of Kentucky, Lexington, KY USA 40506-0045

We outline a theoretical procedure for converting two-dimensional Navier–Stokes codes to three-dimensional ones based on a modification of Douglas & Gunn time splitting with emphasis on exploiting opportunities for parallelization. We then present results computed on several different platforms, employing both compiler directives and MPI to achieve parallelization. We conclude that in the context of structured-grid codes that may be readily time split, our procedure provides an effective way to construct 3-D Navier–Stokes codes from existing 2-D ones, thus enabling continued use of "legacy" codes that have undergone extensive development and testing.

1. INTRODUCTION

Until only recently it has generally been too time consuming to perform full 3-D Navier–Stokes (N.–S.) calculations, at least in the context of often-required "over-night turnaround" needed in an industrial setting. (We note, however, that essentially all commercial CFD software has been 3-D from the beginning, leading to use of often very coarsely-gridded problem domains resulting in often very inaccurate predictions.) An interesting practical question is whether existing 2-D research and/or proprietary codes can be readily converted to 3D, or should they simply be discarded and new 3-D codes built. In many cases considerable time and effort have been invested in development of 2-D codes, and they have been quite thoroughly validated. Thus, if an easy approach to 3-D extension can be found, it would provide an attractive way to construct new 3-D codes.

In this report we will demonstrate in a somewhat specific context that such an extension is readily obtained by utilizing a 2-D code for planar solves within a 3-D domain via a certain interpretation of Douglas & Gunn [1] time splitting. It is easily seen that several parallelization issues arise in the investigation of such a formalism; these will be the focus of the reported research. In Sec. 2 we present the form of the N.–S. equations to be considered, and in Sec. 3 we describe the version of Douglas & Gunn time splitting that lends itself naturally to the 2-D → 3-D conversion we seek. Along with this we briefly outline the overall conversion process, emphasizing the parallelization opportunities. Section 4 contains results of testing parallelization strategies for a particular well-known problem using MPI on three different platforms: *i)* Cray T3E, *ii)* HP SPP-2200 and *iii)* HP N-4000. We have also employed compiler directives on the HP SPP-2200 system. A final section provides our conclusions from the current study and recommendations for future related work. In particular, we will show that MPI has not provided very effective parallelization for the code being considered, especially on Hewlett Packard (HP) hardware.

2. GOVERNING EQUATIONS

In the present study we confine our attention to the incompressible N.–S. equations in 3-D Cartesian coordinates, expressed in dimensionless form as

$$\nabla \cdot \boldsymbol{U} = 0, \tag{1a}$$

$$\boldsymbol{U}_t + \boldsymbol{U} \cdot \nabla \boldsymbol{U} = -\nabla P + \frac{1}{Re}\Delta \boldsymbol{U}. \tag{1b}$$

Here, $\boldsymbol{U} \equiv (u, v, w)^T$ is the velocity vector, and P is pressure; $Re \equiv UL/\nu$ is the Reynolds number based on length scale L, velocity scale U and the kinematic viscosity, ν. The subscript t denotes partial differentiation with respect to time, and ∇ and Δ are the gradient and Laplace operators, respectively. We will not at this time provide specific boundary and initial data, but merely assume such data will be available so as to constitute a well-posed problem on a closed, bounded domain $\Omega \subset \mathbb{R}^3$.

We will assume that Eqs. (1) are discretized via a (structured) staggered-grid finite-volume approximation, as indicated in Fig. 1, and solved by a projection method (see, *e.g.*, Gresho [2]). We further assume that the discretization leads to banded matrices of structure consistent with "compact" banding in each separate spatial direction, at least after possible permutation of indices.

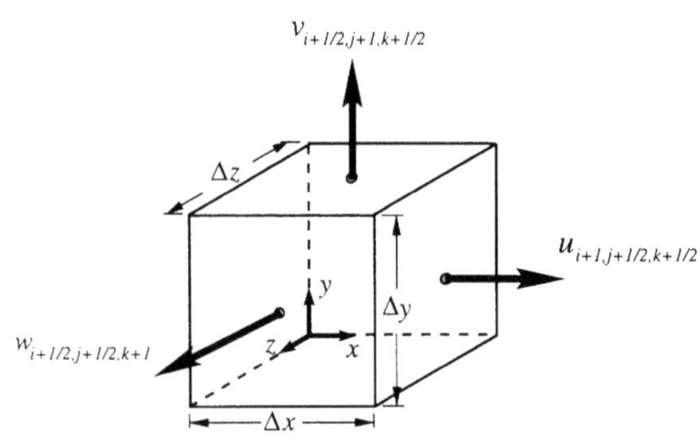

Figure 1. Schematic of 3-D staggered-grid finite-volume cell.

Here we will be concerned only with the momentum equations (1b), and we will employ 'projection 1' of Ref. [2] as the solution procedure, thus eliminating the ∇P term. This reduces the problem to a 3-D system of Burgers' equations for so-called "auxiliary velocities," which in a complete N.–S. calculation would be projected onto a divergence-free subspace of solutions via a Poisson equation solution during each time step. We remark that the Poisson equation presents separate, and in many respects easier, questions associated with the 2-D $\rightarrow$ 3-D conversion; these will not be treated here. Thus, the equations to be studied in the present effort are:

$$u_t + (u^2)_x + (uv)_y + (uw)_z = \frac{1}{Re}\Delta u, \tag{2a}$$

$$v_t + (uv)_x + (v^2)_y + (vw)_z = \frac{1}{Re}\Delta v, \tag{2b}$$

$$w_t + (uw)_x + (vw)_y + (w^2)_z = \frac{1}{Re}\Delta w. \tag{2c}$$

We comment that although this study will be carried out in a Cartesian coordinate system, none of the techniques we employ are restricted to this case; *viz.*, generalized coordinates can also be used. Indeed, our basic numerical procedure consists of trapezoidal integration in time, Δ-form quasilinearization for treatment of nonlinear (and bilinear) terms in Eqs. (2), and centered spatial differencing. This leads to tridiagonal matrices in each spatial direction for each Newton-Kantorovich iteration, resulting in only $\mathcal{O}(N)$ arithmetic operations per time step ($N \equiv N_x N_y N_z$, each factor being the number of grid cells in the indicated spatial direction), and this carries over directly to generalized coordinates.

3. DOUGLAS & GUNN TIME SPLITTING

It is well known (see, *e.g.*, [1]) that in the context of the numerical approach described above, each equation of the system (2) is approximated by an algebraic system of the form (*e.g.*, for (2a))

$$\left(I + A^{n+1}\right) u^{n+1} + B^n u^n = f^n, \tag{3}$$

where A^{n+1} is a $N \times N$ matrix whose elements are evaluated at the advanced time level $n+1$; B^n is a $N \times N$ matrix with elements at time level n; I is the $N \times N$ identity matrix, and f^n is a *known* function (but possibly containing time level $n+1$ information).

Typically, the matrix A has the structure of a discrete Laplacian, and it can be additively decomposed as the sum of tridiagonal matrices: $A = A_x + A_y + A_z$, with the subscripts indicating the direction of spatial differencing. It should be noted that for any given ordering of the indexing of the grid function u only one of A_x, A_y, A_z is compactly banded; but there are separate orderings (induced in practice by nested DO-Loop ordering) leading to compact banding for each matrix. Finally, for purposes of the 2-D $\rightarrow$ 3-D conversion, it is crucial to recognize that $A_{x+y} \equiv A_x + A_y$ is the usual 2-D discrete Laplacian. Thus, the first step in applying Douglas & Gunn time splitting in the 3-D case is to decompose the matrix A as

$$A = A_{x+y} + A_z, \tag{4}$$

while noting that

$$A_{x+y} = A_x + A_y. \tag{5}$$

Then the splitting indicated in Eq. (4) results in one 2-D solve, and one 1-D solve, the former to be accomplished with an existing 2-D code.

The most efficient computational version of Douglas & Gunn splitting is the δ-form. To obtain this, define $u^{(q)}$, $q = 1, 2, 3$ with $u^{(3)} \equiv u^{n+1}$ such that $u^{(q)}$ is the solution from the qth split step calculation, and

$$\delta u^{(q)} \equiv u^{(q)} - u^n. \tag{6}$$

Then the δ-form equations are written as

$$\left(I + A_{x+y}^{n+1}\right) \delta u^{(2)} = f^n - \left(A^{n+1} + B^n\right) u^n, \tag{7a}$$

$$\left(I + A_z^{n+1}\right) \delta u^{(3)} = \delta u^{(2)}. \tag{7b}$$

The first of these equations will be solved by an existing 2-D code on each of the discrete xy planes of the 3-D computational domain, and (7b) will then be solved to obtain $\delta u^{(3)}$. Rearrangement of Eq. (6) then produces

$$u^{n+1} = u^n + \delta u^{(3)}, \tag{8}$$

the advanced time level result.

The preceding analysis demonstrates the theoretical feasibility of the desired 2-D $\rightarrow$ 3-D conversion in the context of Douglas & Gunn time splitting, but there are many implementation details that must be addressed. We will describe some of these here, with an emphasis on aspects of parallelization. The main items to be considered in the conversion process are: *i)* overall code structure, *ii)* array structure and *iii)* parallelization. Clearly, these are not completely independent of one another.

With a 2-D code already in existence, the main changes to the basic code structure involve adding code for the third (z-) momentum equation, and for the additional advective and diffusive terms of the original momentum equations, corresponding to the new third direction. In the context of time-split schemes as we are treating here, and especially for a sequential solution algorithm (one momentum equation solved at a time, as opposed to block-coupled solves), this is completely straightforward. In particular, one simply copies code of one of the original momentum equations and changes names of variables to obtain the new third equation; all required additional terms for the two original equations can be included in the function f^n and matrices A^{n+1}, B^n in Eq. (7a) and/or in the A_z^{n+1} matrix of Eq. (7b), as appropriate.

The array structure must now include time level n and $n+1$ arrays for the 3-D solution components, and arrays containing problem geometry information must now be 3D. The simplest approach, especially with regard to parallelization, seems to be to retain the original array structure of the 2-D code and transfer data between this and the complete 3-D arrays on a plane-by-plane basis, as needed.

Parallelization is crucial in 3-D N.–S. calculations, and we now describe the various ways by which it can be introduced into an algorithm constructed per the above prescriptions. The first, and most obvious, opportunity for parallelization involves the 2-D planar solves corresponding to Eq. (7a), utilizing the existing 2-D code on each plane. The initial step in this process is evaluation of the right-hand side of Eq. (7a). Because of the three dimensionality, evaluation on any given plane requires data from the two nearest-neighbor planes. Thus, to avoid memory-access conflicts it is probably advisable to employ a Red-Green-Blue (R-G-B) ordering of the planes. On a typical SMP architecture with several nodes per hypernode (such as the HP SPP-2200 and SGI Origin 2000 series machines), data would be transferred to and from one red plane per processor, plus data from its immediately adjacent green and blue planes. In many cases all red planes can be processed simultaneously since even on a grid of 10^6 cells one would expect only $\sim \mathcal{O}(10^2)$ cells in any given direction, and this implies $\sim \mathcal{O}(10^1)$ red planes. This process is then repeated for the green and blue planes.

Once the right-hand sides have been evaluated, the solution process can be carried out. At this point all planar calculations are completely independent, so assignment of planes to individual processors should be done in a manner to effect the most efficient memory accesses. This can vary among the various SMPs, but in general for current typical problem sizes ($\sim \mathcal{O}(10^6)$ cells) each processor should operate on at least one plane ($\sim \mathcal{O}(10^4)$ cells) to balance CPU and communication times. As a corollary, parallelization of 1-D line solves within the 2-D planes would not likely be useful because the lines are generally too short ($\sim \mathcal{O}(10^2)$ cells) to keep a processor busy. On a large SMP with say, $\mathcal{O}(10^2)$ processors, it is possible at least in principle to perform all 2-D planar calculations simultaneously, implying that the solution of Eq. (7a) requires the same order of wall-clock time as would a 2-D problem.

We now consider parallelization of the solution process for Eq. (7b). There are N_xN_y individual 1-D problems in the z direction, and several strategies can be envisioned for treating these. Each of these independent problems will typically contain $\mathcal{O}(10^2)$ dependent variables which, as we have already observed, is probably far too few to allow assigning each 1-D equation to a single processor. At the same time, the total number of variables is expected to be $\sim \mathcal{O}(10^6)$, so even though there exists a permutation of A_z^{n+1} rendering it compactly banded, a problem this large will probably not fit in the cache of a single processor. It thus appears that the best strategy might be what is effectively domain decomposition resulting in sending well-defined collections of 1-D problems to separate processors. Indeed, on a SMP this can be done in a hierarchical fashion via, *e.g.*, a R-G-B ordering in which planes of lines would be labeled by color, all planes of a given color sent to a hypernode via MPI, and then individual planes assigned to separate processors within the hypernode utilizing the shared-memory paradigm. We note that a similar strategy might be employed during the second step of the solution of Eq. (7a) as well. In any case, theoretically, if there are sufficient processors to assign a 2-D plane of 1-D problems to each processor, the run time should then be the same as that of a corresponding 2-D problem, modulo communication time.

We can summarize the overall solution/parallelization process as follows. First are the right-hand-side evaluations for Eq. (7a), which are parallelizable but possibly subject to memory-access conflicts. R-G-B ordering is recommended, and the three colors must be processed sequentially—but with significant parallelization within each color. Theoretically, with perfect parallel efficiency, run times for this step could be within a factor of *three* per equation of those for a corresponding 2-D problem, rather than a factor of $N_z \sim \mathcal{O}(10^2)$ without parallelization. Of course, there is one additional equation in the 3-D N.-S. system. Second, the solution of Eq. (7a), once the right-hand-side evaluations are complete, can be done in approximately the *same* amount of wall-clock time as that of a corresponding 2-D problem, per equation (instead of a factor of N_z greater). Finally, the 1-D solves in the third direction can also be done in the *same* amount of wall-clock time as a single 2-D solve in that direction (instead of a factor of, say $\min(N_x, N_y)$ times that of the planar solve).

We conclude from all this that not only can a 2-D code be converted to a 3-D one, but moreover it can be parallelized very effectively, at least in theory.

4. PARALLELIZATION RESULTS

Variants of the parallelization strategies associated with the 2-D → 3-D code conversion described herein have been tested for the standard 3-D lid-driven cavity problem on three different parallel platforms, namely, a Cray T3E at Lawrence Berkeley Laboratories, a HP SPP-2200 at the University of Kentucky and its recent replacement, a HP N-4000. All calculations were performed in 64-bit arithmetic employing a Fortran 77 code. Discretization was done on a 72^3 grid for all machines, except where specifically noted, and computations were run for 100 time steps (actually, for 200 on the T3E, but we report one-half times for these data). MPI was utilized on all three machines. It is worth mention that since we could not obtain dedicated time for runs on the Cray T3E, we did not do so on the HP platforms either. Thus, results reflect parallel performance in a "realistic, production" environment rather than an ideal one.

Figure 2 presents wall-clock times *vs.* number of processors for all three platforms, indicating that although the HP machines are significantly faster than the Cray T3E per processor, they do not provide effective parallelization beyond *at most* four processors. Figure 3 depicts the corresponding speedups from which we see this in a slightly different way.

We also carried out limited calculations employing compiler directives (similar to OpenMP, but predating this) on the HP SPP-2200 to obtain shared-memory performance data. Here, a 64^3 grid was used (because the SPP had insufficient memory per hypernode when in shared-memory mode to allow larger problems), and computations were done with one, two and four processors. Results of speedup are displayed in Fig. 4. Comparing this with Fig. 3 shows that on the SPP-2200 compiler directives provide significantly more effective parallelization than does MPI.

5. SUMMARY

We have demonstrated a formalism for converting existing 2-D Navier–Stokes codes to 3-D, discussed aspects of parallelization in this context and provided results of implementing this procedure via MPI on three different systems, and with compiler directives on one system. These results are not particularly impressive, but it must be emphasized that they were produced in a non-dedicated, production environment. Hence, they are probably quite representative of actual performance in an industrial setting.

There are additional efforts that might be made to improve the parallel performance of the code under study. Especially on the HP N-4000 which has 96 processors, it is possible to attempt implementing combinations of MPI (across nodes) and OpenMP (between multiple processors on a single node). This is the subject of current, ongoing investigations which will be reported elsewhere.

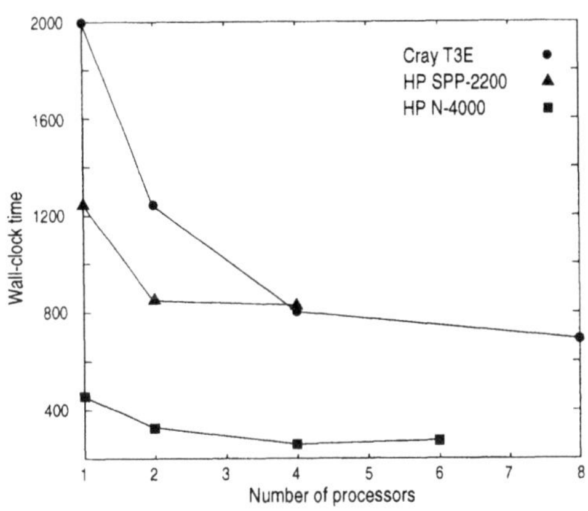

Figure 2. Wall-clock times for MPI parallelization

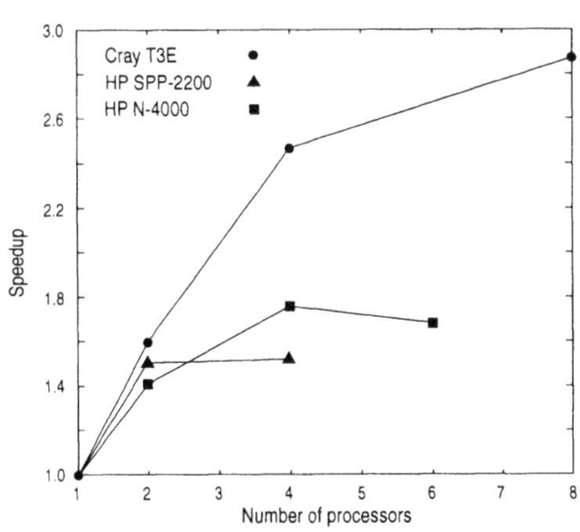

Figure 3. Speedups for MPI parallelization

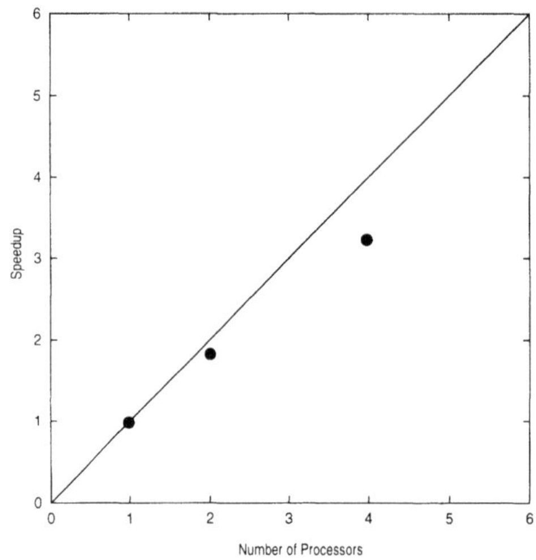

Figure 4. Speedups for 64^3 grid cell calculations with shared-memory parallelization.

ACKNOWLEDGEMENTS

The authors wish to express there gratitude to Lawrence Berkeley Laboratories for the use of their Cray T3E, and JMM also acknowledges the support of General Electric Aircraft Engines for development of the 3-D code studied here.

REFERENCES

1. J. Douglas, Jr. and J. E. Gunn, *Numer. Math.*, **6** (1964) 428.
2. P. M. Gresho, *Int. J. Numer. Meth. Fluids*, **11** (1990) 587.

4. Load Balancing

Parallel Computational Fluid Dynamics – Trends and Applications
C.B. Jenssen et al. (Editors)

Dynamic Load Balancing in International Distributed Heterogeneous Workstation Clusters

T. Bönisch[a], J.D. Chen[b], A. Ecer[b], Y.P. Chien[b], H.U. Akay[b]

[a] High Performance Computing Center Stuttgart, Allmandring 30, D-70550 Stuttgart, Germany

[b] Purdue School of Engineering and Technology, 723 W. Michigan St., Indianapolis, Indiana 46202

This paper describes a dynamic load-balancing scheme developed for improving the efficiency of running parallel CFD applications in a multi-site environment. Target of this approach is the minimization of the total elapsed time of the parallel job. The benefits of load balancing should exceed its costs that consist of the additional expenses for the load redistribution and the time needed by the load-balancer itself.

1. INTRODUCTION

In research collaborations it makes sense to use the resources of the other sites in addition to the local ones. This shortens the elapsed execution time for solving big problems with parallel applications. In such multi-site environments, especially when different owners operate the sites, reservation of dedicated time for these cases becomes difficult. Therefore, the main objective of this effort is to run parallel CFD codes efficiently on all available computer resources. The computer resource includes workstation clusters on different sites (may be on different continents) communicating through complex network topologies.

2. THE LOAD-BALANCER

The basic assumption adopted in our study is that the computers are multi-user machines running Unix or Windows NT operating systems. Each user can access all or a subset of these computers. Furthermore, it is assumed that the problem is divided into a number of blocks. In order to enable load balancing, the number of blocks should exceed by far the number of processors that will be used for this application. The load balancer distributes all these blocks to the available computers. Blocks on all these computers are running concurrently and are also competing for computation power with all other running jobs on the respective host.

2.1. Information for the load-balancer

In order to do load-blancing the load-balancer needs information about the computation environment and its condition, i.e., the load of all available computers and the network speed between these available machines. This information is provided by two tools running in parallel to the execution of the application. The tool "Ptrack" provides the load-balancer with the information about the average load of all available machines excluding our CFD job. "Ctrack" measures the network performance by detecting the time needed by a 1 KB packet between each pair of computers including the local loopback. Since measuring the communication speed this way is not very suitable for networks with highly different speeds, especially, transatlantic network connections compared with local networks, we currently experimenting improved measurement techniques. The described two tools are available for NT and Unix machines.

Furthermore, the load-balancer needs information about the application and its execution on the computers: the execution time of a block with the current application on each used machine and the amount of data exchanged within each iteration. This information is provided by the application itself, e.g. the parallelization library of the application [1]. Based on all the measured information, the cost estimator of the load balancer is able to calculate the expected run time of the next execution cycle for any given new block distribution.

Since some computers use the same file server and other computers use different file servers, the load balancer needs to be able to move block data from one machine to another if they do not share the same file server. We use a message-passing method for block transfer in order to avoid problems of different data representations on various architectures.

As the load balancer is able to deal with all the mentioned CFD applications, the block mover should do so, too. However, load balancer does not have the information about the CFD internal block organization and how to deal with them. To be able to handle all CFD programs, the block mover is build up modularly. The user or the application programmer respectively has to provide only four functions. These functions are for reading a block from and writing a block to disk as well as sending and receiving a block. These functions are normally part of the parallel application program and have to be made available to the block mover via its predefined interface. Using these functions, the block mover is able to redistribute the blocks of all corresponding user applications.

2.2. Optimizer

Besides the mentioned cost estimator, the core part of the load balancer is the optimizer. Optimizer tries to find a good distribution of blocks to the available computers within a short running time. This optimization problem is NP-complete, i.e. it is impossible to find the global optimal solution within an appropriate running time for large problems because there are so many possible solutions to consider.

Therefore, we implemented and compared several different optimization strategies and algorithms. The first algorithm implemented is the Greedy Algorithm. Starting with the current block distribution it tries to move each block, one after the other, to each other machine. It then chooses the block move with the best result, i.e. it moves the block to the specific target machine which leads to the shortest estimated running time of the application in the next cycle. Based on this new (and currently better) solution, the algorithm tries to find another block move using the same steps again. This optimization process is repeated as long as the algorithm is able to find better solutions. A problem of the original Greedy Algorithm used for the optimizer is, that it is only able to move one block from one machine to another within one optimization step. This limits the solution space and inhibits very often to find a good solution, i.e. the Greedy Algorithm sometimes gets stuck in a local minimum ignoring much better solutions.

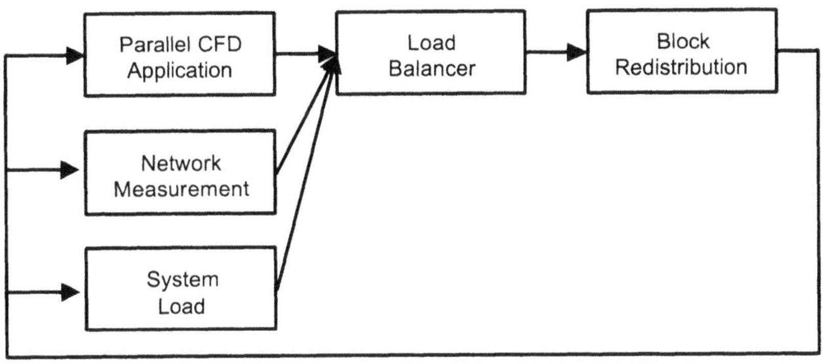

Figure 1. The load-balancing cycle.

We tried other algorithms to improve the quality of optimization. The first three algorithms use statistical approaches, i.e. they all start with one solution and calculate a new load distribution by choosing randomly the move of one block to a target machine. The difference between the three statistical algorithms is the method of handling the new block distribution. The first algorithm is the Threshold Accepting (TA). The algorithm takes a new distribution as a solution even if it is a little bit worse (the threshold in seconds) than the last one [2]. If there is no increase in quality for some time, the threshold is decreased. If there is no increase in quality with threshold zero, the algorithm stops.

The second algorithm is Record-to-Record Travel (RRT). The algorithm works with a so-called record that stores the best solution it has already found [3]. It accepts a new distribution as solution to work further with, if the solution is better or a little bit worse than the record. If the new solution is better than the present record, this solution becomes the new record. If there is no increase in the solution quality for some time, the algorithm stops.

The third algorithm is Great Deluge algorithm. This algorithm works in a completely different way [3]. It uses two parameters, the (water) level and a decreasing number. If a solution is better than the level, the solution is accepted and the level is decreased by the decreasing parameter. If there is no better solution for some time, the algorithm stops.

The fourth algorithm is again of the greedy type, but it gets around the main problem of the first Greedy Algorithm. The new algorithm first moves one block away from each host to a buffer space. Than it moves each block out of the buffer, one after the other, to the host onto which the block fits best. This is done as long as there are improvements with the resulting block distribution. In addition, we also made experiments with different genetic algorithms for the optimization of the block distribution.

After the redistribution of the blocks, the application is run further on with that new block distribution. The resulting load balancing cycle is shown in Figure 1.

3. RESULTS

All measurements and tests of the load-balancer were done in a distributed computation environment that consists of workstation clusters on three sites (figure 2):

- 8 nodes of the IBM RS/6000 SP in Bloomington, Indiana, U.S.
  Power II SC, 160 MHz, 256 MB memory, AIX (b1 – b8)
- 8 nodes of the IBM RS/6000 SP in Stuttgart, Germany
  Power II Thin2, 66 MHz, 128 MB memory, AIX (s1 – s8)
- 6 IBM RS/6000 7043 Workstations Model 260 in Indianapolis, Indiana, U.S.
  Power III , 200 MHz, 512 MB memory, AIX (iw1 – iw6) and
- 10 PC´s in Indianapolis, Indiana, U.S.
  Pentium II, 400 MHz, 256 MB memory, Windows NT (ip1, ip4 – ip13)

The connection between Indianapolis/U.S. and Stuttgart/Germany is the German research network (DFN) with its transatlantic link and Abilene. The connection between Indianapolis and Bloomington is ATM and HSSI.

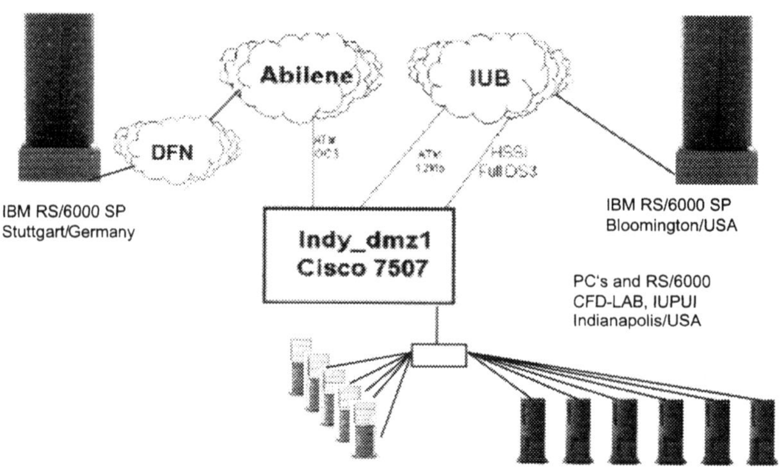

Figure 2. Our distributed computation environment.

3.1. Two Site Example Run

The following experiment demonstrates the effectiveness of load balancing. The CFD-problem [4] consists of 64 equal sized data blocks. These blocks are distributed to 24 of the computers (nodes) described above.

We do load balancing once a while during the execution of the application in order to respond to load changes on the machines. For long running jobs, the load balancing is executed once in approximately half an hour to an hour, depending on the effort to redistribute the blocks, i.e. the data size.

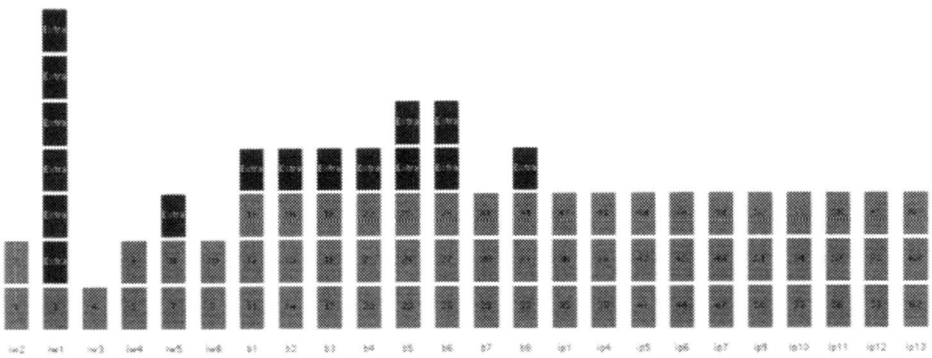

Figure 3. Block distribution for the initial cycle of the application, elapsed time 1379 s.

In the case shown on this page, load balancing is done every 100-iterations of the application program's main loop. In the first execution cycle we use an arbitrary initial distribution with three blocks assigned to each host and only one or two to the workstations known as to be faster (see dark gray blocks in Figure 3).

In these pictures the dark gray and light gray boxes represent the blocks of the CFD application. The black ones represent extraneous load. The extraneous loads are the processes on the machines that are not part of the CFD application. The light gray boxes are blocks that have been moved by the load-balancer before this cycle, whereas the dark gray ones stayed on the same computer.

After the first load-balancing step, elapsed time for the application went down from 1379 to 371 seconds further down to 363 seconds within the following steps. The block distribution for the forth cycle is given in Figure 4. The PC's (ip1 to ip13) are emptied. This is reasonable because the PC's in this case are using a non-optimized version of the executable.

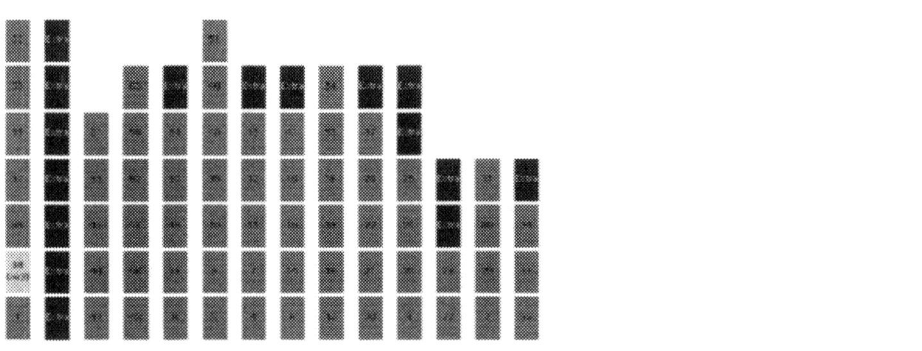

Figure 4. Forth cycle of the application, elapsed time 363 s.

Furthermore, we can see another effect in this case. We considered the eight nodes of the Bloomington SP (b1 to b8) being equal. But surprisingly the load balancer showed unexpected results. It puts fewer loads on node 6 to 8 compared to node 1 to 5. We could not explain this phenomenon because the load measurements and timings showed that the load balancer did a good job. Asking the system administrators, we figured out that node 6 to 8 have a different memory layout (2x128 MB compared with 4x64MB on node 1 to 5) which leads to a worse interleave factor and therefore to a worse performance.

This case demonstrates the advantages of the load balance program that it is able to deal with the efficiency differences in compilers used on different types of machines and with different sustained performance on nodes even if they have been assumed to be equal.

3.2. Optimization algorithms

To compare the aforementioned optimization algorithms, we made tests on 21 of the 32 machines: 5 of the IBM's and 8 of the PC's in Indianapolis and the 8 nodes of the SP in Stuttgart. We executed the load distributions suggested by different algorithms and recorded the actual execution time for each load distribution. The computation time needed by the optimizer itself is also recorded. These values are gathered in table 1. For the genetic algorithm we didn't make an application run because the time needed for running the load-balancer and the expected run time of the application would be much longer than that of the good combinations.

Figure 5 shows the initial distribution of the blocks for one of these runs. The other two initial distributions are quasi similar. In the following figures we give the results after the second load balancing step, i.e. the block distribution for the third cycle of the application and the times needed for that application cycle.

It is obvious that the results with the original Greedy Algorithm are not the best (Figure 6). There are some "wholes" which affect the performance. Additionally, in the second cycle it had a worse result of 800 s. The RRT shows a better result of 623 s.

However, figure 7 also displays a problem of the statistical algorithms. These algorithms move too many blocks around in load balancing. One solution to prevent this would be adding an estimated block movement time to the estimated cost function for load balancing. Consequently, the solution quality for the optimizer would consider block movement costs that should lead to improved results of the statistical algorithms.

Table 1. Comparison of the different optimization algorithms

Optimization Algorithm	Runtime Optimizer	Runtime CFD-Application
Original Greedy	189 s	640 s
TA	56 s	627 s
RRT	56 s	587 s
Great Deluge	33 s	978 s
Greedy + heuristic	22 s	604 s
Genetic Alg.	ca. 550 s	---

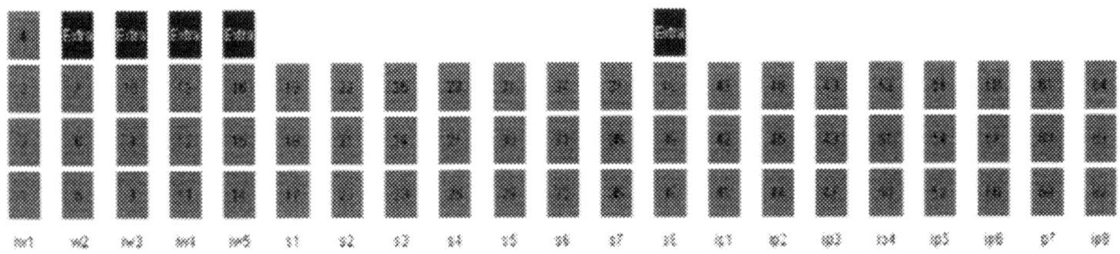

Figure 5. Initial block distribution, elapsed time for the application 1481 s

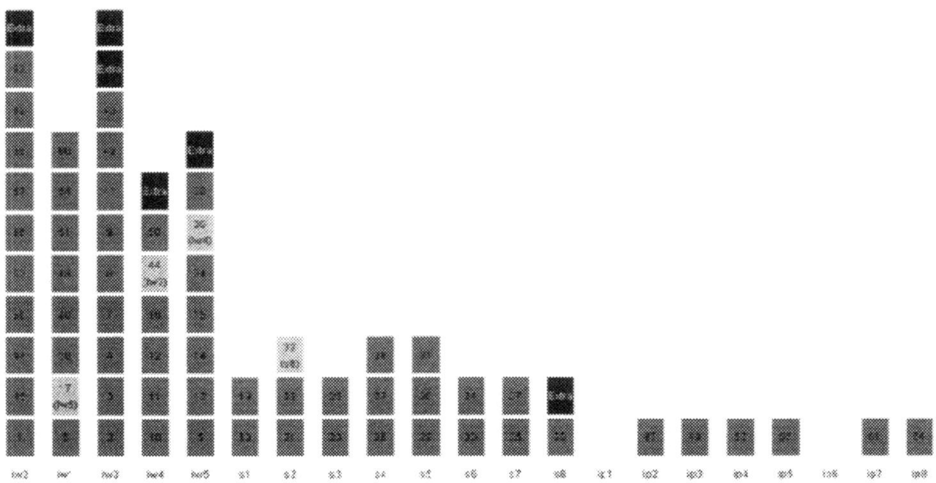

Figure 6. Distribution for the third application cycle (Greedy Alorithm), elapsed time 659 s.

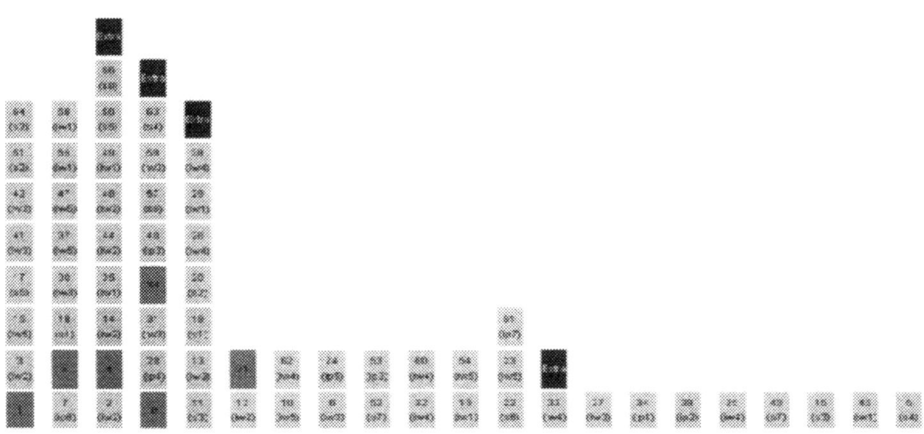

Figure 7. Third application cycle, optimization with RRT, elapsed time 623 s.

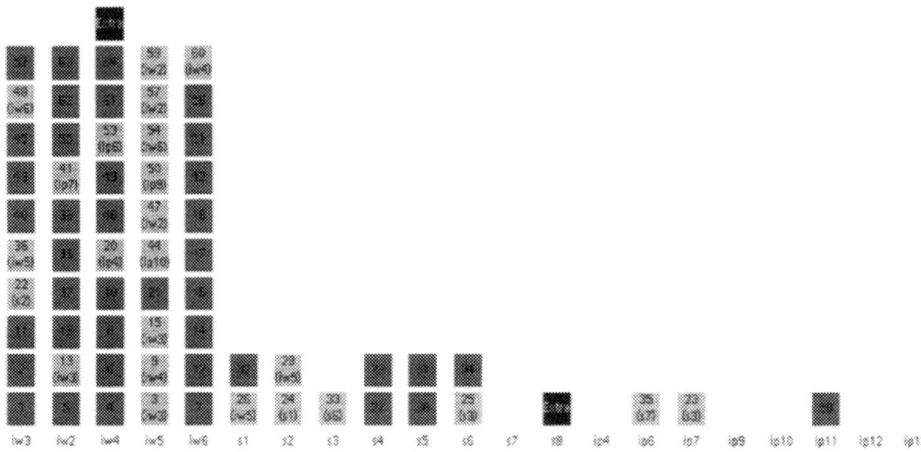

Figure 8. Distribution in the third application cycle (Heuristic Greedy), elapsed time 548 s.

The Heuristic Greedy Algorithm in figure 8 shows that it is able to provide a good result regarding the application time and it did not move so many blocks around in load balancing. Therefore, this algorithm is the best in the moment.

REFERENCES

[1] Y.P. Chien, J.D. Chen, A.Ecer, H.U. Akay: *Dynamic Load Balancing for Parallel CFD On Windows NT*, Proceedings of the Parallel CFD Conference 1999.

[2] G. Dueck and T. Scheuer: *Threshold Accepting: A General Purpose Optimization Algorithm Appearing Superior to Simulated Annealing*, Journal of Computational Physics 90, 161-175 (1990).

[3] G. Dueck: *New Optimization Heuristics: The Great Deluge Algorithm and the Record-to-Record Travel*, Journal of Computational Physics 104, 86-92 (1993).

[4] *ParCFD Test Case*: accessible via http://www.parcfd.org.

[5] Y.P. Chien, F. Carpenter, A.Ecer, H.U. Akay: *Load-balancing for parallel computation of fluid dynamic problems.*

[6] A. Ecer, Y.P. Chien, J.D. Chen, T. Bönisch, H.U. Akay: *Dynamic Load-Balancing for Distributed Heterogeneous Computing of Parallel CFD Problems*, proceedings of NASA High Performance Computing and Communications Computational Aero Science Workshop (CD-ROM), February 15 – 17, 2000, California, USA.

Parallel Computational Fluid Dynamics – Trends and Applications
C.B. Jenssen et al. (Editors)

Dynamic Load Balancing for Unstructured Fluent

N. Gopalaswamy,K. Krishnan and T. Tysinger

Fluent Incorporated, 10, Cavendish Court, Lebanon, New Hampshire, U.S.A.

The paper describes the implementation of a dynamic load balancing algorithm in the parallel unstructured version of the FLUENT code. Since FLUENT uses a domain decomposition method for parallelization, dynamic load balancing is done by migrating parts of the subdomains between processors during execution. An optimization algorithm is used to determine the migration pattern with the help of certain assumptions to reduce the search space of the optimization problem. Some examples are presented demonstrating the performance of the load balancing algorithm on a heterogeneous network of workstations with varying computational speeds and communication parameters.

1. INTRODUCTION

With the increasing emphasis on efficient parallel computation driven by current industry demands to solve ever greater and more complex physical problems, a practical approach to enable dynamic load balancing into the unstructured FLUENT codes is presented. The parallel FLUENT codes use domain decomposition to subdivide the mesh into partitions and distribute the partitions among different processors upon startup. Subsequently, during execution, parallel computation may become unbalanced because of various factors. Some of them could include: a) uneven adaption of the mesh across the partitions, b) a change in the solution parameters, c) extraneous processes, and d) changes in the network traffic.

There has been a significant amount of work in the last few years on this topic [1,2,3,4,5,6]. The range of algorithms presented and their specific application varies depending on the constraints in the load balancing problem and the nature of the code. There could be additional constraints on the load balancing problem in terms of the memory resources of the parallel architecture, utilization policy for the available computers etc. It is the aim of the current work to present an approach to load balancing with the goal of minimizing the wall clock execution time T_{wall} of FLUENT, given a finite set of resources. The resources can be defined in terms of the machines available, number of processors to use per machine, the network connectivity between the machines, and the amount of memory to use per machine. The dynamic load balancing procedure outlined describes the metrics used to define the amount of load imbalance and the steps attempted to minimize the imbalance.

2. APPROACH

Unstructured FLUENT uses a cell-centered finite volume discretization. So computation mostly involves looping over cells and their associated faces. During parallel computation certain parameters required by the load balancing procedure are monitored and stored on each compute-node executing on a processor. There is a periodic check to see if there is a need to perform load balancing. This period can be calculated automatically, or specified by the user. Dynamic load balancing is carried out by redistributing groups of cells between the processors at runtime. The groups of cells, called partitions, are identified by a partitioning algorithm, which may be invoked at the beginning of the procedure if the partitions do not already exist. An optimization algorithm is then used to determine the redistribution pattern between the processors. If the optimization algorithm yields a distribution of partitions which can result in a lower execution time than the current one, a load imbalance exists. The savings in execution time should be above a certain threshold in order to proceed further in the load balancing procedure. A profit analysis is carried out to determine if the savings in execution time from the load balanced distribution outweigh the redistribution costs. Depending upon the outcome of the analysis, a migration module may be invoked which carries out the redistribution of partitions between the processors. Any FLUENT process on a processor with no partitions is then shut down, and the parallel execution resumes.

2.1. Partitioning for Dynamic Load Balancing

If not already done, the computational mesh is partitioned into n partitions such that $n \gg P$, where P is the number of available processors. Fluent has a selection of various geometric partitioners and the Metis partitioner[7]. These are available in both parallel and serial versions of the code. Each partition has an overlapping cell layer with its neighbors. A computational subdomain on each processor can be considered as an agglomeration of these partitions on that processor. Each partition can only be in one subdomain. The graph connectivity $G = (S, E)$ of these partitions is computed at the time of partitioning and stored in Compressed Storage Row (CSR) format. The resulting graph is used during the agglomeration phase to compute the graph connectivity of the subdomains formed by the agglomeration. The vertices of the graph S represent the various partitions and the edge weights E of the graph represent the number of cells in the overlapping cell layer between partitions.

2.2. Balance Parameters

There are time-stamps present in the code which are triggered by communication events. Based on these time-stamps, the following parameters are computed at every iteration:

1. the elapsed computational time for each subdomain on a processor,
2. the elapsed time required for communication of messages between the processors,
3. the number of messages and the number of bytes exchanged between the processors,
4. the order of the communication exchanges,
5. and, the total elapsed time per iteration.

For the first four quantities above a running average is also maintained. The averaging interval is chosen to be the balance monitoring period. At every invocation of the balance monitor, the available memory on the processors is queried. In the case of an SMP machine, the per processor memory is taken to be the ratio of the total available memory on the machine, to the number of processors. Also the latency and bandwidth of the network interconnection between the processors is calculated. If there are P processors, the values are stored in two $P \times P$ matrices.

2.3. Agglomeration of Partitions for Load Balancing

Parallel FLUENT is organized as a "host" process which handles I/O, and one or more "compute-nodes" which carry out the flow solution. The compute-nodes send their portion of the graph corresponding to the partitions they currently own to the host process. The complete graph is built on the host process. The agglomeration is also carried out on the host process. A subdomain is formed by the agglomeration of partitions on each processor. Let S be the set of all partitions. Then, considering a subdomain on processor i, let A_i be the set of partitions on processor i. Thus, $A_i \in S$. It is allowed that $A_i = \emptyset$. This would eliminate processor i for consideration in the load balancing procedure. The agglomeration is constructed by setting the processor id of each partition $a_i \rightarrow \mathrm{id}, \mathrm{a_i} \in \mathrm{A_i}$ to be i on the host process. Due to memory constraints, the set A_i must satisfy:

$$\sum a_i \rightarrow \mathrm{ncells} \leq \mathrm{M_i}, \mathrm{a_i} \in \mathrm{A_i}, \mathrm{a_i} \rightarrow \mathrm{id} = \mathrm{i} \tag{1}$$

where M_i is the total number of cells that can fit within available memory on processor i. Thus, from the graph $G = (S, E)$, various subgraphs are constructed:

$$G_i = (A_i, E_i), i = 1 \dots P \tag{2}$$

If we define a set of edges K_i^j such that

$$K_i^j = E_i \cap E_j; j \neq i \tag{3}$$

this would represent the edges cut between the subdomains formed by agglomeration on processors i and j. The edge weights of these edge cuts would represent the number of the cells in the overlapping layer between subdomains on processors i and j.

2.4. Computing the Estimated Elapsed Time

During each iteration, there are multiple exchanges carried out depending upon the connectivity between the subdomains. There are also some global operations interspersed with the computation. Modeling the entire pattern of computation and communication can become complex so a simplified model is constructed in order to calculate the estimated time.

Considering a subdomain on processor i, the elapsed computation time can be approximated as:

$$C_i = N_c^i t_c^i \tag{4}$$

where N_c^i is the number of cells on processor i and t_c^i is the elapsed time for computation for the cells on processor i. The elapsed communication time spent by each processor i can be computed as:

$$H_i = \sum_{j=1}^{P} \left(b_i^j I_i^j + l_i^j N_i^j \right) \tag{5}$$

where b_i^j is the bandwidth of the network interconnection between processor i and j and similarly l_i^j represents the latency. These values are obtained from the $P \times P$ matrices described previously. The quantity I_i^j represents the number of bytes exchanged between processor i and j and N_i^j is the number of messages. The value I_i^j is proportional to the edge weights of the edges cut in K_i^j. A scaling constant is necessary to translate the number of bytes exchanged for a given number of cells in the overlapping layer. This constant is based on the averaged values of message sizes stored in the time-stamp history for previous iterations. A similar procedure is employed for estimating N_i^j.

Besides communication exchanges, there are many global operations which must also be accounted for. If processor k is chosen as the synchronizing point, then the elapsed time for the global operations can be written as:

$$Q_i = \sum_{j=1}^{P} l_k^j N_g; j \neq k \tag{6}$$

where N_g is the number of global operations per iteration and is also stored as part of the time-stamp information.

The total elapsed time per iteration for each agglomeration can thus be written as:

$$T_{wall} = C_i + H_i + W_i + Q_i \tag{7}$$

where W_i is the waiting time. T_{wall} is the same for all processors. The waiting time on each processor i can be written as:

$$W_i = F(C_j, H_j); j = 1 \ldots P \tag{8}$$

where $F()$ is a complex function since the waiting time on a processor is dependent upon events occurring on other processors, and also the order of communication between the different processors. This function is solved numerically using an advancing front algorithm till convergence is achieved. After convergence the value of T_{wall} can be obtained.

Thus various agglomerations of partitions can be evaluated to provide an estimate of the wall clock time for that particular agglomeration.

2.5. Optimization of Agglomeration

The search space of the optimization problem is P^n, which grows exponentially with the number of partitions and available processors. If all these combinations of partitions were evaluated, it would cover the complete solution space of the load balancing problem. The estimated time for each such combination of partitions could be calculated and the combination corresponding to the minimum estimated time would be the the solution to the load balancing problem. However, this is an NP-complete problem and thus impractical to solve using the above approach for a reasonable number of partitions and processors. A variant of the greedy algorithm is used to solve the optimization problem. A two-pass load balancing procedure is used. The optimization problem is formulated in terms of a single variable, the number of subpartitions agglomerated on a processor. The agglomeration is restricted to a sequential set of partitions. The partitions are numbered in sequence during the partitioning phase. Usually partitions which are close to each other in the sequential order are also geometrical neighbors if a geometric based partitioning is used, though this may not be true for other partitioning methods. These set of assumptions result in the load balancing problem being recast as an integer optimization problem. We can express the problem as find $\bar{T}_{wall}$ such that:

$$\bar{T}_{wall} = \frac{\min}{A_i} \left(T_{wall} = C_i + H_i + W_i + Q_i; A_i \in S \right) \tag{9}$$

for all possible subsets A_i such that, $A_i = X_i$, where X is a vector satisfying the property $\sum_{j=1}^{P} x_j = N_p; x_j \in X_i$. Here N_p is the total number of partitions produced by the partitioner. Thus the elapsed time calculation result can be written as $T_{wall} = f \cdot X_i$ where the operator $f\cdot$ represents the function used to calculate the elapsed time and takes a vector X_i as the input.

The elapsed time difference is calculated for the following combinations in the inner pass for $(x_j \in X_i)$:

$$\Delta T_{wall} = f \cdot X(x_j \pm \Delta n) \tag{10}$$

where Δn is the number of partitions swapped with the neighbor x_{j+1}, and is chosen empirically to accelerate convergence. If there is a reduction in the execution time, this value of x_j is saved.

The second pass repeats the above procedure for other pairs of subdomain interfaces (x_k, x_{k+1}). In each pass the vector corresponding to the local minimum is saved and compared to the new elapsed time calculated. This procedure is repeated until a certain maximum number of iterations has been reached. The vector X_i corresponding to the minimum estimated elapsed time is then used as the load balanced distribution.

2.6. Profit Analysis

Once a given load balanced distribution has been calculated, the estimated time for such a distribution is compared to the existing elapsed time. If there is a savings in time above a certain threshold, the time needed for redistribution is calculated. Based on any past redistribution time, the time is scaled linearly based on the total volume of cells emigrating. The redistribution phase involves exchanging the information associated with the migrating cells, and rebuilding the connectivity on each processor. If the number of iterations to be carried out within the load balancing window is such that:

$$n \geq \frac{T_{rd}}{T_{wall} - \bar{T}_{wall}} \tag{11}$$

where T_{rd} is the estimated redistribution time, then the redistribution is carried out. If any processor ends up with zero cells on it, then if possible, the Fluent process on that processor is shut down. In the current implementation, that processor will not be used again.

2.7. Adaption Event

Adaptive meshing can be carried out to refine and/or coarsen the mesh depending upon some user specified criteria. Before adaption, the cells which will be refined or coarsened are marked and repartitioning is carried out based on the post adaption cell distribution. A simple balancing procedure is carried out based on the difference in the computational load between the different processors. Here the number of processors is fixed to be the current set of processors. If the computational load imbalance is more than a specified threshold, and the redistribution time calculated is less than the savings in execution time, redistribution is carried out. Since the redistribution is carried out before adaption actually takes place, the volume of cell movement is less than or equal to what would be necessary if the same was done after adaption. Thus, adaption itself can happen in a more load balanced fashion. After adaption, the load balancing procedure is invoked during the next interval, which then takes into account both computational and communication costs as described previously in calculating further necessity for load balancing.

3. RESULTS

The hardware used for the following results is a cluster of networked machines consisting of:

- Two single processor Sun Ultra/60 machines running Solaris
- Two dual processor Pentium 450 MHz machines running Linux
- Four dual processor Pentium 550MHz Xeon machines running Linux

The Linux machines are in a cluster connected by a Fast Ethernet switch. The Solaris machines are connected to a separate Fast Ethernet switch, which communicates directly with the Linux cluster switch. Thus, the Solaris machines incur additional latency when communicating with any of the Linux machines, and vice versa. To demonstrate the performance of the load balancing algorithm, the following test cases are used:

1. A case corresponding to a mesh of 248000 tetrahedral and hexahedral cells is used to compute the turbulent flow inside an automotive engine valveport.
2. A case corresponding to a mesh of 32000 hexahedral cells is used to compute turbulent flow in an elbow shaped duct.

First, the effect of variation in the computational load in a heterogeneous cluster consisting of the Sun and Linux machines is studied. The difference between the fastest and the slowest machines running FLUENT is approximately a factor of two. Test case 1 is partitioned equally for 1 to 8 processors and is loaded on the heterogeneous cluster in such a way that there is at least one slow machine in the cluster. The machines are all used in single processor mode. About 25 iterations are carried out, and then balancing is turned on. The time before and after balancing is displayed in Figure 1. It can be seen from the figure that depending upon the variation in the computational load and the number of processors, load balancing can produce significant savings in the execution time.

Next, the effect of differences in the communication speeds of the network connections between the machines in a homogeneous cluster is studied. A homogeneous cluster is used to exclude the effects of variations in computational speed that may be present in a heterogeneous cluster. A network load is imposed in the form of a concurrent data transfer between two machines in the cluster, in such a way that at least one machine with a FLUENT process on it is subject to the network load. Also, Test case 2 is used since it is a small case with a higher communication/computation ratio than Test case

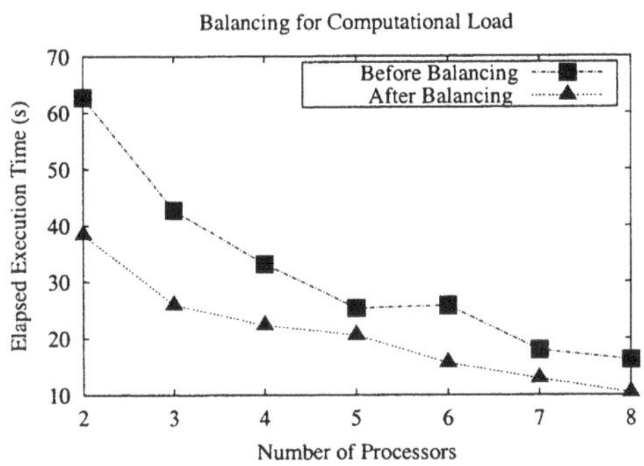

Figure 1. Effect of Load Balancing on a Heterogeneous Cluster

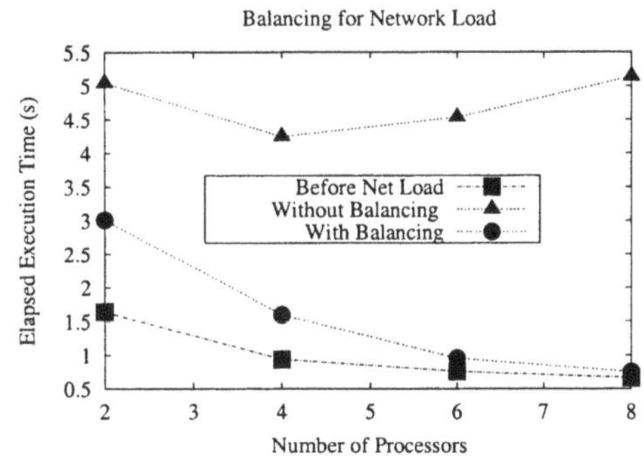

Figure 2. Effect of Load Balancing due to Network Load on a Homogeneous Cluster

1. The case is run for 10 iterations, and then a network load is imposed. After another 10 iterations, the load balancing procedure is invoked. The timings for each phase are displayed in Figure 2. One can see from the figure that there is a marked increase in the execution time because of the network load, and the load balancing procedure ameliorates this effect to a large degree, usually by migrating all partitions off the offending machine and shutting the FLUENT process down on that machine. As the number of processors increases, the effect of shutting down the FLUENT process on the machine results in a smaller increase in the per process computational load.

Finally, Test case 2 is used to demonstrate the effects of uneven adaption on the execution time. A homogeneous cluster of 4 dual processor Pentium 550MHz/Linux machines is used. Approximately 2% of the cells are marked for adaption by selecting a suitable pressure gradient threshold. Most of these cells marked for adaption are located in one part of the domain, and thus would lead to an uneven distribution of cells in the domain after adaption. The cell count after adaption increases to approximately 43000 from 32000 cells. If balancing is not enabled, this leads to load imbalance as shown in Figure 3. With load balancing enabled however, this load imbalance is greatly minimized as can be seen from the same figure.

The implementation of the load balancing algorithm into the FLUENT code can produce significant savings in situations that some users may encounter. Future work in enhancing the optimization algorithm and providing additional usability to fully exploit this new feature is also planned.

4. REFERENCES

1. N. Touheed, P. Selwood, P.K. Jimack, M. Berzins and P.M. Dew, "Parallel Dynamic Load-Balancing for the Solution of Transient CFD Problems Using Adaptive Tetrahedral Meshes," Parallel Computational Fluid Dynamics 1998, Edited by D.R. Emerson et al., Elsevier, Amsterdam, pp. 81-88, 1998.

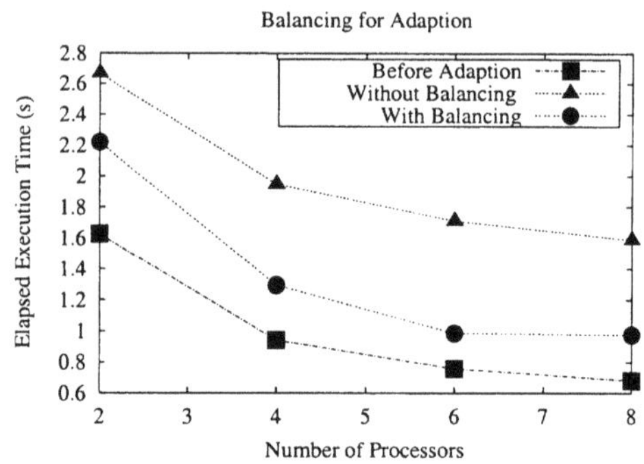

Figure 3. Effect of Load Balancing due to Adaption on a Homogeneous Cluster

2. M. Streng, "Load Balancing for Computational Fluid Dynamics Calculations," *High Performance Computing in Fluid Dynamics*, Kluwer Academic Publishers, Netherlands, pp. 145-172, 1996.

3. A. Vidwans, Y. Kallinderis and V. Venkatakrishnan, "Parallel Dynamic Load-Balancing Algorithms for Three-Dimensional Adaptive Unstructured Grids," *AIAA Journal*, vol. 32 No. 3, pp. 497-505, March 1994.

4. L. Oliker and R. Biswas, "PLUM: Parallel Load Balancing for Adaptive Unstructured Meshes," *Journal of Parallel and Distributed Computing*, vol. 52, pp. 150-177, 1998.

5. N. Gopalaswamy, Y.P. Chien, A. Ecer, H.U. Akay, R.A. Blech and G.L. Cole, "An Investigation of Load Balancing Strategies for CFD Applications on Parallel Computers," Parallel Computational Fluid Dynamics 1995, Edited by A. Ecer et al., Elsevier, Amsterdam, pp. 703-710, 1995.

6. T. Tysinger, D. Banerjee, M. Missaghi and J. Murthy, "Parallel Processing for Solution-Adaptive Computation of Fluid Flow," Parallel Computational Fluid Dynamics 1995, Edited by A. Ecer et al., Elsevier, Amsterdam, pp. 513-520, 1995.

7. G. Karypis, and V. Kumar, "METIS: A Software Package for Partitioning Unstructured Graphs, Partitioning Meshes, and Computing Fill-Reducing Orderings of Sparse Matrices," University of Minnesota, Department of Computer Science, Minneapolis, MN, September 1998.

Parallel Computational Fluid Dynamics – Trends and Applications
C.B. Jenssen et al. (Editors)

Parallel Computing and Dynamic Load Balancing of ADPAC On a Heterogeneous Cluster of Unix and NT Operating Systems

H.U. Akay, A. Ecer, E.Yilmaz, L.P. Loo, and R.U. Payli

Computational Fluid Dynamics Laboratory
Department of Mechanical Engineering
Indiana University-Purdue University Indianapolis (IUPUI)
Indianapolis, IN 46202, USA

1. INTRODUCTION

The current research involves parallel processing of ADPAC (An Advanced Ducted Propfan Analysis Code) on local area and wide area networks formed by combined Unix and Windows/NT environments. ADPAC was originally developed at NASA and Rolls-Royce (formerly, Allison Engine Company) to run on Unix-based system computers [1, 2]. The results obtained from each workstation are compared and verified by using a CFD test case. Timing evaluation and dynamic load balancing studies are conducted. The dynamic load balancing tools developed at the IUPUI's CFD laboratory are utilized in conjunction with a 500,000 grid points CFD test case to perform this task. All the work involving timing evaluation and dynamic load balancing are done by using a 100Mb switch as the communication network. Special emphasis is given to running the code in heterogeneous workstation environments. In order to evaluate the performance of each workstation in heterogeneous environments, the computational and communication times obtained are analyzed. The speedup and efficiency of each workstation in parallel computation are also compared. The effects of load balancing on the total elapsed time and for achieving balanced load in heterogeneous environments are demonstrated in this document.

2. GOVERNING EQUATIONS

The numerical solution for ADPAC uses the conservation law form of the Navier-Stokes equations. In the present study, the inviscid version of ADPAC, i.e., Euler equations, are used. For a rotating frame of reference in cylindirical coordinates, the control volume form of the Euler equations is defined as:

$$\int_V \frac{\partial}{\partial t}(\mathbf{Q})dV + \int_A [\mathbf{F}_{inv} dA_z + \mathbf{G}_{inv} dA_r + (\mathbf{H}_{inv} - r\omega\mathbf{Q})dA_\theta] = \int_V \mathbf{K}\, dV \tag{1}$$

where ω is the rotational speed and

$$\mathbf{Q}=\begin{bmatrix}\rho\\ \rho v_z\\ \rho v_r\\ \rho v_\theta\\ \rho e_t\end{bmatrix},\ \mathbf{F}_{inv}=\begin{bmatrix}\rho v_z\\ \rho v_z^2+p\\ \rho v_z v_r\\ r\rho v_z v_\theta\\ \rho v_z H\end{bmatrix},\ \mathbf{G}_{inv}=\begin{bmatrix}\rho v_r\\ \rho v_z v_r\\ \rho v_r^2+p\\ r\rho v_r v_\theta\\ \rho v_r H\end{bmatrix},\ \mathbf{H}_{inv}=\begin{bmatrix}\rho v_\theta\\ \rho v_z v_\theta\\ \rho v_r v_\theta\\ r(\rho v_\theta^2+p)\\ \rho v_\theta H\end{bmatrix},\ \mathbf{K}=\begin{bmatrix}0\\ 0\\ \dfrac{\rho v_\theta^2+p}{r}\\ 0\\ 0\end{bmatrix} \quad (2)$$

In Equation 2, ρ, p, v_z, v_r, and v_θ denote the density, pressure, and axial, radial and circumferential velocity components, respectively, relative to the coordinate system used in ADPAC; $\mathbf{Q}$ is the vector of conservation variables; $\mathbf{F}_{inv}, \mathbf{G}_{inv}$, and $\mathbf{H}_{inv}$ are the inviscid flux vectors; and $\mathbf{K}$ is the source vector due to rotating frame of reference. The total internal energy for a perfect gas, ρe_t, is defined as (γ is the specific heat ratio):

$$\rho e_t = \frac{p}{(\gamma-1)} + \frac{1}{2}\rho(v_z^2+v_r^2+v_\theta^2) \quad (3)$$

3. COMPUTING ENVIRONMENTS

In the present study, two clusters, formed by Unix and Windows NT operating systems, are used. The first cluster, located at the CFD laboratory of IUPUI, consists of six Unix-based IBM workstations and 16 Windows/NT-based Pentium-II PCs. The second cluster is created by connecting the local network through the Internet to IBM SP2 parallel computer with 139 compute nodes located in Indiana University at Bloomington, Indiana. The first cluster is considered as Local Area Network (LAN) whereas the second one is considered as Wide Area Network (WAN). The Unix-based workstations are IBM RS/6000 (RS6K) 43P-260 workstation with 512 MB RAM memory, 4MB of cache memory, and Power3 processors. The PCs are single processor computers with Pentium II, 256MB RAM memory, and 512K of cache memory. The IBM SP2 system consists of P2SC and Power3 processors with varying 256 MB and 512 MB of RAM memory options. Computers in the LAN cluster are connected to each other via 100Mb switches, while the Internet is used to connect to SP2 system in Bloomington. A grid-based parallel data base management tool, GPAR [2], developed in CFD Laboratory at IUPUI, is used as the tool for exchanging information between domain blocks of interest. The PVM library is used for communicating across different processors.

4. TIMING EVALUATIONS

A test case was considered for timing evaluations. Timing results were based on the steady-state solutions with the inviscid version of the ADPAC code on the LAN cluster. The total elapsed time for one processor was estimated based on the average elapsed time per iteration per node obtained for the two processors. In obtaining the timing results, each case was run both on NT and the heterogeneous environment (consisting of NT and Unix RS6K), by varying the number of machines or processors from two to twelve. However, the timing results for RS6K were done up to six machines because there are only six RS6K machines available in the laboratory. Hence, in some cases, the performances of each computing environment was evaluated based on two, four and six machines only.

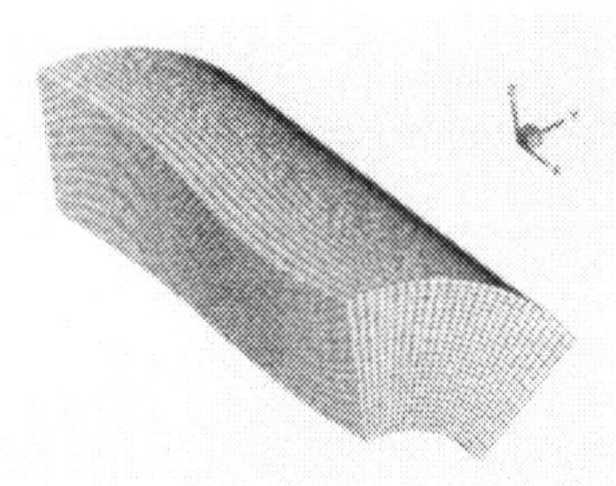

Figure 1. Grids for S-duct geometry.

The timing recorded includes the total elapsed and CPU time. A series of elapsed times for various computing environments is obtained statistically. Based on these timing results, the evaluation is assessed from three aspects: relative computational speed, parallel speedup, and parallel efficiency. The full-optimized Fortran compiler options were used for both Unix and NT machines.

The test case, given in Figure 1, is a swirling S-duct problem with a 765x25x25 grid in the axial, radial, and circumferential directions, respectively. Division of the domain into blocks was done along the axial direction. Shown in Figure 2 are the total elapsed times of two, four, and six-block solutions of the problem using Unix/RS6K and NT on LAN cluster. Also shown in the same figure are the interface solver times, which are due to information exchange between the blocks. It is observed that the computational speed of RS6K is higher than NT by a factor of 1.85 to 1.45 from 2 to 6 machines, respectively. For the same number of machines used, the elapsed time for the heterogeneous environment falls between NT and Unix/RS6K.

The corresponding speedup and efficiency curves versus the number of machines are plotted in Figure 3. It is apparent that the parallel efficiency for RS6K drops significantly when the number of machines increases. On the other hand, the efficiencies for NT and heterogeneous workstation are higher. The efficiency for RS6K and heterogeneous drop below 80% when 12 processors are used. This happens at the 6-processor case for RS6K. This is because the computational grids in each machine are relatively small and the time spent for message passing is becoming more dominant. These phenomena imply that as the computing power of computers are increased, the benefit of using more than a certain number of processors for a certain number of data blocks would be diminished.

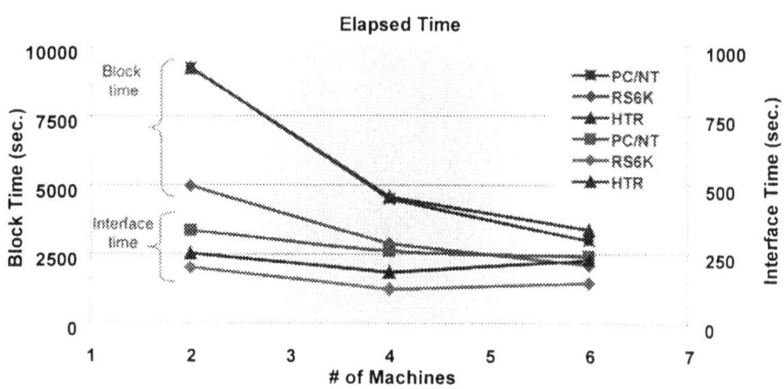

Figure 2. Elapsed time comparison for network of workstations (HTR = heterogeneous).

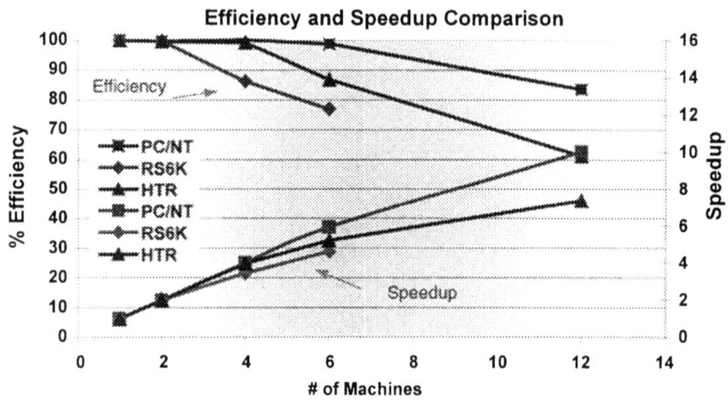

Figure 3. Efficiency and speedup comparison for network of workstations.

As mentioned in the foregoing, the poor speedup and efficiency could be attributed to the increase of communication time. Although, the increase in communication time is not significant for these cases, which is normally the case for steady flow, any decrease in computation time was quickly offset by the increase of communication time. Generally, as the number of machines is increased, communication within processors is increased. In this case, proximity of machines could be the major contribution for the increase of communication time. For example, the relatively low speedup and efficiency in heterogeneous workstation are the consequences of higher communication due to machine locations.

5. DYNAMIC LOAD BALANCING

Dynamic Load Balancing, DLB, was applied to the same test case for the ADPAC solutions both on LAN and WAN clusters. As given in the flow diagram of DLB task in Figure 4, communication tracking and process tracking programs are run with ADPAC first to get stable and long running loads on the the network. Sampling time for process tracking is user defined. Based on this sampling of computer and network loads a new distribution of blocks is obtained by using other DLB tools based on the Greedy optimization algorithm [3, 4]. Assuming this new load distribution is stable and constant, the application program is run as many as needed for a DLB cycle. Generally 3-4 cycles of DLB are found sufficient in the present problem.

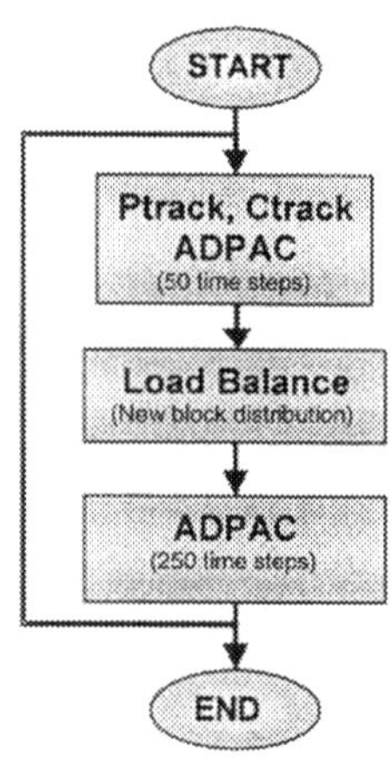

Figure 4. Flow diagram of DLB tasks.

Shown in Figures 5 and 6 are the load

distributions obtained by dynamic load balancing on the LAN environment. In this case, all computers were equally loaded assuming that the user has no idea about efficiency and power of computers and network at the beginning. Extraneous loads (i.e., other users) exist on computers in the Unix side during the ADPAC simulation. It is very obvious that extraneous loads were taken into account and very smooth and equal distribution was obtained at the third cycle. Three cycles were enough to reduce the total elapsed time by 25%. After that, further load balancings do not provide any time reduction. This is because a local minimum, which is a criterion used to determine the minimum time, has been achieved. The total amount of time reduced is substantial. Further experiments to illustrate a better time reduction by avoiding the local minimum, using improved optimization algorithms are in progress.

In the case of running ADPAC on the WAN environment, the same test case was used with and without DLB. Thirteen IBM SP2 nodes were introduced to the present cluster in the CFD laboratory. Therefore, a total of 32 compute nodes were involved to get solutions for 64 blocks. It was assumed that no information about loads and performance of remote resources were available at the time of submitting initial run, but just availability of computers itself. Therefore, all blocks were initially distributed on locally available computers almost equally. The initial load distribution is given in Figure 7. After three cycles of DLB, the block distribution is shown in Figure 8. Some of the blocks were moved to the SP2 to balance the overall distribution. The local cluster is almost equally loaded. Exraneous loads are changing at each cycle. Table 1 gives the elapsed time of the ADPAC simulation with and without DLB on both the LAN and WAN environments.

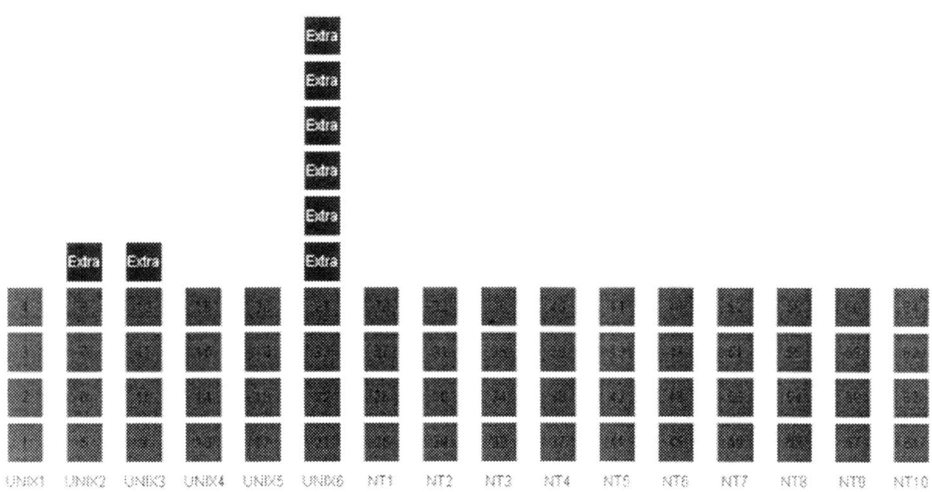

Figure 5. Initial load distribution on 16 computers for 64 blocks on the LAN cluster.

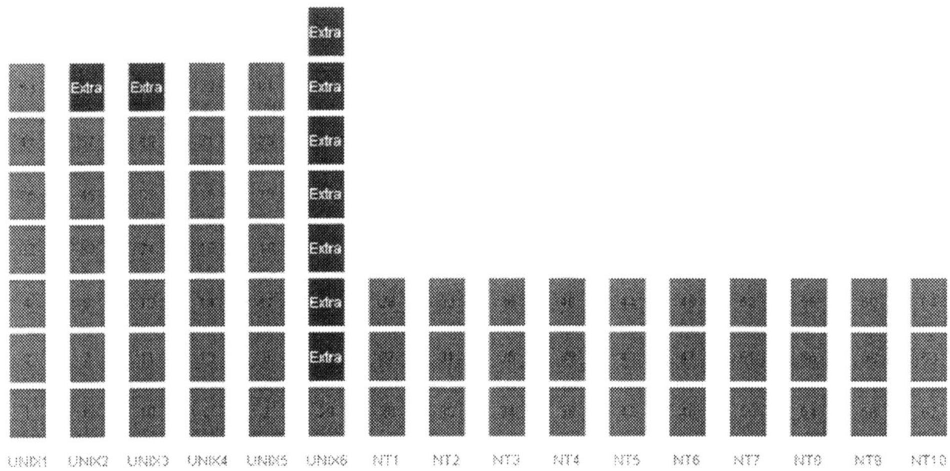

Figure 6. Load distribution on 16 computers for 64 blocks on the LAN cluster after 3 cyles of DLB.

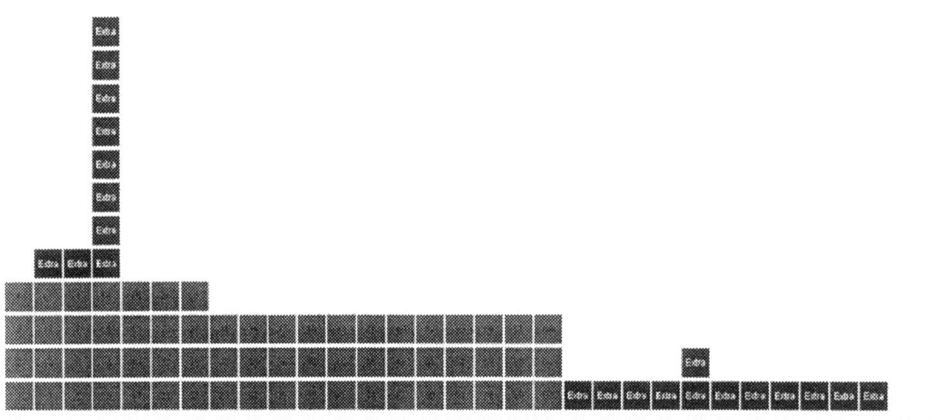

Figure 7. Initial load distribution on 32 computers for 64 blocks on the WAN cluster.

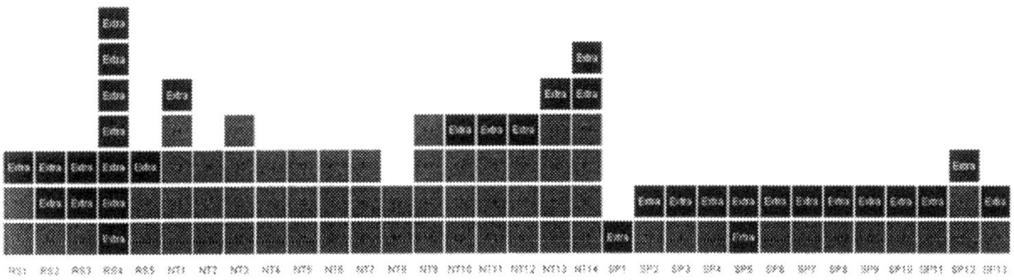

Figure 8. Load distribution on 32 computers for 64 blocks on the WAN cluster after 3 cyles of DLB.

Table 1. ADPAC elapsed time for DLB on LAN and WAN clusters for 64 blocks.

	LAN	*WAN*
Cost without DLB for 900 ADPAC time steps, hours	2.22	4.58
Cost with 3 cycles of DLB for 900 ADPAC time steps, hours	1.67	2.43
Improvement, %	25	47

6. CONCLUSIONS

From the timing results and speedups obtained, it is evident that parallel computations are less satisfactory than using either Pentium/NT or RS6K machines individiully as computational platform. Although, in some cases, the NT cluster has achieved better speedup, its computational speed is about two times slower than the RS6K cluster. On the other hand, the RS6K workstation is less efficient compared to the NT workstation for increased number of blocks, mainly due to the inability of the network speed to keep up with the faster processing speed of RS6K. This tradeoff between elapsed time and efficiency is typical when solving parallel codes as the number of blocks increases and the block sizes decreases for a given computational grid. Since the computational speed of each processor may be different, the parallel performance has substantially increased by employing the heterogeneous computing environment approach. In heterogeneous workstation, the communication speed may also vary due to the difference in communication bandwidth for different computer architecture. From our previous studies, other factors that affect the parallel performance could be attributed to the proximity of machines and block and interface sizes. In this study, these factors are found to have a different degree of impact on the parallel performance. The proximity of machines and blocks and interface sizes have added some communication weights to the existing communication routines in ADPAC, causing inefficient parallel performance to occur beyond 4 processors.

As for the DLB test runs, it is obvious that the load balancing leads to an improvement in total elapsed time of ADPAC computations. In the LAN case, balanced block distribution reflects performance and computational power of computers involved in this run. Since IBM RS/6000 computers are faster than PC/PII computers, there are more loads on these computers accordingly. Extraneous loads are taken into account in the new block distribution.

An improvement of 25% in elapsed time for the ADPAC computation was observed. In the case of the WAN environment, similar improvement in the ADPAC computation times was observed. It is obvious that the total improvement depends on the initial distribution. If it is already a balanced distribution, then one cannot expect any substantial improvement from the DLB. However, if no accurate information is available for the power of computers a priori, the best way is to use either an equal distribution or load only the local cluster. In the WAN application, other important issues that should be considered are the communication bandwidth and speed. In the present application, this is also taken into account in each new block distribution obtained via the DLB.

ACKNOWLEDGEMENTS

This research was supported by the NASA Glenn Research Center under Grant No: NAG3-2260. The authors would like to express their gratitude to NASA and Rolls-Royce Corporation for providing inviscid version of the ADPAC code in this research.

REFERENCES

1. E.J. Hall, R.A. Delaney, and J.L. Bettner, *Investigation of Advanced Counterrotation Blade Configuration Concepts for High Speed Turboprop Systems*, NASA Contractor Report CR-187106, May 1991.
2. A. Ecer, H.U. Akay, W.B. Kemle, H. Wang, D. Ercoskun, and E.J. Hall, "Parallel Computation of Fluid Dynamics Problems," *Computer Methods in Applied Mechanics and Engineering*, Vol. 112, 1994, pp. 91-108.
3. Y.P. Chien, A. Ecer, H.U. Akay, and F. Carpenter, "Dynamic Load Balancing on Network of Workstations for Solving Computational Fluid Dynamics Problems," *Computer Methods in Applied Mechanics and Engineering*, Vol. 119, 1994, pp. 17-33.
4. Y.P. Chien, J.D. Chen, A. Ecer, and H.U. Akay, "Dynamic Load Balancing for Parallel CFD on NT Networks," *Proceeding of Parallel CFD'99*, May 23-26, 1999, Edited by. D. Keyes, et al., Elsevier Science, Amsterdam, The Netherlands, (in print).

Parallel Computational Fluid Dynamics – Trends and Applications
C.B. Jenssen et al. (Editors)

Efficient techniques for decomposing composite overlapping grids

Stefan Nilsson[a*]

[a]Department of Naval Architecture and Ocean Engineering,
Chalmers University of Technology,
Chalmers tvärgata 8, 412 96 Göteborg, Sweden

Several different schemes for decomposing and distributing composite overlapping on parallel computers grids are described. Some of the difficulties connected with decomposing overlapping grids and guidelines for overcoming them are given. Some early experiences from implementing and testing the implemented schemes are reported.

1. INTRODUCTION

Composite overlapping grid methods are nowadays used routinely to solve problems from fluid mechanics in complex geometries, e.g. [1] and [2]. The perhaps most attractive feature of these methods is the relative ease with which moving and locally deforming geometries can be treated. This ease stems from the fact that fields defined on the component grids making up the composite grid are coupled only through interpolation operators, and hence the component grids can be moved more or less independently. Figure 1 shows an example of a composite overlapping grid. An excellent review of computations on composite overlapping grids is given in [3].

With the increasing dominance of distributed memory machines in the High Performance computing society of today, efficient algorithms are needed for decomposing the computational domain into sub-domains and distributing work evenly when using composite overlapping grids (hereafter referred to as overlapping grids). This paper describes some work done by us in this area of research.

2. PARALLEL COMPUTING ON OVERLAPPING GRIDS

Although several papers report successful uses of parallel computing on overlapping grids, e.g. [2,4,5], and several different decomposition schemes have been proposed, questions still remain regarding when and where each scheme should be applied. In this section we start by describing some of the difficulties encountered when working on this problem, and then present some solutions offered by other authors. Last we define the application domain of our use of overlapping grids and try to define the conditions to be fulfilled by a good decomposition scheme.

*The majority of this work was carried out while visiting the Edinburgh Parallel Computing Centre under the TRACS programme, sponsored by the EU's Training and Mobility of Researchers Programme under grant number ERB-FMGE-CT95-0051.

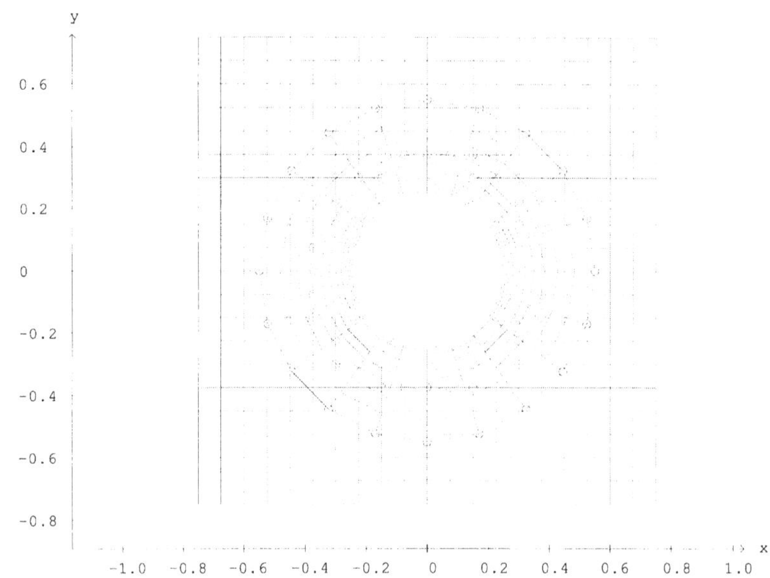

Figure 1. Example of simple overlapping grid consisting of two component grids. Circles indicate points where function values need to be interpolated from another grid

2.1. Inherent difficulties

When solving finite-difference problems on overlapping grids a mix of structured and unstructured operations are needed at each time-step/iteration. This mix is what stops the decomposition of the grids from being a straightforward procedure.

On the component grid level all grid points are either discretization points where structured finite-difference operators are applied or hole points where no computing is necessary, giving structured operations on a irregular regions. On the boundaries of discretization regions, interpolation operators are used to compute needed values from neighboring component grid, this is a totally unstructured operation.

Thus, the decomposition of overlapping grids presents some extra problems compared to the multi-block case which can be seen as a special case of overlapping grid with zero overlap between component grids (no hole points) and trivial interpolation operators.

2.2. Suggested strategies

Overlapping grid methods are basically ways to decompose a domain. Therefore, a natural coarse grained parallelism exists from the start, where one might partition entire component grids onto separate processors one by one. As the component grids have varying number of discretization- and hole points this easily leads to load imbalance among processors. However, this strategy can give excellent results when the number of component grids is very large. In [2] a code using this method is described, furthermore the decomposition algorithm used there tries to group component grids belonging to the same processor together geometrically, thereby reducing the amount of communication in the interpolation step. Modifications of this basic algorithm have been proposed by several authors to improve the performance, some of these are described in [6], where they are applied to multi-block grids.

At the other extreme we might partition each component grid individually over all processors. This automatically gives good load balance, but increases the amount of necessary communication among processors. To alleviate the latter problem, generalizations where the component grids and/or the processors are grouped together before partitioning can be used. The basic method as well as some variants are thoroughly analyzed, although mostly in a zero overlap context, in [7] and [8].

The two basic methods described above illustrates two different levels of obvious parallelism that can be made exploited when decomposing overlapping grids. Before we can discuss their respective merits and compare with other less obvious techniques, we need to decide what we should look for in an algorithm for overlapping grid decomposition.

2.3. What is a good decomposition scheme?

The applications of interest to us are computations of unsteady flows using finite-difference discretizations in geometries that might be dynamic. Linear equation systems appearing during in the solution process are solved using preconditioned Krylov sub-space iterative solvers.

Necessary communication among processors therefore involves: (a) global reduction operations needed for iterative solvers and global time-step determination, and (b) neighbor-neighbor communication needed for finite-difference- and interpolation operators. The overhead for doing (a) cannot be influenced by our decomposition algorithm. The overhead for the communication in (b) is dependent on the number of neighbors of each processor's sub-domain and the "length" of the border of the sub-domain, where the length is equal to the number of grid points along the border.

From (b) we see that in addition to the demand that a decomposition algorithm should balance the computational work among the processors we also need to minimize the number of grid points lying on the borders of each sub-domain and the number of bordering sub-domains.

Furthermore, for the cases where moving geometries are involved the partitioning algorithm used should be reasonably fast as we might need to recompute the partitioning when the grids move relative to each other. Algorithms which can take advantage of a good initial guess would be especially valuable in these cases.

3. IMPLEMENTATION AND STUDIES OF PERFORMANCE

A library written in C++ and including classes supporting computations on overlapping grids was coupled with the MPI-based PETSc-library [9], the latter used to ease the implementation of several different decomposition schemes. We also intend to use PETSc in the future to test different preconditioners in combination with the techniques described here.

3.1. Implemented decomposition schemes

Several decomposition schemes have been implemented and tested. The four examples given here represent techniques considered promising for different reasons. They are of two basic types:

Generalized single-grid partitioning This name is taken from [8]. Each component grid is decomposed separately among all processors. This ignores coupling due to the interpolation operators and is therefore easier to compute, but for the same reason we risk ignoring important information when doing our decomposition;

Composite grid partitioning Coupling between component grids is taken into account to some degree. Doing this should provide better quality decompositions, but could prove too costly in the cases where we have dynamic geometries.

The schemes implemented where:

1. **Normal partitioning :** Every component grid is divided into a number of blocks containing an equal number of grid points (hole and non-hole) so as to minimize the length of borders of the blocks, and these blocks are then partitioned over all processors [7]. Typically the number of blocks is equal to the number of processors. This is very efficient for moving geometries as the partitioning remains unchanged as the grids move. However, as we are not explicitly making sure that the number of non-hole grid points in every processors sub-domain is the same we could end up with badly load balanced decompositions.

2. **Recursive k-section partitioning :** To solve the problem with load imbalance that might occur using Normal partitioning, a recursive k-section was used on the component grids so that all partitions contained the same number of non-hole points [10,11]. Figure 2 shows an example. This method ensures good load balance, but introduces more complicated communication patters and is somewhat less straightforward to compute. When the grids move or deform the decomposition needs to be recomputed, [10] contains some suggestions on how this work can be minimized.

3. **Unstructured graph partitioning :** The complete overlapping grid (except for hole points) is treated as one totally unstructured grid with (directed) connectivity graph A. Then $(A+A^T)$ was decomposed using the METIS library [12] with weights on the graph edges used to simulate the un-symmetric connections between the component grids (due to interpolation). This is expensive, especially in the dynamic geometry case, but usually gives good decomposition quality. METIS also exists in a parallel version called ParMETIS, but this has not yet been tried.

4. **Block graph partitioning :** This can be viewed as a generalization of the unstructured graph partitioning scheme or an extension of the normal partitioning scheme, it was proposed in [6]. Each component grids is decomposed into a number of blocks, not necessarily the same as the number of processors. Then the computational weight, equal to the number of non-hole grid points, of every block is computed. Connectivity resulting from discretization and interpolation between

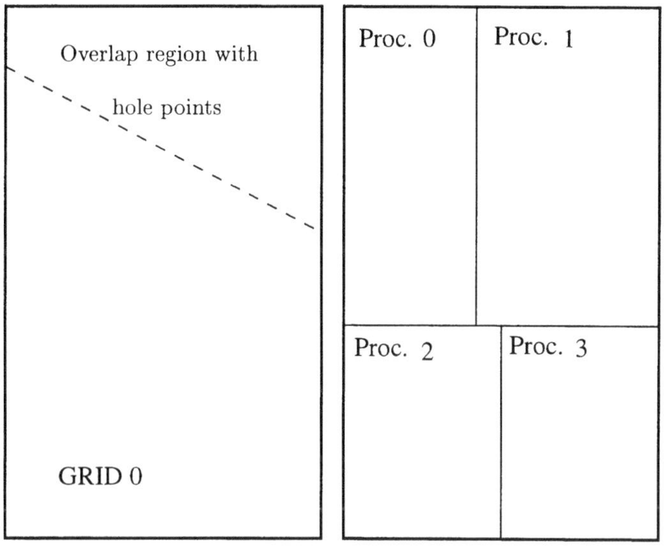

(a) Overlap situation for Grid 0

(b) Resulting decomposition for four processors

Figure 2. Component grid 0 is overlapped by other component grids with higher priority (a). In the overlap region, hole points occur. Decomposing component grid 0 among four processors using recursive k-sectioning can be done in two successive bi-sections in different directions (b).

blocks is also computed and the result is a graph with weights on both edges and vertices. This graph is then decomposed. Good approximations for the weights needed can be obtained quite cheaply, making the computational cost for this scheme tractable. In figure 3 a block graph is shown.

3.2. Measuring performance

To estimate the efficiency of each scheme we measured the amount of message passing going on during the solution process of our flow problem and then computed two numbers relevant to the discussion in 2.3. These numbers are:

1. computational load balance $\mathcal{C}$, defined as $\mathcal{C} = \lambda/\lambda_{max}$, where λ is the mean number of non-hole points in the sub-domains and λ_{max} the maximum number across all processors;

2. communication overhead $\mathcal{M}$, $\mathcal{M} = (\#_{neigh.}\alpha + \beta m)_{max}$, where $\#_{neigh.}$ is the number of neighboring sub-domains, m the length of the border as defined in 2.3, α and β the

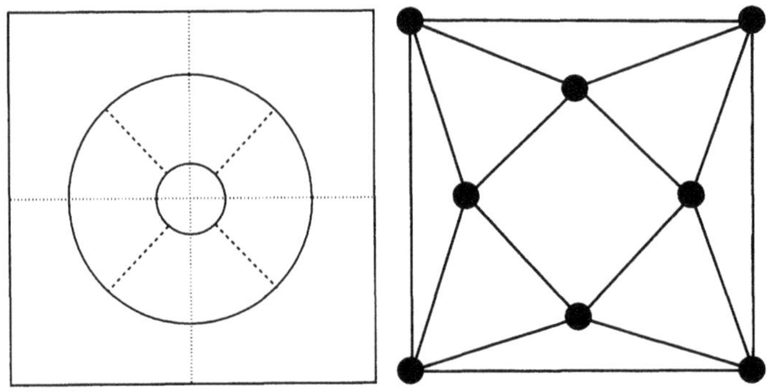

(a) Blocking of component grids (b) Resulting connectivity graph

Figure 3. If the component grids from figure 1 where divided into four blocks each (a), the graph shown in (b) would be the result.

communication latency and time to transfer one floating-point number respectively, and finally $(\ldots)_{max}$ the max value across all sub-domains.

The second number is admittedly a very naive estimate of the communication cost as it assumes there are no memory conflicts or contention for message passing resources in the network, and also that all communication needed for the complete overlapping grid is done simultaneously and not e.g. once for every component grid. Furthermore α and β are hardware-dependent, something which shows that the ideal scheme might vary depending upon our communication network. Nevertheless, this gives us some information on the performance of the different schemes. We used the values $(\alpha = 50, \beta = 0.01)$ μs, these being typical values for an MIMD type machine a few years ago [13].

3.3. Results

As a test case we used a simple overlapping grid containing four component grids and used for internal flow simulations. The total number of grid points was 171246, and of these 16071 where hole points. Computations where done for 16 and 45 processors. Table 1 shows the results.

As can be seen scheme 1 has the worst computational load balance, and also performs badly when communication overhead is concerned. The computationally most expensive scheme, number 3, has the overall best performance.

The reason scheme 1 and 2 have high $\mathcal{M}$ values is the big number of neighboring sub-domains they end up with because of the neglection of coupling between component grids done when computing them. These values are very dependent upon the type of network used in our parallel computer, and the figures given here should be taken only as examples. However, when the communication cost is broken down into separate pieces it is seen that

Table 1
Results using the four decomposition schemes from 3.1 Values for $\mathcal{M}$ have been normalized against the value for scheme 1

Decomposition	16 processors		45 processors	
scheme	$\mathcal{C}$	$\mathcal{M}$	$\mathcal{C}$	$\mathcal{M}$
1	0.904	1	0.880	1
2	0.989	0.919	0.966	1.19
3	0.971	0.434	0.971	0.377
4*	0.969	0.488	0.926	0.426

*Number of blocks per component grid was equal to ten times the number of processors

schemes based on composite grid partitioning have better behaviour for both $\#_{neigh.}$ and m.

4. CONCLUSIONS AND REMAINING WORK

We have implemented and tested a few different decomposition schemes for composite overlapping grids. To complete the evaluation more tests need to be run on a range of different (two- and three-dimensional) geometries.

In the case of dynamic geometries we need to decide the allowable amount of overhead connected with the computation of the decomposition. This will vary with the type of time-stepping used, (explicit or implicit), the physics simulated, etc.

Another interesting research area is the investigation of different "domain decomposition preconditioners" in combination with the schemes implemented so far.

REFERENCES

1. D. L. Brown, W. D. Henshaw and D. J. Quinlan: Overture: Object-Oriented Tools for Overset Grid Applications. AIAA conference on Applied Aerodynamics, UCRL-JC-134018 (1999)
2. R. L. Meakin, A. M. Wissink: Unsteady aerodynamic simulation of static and moving bodies using scalable computers. AIAA-99-3302-CP, 14th AIAA Computational Fluid Dynamics Conference, Norfolk, VA. (1999)
3. G. Chesshire, W. D. Henshaw: Composite Overlapping Meshes for the Solution of Partial Differential Equations. J. Comput. Phys. (1990) **90**
4. G. Chesshire, V. K. Naik: An environment for parallel and distributed computation with application to overlapping grids. IBM J. Res. Develop. (1994) **38**
5. C. S. Kim, C. Kim, O. H. Rho: Parallel Computations of High-Lift Airfoil Flows Using Two-Equation Turbulence Models. AIAA Journal (2000) **38**
6. J. Rantakokko: Data Partitioning Methods and Parallel Block-Oriented PDE Solvers. Ph. D. thesis, Uppsala University, Uppsala, Sweden. (1998)
7. M. Thuné: Straightforward partitioning of composite grids for explicit difference methods. Parallel Computing. (1991) **17**

8. M. Thuné: Partitioning strategies for composite grids. Parallel Algorithms and Applications. (1997) **11**
9. S. Balay, W. Gropp, L. C. McInnes, B. Smith: PETSc 2.0 Users Manual. ANL-95/11 - Revision 2.0.24, Argonne National Laboratory.
10. M. J. Berger, S. H. Bokhari: A Partitioning Strategy for Nonuniform Problems on Multiprocessors. IEEE Transactions on Computers. (1987)
11. M. Ashworth, R. Proctor, J. I. Allen, J. C. Blackford: Coupled Marine Ecosystem Modelling on High-Performance Computers. Presented at the Parallel CFD '99 conference , 23rd-26th May 1999, Williamsburg, Virginia, USA.
12. G. Karypis, V. Kumar: METIS: A Software Package for Partitioning Unstructured Graphs, Partitioning Meshes, and Computing Fill-Reducing Orderings of Sparse Matrices. Tech. Rep., University of Minnesota, Dep. Computer Science (1998)
13. L. Eldén, H. Park, Y. Saad: Scientific Computing on High Performance Computers. Working Manuscript (1998)

5. Tools and Environments

Parallel Computational Fluid Dynamics – Trends and Applications
C.B. Jenssen et al. (Editors)

Computer Load Measurement for Parallel Computing

Y.P. Chien, J.D. Chen, A. Ecer, and H.U. Akay

Computational Fluid Dynamics Laboratory
Purdue School of Engineering and Technology
Indiana University-Purdue University Indianapolis (IUPUI)
Indianapolis, IN 46202, USA

When multi-user computers are used for running parallel jobs, load balancing the computers is essential for taking the advantage of parallel and distributed processing. One important factor used in load balancing is the current load of the computers. The computer load information is commonly provided by operating systems (OS) and by load management systems. However, we found that the computer load information provided by the operating systems is misleading when there are tightly coupled parallel processes running on the system. The misleading information can cause a load balancer to suggest very poor load distribution for newly added parallel jobs. This paper explains why the OS provided load information may not be useful and provides a method to get the modified computer load information based on the OS provided information. The modified computer load information has been successfully used for load balancing of multiple parallel jobs on RS6000 workstations and Windows NT based computers. A load-balancing example based on the modified computer load information is presented to demonstrate the usefulness of the method.

1. INTRODUCTION

As parallel and distributed computing becomes common, the efficient use of shared computer resources becomes an important issue [1, 2]. Many factors need to be considered for load scheduling and balancing of parallel processes. One important factor is the current load of the computers. The computer load is usually defined as the runnable and non-waiting processes. All multi-user operating systems (e.g., UNIX, Windows/NT) provide the current and past computer load information [3,4]. Most parallel computing load management systems (e.g., LSF) also provide the load information of all computers. The method used in all these load measurements is the statistical average of periodically measured load in different time periods [5].

In a recent load balancing study, we noticed that the statistical computer load information is useful for load balancing of a parallel job only under the condition that there is no other mutually dependent parallel jobs running concurrently. When there are mutually dependent parallel processes running on the computers, the OS provided load information of some computers is misleading and cause the load scheduling and balancing program to suggest poor load distributions for parallel jobs. In this paper, we analyze this problem and suggest a method to modify the computer load information so that it can be useful for future load balancing [6,7]. A load-balancing example for three concurrently running but independent parallel CFD jobs (each job has multiple processes) is described to demonstrate the effectiveness of the new computer load information.

2. PROBLEM DESCRIPTION

The problem of OS provided computer load information can be demonstrated by the following experiments: A parallel computation fluid dynamics (CFD) problem has 30 approximately equal sized data blocks. These data blocks are distributed to 5 identical IBM RS6000 workstations (as shown in the second line of Table 1). These computers are dedicated for this experiment and there are no other users or processes on these computers. The parallel CFD program is executed for 100 time steps and the computer load is measured during the program execution. The method used in the load measurement is the statistical average of periodically measured load. The result of load measurement is depicted in the third row of Table 1. After adding an independent sequential job to the lightly loaded computer 1, the parallel CFD program is executed again for 100 time steps and the computer load is again measured during the program execution. The measurement result is listed in the fourth row of Table 1.

Table 1. Computer load measurements.

Computer Number	**1**	**2**	**3**	**4**	**5**
Actual number of processes	6	4	5	5	10
Measured load without extraneous load on any computer	1.50	0.57	0.96	2.68	9.12
Measured load after adding an extraneous load on computer 1	4.42	1.52	0.78	1.82	9.24

Table 1 shows that the actual numbers of processes running on most computers are far different from the measured computer loads during the first execution of the parallel CFD program. The load information is obtained by using *ps* command in UNIX and by reading registry table in Windows NT. The measured load is approximately the same as the actual load only for the most heavily loaded computer 5. In the second execution of the parallel CFD program, although only one extraneous process is added to computer 1, the measured number of runnable load jumped from 1.5 to 4.42. The worse part is that the load measurements of other computers are also severely affected. However, the measured computer load never exceeds the actual number of processes on the computer in all of our test experiments. After making sure that there is no error in the load measurement, we are convinced that the dependency between the parallel CFD processes caused such phenomena.

This phenomenon can be explained as follows. In this test case, the sizes of data blocks are similar, the speeds of computers are the same, and the load distribution is unbalanced. Therefore, the computation of the processes on the lightly loaded computers finishes earlier than that on the heavily load computers. During the parallel CFD execution, each process exchanges boundary conditions with neighbor processes in every time step. Therefore, the processes on the lightly loaded computers wait for the boundary information from the processes on the heavily load computers in every time step. When the processes are on the waiting queue, they are not counted as the computer load by the statistical periodic load measurement. On the most heavily loaded computer, the parallel processes do not wait for information from neighbor processes on other computers so that the parallel processes on the most heavily loaded computer always stays in the running queue and do not go to the wait queue.

This computer load measurement problem does not affect scheduling of sequential jobs. In most OS supported job-scheduling systems, each job is treated as a sequential job. The job scheduler adds new processes to the lightly loaded computers one by one. Even though the measured computer load does not change linearly to the actually added new processes, it does show which computer is lightly loaded. Therefore, the measured load information can still be used.

This computer load measurement problem only affects the load balancing of multiple parallel jobs. When scheduling parallel jobs, several processes can be added to an apparently lightly loaded computer. Since the actual load change does not correspond to the actual number of processes added, the scheduler can severely overload the lightly loaded computer. Without knowing exactly how the parallel CFD processes exchange information during execution, it is difficult to predict how the newly added processes affect the computer load just based on the currently measured loads of all computers.

3. SOLUTION

To prevent load balancer of parallel jobs from overloading lightly loaded computers, the reliable load information should be obtained. Since there is no inter-process communication for sequential processes, the actual number of sequential load should be the same as the measured load of the sequential jobs (assuming the sequential load is computation bound). Since the measured computer load corresponding to the parallel processes is unreliable, we choose to record the actual number of parallel processes on each computer. A parallel load registry mechanism is developed and installed on each computer. Whenever a parallel job is started, the actually number of parallel processes on each computer is registered on that computer. Whenever a parallel job is stopped, the registration of all of its processes on any computer is deleted. The load of each computer is modified through the following steps:

1. Obtain the list of all computer loads on each computer using OS supported tools.
2. Count the sequential loads from the list (the names of the parallel loads are known)
3. The number of loads on each machine is the sum of the counted sequential loads and recorded parallel loads.

This modified load information takes into the account the dependencies of tightly coupled parallel processes. In fact, it assumes that all parallel processes do not need to wait for information from other processes. This assumption is valid when the parallel process load on all computers are balanced. Since the modified load information never under count the load on the computer, the load balancer will not overload any computer due to the input load information of computers. This modified load information should not be used to describe what the computer load was in the past. It is derived used for future load balancing.

4. IMPLEMENTATION

In order to keep tracking computer and network load information, we developed a JAVA program called the System Agent. One component of the System Agent is to find out the number of sequential jobs running on the computer. We call this component PTrack. One System Agent is executed on each computer. To keep track of the number of parallel processes on each computer, we also developed a parallel job load balancer and dispatcher, called the DLB Agent (Dynamic Load Balancing Agent) in JAVA. A DLB Agent is associated with each parallel job. Users must submit their parallel jobs to the computers through a DLB agent. The relationship between the DLB Agent, the System Agent, and PTrack is depicted in Figure 1.

The DLB agent of a parallel job uses the AddJob function to tell the System Agent of all the computers used and how many parallel processes are added to that computer. The name and number of all parallel processes are stored in the JobTable of the System Agent. PTrack periodically reads the JobTable and obtains the names of all parallel processes running on the computer. PTrack also periodically gathers the list of computer loads by using OS provided utility functions. By removing the known parallel processes from the list of measured computer load, PTrack counts the number of sequential processes on the computer. The pseudo code of PTrack is shown on the next page.

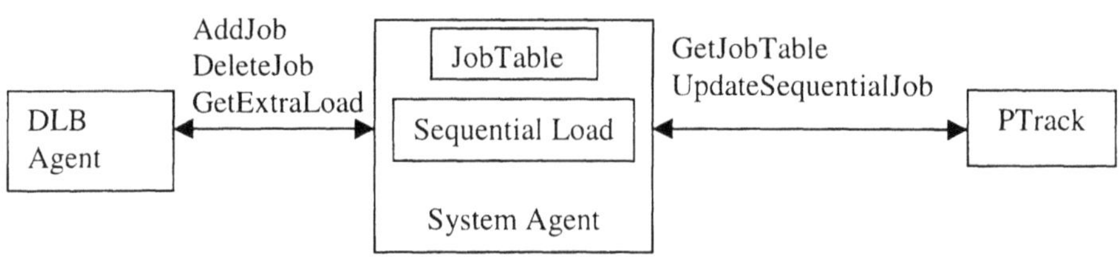

Figure 1: Software tools for load measurement.

```
class PTrack extends Thread
{
   Vector jobTable = new Vector();
    public void run()
    {
        int sLoad = 0;
        int i = 0;
        for(;;)
        {
            i++;
            //Get the names of the parallel processes on this computer
            jobTable = SystemAgent.GetJobTable();

            if(i==50)  //Report load measurement in every 50 iterations
            {
                SystemAgent.UpdateSequentialLoad((float)sLoad/i);
                i = 0;
                sLoad = 0;
            }
            else  //make statitical average
            {
                    sLoad = sLoad + sequentialLoadCounter();
            }

            //wait a while before next measurement
            Thread.sleep(1000);
        } // end of for loop
    }

    private int sequentialLoadCounter()
    {
        - Use system related method, such as ps command in UNIX system,
        to find all runnable processes.
        - If there are some parallel load included (parallel load can be found in JobTable),
        don't count them.
        - Return the number of the sequential load;
    }
}
```

The DLB Agent and the System Agent have been successfully tested on IBM RS6000 and Windows/NT based Pentium computers.

5. DEMONSTRATION EXAMPLE

To demonstrate that the modified computer load measurement method is useful for computer load balancing, a dynamic load balancing case is shown here. Three parallel jobs (job 1, 2, and 3) use the same set of five IBM RS6000 computers. These jobs are load balanced independently by three DLB Agents. The start and stop of any parallel job is independent of other parallel jobs. In the experiment, each parallel job is load balanced 10 times. Figure 2 shows the elapsed execution time of all three parallel jobs after each balance cycle.

In Figure 2, the horizontal direction is the time axis. The vertical direction is the elapsed execution time (in the unit of seconds) per 100 time steps. As may be observed in Figure 2, each job initially tried to distribute its processes to all computers evenly. Job 1 starts at time 16:45, job 2 starts at 16:46 and Job 3 starts at 17:00. Before the second job was added to the computers, the load balancer for the first job tries to load balance Job 1. After each new job was dispatched to the computers, the elapsed execution time of currently running jobs increases considerably. After Ptrack detects the new processes on the system, the new load information is derived. The new load information is used for the next load balancing cycle. The load balancer of each job load balances its responsible parallel job. Since the load balancer works well, the new PTrack does provide useful computer load information to the load balancer in an environment of multiple parallel jobs on the same computers.

6. CONCLUSION

It has been determined that the computation load information provided by the operating systems is not useful for load balancing of parallel jobs. A new method is developed to provide meaningful computation load information for computers running parallel jobs. The output of the new load measuring method was successfully used in the load balancing of parallel CFD cases.

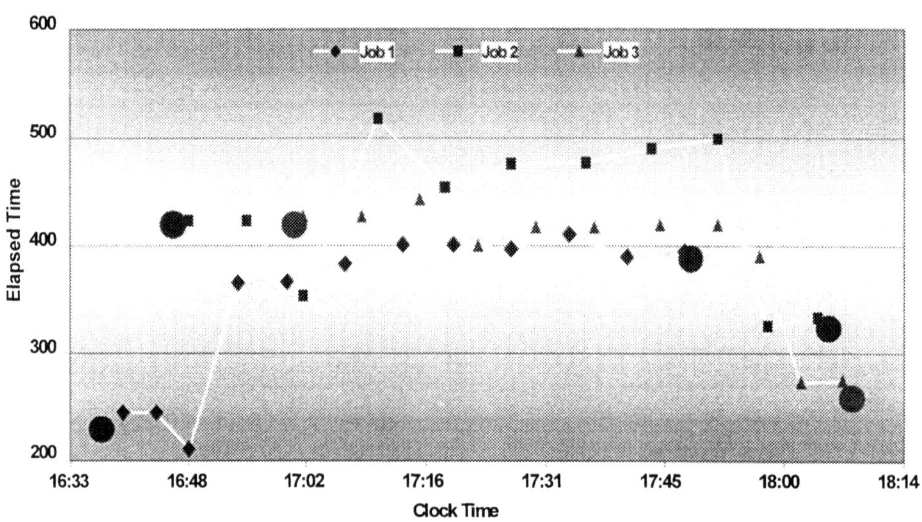

Figure 2. Dynamic load balancing based on the modified computer load information.

ACKNOWLEDGEMENT

Financial support provided for this research by the NASA Glenn Research Center (Grant No. NAG3-2260) is gratefully acknowledged. The authors are also grateful for the computer support provided by the IBM Research Center throughout this study.

REFERENCES

[1] T.E. Tezduyar, S. Aliabadi, M. Behr, A. Johnson, and S. Mittal, 'Parallel Finite Element Computation of 3D Flows,' *IEEE Computer*, 1993, pp. 27-36.
[2] D. Williams, 'Performance of Dynamic Load Balancing Algorithms for Unstructured Grid Calculations,' *CalTech Report*, C3P913, 1990.
[3] K.H. Rosen, R.R. Rosinski, and J.M. Farber, *Unix System V Release 4: An Introduction*, Osborene McGraw-Hill, 1990.
[4] E. Frisch, *Essential Windows NT System Administration*, O'Reilly, 1998.
[5] *LSF User's and Administrator's Guide*, Platform Computing Corporation, 1993.
[6] Y.P. Chien, F. Carpenter, A. Ecer, and H.U. Akay, 'Computer Load Balancing for Parallel Computation of Fluid Dynamics Problems,' *Computer Methods in Applied Mechanics and Engineering*, Vol. 120, 1994, pp. 119-130
[7] Y.P. Chien, A. Ecer, H.U. Akay, and F. Carpenter, 'Dynamic Load Balancing on Network of Workstations for Solving Computational Fluid Dynamics Problems,' *Computer Methods in Applied Mechanics and Engineering*, 119, 1994, pp. 17-33.

Parallel Computational Fluid Dynamics – Trends and Applications
C.B. Jenssen et al. (Editors)

Numerical Algorithms and software tools for Efficient Meta-Computing

M. Garbey [a] M. Hess [b], Ph. Piras[a,b], M. Resch[b], D. Tromeur-Dervout[a] *

[a] Center for the Development of Scientific Parallel Computing, CDCSP/ISTIL - University Lyon 1, 69622 Villeurbanne cedex,France

[b] High Performance Computing Center Stuttgart (HLRS), Allmandring 30, D-70550 Stuttgart, Germany

This paper reports the development of a numerical Aitken-Schwarz domain decomposition method associated with the PACX-MPI message passing library, to perform the computation of a Navier-Stokes model in velocity-vorticity formulation in a metacomputing framework.

1. Introduction

The development of cost affordable parallel computer hardware architecture evolves in less and less integration of the components in a same location (CPU, cache, memory). From the shared memory architectures of the early 80s (Cray XMP), the architecture went to the distributed memory in the 90s (Intel Paragon, Cray T3E), and has been followed in the end of 90s by a hybrid hardware architecture as clusters built of shared memory systems linked by a dedicated communication network (Dec Alpha architecture with memory channel). One can further extend the concept by considering metacomputing environments, where the architecture can be a constellation of supercomputers linked through national or international communication networks. Indeed, large scale computing on a network of parallel computers seems to be mature enough from the computer science point of view to allow experiments for real simulations. Nevertheless, the old but readily problem of designing efficient parallel programs for such architectures is still: to reduce or to **relax the time to access to the distant required data**. This constraint is quite strong because there is at least an order of 2 of magnitude between the communication bandwidth inside the parallel computers and the communication bandwidth of the network linking the distant parallel computers. One can argue that the real improvement of the distant network will overcome this constraint. Actually dedicated links of some hundred megabyte/s can be used and some research projects expect a dedicated bandwidth of some gigabyte/s[1]. But, as the final objective is to run real applications involving hundred of processors in a non-dedicated environment, the communication bandwidth will be typically shared and so reduced and in fact the problem of the difference of the communication bandwidth performance for intercommunication and intracommunication will be still persistent.

*The work of CDCSP was supported by the Région Rhône Alpes.

Obviously, this kind of computation that requires huge computational and network resources will be useful for time critical applications where the simulation is conducted to provide data as a basis to make the right decisions. Concentration of human forces and sea barrier resources on seaboard in case of pollution by oil, computing of risks to choose burying sites for nuclear waste, determining risks of contaminated areas in case of air dead pollution to save the population, are examples for such critical applications.

This paper reports some first findings on how to realize this kind of metacomputing application, the problems encountered, and the users environments that need to be developed. Efficient metacomputing needs the development of communication tools as well as development of numerical algorithms in order to take advantage of the hardware of each MPP and to minimize the number and the volume of communications on the slow network.

We address as an example the challenge of metacomputing with two distant parallel computers linked by a slow affordable network providing a bandwidth of about 2-10 Mbit/s and running the numerical approximation of a Navier-Stokes model. By solving an elliptic problem we show that efficient metacomputing needs to develop **new parallel numerical algorithms** for such multi-cluster architectures. On top of the local parallel algorithms within each cluster, we develop new *robust and parallel* algorithms that work with several clusters linked by a slow communication network. We use a two level domain decomposition algorithm that is based on an Aitken or Steffensen acceleration procedure combined to Schwarz for the outer loop and a standard parallel domain decomposition for the inner loop [7].

2. Coupling tools for metacomputing

As far as we know several tools exist to couple applications on different hardware architectures [2,3]. Most applications run one physical model on each architecture and exchange the needed fields between the applications. Several coupling softwares have been developed based on the Corba technology [4]. Although this technology has been successful in industrial applications to integrate several components of the industrial process, we do not use this approach because of the following reason: first we are in this case interested to run one single application, and to distribute it on several distant computers. The Corba technology uses a quite complex and time consuming protocol for security that is not essential when doing parallel computation on a single scientific application where both data type and pattern are quite regular even on irregular mesh. Second, this approach needs to develop an overlayer of communications of the original code that can be quite difficult to program. A large part of people that deal with huge computation, as mechanical engineers, applied mathematicians and physicians, know quite well about procedural language like C or Fortran and about communication libraries like PVM or MPI. However, they are typically not so familiar with object oriented concepts as used in Corba.

Although these coupling tools are available to couple applications, some numerical developments are still necessary in order to obtain good performances in a metacomputing environment. We develop for this purpose algorithmic schemes called $C(p, q, j)$ to solve coupled system of PDE's [9].

In the case of one single application on which we focus here, some software as LSF allows to manage an application running on several distant resources. Although these facilities to distribute processes on physical distant processors are available, it is not possible at the present time to specify the location of a specific process on a processor of a specific parallel computer involved in the computing pool. This capability is of leading importance in the metacomputing framework in order to relax the constraints on communication of the numerical algorithm between distant processors. The users must be able to distribute processes precisely according to his needs.

The HPC Center of Stuttgart (HLRS) provides a communication library called PACX-MPI [13][10] that allows to run MPI applications on meta-computers [11][12]. PACX-MPI concentrates on expanding MPI for usage in an environment that couples different platforms. With this library the communication inside each MPP relies on the highly tuned MPI-version of the vendor, while communication between the two MPPs is based on the TCP/IP standard protocol. For the data exchange between the two MPPs each side has to provide 2 extra nodes for the communication, one for each direction. While one node is always waiting for MPI commands from inner nodes to transfer to the other MPP, the other node is executing commands received from the other MPP and hands data over to its own inner nodes. This tools is well adapted to multi-cluster parallel computer architectures.

3. An application example : elliptic solver of a Navier-Stokes code

As an example of computation, we focus on the numerical solution of a Navier-Stokes model written in velocity-vorticity ($V-\omega$) formulation. From the numerical point of view, we have to solve six equations with 3D fields of unknowns: three equations are devoted to the transport of the vorticity and three other are equations devoted to the computation of the divergence free velocity field. Computations of the three velocity fields equations are the most time consuming part of the elapsed time because corresponding residuals must be kept very low in order to satisfy the divergence free constraint. Mathematically the PDE problem is written as follows:

$$\frac{\vec{\omega^{n+1}} - \vec{\omega^{n}}}{\Delta t} - \vec{\nabla} \times (\vec{V^{n}} \times \vec{\omega^{n+1}}) = \frac{1}{Re} \nabla^2 \vec{\omega^{\mathbf{n+1}}} \tag{1}$$

$$\nabla^2 \vec{\mathbf{V}^{\mathbf{n+1}}} = \vec{\nabla} \times \vec{\omega^{n}} \tag{2}$$

We used 2nd order finite differences in space and 2nd order in time. The time marching scheme for the vorticity is the ADI scheme with correction of Douglas [5]. The velocity is solved with a classical Multi-grid method but with a dual Schur complement domain decomposition to solve the coarser grid problem [6] for parallel efficiency purpose.

The 3D lid driven cavity is the target application. The geometry under consideration now is rectangular ($(0,3) \times (0,1)^2$) but one might extend the methodology to complex geometries with fictitious domain methods. The goal is to solve this problem between distant parallel computers.

As we already have an MPI code that treats this lid driven cavity problem, we want to split the domain of computation in macro domains with a variable overlap area. Each macro domain is associated with one pool of processors belonging to the same parallel computer. We apply Schwarz additive domain decomposition to solve the velocity equation, using a Multigrid solver as the inner solver on each macro domain. Thus the intercommunication corresponds to the communications involved in the Schwarz algorithm. The other communication for the inner solver on each macro domain is only between processors belonging to the same computer. We use PACX-MPI to provide intercommunication in order to use the best communication network where possible. An overlap area of one mesh size is numerically sufficient for the Aitken Schwarz methodology under consideration that is developed belows.

4. Aitken-Schwarz Domain Decomposition

It is well known that Schwarz algorithm is a very slow but robust algorithm for elliptic operators. The linear convergence rate depends only on the geometry of the domain and on the operator. We develop a numerical procedure, called Aitken-Schwarz, to accelerate this algorithm based on the linear convergence property of the method and the separability of the elliptic operator . This Aitken-Schwarz Domain Decomposition was first presented at the Parallel CFD 99 conference [7] and was extended to more general linear and non linear problems at Domain Decomposition DD12 conference [8]. We keep the sequel of traces of few Schwarz iterates solution at the artificial interfaces. As the convergence of the error between these traces of iterate solutions and the exact solution is purely linear, Aitken acceleration may compute the trace of the exact solution at the artificial interfaces between the subdomains of the domain decomposition.

Let us exemplify the algorithm for the velocity solver of Navier-Stokes in velocity-vorticity formulation. Our Poisson problem writes $u_{xx} + u_{yy} + u_{zz} = f$ in the square $(0,3) \times (0,1)^2$ with homogeneous Dirichlet boundary conditions. We partition the domain into two overlapping stripes $(0,a) \times (0,1)^2 \bigcup (b,3) \times (0,1)^2$ with $b > a$. We introduce the regular discretization in the y and z direction $y_i = (i-1)h_y$, $h_y = \frac{1}{N_y - 1}$, , $z_j = (j-1)h_z$, $h_z = \frac{1}{N_z - 1}$, and central second-order finite differences in the y and z directions. Let us denote by $\hat{u}_{ij}$ (respect.. $\hat{f}_{ij}$) the coefficient of the sine expansion of u (respect.. f). The semi-discretized equation for each sinus wave is then

$$\hat{u}_{ij,xx} - (4/h_y^2\, sin^2(i\frac{h_y}{2}) + 4/h_z^2\, sin^2(j\frac{h_z}{2}))\ \hat{u}_{ij} = \hat{f}_{ij}, \tag{3}$$

The algorithm for this specific case is the following: First, we compute three iterates with the additive Schwarz algorithm. Second, we compute the sinus wave expansion of the trace of the iterates on the interface Γ_i with fast Fourier transforms. Third, we compute the limit of the wave coefficients sequence via Aitken acceleration separately for each interface:

$$\hat{u}^{\infty}_{kl|\Gamma_i} = \frac{\hat{u}^0_{kl|\Gamma_i}\hat{u}^3_{kl|\Gamma_i} - \hat{u}^1_{kl|\Gamma_i}\hat{u}^2_{kl|\Gamma_i}}{\hat{u}^3_{kl|\Gamma_i} - \hat{u}^2_{kl|\Gamma_i} - u^1_{kl|\Gamma_i} + u^0_{kl|\Gamma_i}}. \tag{4}$$

A last solution step in each subdomain with the new computed boundary solution gives the final solution. We must notice that a better implementation of the method can be achieved: three Schwarz iterates are needed if one takes into account the coupling $(\hat{u}_{kl|\Gamma_1}, \hat{u}_{kl|\Gamma_2})$ between interfaces and accelerates globally $(u_{kl|\Gamma_1}, u_{kl|\Gamma_2})$. Two Schwarz iterates are necessary if one computes analytically the amplification vector of each wave $\hat{u}_{kl|\Gamma_i}$ [8]. In the lid driven cavity problem, we adapt the Aitken-Schwarz method to solve the three elliptic equations with non homogeneous boundary conditions of the velocity . In this problem we have one non homogeneous boundary condition for the 2nd velocity component on the boundary $z = 1/2$ corresponding to the moving boundary. We shift the solution by Φ analytical solution of :

$$\Delta\Phi = 0 \ on \ \Omega, \quad \Phi_{|\partial\Omega-\{z=1/2\}} = 0, \quad \Phi_{|z=1/2} = 1. \tag{5}$$

and retrieve a numerical problem with homogeneous boundary conditions.

5. Numerical results and performances

The target computers that we used for these firsts experiments are 4 alpha servers 4100 with 4 processors ev5 400 Mhz gathered in a Digital Tru64 cluster named MOBY-DECK with an internal latency of $5\mu s$ and a bandwidth of 800 Mbit/s.

5.1. Performance of the code in a metacomputing context

First we report on an experiment that actually demonstrates the need of new algorithms like Aitken-Schwarz for metacomputing. We consider 3 Laplace solves and a total of 197000 unknowns. Two first rows of Table 1 show that with a good spreading of the macro domains, that is one macro domain on each parallel computer, we can lower the communication bandwidth between these two computers without loosing too much efficiency. On the other hand if we split the macro domain between two distant computers then the elapsed time increase by a factor of 10. As a matter of fact, the communication of the inner solver (multigrid) much more than 10 Mbit/s to be efficient.

Second, we report on an experiment to solve with Aitken-Schwarz the 3 components of velocity using MPICH between a SUN Enterprise 10000 located at ENS Lyon, MOBY-DECK and a Compaq DS20 located at University of Lyon, connected with a 2 Mbit/s non dedicated link between these 3 computers. The total number of unknowns in this last experiment is 288000. Processors of MOBY-DECK are 30% faster than the SUN Enterprise 10000 for this application. Table 2 shows that the elapse time increases only by 7.4% when degrading the communication bandwidth by a factor of 50.

5.2. Overhead induced by PACX on a single MPP

The MPICH drawback is to do not use the best MPI implementation on the machine. For example on MOBY-DECK, MPICH used the 100 Mbit/s (theoritical) communication bandwidth on FDDI communication network but not the 800 Mbit/s (theoritical) communication bandwidth of memory channel communication network. In this section, we show that PACX-MPI communication software allow us to overcome this difficulty and we do not consider the overall performance of the numerical method. The experiment has been run with the full NS code describe in section 3. In PACX-MPI each parallel computer is considered as a partition and two added processors per partition are devoted to the communication between partitions. We call in the following such communications as external communications and communications inside a partition as internal communications.

We timed on MOBY-DECK point to point communications with one partition of the machine. The overhead introduced by PACX-MPI for internal communication is around 3 microseconds for the latency and the bandwidth is reduced of about 3%.

We next considered two partitions of the machine and a communication from one processor belonging to one partition to a processor belonging to the other partition. Then an additional overhead for communication is induced because of the access to the TCP protocol. An effective communication bandwidth of 60 Mbit/s and a latency of 1,1 ms as been measured for these inter partitions communications with PACX-MPI.

Table 3 gives the time in seconds of external communications and the time for the total communication (the sum of internal and external communications) for a domain split into two macro domains of 2×2 processors. The MPI row considers one partition of the machine and the vendor MPI communication library with 8 processes. The PACX-MPI row considers two partitions of the machine with PACX-MPI library software. In this last case 4 processes for the communication between partitions have been added to the 8 processes needed. The overhead induced by PACX on the total communication is almost the overhead of the external communication. We can hardly see any difference for the internal communication. The low overhead in the time of resolution between the 2 tests demonstrate the suitability of the use of PAXC-MPI in a metacomputing context.

Table 4 compares the communication elapsed time of the classical multigrid solver on one macro domain (without domain decomposition for the coarser grid problem) with the communication elapsed time of the Aitken-Schwarz domain decomposition method to solve the three velocity equations. The first method used MPI Vendor on one partition of the machine, the second used PACX-MPI communication library on two partitions of the machine with one macro domain each. Two global sizes of the computational domain $125 \times 32 \times 64$ and $125 \times 63 \times 64$ have been considered. The time of communication for the velocity is more important for the Aitken-Schwarz code than for multigrid solver on one macro domain ($125 \times 64 \times 64$) case. The communications time of one inner solver iterate in the whole domain competes with the communications time of 4 inner solver iterates in reduced domains. When the number of communications in the inner solver increases (two times more for the ($125 \times 63 \times 64$) case than for the ($125 \times 32 \times 64$) case) Aitken Schwarz seems to be less competitive because of the 4 solves of Aitken-Schwarz. Nevertheless, this implementation with (4) is not the optimal one [8].

6. Conclusions

This paper demonstrates the need for new algorithms associated with efficient communication software to solve numerical PDEs in a metacomputing framework. The experiments shown still use only a few processors, but experiments using more than 1000 processors are under investigation with the Aitken-Schwarz methodology and PACX-MPI.

We must report some difficulties that occur in the metacomputing framework, in order to indicate some tracks of development. The main problem encountered in the development of this experiment is the code validation on different platforms due to the difference of the MPI vendor capability depending on the hardware resources. For example, we developed a version of the code using the buffered non-blocking communication code, that runs without problems on a DEC Alpha but is limited to small problem sizes on the SUN Enterprise 10000 of ENS Lyon. At the present time, our metacomputing experiments require quite heavy management of source files, data files and makefiles on several distant computational sites. The software environment of each machine, scientific library, compiler optimization, development tools and batch system lead to a capitalization of the know-how on several architectures. However, we need to develop :

• generic procedures of compiling that adapt themselves to the software and hardware environment.

• automatic management of source codes, data files, and job submissions as provided in TME [14].

• tools that give an accurate image of the network or computer resources availability. Notably, the time of latency, the average value and real time value of the communication bandwidth between two distant computers, the load of the network.
We are currently investigating the development of such tools.

REFERENCES

1. T. Eickermann, H. Grund and J. Henrichs: "Performance Issues of Distributed MPI Applications in a German Gigabit Testbed", In: Jack Dongarra, Emilio Luque, Tomas Margalef, (Eds.), 'Recent Advances in Parallel Virtual Machine and Message Passing', Lecture Notes in Computer Science, 180-195, Springer, 1998.
2. Ian Foster, Carl Kesselman: "GLOBUS: A Metacomputing Infrastructure Toolkit", International Journal of Supercomputer Applications, 11, 115-128, 1997.
3. Andrew Grimshaw, Adam Ferrari, Fritz Knabe, Marty Humphrey: "Legion: An Operating System for Wide-Area Computing", Technical Report, University of Virginia, CS-99-12, 1999.
4. P. Beaugendre, T. Priol, G. Allon, and D. Delavaux. A Client/Serveur Approach for HPC Applications within a Networking Environment. In Proc. of HPCN'98, number 1401 in LNCS, Springer Verlag, pages 518–525, Amsterdam, Pays-Bas, April 1998.
5. J. Douglas Jr: " Alternating direction methods for three space variables ", Numerische Mathematik 4, pp. 41-63, 1962.
6. Edjlali G., Garbey M., Tromeur-Dervout D.: Interoperability Parallel Programs Approach To Simulate 3d Frontal Polymerization Processes . Journal of Parallel Computing, Vol 25, pp.1161-1191, 1999
7. Garbey M., Tromeur-Dervout D.: "Operator splitting and Domain Decomposition for Multicluster", 11th International Conference Parallel CFD99, D. Keyes, A. Ecer, N. Satofuka, P. Fox, J. Periaux Editors, North-Holland publisher ISBN 0-444-50571-7, Williamsburg, pp. 27-36, 1999.
8. Garbey M., Tromeur-Dervout D.: "Two level Domain Decomposition for Multiclusters", 12th International Conference on Domain Decomposition Methods DD12, T. Chan, T. Kako, H. Kawarada and O. Pironneau Editors, DDM org publisher, 2000, http://applmath.tg.chiba-u.ac.jp/dd12/proceedings/Garbey.ps.gz
9. Garbey M., Tromeur-Dervout D.: "A Parallel Adaptive Coupling Algorithm for Systems of Differential Equations", J. Comp. Phys., Vol 161, pp.401-427, 2000
10. Edgar Gabriel, Michael Resch, Thomas Beisel, Rainer Keller: "Distributed Computing in a heterogenous computing environment", in Vassil Alexandrov, Jack Dongarra (Eds.) 'Recent Advances in Parallel Virtual Machine and Message Passing Interface', Lecture Notes in Computer Science, Springer, 1998, pp 180-188.
11. Matthias A. Brune, Graham E. Fagg, Michael Resch: "Message-Passing Environments for Metacomputing", Future Generation Computer Systems (15)5-6 (1999) pp. 699-712.
12. Michael Resch, Dirk Rantzau and Robert Stoy: "Metacomputing Experience in a Transatlantic Wide Area Application Testbed", Future Generation Computer Systems (15)5-6 (1999) pp. 807-816.
13. Thomas Beisel , Edgar Gabriel, Michael Resch: "An Extension to MPI for Distributed Computing on MPPs", in Marian Bubak, Jack Dongarra, Jerzy Wasniewski (Eds.) 'Recent Advances in Parallel Virtual Machine and Message Passing Interface', Lecture Notes in Computer Science, Springer, 1997, 75-83.
14. "TME (Task Mapping Editor), Japan Atomic Energy Research Institute (JAERI), JAERI-Data/Code 2000-010, Tokyo, 2000.

Cluster 1 3 or 4 processors	Cluster 2 2 or 3 or 4 processors	Elapse time in second	Bandwidth of network
4 ev5 CDCSP-MOBY	4 ev5 CDCSP-MOBY	28.4s	100 Mbit/s
4 ev5 CDCSP-MOBY	2 ev6 CDCSP-DS20	29.4s	10 Mbit/s
2 ev5 CDCSP-MOBY 1 ev6 CDCSP-DS20	2 ev5 CDCSP-MOBY 1 ev6 CDCSP-DS20	220.7s	10 Mbit/s

Table 1
Performance of the analytical Aitken Schwarz algorithm on intranet.

Cluster 1 4 processors	Cluster 2 4 processors	Cluster 3 4 or 2 processors	Elapse time in second	Bandwidth of network
4 PSMN-SDF1	4 PSMN-SDF1	4 PSMN-SDF1	28.8 s	not available
4 ev5 CDCSP-MOBY	4 ev5 CDCSP-MOBY	4 ev5 CDCSP-MOBY	20.7s	100 Mbit/s
4 PSMN-SDF1	4 ev5 CDCSP-MOBY	2 ev6 CDCSP-DS20	31.2s	2 Mbit/s

Table 2
Performance of the analytical Aitken Schwarz algorithm on City's Network.

	Vorticity		Velocity	
	External com	Total com	External com	Total com
MPI	0.88	1.36	0.29	10.0
PACX+MPI	1.00	1.5	0.7	10.4

Table 3
Overhead induced by PACX on a single MPP

Velocity	$Macro_x(p_x \times p_y) = 2(2 \times 1)$ $(Ng_x \times Ng_y \times Ng_z) = (125 \times 32 \times 64)$		$Macro_x(p_x \times p_y) = 2(2 \times 2)$ $(Ng_x \times Ng_y \times Ng_z) = (125 \times 63 \times 64)$	
	External com	Total com	External com	Total com
MGM+MPI		4.66		7.1
Aitken-Schwarz + MGM + PACX-MPI	0.28	4.3	0.7	10.4

Table 4
Comparison of communication elapse time for the velocity problem .

Parallel Computational Fluid Dynamics – Trends and Applications
C.B. Jenssen et al. (Editors)

Mixed C++/Fortran 90 Implementation of Parallel Flow Solvers

Malin Ljungberg and Michael Thuné [a]*

[a]Department of Scientific Computing, Uppsala University, Box 120, S-751 04 Uppsala, Sweden. Email: malinl, michael@tdb.uu.se.

The following article gives a short presentation on how to wrap user defined modules in Fortran 90 in C++ classes, so as to make selected parts of the functionality available in a C++ shadow class. This approach is demonstrated on a relatively large object-based Fortran 90 library. For validation we consider flow solvers implemented via calls to this library.

1. INTRODUCTION

With the increasing performance of computers has come an ability treat more complex flow cases, with complicated geometries. In flow solvers based on finite difference approximations, such geometries are handled via the insertion of composite, structured grids (typically so called multi-block grids). Fig. 1 shows a simplified model problem to illustrate this. It is an expansion pipe, and the geometry requires a five-block grid (see Fig. 2).

The traditional way of constructing a flow solver is to write it from scratch, usually in Fortran, using the built-in data structures of that language. However, the complicated data structures of modern flow solvers, and the additional aspect of parallelization, has made this approach to software construction inadequate. There is an apparent need for software tools that raise the level of abstraction considerably [1].

To this end, we developed Cogito [2], a Fortran 90 library supporting implementation of parallel PDE solvers. Cogito has an object-oriented design. The core classes are Grid, Composite Grid, and Grid Function. Each individual Grid Function object is automatically distributed over several processors. The parallelism is of SPMD type, and uses message passing via MPI. The message passing takes place within the object and is invisible to the user.

The present classes in Cogito are essentially data stores with associated operations. They do not initiate interaction with other objects. The starting point for the present work was a wish to extend Cogito with, among other things, support for preconditioned Krylov subspace methods. Also, we would like to integrate Cogito with Compose, which is a related effort by our group [3][4], implemented in C++. Compose contains classes on a higher level of abstraction, representing, for example, partial differential equations (PDEs) as stand-alone objects. The combination of the flexibility of a framework

*This research was supported by the Swedish Research Council for Engineering Sciences.

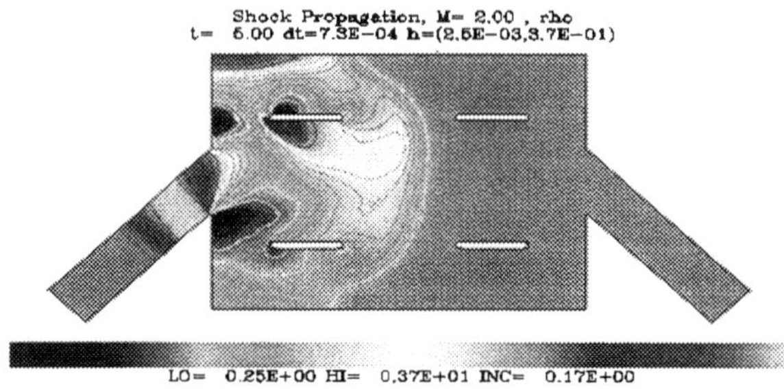

Figure 1. Simulation of airflow through an expansion pipe.

like Compose with the high parallel efficiency of the Cogito package would give us an efficient and flexible parallel PDE solver framework.

These extensions of Cogito will lead to a fully object-oriented software architecture, where a complete algorithm is composed of different interacting objects in a "plug-and-play" manner. The limited object-oriented facilities of Fortran 90 are insufficient for such an approach. At the same time, we want to reuse the existing Fortran 90/MPI implementation of Cogito, since considerable effort was invested in it (~20000 lines of code), and since it contains numerically intensive operations, for which Fortran gives good efficiency. As the handling of multi dimensional arrays in Fortran is superior to that of C++ both from a coding and from an efficiency point of view it is desirable to to keep these parts of the code in Fortran, particularly when they are used in time critical parts of the code.

Our approach to this problem is to put a C++ wrapper class around each relevant Fortran class. This will allow for extensions to the simple Fortran objects. It will allow for the use of templates, inheritance and virtual functions, features which can greatly enhance the structure and readability of the code. The C++ standard template library

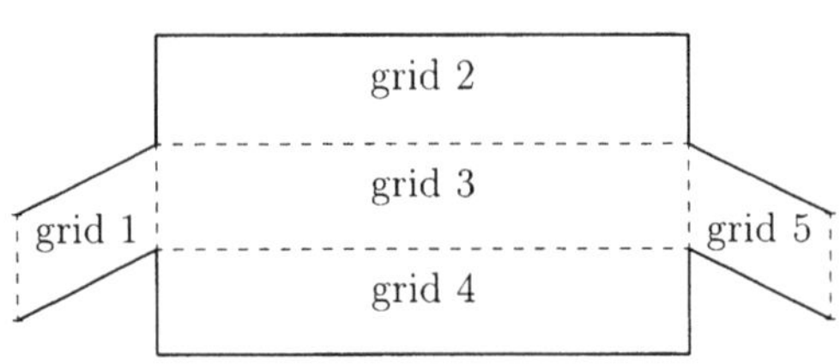

Figure 2. A schematic view of the multi-block grid used in the airflow simulations.

can also be used, which contains vectors, lists and other container objects for general classes. Being able to use existing and well tested code will allow the implementation to focus on the numerical problems being solved.

The aim with the present work is to verify the viability of this mixed language solution. In particular we need to check that efficiency is not compromised, and that the upper C++ layer will provide us with the increased flexibility we need to in order to be able to extend Cogito in the directions discussed above.

Our contribution relates to a growing literature on how to use Fortran 90 in an object-oriented fashion (see, e.g., [5], [6], and references therein). The issue of how to use preexisting C++ packages from within Fortran 90 has also attracted attention [7]. Vice versa, calling Fortran 77 from C/C++, and using calls to Fortran 77/Fortran 90 from within methods in C++ classes, is relatively straightforward [8] [9].

Our situation is more complicated, since we wrap C++ around an *object-based* Fortran 90 code. This was discussed in [10]. Our approach is similar, but avoids one of the steps in their method of wrapping. Moreover, we demonstrate our approach on a relatively large object-based Fortran 90 library. Finally, we validate the result by considering flow solvers implemented via calls to this library.

2. INTERFACING BETWEEN C++ AND FORTRAN-90

In order to wrap our Fortran 90 pseudo objects we first add a flat interface, i.e. Fortran methods for performing all the relevant "member functions" of the Fortran 90 pseudo object, but without explicit reference to the particular data type. On the C++ side of things pointers to Fortran 90 pseudo objects are represented as pointers to void.

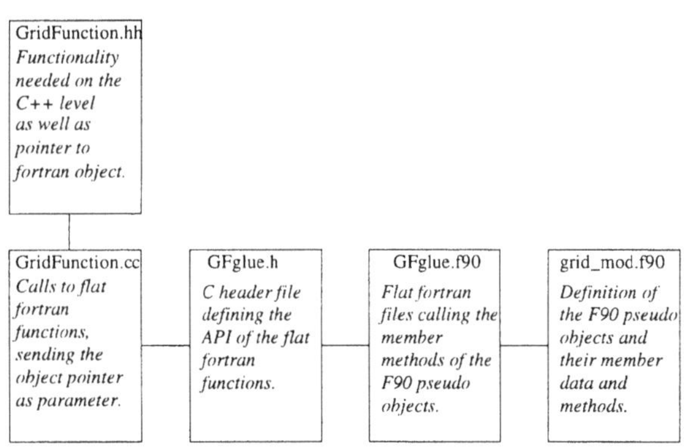

Figure 3. File structure for the C++/Fortran 90 interface.

The C++ shadow object will have all the functionality of the Fortran pseudo object which is made available through the flat interface, and it is also possible to add functionality directly to the C++ shadow, which will then become slightly more than a shadow.

The method we have used for dealing with the above issues has much in common with the method described in [10], the main difference being that we avoid the translation between object pointers and the so called opaque pointers by letting the the C++ shadow of the Fortran 90 object keep a void pointer pointing straight to the memory allocated by the Fortran routine. This works since this piece of memory is only kept for reference by the C++ object, and any actual action taken will only be by Fortran 90 code. Using a void pointer clearly indicates that this chunk of memory is in a format which is not directly usable by the C++ shadow object.

3. VALIDATION

As a demonstration of the viability of the mixed language approach discussed above, we consider two examples involving flow solvers. One is a simple advection problem, the other is a compressible Navier-Stokes code. Both have been implemented via calls to the Cogito library and are executable on parallel platforms supporting MPI.

The points we make are:

- that the C++/F90 implementation gives the same execution speed as the original, pure Fortran 90 code.
- that the C++/F90 version allows for a much more convenient extension of Cogito into a fully object-oriented framework.

3.1. Execution Speed: 2D Advection

Our first example is an implementation in Cogito of a Leap-Frog solver for a simple two dimensional advection equation, $u_t + u_x + u_y = F(x, y)$, using centered second ordered differences in the spatial directions. The problem is solved on a multi-block grid, and works in parallel. The parallelism is of SPMD type. Internally, Cogito uses MPI for message passing.

In order to test the efficiency of the C++/F90 version of Cogito, we have translated the construction of the Leap-Frog method into C++. Two of the Fortran classes, GridFunction and CompositeGrid, have been wrapped with C++ shadows which allow us to perform this construction purely in C++, using the shadow objects.

A series of test runs were performed on a parallel computer consisting of two Sun Ultra Enterprise 6000 servers connected through the Sun WildFire interconnect, cf [11]. The compilers used were the Sun Fortran 90 and hpc as well as Sun C++ compilers versions 4.2. In each of the test runs 8 processors were used.

In order to compare the performance of the C++ wrapped application to that of the pure Fortran version, tests were run with different optimization flags. Five timed runs were performed for each of the versions for five different settings of the optimization flag. The -fast flag[2] was used for all cases where the optimization flag was greater than two.

---

[2]The -fast option is short for a number of options for both the CC and the F90 compiler, i.e. native hardware target, in-lining certain math library routines, optimization of floating-point operations, memory alignment to allow generation of faster double word load/store instructions, linking to the optimized math library, etc.

It is not possible to combine the lower optimization flags with the -fast option. No other compiler flags were used besides the -o and -fast flags.

The differences between the two cases measured as ratios (time for C++ divided by time for Fortran) show that we never have more than 3 percent discrepancy. The fact that the C++ version is actually faster for optimization levels 0 and 2 leads us to believe that the variations are statistical, i.e. due to the fact that the processors are not totally dedicated, and we sometimes get more and sometimes less processor time. This is also indicated by the fact that for all test series the measurements for the mixed version were overlapping with those from the pure Fortran 90 version.

Thus for the 8 processor case our measurements show that the time differences between the two versions are insignificant and also independent of the optimization level. We can add the flexibility of having a fully object-oriented interface on top of our Fortran pseudo objects without having to pay a price in performance.

3.2. Execution Speed: Navier-Stokes

As a second example we converted an existing compressible Navier-Stokes code to C++ by using our new C++/Fortran 90 wrapper layer. Time stepping is by a n step Runge-Kutta method, where n is tunable. In our test runs n was set to 5.

A Sun Ultra Enterprise 6000 was used for these test runs, with 14 processors available for the tests. The compilers were now the Sun Fortran 90 and hpc, as well as Sun C++ compilers versions 5.0.

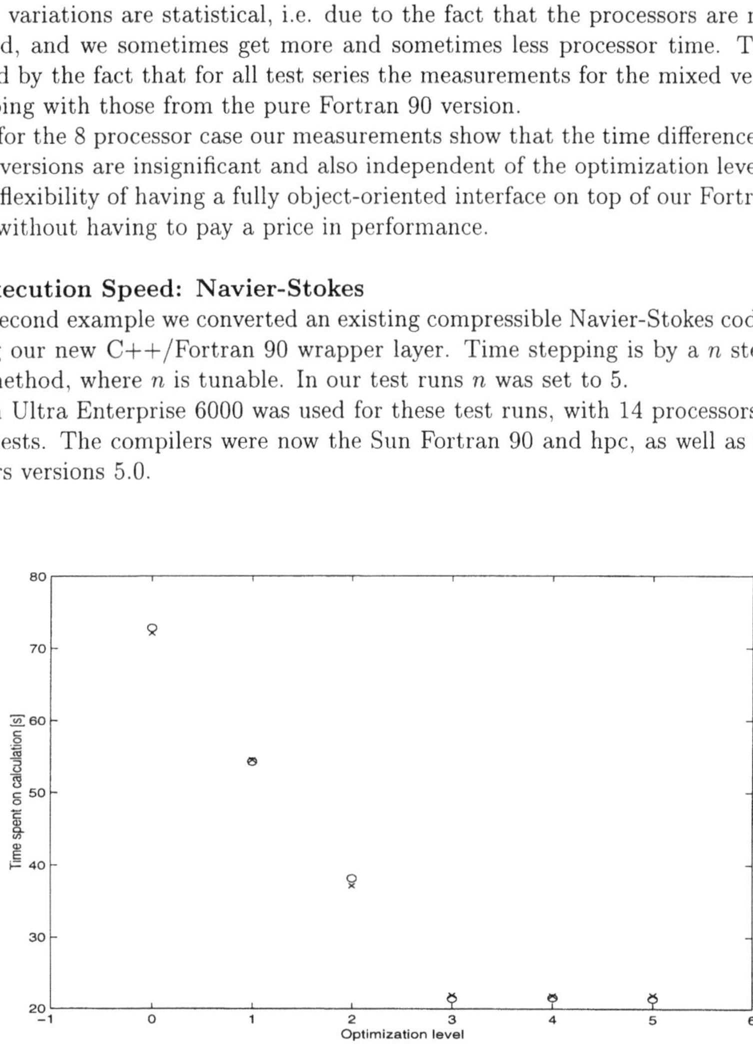

Figure 4. Comparison of computation times for C++ wrapped version (x) and pure Fortran 90 version (o) of the Advec program. The variations go both ways, and in no case do the results differ by more than 3%.

The geometry on which the solution was calculated was a composite grid consisting of 5 sub-grids, as shown in Fig. 2. In all runs the 5 sub-grids were set to have the same number of grid points. The number of time steps for which the problem was run was also kept fixed during the runs.

In order to compare not only the speed, but also the scalability of the Fortran 90 versus the mixed implementation, we made measurements of both the speedup and the size-up of the two.

Speedup, $sp(nproc)$, was measured as the time to run the problem on one processor divided by the time to run the same problem on $nproc$ processors, where $nproc$ is the number of processors; $sp(nproc) = time(1)/time(nproc)$.

The speedup measurements were performed by running the problem with the grid size set to 5 by 17 by 21, and the number of time steps set to 1000. The number of processors used ranged from 1 to 12, for both the pure Fortran 90 implementation and for the C++/F90 version of the code.

Size-up, $sz(nproc)$, was measured as the work done one one processor during a specific time divided by the work done on $nproc$ processors during the same amount of time. The work was assumed to be a constant times the number of grid points treated, $ngrid$. The number of time steps were kept constant.

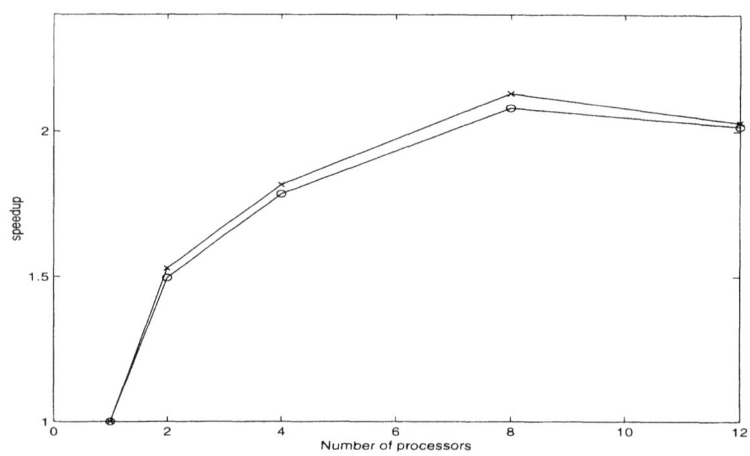

Figure 5. Comparison of fixed size speedup for a small problem. o indicates the mixed C++/F90 version, while x indicates the pure Fortran 90 version of the code. The maximum discrepancy between the two versions is 2.5%, and there is no trend of increasing differences as the number of processors goes up.

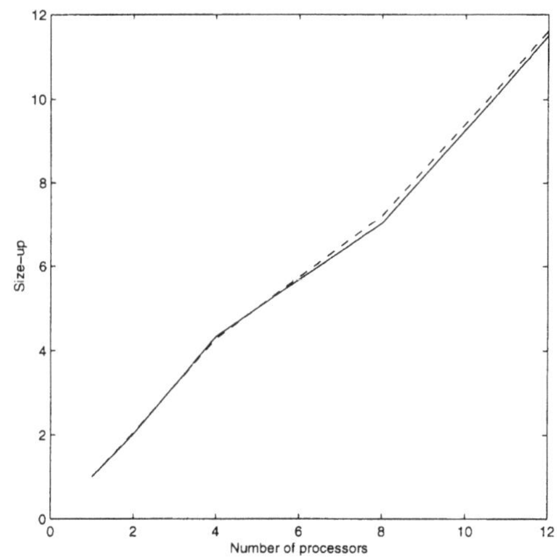

Figure 6. Comparison of size-up for the two versions of the code. Dashed represents the pure Fortran 90 version while solid is the mixed C++/F90 implementation of the code. The maximum discrepancy between the two cases is 4.5%, and there is no trend of an increasing discrepancy as the number of processors goes up.

Since it was impossible to tune the grid sizes perfectly so that all runs took exactly the same time a small time correction was then also introduced when calculating the work done for a fixed time. This time correction was simply the time used to run the problem on *nproc* processors divided by the time used on one processor. The size-up was thus calculated as $sz(nproc) = ngrid(nproc)/ngrid(1) * t(1)/t(nproc)$, where *ngrid* is the number of grid points used and t is the time the run took for a certain number of processors.

For both the speedup and size-up measurements the differences between the two versions of the code were as small as, or smaller than, expected. The maximum difference we saw was that the mixed C++/F90 version was 4.5% slower for 8 processors in the size-up measurements. Of particular importance is the fact that there was no trend towards increasingly bad performance as the problem size and/or number of processors was scaled up.

3.3. Benefits of the Added Flexibility

In the original Fortran 90 version of the advection program the stencils used in the difference scheme are hard coded in the main program. By introducing an Operator class we will get a natural spot to keep references to these. They will be kept as member data of the Operator class. Furthermore, as a next step, thanks to the true inheritance available in C++ we can introduce a general Stencil class, and have selectable stencil types available in the Operator class. The specific Stencil types will be naturally implemented as derived classes from the Stencil base class. This type of plug and play is easily done in C++, while it requires complicated workarounds when using Fortran 90 pseudo objects, which lack inheritance.

In the Navier-Stokes code the calculation of the boundary conditions is implemented as a separate part of the code. In the future object-oriented version of the code we will be able to integrate the boundary conditions as a class associated with the Operator class. Inheritance will allow us to use any of several different types of boundary conditions, all of which will inherit ("is a") from a BoundaryCondition base class. In this way the specific behavior of the particular boundary conditions being used will not have to be visible at higher levels of the code, but will rather be seen as an implementation detail of the boundary condition object in question. This would be possible, but much more inconvenient to implement in the original, pure Fortran 90, setting.

REFERENCES

1. A. H. HAYES *The changing face of high-performance computing in the United States*, in Wuhan Journal of Natural Sciences, Keynote Lecture at the International Conference on Parallel Algorithms, 1995, pp. 420–429, Wuhan University, Wuhan, P. R. China, 1996.
2. MICHAEL THUNÉ, EVA MOSSBERG, PETER OLSSON, JARMO RANTAKOKKO, KRISTER ÅHLANDER, KURT OTTO *Object-Oriented Construction of Parallel PDE Solvers*, in Modern Software Tools in Scientific Computing (E. Arge, A. M. Bruaset and H. P. Langtangen, eds.), pp. 203–226, Birkhäuser, 1997.
3. KRISTER ÅHLANDER, *An Oject-Oriented Framework for PDE Solvers*, Acta Universitatis Upsaliensis, Comprehensive Summaries of Uppsala Dissertations from the Faculty of Science and Technology 423, ISBN 91-554-4379-6, 1999
4. KRISTER ÅHLANDER, *An Extendable PDE Solver with Reusable Components*, Computational Technologies for fluid/thermal/structural/chemical systems with industrial applications, PVP-Vol. 397-1, 1999
5. MARK G. GRAY AND RANDY M. ROBERTS, *Object-Based Programming in Fortran 90*, Computers in Physics, Vol 11, No 4, Jul/Aug 1997, pp. 355-361.
6. *http://www.cs.rpi.edu/~szymansk/oof90.html*
7. CHARLES D. NORTON, VIKTOR DECYK, AND JOAN SLOTTOW, *Applying Fortran 90 and Object-Oriented Techniques to Scientific Applications*, LNCS 1543, Object-Oriented Technology, ECOOP'98 Workshop Reader, (1998), pp. 462–463.
8. PAUL F. DUBOIS, LEE BUSBY, , *Portable, Powerful Fortran Programs*, Computers in Physics, Vol 7, No 1, Jan/Feb 1993.
9. *http://www-zeus.desy.de/~burow/cfortran/.*
10. MARK G. GRAY, RANDY M. ROBERTS, AND TOM M. EVANS, *Shadow-Object Interface Between Fortran 95 and C++*, Computers in Science and Engineering, Vol 1, No 2, Mar/Apr 1999, pp. 63-70
11. *http://www.tdb.uu.se/~hpc/albireo/albireo.html.*

Parallel Computational Fluid Dynamics – Trends and Applications
C.B. Jenssen et al. (Editors)

COUPL+ : Progress towards an integrated parallel PDE solving environment

Mike Rudgyard[a], David Lecomber[b] and Thilo Schönfeld[c]

[a]Department of Computer Science
University of Warwick
Coventry CV4 7AL, UK

[b]Oxford University Computing Laboratory
Oxford OX7 4NN

[c]CERFACS, Toulouse 31057, France

COUPL+ is a Fortran/C software library that simplifies the task of programming grid-based applications on distributed memory parallel machines. We discuss the basic concepts used by the library, and indicate how a typical application, such as an unstructured CFD solver, can make use of these.

1. Introduction

In recent years, the effective use of parallel computers has become of major importance to numerical simulation for research and design. Simulations that previously required expensive vector supercomputers may now be tackled on a clusters of PCs or workstations at little cost, while the advent of massively parallel machines with high bandwidth networks and high-performance processors has meant that many previously intractable problems may now be tackled.

Despite the rapid advances in hardware technology, there are few software tools that enable non-expert users to make good use of parallel computers, especially when we consider problems that involve a substantial amount of indirection, or with dynamic or irregular data-access patterns. In our view, it is highly unlikely that compiler technology alone will provide the means to automatically parallelize such complex code with any degree of scalability, at least in the short to medium term. Numerical libraries offer a more realistic alternative, although their development is often driven by the specific needs of particular communities and to date relatively few attempts have been made to merge ideas within a common framework. Nonetheless common communication standards such as PVM, MPI and BSP have emerged, and many more projects that aim to create more general purpose tools have now begun; examples include PETSc [1] and Aztec [11]. Here we discuss one other such high-level approach, namely the COUPL+ parallel library.

2. COUPL+ and its Generic Parallel Tools

COUPL+ is a direct descendent of the Oxford Library for Unstructured Solvers (OPlus) [3] and the CERFACS and Oxford University Parallel Library (COUPL) [9], both of which attempted to create a software framework suitable for large-scale parallel simulations based on the solution of PDEs on unstructured or hybrid grids. By concentrating on the simple data structures that are associated with an underlying mesh, the library aims to provide an environment in which the whole simulation may be viewed consistently, from mesh manipulation and refinement, to the discretisation of the problem and its eventual solution and visualisation. For many non–linear problems of interest (where the construction of residuals or matrices and the solution of the resulting problem is strongly coupled), we believe that this integrated approach is necessary in order to provide a simple interface to the end user.

The central difference between COUPL+ and its predecessors is that every operation in COUPL+ is fully distributed, including the I/O, the data partitioning and the grid manipulation. This means that COUPL+ does not require a *server* process with large amounts of memory, and so it is relevant to very large-scale simulations such as Navier-Stokes computations on complex three dimensional grids. For similar reasons, COUPL+ is applicable to dynamic applications, such as codes that include grid refinement or movement. Many of the distributed tools that underpin the library are described in [6].

2.1. Basic Principles

COUPL+ makes use of a straightforward data structure based on the concepts of distributed **sets** (such as the nodes or cells of a mesh), *distributed data* that span the sets (such as the nodal solution), and *connectivities* between the sets (such as the cell-to-node pointer, or a node adjacency graph).

Typically, sets are declared at the beginning of the application programme, and information about these (such as the global number of elements of each set) is provided by the user. The library then provides *set identifiers* or handles, which are subsequently required by a number of COUPL+ functions and subroutines.

Any distributed data that spans a valid set is dynamically allocated using the libraries memory management functions (note that this requires the pointer extension to be used from Fortran 77). The data may either be logically equivalent to a two dimensional array (so-called *regular data*), or it may be stored as *irregular data*, indexed by a pointer array that spans the set. The latter form is particularly useful for sparse matrix storage schemes, or for graph structures. Fortunately, these two data types (or close variants of these) are commonly the only types used within the majority of existing simulation codes.

2.2. Restrictions

Like the OPlus library [3], COUPL+ is relevant to those algorithms for which the result is independent of the order in which the operations are performed. As a result, iterative techniques such as point or line Gauss–Seidel, as well as global solution (or preconditioning) techniques such as ILU, are not straightforward to implement using the model. Examples of *order–independent* algorithms include explicit time–stepping, many fixed–point iteration techniques, Krylov subspace methods and Multi-grid (with suitable smoothers). If we generalise the above definition to include algorithms where we work on blocks of data that correspond to individual *partitions*, block-iterative, block-preconditioned and domain decomposition methods all fit naturally within the COUPL+ framework.

2.3. Parallel I/O Models

There are many possibilities for file I/O on shared and distributed memory parallel architectures. Unfortunately, there is no single model that works best on all architectures. Issues such as whether I/O must take place via a server process, whether each processor has a local disks, or whether an auto-mounted file system exists, means that a portable library requires a flexible treatment of this issue.

COUPL+ defines four I/O types that may be associated with a file when this is opened: A Shared I/O file is stored on the same processor as the COUPL+ server, although it will be accessed by all *client* or *compute* processes; any distributed data that is stored in such files is stored in its scalar form (ie. for the unpartitioned problem). An Ordered I/O file is stored on the same processor as the COUPL+ server, and is also accessible from all client processes; distributed data within these files is stored by partition: the values for partition 1 being stored first, followed by the values for partition 2, etc. If a file is defined as being Single I/O, then the file is physically opened by the COUPL+ server, although only the client process that initiated the opening of the file will be allowed to access that file in subsequent actions. Local I/O files reside locally to the client processes, and data may be written or read in a locally 'scalar' fashion.

Typically, Shared I/O files are the most useful since preprocessing tools such as grid generators usually provide data in 'scalar form'; data written to these files may subsequently be used by the same visualization or post-processing tools, irregardless of the number of processes used by the computation itself. Ordered I/O files are useful for reading pre-partitioned data, and permit COUPL+ to make use of parallel grid generators or external partitioning tools (should the user wish to do this). Single I/O files are useful for debugging purposes, while Local I/O files may be used for dumping or retrieving large amounts of data from disk during a computation.

2.4. Reading and Writing Data in Parallel

There are three mechanisms for reading and writing data within COUPL+: *Read* and *Write* functions; *Input* and *Output* functions; and *Import* and *Export* functions.

The basic Read and Write functions are used to read and write scalar data to and from the server processes. On reading data, each processor receives a copy of the data that is stored on file. On writing data, the copy of the data that is held by the *root* processor is stored on file.

The Input and Output functions are used for distributed data. On reading, the data is automatically partitioned in a manner that is consistent with the set it spans. On writing, the data is renumbered so as to be consistent with the file-type that it is to be written to (ie. it is written by partition for distributed files, or according to the ordering of the initial scalar data for shared files).

Finally, the Import and Export functions are used for (integer) connectivity arrays and will ensure that any new sets that are referenced during a read operation are automatically partitioned (by inheritance); at the same time any interface lists that are used internally by COUPL+ for the execution of parallel loops is updated.

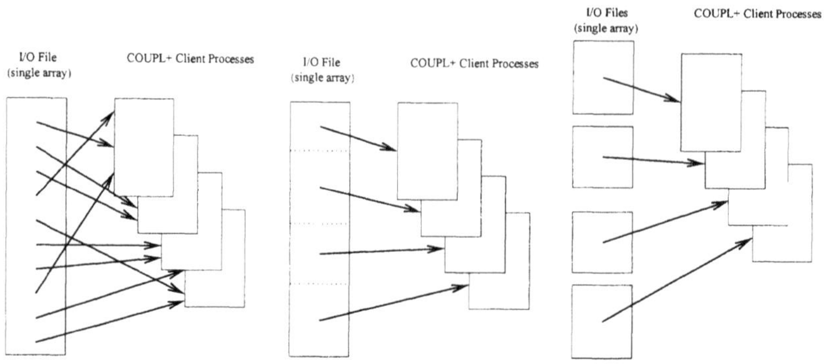

Figure 1. Schematic of shared, ordered and local I/O for reading COUPL+ set data

2.5. Distributed Partitioning

An integrated partitioning facility was one of the major design criteria for COUPL+. This avoids the need to pre-process data before the parallel execution of the code, and (as noted above) is paramount when we consider large-scale problems or require dynamic repartitioning.

The library includes distributed partitioning based on geometric techniques such as the Recursive Co-ordinate and Recursive Inertial Bisection algorithms. For a typical grid-based problem, nodal co-ordinate information is read from file using the inbuilt *shared* I/O functions. As the library recognizes internally that the set of nodes has yet to be partitioned, the coordinate data is distributed equally between the processors in an ad-hoc manner. A subsequent call to the geometric partitioning tool ensures that this data is then repartitioned sensibly.

When the Import function is used to read connectivity information from disk, the library makes use of information that it holds concerning the *spanned* set as well as the set into which the connectivity points, or the *target set.* If the target set has been partitioned, then the spanned set inherits this partitioning — for example, in a typical Finite Element or Finite Volume application where the nodes have been partitioned geometrically, the cells will be partitioned automatically through a subsequent Import operation. Similarly, if the target set has already been partitioned, then the spanned set will be partitioned accordingly. If neither the spanned nor the target set has been partitioned, then COUPL+ automatically makes use of a distributed hierarchical graph bisection algorithm (similar to that described in [5]) to partition both.

When sets are partitioned, any data or connectivity that spans this set is automatically repartitioned, reallocated and renumbered by the library. Similarly, the interface information that is required internally by COUPL+ in order to execute the parallel loop syntax is updated behind the scenes.

2.6. The Parallel Loop Syntax

Following the approach taken in OPlus, a simple parallel loop syntax is used for maintaining data coherency at partition interfaces. This ensures no low-level message-passing is visible to end-user applications, so that his code closely resembles scalar code.

The technique is best explained by a simple Fortran example (the C code is of similar form, except using `while` and `for` loops). In the following, we consider the computation of dual volumes of a triangular mesh through a gather/compute/scatter operation that is typical of unstructured CFD applications (for finite element or finite volume approximations). We note that the parallel code only differs from its scalar counterpart due to the `do while` and `kpl_access` functions that are defined below.

```
c
c     Compute the dual volumes of a triangular mesh;
c
      third = 1.d0/3.d0

      do while ( kpl_par_loop(itri_set, ifrom, ito, ierr) )

        call kpl_access( kpl_data_access, kpl_no_access, x,
     &                   ielno, 1, 2, ierror)
        call kpl_access( kpl_data_access, kpl_data_access, voln,
     &                   ielno, 1, 1, ierror)

        do n = ifrom, ito

          n1 = ielno(1,n)
          n2 = ielno(2,n)
          n3 = ielno(3,n)
          volcell = 0.5d0 * ( (x(2,n2) - x(2,n1))*(x(1,n1) - x(1,n3)) -
     &                        (x(1,n2) - x(1,n1))*(x(2,n1) - x(2,n3)) )
          voln(n1) = voln(n1) + third * volcell
          voln(n2) = voln(n2) + third * volcell
          voln(n3) = voln(n3) + third * volcell

        end do

      end do
```

In the above, `kpl_par_loop` is a logical function that returns the loop indices `ifrom` and `ito` given the set identifier relating to the set of triangles, `itri_set`, over which the loop is performed.

`kpl_access` is a function that defines how the distributed data within the loop is accessed (using the flags `kpl_data_access` and `kpl_no_access`). In this case the coordinate array `x()` is read from memory through an indirection using an integer connectivity array, `ielno()`; the latter stores the vertex numbers of the triangular elements. Note that the array `x()` is not written back to memory within the loop. Similarly, the array `voln()` is read from memory and then written back to memory through an indirection `ielno()`.

On the first pass of the outer loop, `kpl_par_loop` is true, although `ito` is given a value less than `ifrom`. The library then analyses the loop in order to decide how to copy the minimum amount of information so as to ensure that owned values of each set are up-to-date following the execution of the loop; during this or subsequent passes, the values of `ito` and `ifrom` are set accordingly, and information to be copied from remote processors is scheduled as required.

Certain other optimizations can also be performed by the library if the user details a particular reduction operation instead of the standard `kpl_data_access` flag.

3. Grid-based and Related tools

In addition to the generic tools that have been detailed above, COUPL+ includes a number of functionalities that are specific to unstructured and hybrid grid applications.

3.1. Connectivity-on-demand

The connectivity-on-demand capabilities offered by COUPL+ may be used to compute derived connectivities such as the edge-to-node, edge-to-cell, face-to-cell and node-to-node indirections *on the fly*, assuming that the mesh has been stored on file in a common format, such as through the cell-to-node and boundary facet-to-cell pointers. This greatly simplifies the task of writing parallel edge-based or cell-based finite volume codes, or the creation of sparse matrix or graph representations of the mesh.

The library makes internal use of distributed merge algorithms to ensure that any new sets created when *extracting* derived connectivities may subsequently be used in parallel loops. Because of this, these tools are extremely powerful and very easy to use.

3.2. Parallel Multigrid and Inter-grid Communication

Recent work on COUPL+ has concentrated on the development of tools for parallel multigrid and multi-physics. These include tools for agglomerating edge weights and for creating valid coarse meshes in parallel, as well as tools for determining the connectivities between distinct overlying grids.

The agglomeration tool makes use of a local (processor-wise) agglomeration algorithm; this is combined with the same distributed merge techniques that are used by the connectivity-on-demand capabilities outlined above. The interface for the user is straightforward: he provides a list of edges with corresponding weights and then coarser edges are determined with appropriate symmetric or asymmetric edge weights. The coarse edge sets are defined internally, as are the mechanisms for copying interface data between processors: once again, this means that parallel loops over these new sets may be performed immediately.

Automatic mesh derefinement is presently based on the extension of edge collapsing algorithms that have been successfully used for both tetrahedral and hybrid grids [7]. A distributed prototype of this tool has now been implemented for tetrahedral grids and improvements of this are currently under development. As the scalar technique used in [7] is inherently sequential, the distributed variant requires an independent set finding algorithm similar to that used in the distributed graph partitioning.

Finally, COUPL+ allows several independent grids to be used within the same parallel application and now provides some limited support for interpolation of data between these. By making use of consistent geometric partitioning as well as distributed octree and directed graph searches, the library can automatically extract predefined connectivity arrays that define relationships between any two grids. This capability is obviously relevant to both multi-grid and multi-physics applications.

3.3. Compatibility with Other Libraries

As well as making use of standard message , passing software at a low level (COUPL+ versions exist for MPI, PVM and BSP), the library has interfaces to other well-known software libraries. In particular, real-time parallel visualization is offered by pV3 [4], and a simple interface is provided that takes advantage of COUPL+'s predefined grid-based data structure. The parallel I/O capabilities have also been extended by offering a distributed front-end to the data-base standard ADF [8].

4. Applications and Examples

COUPL+ is presently being used by the *AVBP* CFD code, for both finite volume [10] and finite element [2] versions with turbulence, combustion and LES capabilities. Typical applications, with examples are given in [10]. We are also porting the Rolls Royce *Hydra* code [7], an edge-based multi-grid Finite Volume code, although details of this will also be reported elsewhere.

REFERENCES

1. S. Balay, W. D. Gropp, L. C. McInnes, and B. F. Smith. Efficient management of parallelism in object-oriented numerical software libraries. In E. Arge, A. M. Bruaset, and H. P. Langtangen, editors, *Modern Software Tools in Scientific Computing*, pages 163–202. Birkhauser Press, 1997.
2. O. Colin and M. Rudgyard. Development of high-order Taylor-Galerkin schemes for LES. *J. Comp. Phys.*, 162:338–371, 2000.
3. P.I. Crumpton and M.B. Giles. Aircraft computations using multigrid and an unstructured parallel library. AIAA Paper CP-95-0210, 1995.
4. R. Haimes. pV3: A distributed system for large scale unsteady CFD visualisation. AIAA Paper 94–0321, 1994.
5. George Karypis and Vipin Kumar. Parallel multilevel k-way partition scheme for irregular graphs. *SIAM Review*, 41(2):278–300, 1999.
6. D. Lecomber and M. Rudgyard. Algorithms for generic tools in parallel numerical simulation. In *High Performance Computing and Networking 2000*. Springer, 2000.
7. J-D. Muller P. Moinier and M.B. Giles. Edge-based multigrid and preconditioning for hybrid grids. AIAA Paper CP99-3339, 1999.
8. D. Poirier, S. Allmaras, D.R. McCarthy, M.E. Smith, and F.Y. Enomoto. The CGNS system. AIAA Paper CP98-3007, 1998.
9. M. Rudgyard, T. Schönfeld, and I. d'Ast. A parallel library for CFD and other grid-based applications. In *High Performance Computing and Networking 1996*. Springer, 1996.

Figure 2. Example of a partitioned mesh for a 4M element F15 fighter configuration

10. T. Schönfeld and M. Rudgyard. Steady and unsteady flow simulations using the hybrid flow solver AVBP. *AIAA journal*, 37(11):1378–1385, 1999.
11. R. S. Tuminaro, M. Heroux, S. A. Hutchinson, and J. N. Shadid. Official Aztec user's guide: Version 2.1. Sandia Labs Technical Report, 1999.

Parallel Computational Fluid Dynamics – Trends and Applications
C.B. Jenssen et al. (Editors)

Implementations of a Parallel 3D Thermal Convection Software Package

P. Wang[a]

[a]Jet Propulsion Laboratory
California Institute of Technology
MS 168-522, 4800 Oak Grove Drive
Pasadena CA 91109-8099, U.S.A.
wangp@azalea.jpl.nasa.gov

Thermal convective motions driven by temperature gradients often play an essential role in the behavior of geophysical and astrophysical systems, and obtaining a detailed understanding of their role is often at the core of many important problems in the Planetary sciences. Applications include the dynamics of atmospheres, stellar convection, and convection in gaseous protostellar disks. In this paper, the most advanced numerical algorithm, the finite volume method with an efficient multigrid scheme, to solve the above problems has been investigated, and the design and implementation of a general and efficient parallel version of state-of-the-art software has been discussed in various computer environments. The software can be used by scientists and engineers in the Space field for solving the thermal convection problem for a wide variety of systems characterized by different geometries and dynamical regimes.

1. Introduction

Studies of the natural thermal convection and its influence on the heat transfer and mixing have been pursued vigorously for many years. An example is the convection driven by a horizontal or vertical thermal gradient. A typical model consists of a 2D or 3D rectangular cavity with two horizontal or vertical end walls held at different constant temperatures. In order to determine the flow structure and heat transfer across the cavity, numerous analytical, experimental, and computational techniques have been used.

The most numerically studied form of this problem is the case of a rectangular cavity with differentially heated walls. Such problem in two dimensions has received considerable attention, but very few results have been obtained in three dimensions. For large Rayleigh number 3D flows, there are seldom numerical results available. The existing 3D numerical simulations are still in a rudimentary stage. Most existing numerical studies have suffered from insufficient resolution, and the prominent characteristics of complicated 3D flows have not been discussed in sufficient depth because the accuracy of the solution is not good enough. In particular, at large Rayleigh numbers, greatly enhanced numerical capabilities are essential in order to catch the significant dynamic features in thin boundary layers. The main obstacles have been the limitation of computing power and memory size. The situation gets only worse if one is interested in high Rayleigh number flows in the presence of rotation, that is the case for planets, stars, and protostellar disks. The interest and applications for planetary sciences and astrophysics increase as well as the numerical challenge for such cases, yet our present knowledge of the properties of such flows is even poorer than in non-rotating convection.

Recent advances in the computing hardware have dramatically affected the prospect of modeling 3D time-dependent flows. The significant computational resources of massively parallel supercomputers improve the feasibility of such studies. In addition to using an advanced hardware, designing and implementing a well-optimized parallel code will significantly improve the computational performance and reduce the total research time to complete these studies. Based on the finite volume method, a numerical study for the 3D time-dependent problem has been recently investigated by using massively parallel computing systems [1] [2]. In those studies, the introduction of a parallel multigrid linear solver scheme with a general and portable parallel implementation has been the key to producing dramatically improved performance and to allowing 3D numerical solutions with Rayleigh numbers up to 5×10^7 to be obtained.

In order to use the code for various scientific applications, the development of a general parallel 3D time-dependent thermal convection software package is necessary, such as dealing with various boundary conditions, adding rotation forces, and other issues for different thermal flow problems. In this paper, the design and implementation of such a package is discussed. The reusability, scalability, and portability of the package are addressed. Efficient numerical methods with a general parallel implementation on massively parallel supercomputers is presented. Numerical experiments for different thermal flow problems are given at the end of the paper.

2. The Mathematical Formulation

The flow domain is a rectilinear box of $0 < x < l$, $0 < y < d$, and $0 < z < h$. The appropriate governing equations, subject to the Boussinesq approximation, can be written in non-dimensional form as

$$\left\{ \begin{array}{rcl} \nabla \cdot \mathbf{u} & = & 0, \\ \frac{1}{\sigma}\left(\frac{\partial \mathbf{u}}{\partial t} + \mathbf{u} \cdot \nabla \mathbf{u}\right) & = & -\nabla p + RT\hat{\mathbf{z}} + T_a^{\frac{1}{2}} \mathbf{u} \times \hat{\mathbf{z}} + \nabla^2 \mathbf{u}, \\ \frac{\partial T}{\partial t} + \mathbf{u} \cdot \nabla T & = & \nabla^2 T. \end{array} \right. \tag{1}$$

Dimensionless variables in these equations are the fluid velocity $\mathbf{u}$ with components u, v, and w, the temperature T, and the pressure p, where the $\sigma = \nu/\kappa$ is the Prandtl number; the $R = g\beta\Delta T h^3/\kappa\nu$ is the Rayleigh number; the $T_a = (2\Omega d^2/\nu)^2$ is the Taylor number. Here, the ν is the kinematic viscosity; the κ is the thermal diffusivity; the β is the coefficient of thermal expansion; the Ω is the rotational rate of the reference frame in the vertical direction; the $\hat{\mathbf{z}}$ is the unit vector in the vertical direction; and g is the gravitational acceleration. Several general boundary conditions for the temperature and velocity are considered which include the Dirichlet, Neumann, or periodical conditions. In general the motion is controlled by parameters σ, R, T_a, and the flow domain.

3. The Numerical Approach

The numerical approach is based on the widely used finite volume method [3] with an efficient and fast elliptic multigrid solver for predicting incompressible fluid flows. This approach has been proven to be a remarkably successful implicit method. A normal, staggered grid configuration is used and the conservation equations are integrated over a macro control volume. Local, flow-oriented, upwind interpolation functions are used in the scheme to prevent the possibility of unrealistic oscillatory solutions at high Rayleigh numbers. The discretized equations arising from this scheme, including a pressure equation which consumes most of the computation time, are solved by a parallel multigrid method. This method acts as a convergence accelerator and reduces the CPU time significantly for the entire computation.

The main idea of the multigrid approach is to use the solution on a coarse grid to revise the required solution on a fine grid. In view of this, a hierarchy of grids of different mesh sizes is used to solve the fine grid problem. It has been proven theoretically and practically that the multigrid method has a fast convergence rate independent of the problem size [4] [5]. In the present computation, a V-cycle scheme with a flexible number of grid levels is implemented with the Successive Over-Relaxation (SOR) as the smoother. The injection and linear interpolation are used as restriction and interpolation operators, respectively.

4. The Parallel Implementation

The scientific application software design on the modem parallel system produces many challenges. Like its sequential model, the parallel software has a lifecycle comprising such phases as the requirement analysis and specification, the algorithm design, the program language and model, and the code implementation and testing. However, although these phases have similar objectives as those in the sequential software lifecycle, they have many problems of their own that are not encountered in sequential software development, such as code performance, scalability, portability, and importantly, reusability. The design of parallel codes is a very complicated and time-consuming process, and the maintenance and modification of some exiting scientific application codes are even painful if these codes have a poor user interface and poor documentation. Reusing the products of the software development process is an important way to reduce software costs. In the present software design, the reusability, portability, and scalability are discussed.

4.1. The Reusablity

Various levels of the design hierarchy for the software reusability are considered in the present work. At the top level, a friendly user interface module and a flow solver kernal are designed which allow users to solve a range of thermal convection problems with minimum changes in the user interface module and keep the flow solver kernal untouched. So the entire software can be reused for solving many different thermal convection flows, such as side-heating flows, Rayleight-Benard flows, periodical flows, and other flow problems. At the lower levels, there are five equations (u, v, w, p, T) which require a fast algebraic solver at each time step for a general thermal convection problem. Typically, this solution step dominates the costs associated with the finite volume implementation, especially the pressure equation has the highest cost. Therefore, the choice of solution methodology is perhaps one of the most important implementation issues addressed here. The design of a reusable and efficient parallel multigrid kernel is necessary. Besides present thermal convection problems, many other applications should be able to make use of the parallel multigrid kernel and reduce the total implementation time. There are six major modules implemented in our parallel multigrid kernel: a V-cycle and error control module, a smoother-SOR module, a restriction module, an interpolation module, a coarse grid solver module, and an user interface. Based on the above approach, the software is capable for various applications and the reusability of the software is fully explored.

4.2. The Portability

Given the short life cycle of the massively parallel computer, usually on the order of three to five years, the portability of the software across different computing platforms needs to be addressed. The principal types of the portability usually considered are the binary portability (porting the executable form) and source portability (porting the source language representation). The binary portability is clearly desirable, but it is possible only across strongly similar environments. The source portability assumes the availability of the source code, but it provides opportunities to adapt a software unit to a wide range of environments. In the present study, we focus on the source portability. We address this problem at the application level rather than the system level by designing programs that can sense and adapt to a range of environments. The entire software is designed by C computer language, which is easy to handle complicated data structures, and MPI (Message Passing Interface) communication API (Application Programming Interface) for communications, which is supported by most parallel systems, such as distributed-memory systems, shared-memory systems, and cluster systems. Now the software runs on the Cray T3E, the HP X-class and V-class, the Origin 2000, and the Beowulf system without any modification of the source code.

4.3. The Scalability

Using domain decomposition techniques and the MPI, the entire software package is implemented on distributed-memory systems or shared-memory systems capable of running distributed-memory programs. In order to achieve load balance and to exploit parallelism as much as possible, a general and portable parallel structure based on domain decomposition techniques was designed for the three dimensional flow domain. It has 1D, 2D, and 3D partition features which can be chosen according to different geometry requirements. Those partition features provide the best choice to achieve load balance so the communication will be minimized. For example, if the geometry is a square cavity, the 3D partitioner can be used, while if the geometry is a shallow cavity with a large aspect ratio, the 1D partitioner in x direction can be applied. This optimal design of the partitioner allows users to minimize the communication part and maximize the computation part to achieve better scalability.

5. Examples

Various numerical tests have been performed by the package. The numerical scheme is robust and efficient, and the general parallel structure allows us to use different partitions to suit various physical domains. Two problems are considered here. The first example is

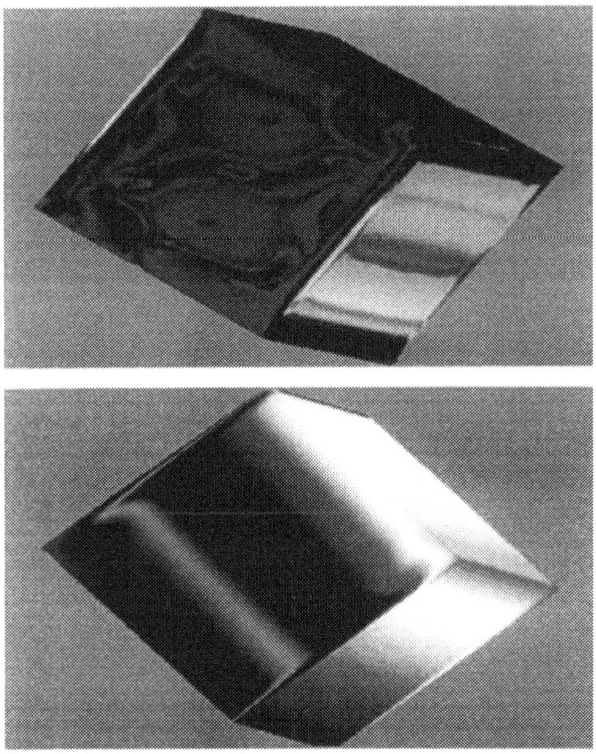

Figure 1. 3D velocity field (top) and temperature field (bottom) for $Ra = 5 * 10^7$ with $\sigma = 0.733$.

the natural convection without rotations in a rectangular cavity of $0 < x < 1$, $0 < y < 1$, and $0 < z < 1$ with differentially heated sidewalls. All other sides use insulating boundary conditions. Here 3D numerical solutions for Rayleigh number 5×10^7 with $\sigma = 0.733$ in $0 \leq x, y, z \leq 1$ are presented, and the solution for $R = 10^7$ was used as the initial state. A grid size $128 \times 128 \times 128$ with a 3D partition was used for the computation. The results are displayed in terms of velocity and temperature. In these figures the hot wall is on the right. All numerical results and visualization have been carried out on the Cray T3E. Figure 1 illustrates the velocity field and the corresponding temperature field. For such a high Rayleigh number, the flow structures become very complicated, and regions of the reverse flow on the horizontal walls are present. Very thin thermal boundary layers are formed on the two sidewalls. The y-variations are very strong near the corners, and eventually affect the main flow. So the overall motion of the flow becomes three-dimensional. The solutions for such high Rayleigh number become strong convective, time-dependent, and periodical.

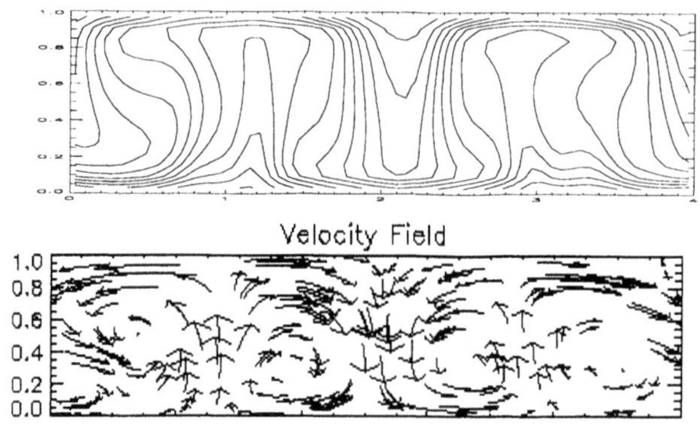

Figure 2. Temperature (top) and velocity (bottom) for $Ra = 10^5$ with $T_a = 10^4$ and $\sigma = 0.733$ at $y = 0.5$.

The second example is the Rayleigh-Benard flow with a rotation in a rectangular cavity of $0 < x < l$, $0 < y < d$, and $0 < z < h$. Numerical results are obtained for a wide range of the Rayleigh number and Tayler number. Here 3D numerical solutions for Rayleigh number 10^5 with $\sigma = 0.733$ and $T_a = 10^4$ in a rectangular cavity of $0 < x < 4$, $0 < y < 1$, and $0 < z < 1$ are given. A 2D partitioner is used for the computation domain, and a small grid size $64 \times 16 \times 16$ is used for the testing problem . All numerical results and visualization have been carried out on a Beowulf system because of the smaller problem size which is an excellent application for the cluster computing. Figure 2 illustrates the temperature and velocity at $y = 0.5$. The figures clearly show the four-rolls convection pattern and thermal plumes.

6. Discussion and Conclusions

We have successfully implemented the finite volume method with an efficient and fast multigrid scheme to solve for three-dimensional, time-dependent, incompressible thermal convection flows on parallel systems. The parallel software is numerically robust, computationally efficient, and portable to any architectures which support MPI for communications. It also has a very flexible partition structure which can be used for any rectangular geometry by applying a 1D, 2D, or 3D partition to achieve load balance. This feature allows us to study various thermal convection flows with different geometries on parallel systems. Because of the optimal implementation of the entire software, the scalability, reusability, and portability are fully explored.

In spite of the difficulties associated with the large Rayleigh number simulation, our results presented here clearly demonstrate the great potential for applying this approach to solve high resolution, large Rayleigh number flow in realistic, 3D geometries using parallel systems. Much higher Rayleigh numbers computations of thermal convection in 3D for various applications are under investigation, which include the deep, rapidly rotating atmospheres of the outer planets, such as the Jupiter and the Saturn in planetary atmospheres science, and the 3D thermal convection in ocean science.

The research described in this paper was performed at the Jet Propulsion Laboratory, California Institute of Technology, under contract to the National Aeronautics and Space Administration. The Cray Supercomputers used in to produce the results in this paper were provided with funding from the NASA offices of Mission to Planet Earth, Aeronautics, and Space Science. The author would like to acknowledge Dr. Peggy Li in JPL for providing the parallel visualization software-ParVox for the graphic work.

REFERENCES

1. P. Wang. Massively parallel finite volume computation of three-dimensional thermal convective flows. *Advances in Engineering Software*, 29:451–461, 1998.
2. P. Wang and R.D.Ferraro. Parallel multigrid finite volume computation of three-dimensional thermal convection. *J. Computers and Mathematics with Applications*, 1999.
3. S.V. Patankar. *Numerical Heat Transfer and Fluid Flow.* Hemisphere, New York, 1980.
4. A. Brandt. Multi-level adaptive solutions to boundary-value problems. *Math. Comp.*, 31:333–390, 1977.
5. S.F. McCormick. *Multilevel Adaptive Methods for Partial Differential Equations.* Frontiers on Applied Mathematics, SIAM, Philadelphia, 1989.

Parallel Computational Fluid Dynamics – Trends and Applications
C.B. Jenssen et al. (Editors)

Development of a Common CFD Platform -UPACS-

Takashi Yamane[a], Kazuomi Yamamoto[b], Shunji Enomoto[b],
Hiroyuki Yamazaki[c], Ryoji Takaki[c], Toshiyuki Iwamiya[c]

[a]Aircraft Propulsion Research Center, National Aerospace Laboratory,
7-44-1 Jindaiji-higashi, Chofu, Tokyo 182-8522, Japan
E-mail:yamane@nal.go.jp

[b]Aeroengines Division, National Aerospace Laboratory

[c]Computational Science Division, National Aerospace Laboratory

UPACS, **U***nified* **P***latform for* **A***erospace* **C***omputational* **S***imulation*, is a project to develop a common CFD platform since 1998 at National Aerospace Laboratory. The project aims not only to overcome the increasing difficulty in recent CFD code programming on parallel computers for aerospace applications which includes complex geometry problems and coupling with different types of simulations such as heat conduction in materials and structure analysis, but also to accelerate the development of CFD technology by sharing a common base code among research scientists and engineers. Currently the development of UPACS for compressible flows with multi-block structured grids is complete and further improvements are planned and being carried out.

1. INTRODUCTION

The progress in CFD and parallel computers during 1990s enabled massive numerical simulations of flow around realistic complicated aircraft configurations, direct optimization of aerodynamic design including structure analysis or heat transfer, complicated unsteady flow in jet engines, and so on. However, the increased complexity in computer programs due to the adaptation to complex configurations, the parallel computation, and the multi-physics couplings brought various problems. One of them is a difficulty in sharing analysis codes and know-how among CFD researchers, because they tend to make their own variations of the programs for their own purposes. The parallel programming is another problem that it made the program much more complicated and the portability was sometimes lost.

In order to overcome the difficulties in such complicated programming and to accelerate the code development, there are some researches to separate flow solvers from parallel processes. DLR has developed a simulation program "TRACE"[1] for turbomachinery flows since early 1990s, while an object oriented framework for parallel CFD using C++ has been proposed by Ohta of High Energy Accelerator Research Organization (KEK)[2].

NAL also has started a project **UPACS**, **U***nified* **P***latform for* **A***erospace* **C***omputational* **S***imulation* in 1998. UPACS is expected to free CFD researchers from the parallel pro-

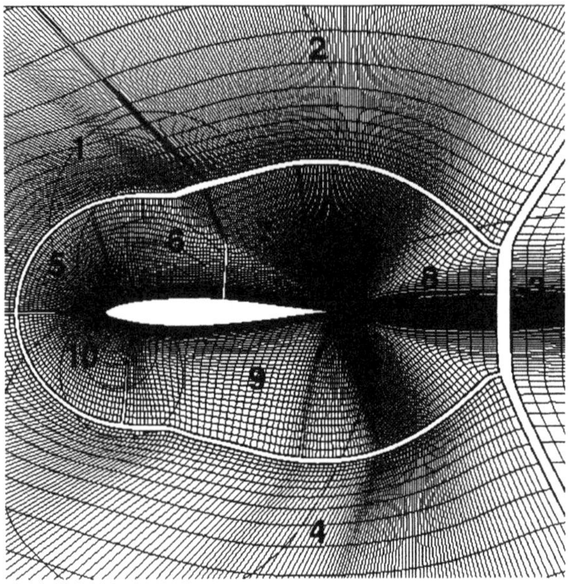

Figure 1. 10 Block Grids around an Aerofoil

gramming, but the biggest advantage would be that both CFD users and code developers can easily share various simulation codes through the UPACS common base.

2. CHRACTERISTICS OF UPACS

The UPACS is based on the following concepts;

- Multi-block structured grid methods
- Separation of CFD solvers from parallel and multi block treatments
- Minimize dependencies on computer archtechtures
- Hierarchical structure of programs and encapsulation of data and calculation procedure
- Support tools for execution of UPACS solver
- Smooth shift from old programs and properties written in Fortran77

2.1. Multi-block sructured grid methods

Multi-block structured grid methods can adapt grid to complex geometric configurations by combination of many structured grids. The UPACS has been designed to be able to handle up to 1000 blocks at this moment. Figure 1 shows an example of 10 block grids around a 2-D aerofoil.

This method should have better reliability and accuracy for aerospace applications, but the difficult and troublesome point is the definition of block-to-block connection information. In order to help this, a pre-process program which searches connection informations among blocks auotmatically has been developed.

2.2. Separation of CFD solvers from parallel and multi block treatment

Based upon the informations obtained from the pre-process program which mentioned above, necessary data for calculation of each block are transfered across processors and the flow calculation inside each block can be treated like a single block solver. With this framework, numerical schemes can be changed by blocks (Fig.2).

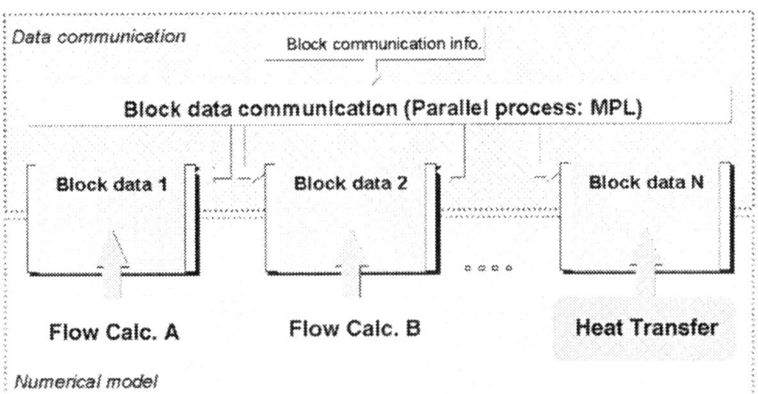

Figure 2. Concept of the separation of parallel multi-block process from numerical solvers

For example, by assigning some blocks to the object material for heat conduction calculation, the external flow field and temperature distribution in the object can be solved at the same time.

Number of blocks in each processor is defined by users freely, but each block must be in one processor. There are two types of data transfer for block communication, processor-to-processor and inside-one-processor, but users need not mind the difference among them.

2.3. Hierarchical structure of programs and encapsulation of data and calculation procedure

The UPACS has been designed by a concept of a hierarchical structure of three layers (Fig.3).

The main loop level corresponds to main program of general CFD codes and controls the iteration process. Some of the parallel processes such as initialization and timing control

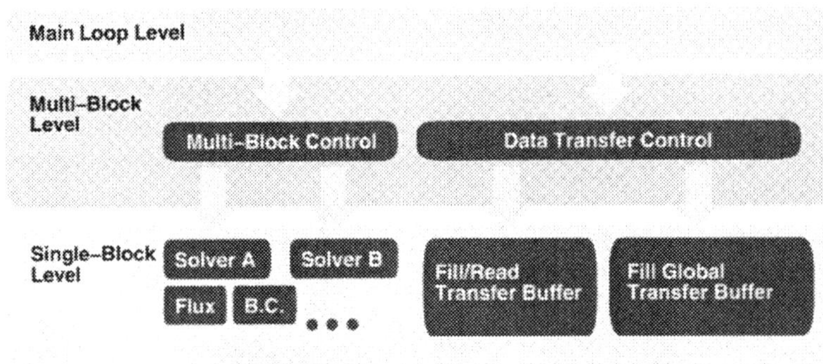

Figure 3. Hierarchical structure of UPACS

of data transfer between blocks are written in the main loop level rouitne, but actual data transfers are treated by the multi-block level in the middle layer which separates the main loop level from the single block level.

The multi-block level takes charge of the assignment of blocks to processors and the communications between blocks. Connecting boundary values which are necessary for the calculation in each block are transfered by this multi-block level, thus calculations inside each block can be made independently and subroutines for single block can be written as if they are conventional CFD programs for one block grid. Variables which are defined for calculations are collected to structures and the multi-block level programs refer only to structure names, so variables can be modified without changing the multi-block level routines.

2.4. Fortran90 and MPI

The concept of "Hierarchical structure of programs and encapsulation of data and calculation procedure" is partly based on the object oriented thinking. C++ is one of the most popular languages for object oriented programs, however, Fortran90 has been chosen for the UPACS. One of the reasons is that the most CFD programs have been developed using Fortran77 in Japan and smooth shift from old programs and property is indispensable. It is advantageous that Fortran usually shows better performance on vector computers than C or C++ and Fortran90 is enough functions for structured programing which is necessary to the UPACS.

MPI[3] has been adopted for parallel process in order to minimize dependencies on computer archtechture but most users and CFD reseachers who want to modify the single block level of the UPACS need not know about MPI.

Figure 4 shows a simplified example of the UPACS program. Subroutines `MPL_init`, `doAllBlocks`, `doAllTransfer` and `MPL_end` which are called in the main program control the multiblock level processes. Subroutines of the single block level are defined in a module (right hand side of the figure) and the subroutine name for single block level is given to multiblock level subroutines. The single block level subroutines can be easily modified or

```
program main
  use upacsManager
  use singleBlockSolver
  ........
  call MPL_init
  call doAllBlocks(initialize1)
  call doAllTransfer(transfer, "grid")
  call doAllBlocks(initialize2)
  ........
  do i=1,imax
     call doAllBlocks(step)
     call doAllTransfer(transfer,"phys")
  end do
  ........
  call doAllBlocks(finalize)
  call MPL_end
end program
```

```
module singleBlockSolver
  use blockDataTypeModule
  private
  public initialize1, ...., step, ...

contains
  subroutine initialize1(block)
    type(blockDataType)::block
    ....
  end subroutine
  ....
  subroutine step(block)
    type(blockDataType)::block
    ....
  end subroutine

end module
```

Figure 4. Structure of UPACS solver program

added without touching the multiblock level routines.

3. CALCULATION PROCEDURE OF UPACS

The actual procedure of flow simulations with the UPACS is shown in Fig.5. In order to execute the solver program "`upacsSolver`", five kinds of files in the following should be prepared;

- Grid data files
- Block connection information file
- Boundary condition definition file
- Parameter database file (PDB file)
- Block-to-processor allocation file

Grid coordinates of each block should be stored in each file which name has a sequential number. Grid points on boundaries must correspond to the grid points on the boundaries of the neighbouring blocks in the current version of the UPACS.

The block connection information file is automatically created by a pre-process program "`createConnect3D`". It also gives a warning which tells the undefined boudary areas, thus users can easily check and write the boudary condition information file. `createConnect3D` plays an important role in making the UPACS for general-purpose flow problems and it enables to write solver programs which are independent of specific shapes and flow conditions.

The parameter database file (PDB file) is a simple text file which contains parameters such as number of iterations, time step, convergence criteria, and so on. A parameter

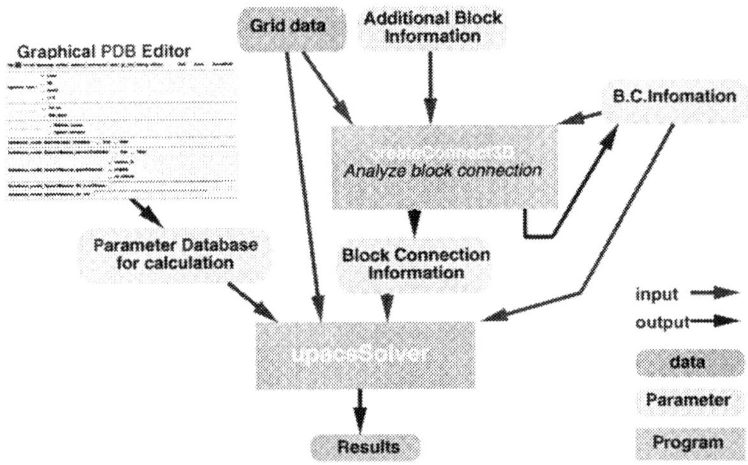

Figure 5. Procedure to use UPACS

name and its value are written in each line of the PDB file but the parameter names need not correspond to variable names in the solver program "`upacsSolver`". `upacsSolver` searches values using parameter names as keys only when the values are required in the program, so parameters which are not necessary for the solver may exist in the PDB file.

Number of parameters in PDB file gets larger as functions of `upacsSolver` increases and it becomes difficult and complicated to set appropriate parameter values. Therefore the PDB editor which is a GUI tools has been developed. It sorts parameters in some groups, shows selections and dependencies between parameters, suggests value ranges, and check inputted values. Currently the PDB editor is being modified that it can easily changed as `upacsSolver` is developed. The new PDB editor will be re-programmed by a definition file which can be written by developers of `upacsSolver`.

Figure 6 is a screen capture image during `upacsSolver` is running. Some physical values on a section can also be displayed simultaneously. A large scale simulation results[4] is shown in Fig.7. This is the supersonic flow around the unmanned, scaled model plane which is under development in the supersonic experimental airplane project of National Aerospace Laboratory. 105 blocks are used for this shape but there is no special modification in UPACS programs for this calculations.

4. CURRENT STATUS AND FUTURE PLAN

The UPACS currently solves compressible flows of perfect gas with the selection of the following numerical schemes.

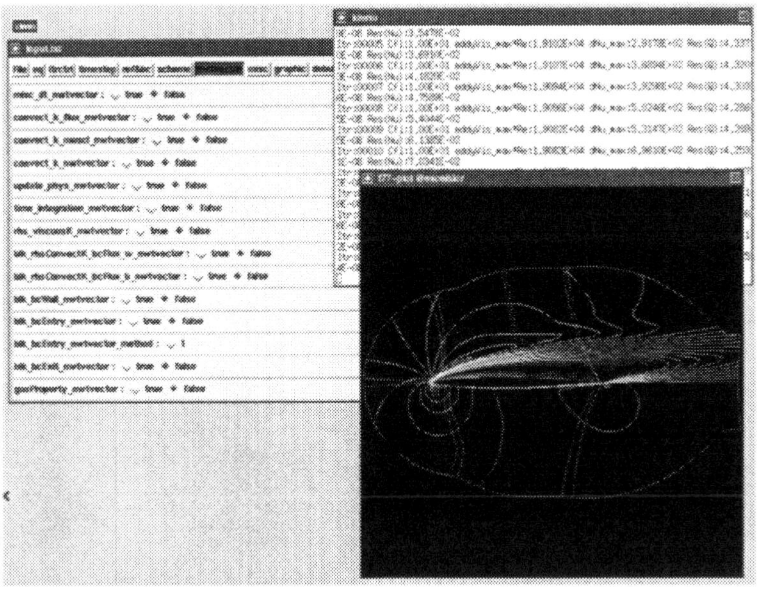

Figure 6. Screen capture of running UPACS

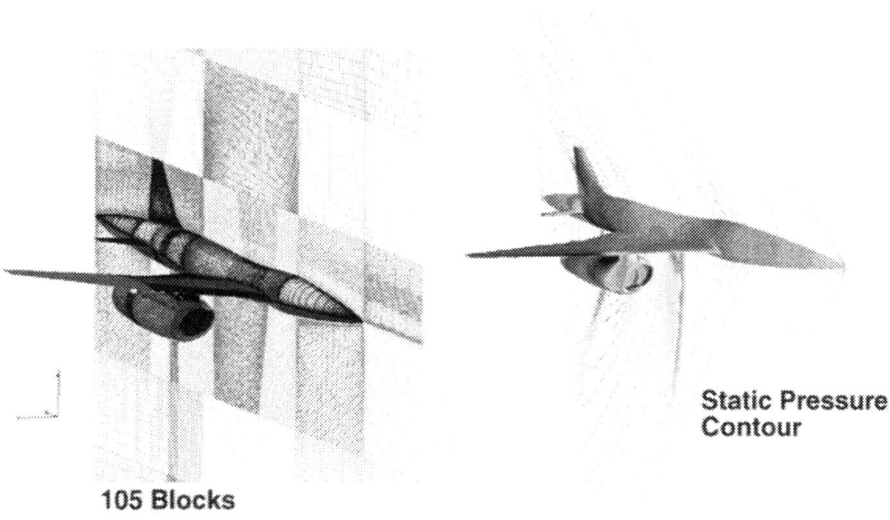

Figure 7. Example of simulation result by UPACS

Convective term	Roe scheme, AUSMDV
Time integration	Runge-Kutta, Matrix Free Gauss Seidel
Turbulence	Baldwin-Lomax, Spalart-Allmaras

Furthermore, the following new functions and modifications are planned. The item marked with ⊙ will be completed during the fiscal year 2000.

- ⊙ Turbomachinery flows
- ⊙ Add various numerical schemes developed in NAL
- • Two equation turbulence models and LES
- • Unstructured Grids
- ⊙ Overset Grids
- ⊙ Conjugate simulations with heat conduction in objects
- • Flows with chemical reactions
- • Conjugate simulations with structure analysis
- • Low Mach number flows
- • Optimization problems using adjoint method

5. CONCLUSION

A common CFD platform "UPACS" which separates flow solvers from parallel processes has been developed in order to accelerate the CFD research and simulation program development, get better portability on various computers, then easily share simulation programs and knowledges among CFD researchers and users. The UPACS solver can effectively simulate flows around complicated geometries with multi-block structured grids using some assisting tools for the UPACS. The UPACS continues its evolution to become a better CFD based tool.

REFERENCES

1. Engel,K., et.al., "Numerical Investigation of the Rotor-Stator Interaction in a Transonic Compressor Stage", AIAA Paper 94-2834
2. Ohta,T., "An Object-Oriented Framework for Parallel Computational Fluid Dynamics" (in Japanese), Proceedings of Aerospace Numerical Simulation Symposium 1998, Special Publication of National Aerospace Laboratory SP-41, 1999
3. "The Message Passing Interface (MPI) stardard", http://www-unix.mcs.anl.gov/mpi/
4. Enomoto,S., "A Structured Grid Method in Simulating Flow around Supersonic Transports", Proceedings of 2nd SST-CFD Workshop, 2000

6. Numerical Schemes and Algorithms

Parallel Computational Fluid Dynamics – Trends and Applications
C.B. Jenssen et al. (Editors)

Numerical investigation of viscous compressible gas flows by means of flow field exposure to acoustic radiation

A.V. Alexandrov [a], B.N. Chetverushkin[b] and T.K. Kozubskaya[b]

[a] Lomonosov Moscow State University, Physical Dept., Vorob'evi Gory, Moscow 119899, Russia

[b]Institute for Mathematical Modelling of Rus.Ac.Sci., 4-A, Miusskaya Sq., Moscow 125047, Russia

In the paper, a new numerical technique of mean flow exposure to acoustic radiation is proposed. The method concludes in the realization of numerical experiment consisting in the examination of noise responses from stochastic acoustic radiation. The technique of such numerical experiment was applied to solving a problem on acoustic noise propagation in a 2D rectangular cavity as well as in a 3D rectangular groove.

1. INTRODUCTION

As it is well known, the acoustic noise arising within gas flows can significantly influence the whole gasdynamic process. For instance, it may negatively affect the structure, may cause a great discomfort both for the airplane (or car) passengers and the people around. So an adequate simulation of acoustic noise becomes a problem of high importance in engineering. In this sense, it is particularly needed to predict not only the mean flow field, but spectral characteristics of noise and the frequencies of higher intensity. In practice, this knowledge provides the detection of critical parameters of a system when a possible destructive influence reaches its fullest value.

In this paper, a numerical tool for the determination of spectral properties of a flow is presented. It is based on the exposure of a mean flow field to stochastic acoustic radiation. In this case, all the corresponding processes are simulated numerically. This technique could be considered as a flow "computer tomography". Besides the detection of flow spectral characteristics, the technique is proposed to be used at the modeling of acoustic sources distributed within flow shear layers and separated zones, namely, for finding the source frequencies.

The difficulty in numerical prediction of aeroacoustics problems results from a small scale of acoustic pulsation especially in comparison with large scale oscillations of gasdynamic parameters. That small scale places strict constrains on the numerical algorithms in use and requires powerful computer facilities. These requirements are further strengthened due to the necessity of using highly refined computational meshes for resolving small scale acoustic disturbances. Actually, higher refined meshes allow the production of flow pattern in more detail. That's why parallel computer systems appear particularly appealing for use in the prediction of aeroacoustic phenomenon.

Parallel distributed memory computers combined with the domain decomposition parallelization technique make more effective the explicit schemes. In particular, the linearized kinetically consistent finite difference (KCFD) schemes [1] are used in the work presented. Such properties of these schemes as homogeneity and soft stability condition provide their robustness and reliability at rather high efficiency especially at using multiprocessor computer systems. The potentials of the usage of linearized KCFD schemes are first shown in [3].

2. MATHEMATICAL FORMULATION

The linearized Navier-Stokes equations in Cartesian coordinates with respect to conservative pulsation variables are used for the mathematical description of acoustic noise propagation. The corresponding linearization procedure is detailed in [2]. Here it is presented in the following compact vector form

$$\frac{\partial \mathbf{Q}'}{\partial t} + \frac{\partial(\mathbf{A}^x\mathbf{Q}')}{\partial x} + \frac{\partial(\mathbf{A}^y\mathbf{Q}')}{\partial y} + \frac{\partial(\mathbf{A}^z\mathbf{Q}')}{\partial z} = \frac{\partial \mathbf{R}'_{NS}}{\partial x} + \frac{\partial \mathbf{S}'_{NS}}{\partial y} + \frac{\partial \mathbf{T}'_{NS}}{\partial z} + \mathbf{S}', \tag{1}$$

where $\mathbf{U}' = (\rho', (\rho u)', (\rho v)', E')^T$ - the linearized conservative variables vector, $\mathbf{S}'$ presents acoustic noise sources (point or distributed) . The above equation system is taken as a base for the acoustic noise prediction.

3. NUMERICAL TECHNIQUE

A numerical technique taken to predict noise propagation phenomena is the linearized KCFD schemes. The details on this numerical algorithm can be found in [2]. In a vector form, it can be written as

$$\mathbf{Q}'_t + (\mathbf{A}^x\mathbf{Q}')_{\overset{\circ}{x}} + (\mathbf{A}^y\mathbf{Q}')_{\overset{\circ}{y}} + (\mathbf{A}^z\mathbf{Q}')_{\overset{\circ}{z}} =$$
$$\left(\frac{\tau^x}{2}(\mathbf{B}^x\mathbf{Q}')_x\right)_{\bar{x}} + \left(\frac{\tau^y}{2}(\mathbf{B}^y\mathbf{Q}')_y\right)_{\bar{y}} + \left(\frac{\tau^z}{2}(\mathbf{B}^z\mathbf{Q}')_z\right)_{\bar{z}} + \langle \mathbf{H}'_{NS} \rangle + \langle \mathbf{S} \rangle . \tag{2}$$

where $\langle \mathbf{S} \rangle$ - linearized source term, $\langle \mathbf{H}'_{NS} \rangle$ -linearized difference Navier-Stokes terms.

All the coefficients of this scheme are space and time variables consisting of mean flow parameters which are considered known in advance.

Owing to the explicitness, the KCFD schemes appear ideally adapted to the distributed memory architecture of multiprocessor computer systems. This property combined with the acceptable soft stability condition of Courant type [1] as well as with the homogeneity of these schemes make the KCFD schemes a robust and efficient numerical algorithm for solving gasdynamics problems on massive parallel computers. This fact comes into importance especially taking into account the necessity of using highly refined meshes for the adequate prediction of small-scale acoustic fluctuations [2].

4. PARALLEL REALIZATION

The domain decomposition technique, namely, its simplest variant is used for parallelizing. In doing so, the whole computational domain is divided into a set of subdomains in accordance with a number of processors available. The data exchange is conducted only along the subdomain boundaries. In order to provide processor load balancing, the subdomain configuration is chosen by requiring an approximate equality of nodes number. Besides, to reduce the time of data exchange, the boundary lengths have to be minimal. This approach provides a good scalability of parallel algorithm as well as its portability. The corresponding parallel codes are written in C for the computation on the parallel distributed memory computer systems.

5. NUMERICAL RESULTS ON FLOW EXPOSURE TO ACOUSTIC RADIATION

The proposed procedure of mean flow exposure to acoustic radiation is implemented in the following way. Firstly, it is assumed that the mean flow field is known in advance. It may be recovered from the experimental data, or it may be predicted by the Reynolds averaged Navier-Stokes equations closed by some turbulence model. To reveal the spectral characteristics of a mean flow, it is proposed to simulate numerically the process of acoustic noise propagation from artificial stochastic noise sources specially prearranged at the boundaries. In doing so, the spectral analysis of acoustic noise detected in the internal points gives the information about characteristic frequencies.

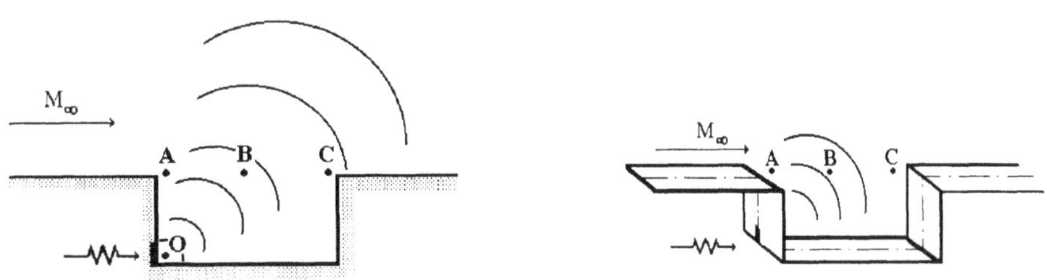

Figure 1. Propagation of acoustic noise from artificial sources in rectangular 2D cavity (left) and 3D rectangular groove (right)

The technique of such numerical experiment is applied to solving a problem on acoustic noise propagation in a 2D rectangular cavity as well as in a 3D rectangular groove formed as a result of 2D cavity extension in z-direction at a terminal distance (see Fig.1). Such 3D problem formulation allows to avoid the need in 3D mean flow field since in this case, it is possible to use the known 2D field at each transverse section. Besides, this 3D formulation completely corresponds to the 2D cavity problem, and this fact seems preferable for the correlation of the numerical results.

Both the 2D cavity and 3D groove are flowed around by a supersonic viscous gas flow as it is shown in Fig.1. The following geometric and gasdynamic parameters of the problems considered are chosen: the ratio of cavity length (groove width) to its depth $L/D = 2.1$, free stream Mach number $Ma = 3.7$, Reynolds numbers $Re = 33000$ and $Re = 3300$ (for cavity) and $Re = 33000$ (for groove).

The artificial noise source simulating surface oscillations is prearranged at the left wall in the immediate vicinity of a bottom. It has a form of a vertical segment of length approximately equal to the tenth part of a wall height. Only pressure pulsation and the corresponding density pulsation at the normal temperature are prescribed. The acoustic pulsation amplitude is taken two order less than the corresponding mean flow value. The stochastic character of acoustic noise emitted from the artificial sources are prescribed as random on time according the law

$$A'_{\text{source}}(y,t) = a(y)\sin(2\pi f_{\text{rand}(t)}t)$$

The random frequencies uniformly varies from 1 to 200 KHz, to cover the acoustic diapason. The points O (close to the source), A (at a left edge), B (in the middle of shear layer formed) and C (at a right edge) (Fig.1) are chosen as specific to study the spectral flow properties. In the case of 2D rectangular cavity, the computation is performed on a set of three condensable meshes '1x1', '2x2' and '4x4'. The mesh refinement is performed on the base of the most coarse mesh '1x1' by a sequential bisection of its cells.

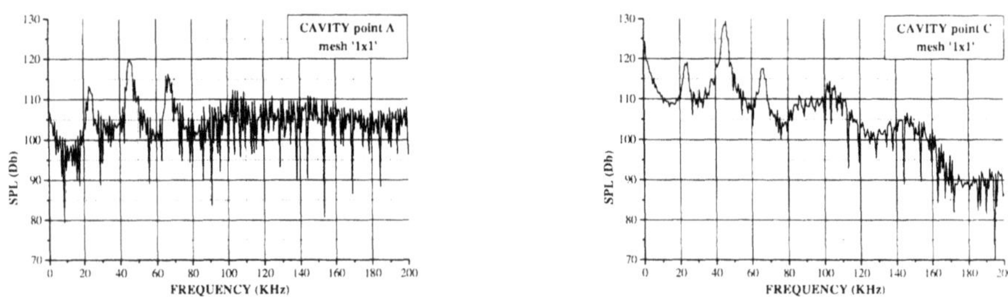

Figure 2. Sound pressure levels in cavity points A (left picture) and C (right picture) - mesh '1x1'

The point C placed at the right cavity edge appears the most responsive to the acoustic radiation affect since the modes of the maximal intensities are detected in it. The peaks found with the help of the above numerical technique quite adequately corresponds to the characteristic frequencies of the system which are detected by natural experiments and confirmed by numerical results at the usage of nonlinear Navier-Stokes equations.

In the case of refined meshes the maximal peaks appear lower and the amplified frequency modes are slightly moved to the right. This fact has to be investigated more carefully. In any case, it testifies that the numerical technique used for the acoustics predictions deserves a great attention.

As for the 3D groove, the spectra appear more "smooth", that is the amplified modes are much worse seen. The Fig.(5) demonstrates the pressure pulsation spectra in the

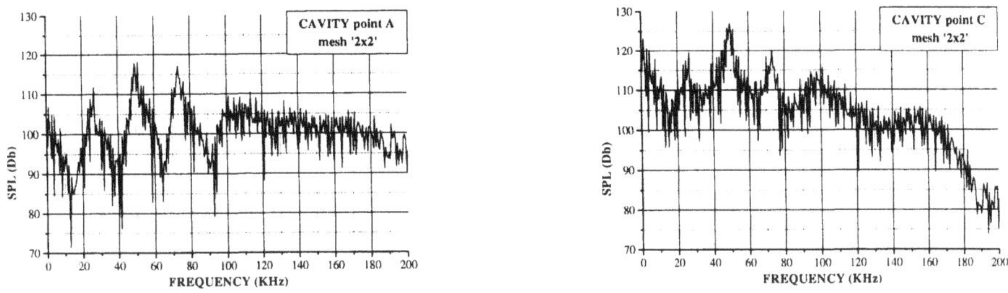

Figure 3. Sound pressure levels in cavity points A (left picture) and C (right picture) – mesh '2x2'

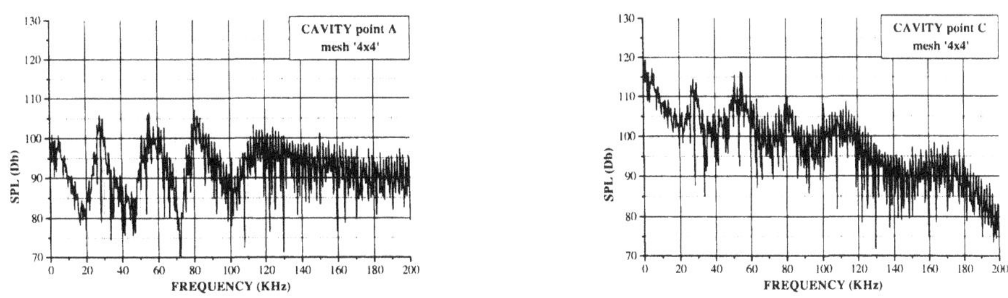

Figure 4. Sound pressure levels in cavity points A (left picture) and C (right picture) – mesh '4x4'

point C (upper diagram) compared with the spectrum in the point CL moved from C along z-direction (lower diagram). It is obvious that the maximal sound pressure level corresponds to the point C. The predictions are carried on the mesh '1x1x1' constructed on the base of coarse 2D mesh '1x1'.

The numerical procedure of mean flow exposure to stochastic acoustic radiation is intended to the study of internal flow properties. It can be applied to about any problem configuration. An important application of this technique is the determination of frequencies of acoustic noise sources distributed within shear layers and separated zones. The main idea here is the supposition that a real flow produce its real noise on the characteristic frequencies, pulsation on which is amplified at the flow exposure to stochastic noise radiation.

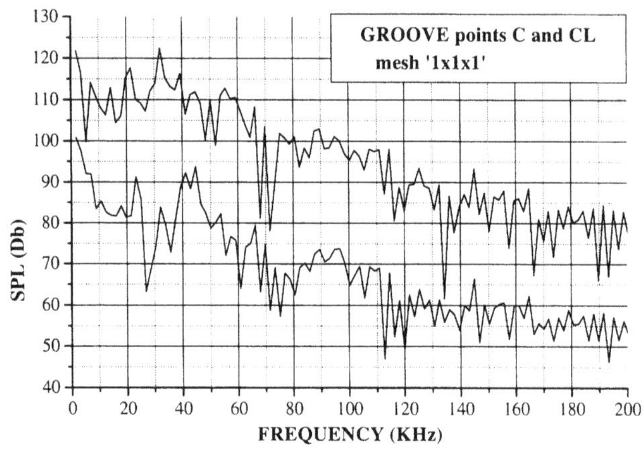

Figure 5. Sound pressure levels in groove points C (upper diagram) and CL (lower diagram) - mesh '1x1x1'

6. CONCLUSION

In the paper, a new numerical technique of mean flow exposure to acoustic radiation is proposed. The method concludes in the realization of numerical experiment consisting in the examination of noise responses from stochastic acoustic radiation emitted from artificial sources in different internal points of a flow. The spectral flow characteristics revealed in such a way can be helpful in engineering, for instance, at the aircraft designing as well as they can give the information about natural distributed noise sources needed for the further simulation of acoustic fields of turbulent flows. The algorithm developed shows its efficiency at solving 2D and 3D test problems for standard aerodynamic configurations.

The work reinforces the statement that it is of singular importance to use parallel computer systems for solving aeroacoustics problems. This is strongly needed because of the necessity to use maximum refined meshes, in order to resolve small-scale acoustic disturbances. Nowadays, powerful multiprocessor computers seem the most preferable to satisfy this requirement.

ACKNOWLEDGEMENT

This work was supported in part by Russian Foundation for Basic Research (Grants No. 99-07-90388, 00-01-00263) and French-Russian A.M. Lyapunov Institute of Applied Mathematics and Informatics (Project 99-02). The computations were performed with the use of German Parsytec PowerExplorer and Parsytec CC-12 (processors PowerPC 601 and PowerPC 604 correspondingly) parallel computer systems.

REFERENCES

1. Elizarova, T.G., and Chetverushkin, B.N. Kinetically consistent schemes for simulation of viscous heat-conductive gas flows. — Journal of Computational Mathem. and Math. Phys. 1988, Vol. 28, N 11, pp. 1695-1710.
2. Antonov, A.N., and Kozubskaya, T.K. Acoustic noise simulation for Supersonic compressible gas flows. — In the Proceedings of ECCOMAS'98 Computational Fluid Dynamics Conference, Athens, Greece, September, 1998, John Wiley & Sons, Chichester, 1998, Vol. 1, Part 2, pp. 48–53.
3. Alexandrov, A.V., Antonov, A.N. and Kozubskaya, T.K. Simulation of Acoustic Wave Propagation through Unsteady Viscous Compressible Gas Flows. — In the Proceedings of Parallel CFD'97 Conference, Manchester, England, May, 1997, Elsevier, Amsterdam, 1997.

Parallel Computational Fluid Dynamics – Trends and Applications
C.B. Jenssen et al. (Editors)

A new low communication parallel algorithm for elliptic partial differential equations

A. Averbuch [a], E. Braverman [b] and M. Israeli [b]

[a]School of Mathematical Sciences, Tel Aviv University, Tel Aviv 69978, Israel

[b]Technion-Israel Institute of Technology, Computer Science Department, Haifa 32000, Israel

We present a low communication, non-iterative, parallel algorithm for a high order (spectral) solution of constant coefficient elliptic equations like the Poisson or the modified Helmholtz equation. The domain is decomposed into non overlapping subdomains, which can be flexibly assigned to distinct processors. Particular solutions are found in subdomains and subsequently matched by the introduction of doublet or singlet layers at the interfaces, resulting in a smooth global solution. The effects of the matching layers must be computed at all interfaces. This is effectively done using multiwavelets bases for computation and compression. Taking into account the smoothness and decay properties of the kernels under consideration. A considerable reduction in global communication is achieved.

1. INTRODUCTION

Solution of initial boundary value problems for large scale nonlinear evolution equations is often required in engineering and scientific applications. Some examples are: incompressible Navier-Stokes equations, problems in elasticity, cosmology, material science, semiconductor device simulation. In order to solve effectively evolution equations with parabolic terms (in the range of parameters of interest for many applications), we resort to the use of semi-implicit discretization in time [1], this device removes the stringent diffusive stability limit. For example, consider the incompressible Navier-Stokes equation

$$\frac{\partial u}{\partial t} = \nu \nabla^2 u - (u \cdot \nabla) u - \nabla p, \tag{1}$$

$$\nabla \cdot u = 0, \qquad \text{in } \Omega \subset R^2 \tag{2}$$

with the boundary conditions

$$u = \phi(x, y) \qquad \text{on } \partial\Omega. \tag{3}$$

The semi-implicit discretization of (1),(2), (3) give rise to one Poisson for the pressure p and two (in 2D) or three (in 3D) modified Helmholtz equations for the velocity components [1] to be solved at each time step. Thus a robust and efficient solution of elliptic equations becomes an important tool for the solution of time dependent problems in computational fluid dynamics.

In this paper the parallelization of the serial algorithm is achieved by decomposition of the computational domain into smaller domains, mainly of rectangular (cubic) shape. The solution in each rectangular domain is based on the fast and accurate algorithm developed in [2,3] for the 2-D case, and in [4] for the 3-D case. The particular solutions, obtained by this approach, are not smooth across domain interfaces. The discontinuities can be removed by adding singularity layers. The effect of a layer at any interface on all other interfaces must be computed. The computational work can be considerably reduced if the potential operators can be represented by sparse matrices. A basis that was built in [5] was chosen due to its favorable localization characteristics which avoid Gibbs phenomena. The resulting sparse representation reduces both the computation and the communication time required for the parallel processing.

An algorithm for the fast solution of the Poisson equation by decomposition of the domain into many small square domains and the subsequent matching of these solutions using the fast multipole method was developed in [6]. However the fast implementation of the algorithm in [6] was stipulated by the division of the global domain into squares with a small number of points in each subdomain. Because a $O(N^3)$ algorithm is applied for the solution of the Poisson equation in the subdomains (here N is a number of points in each direction). Our algorithm has complexity of $O(N^2 \log N)$ in each subdomain. Thus larger subdomains can be efficiently resolved and the number of subdomains is determined by the number of processors in the multiprocessor environment and the behavior of the RHS (in the adaptive version).

The construction of the global solution by matching of the independent local solutions in subdomains is in principal of the same complexity as the computation of the solutions inside the subdomains. The matching procedure involves the computation of the influences at a distance of singular layer. For a moderate number of interfaces we thus achieve an efficient solution for the matching step.

As the number of subdomains increases however the work in the matching step grows approximately as the square of the number of interfaces to be matched. An efficient matching step in the case of many interfaces requires therefore a drastic reduction of communication and computation volume at especially at large distances from the source (the singular layer). The usage of multiwavelets in the discretizations of the various kernels and for compression of all the source arrays (due to the jumps) gives rise to such an economical algorithm. Here the general theory indicates sparsity of the matrices representing the pseudo-differential operators we use. Furthermore the matrices become more sparse as their size grows.

The algorithm has an inherently parallel structure. Therefore, we do not have to modify the serial algorithm when it is ported onto a parallel computer.

2. SOLUTION OF POISSON AND MODIFIED HELMHOLTZ EQUATIONS

2.1. Statement of the problem

We solve the Poisson equation

$$\Delta u = f(x, y) \quad \text{in} \quad \Omega, \tag{4}$$

and a modified Helmholtz equation

$$\Delta u - \lambda^2 u = f(x,y) \quad \text{in} \quad \Omega, \tag{5}$$

with the Dirichlet

$$u = \Phi(x,y) \quad \text{on} \quad \partial\Omega \tag{6}$$

boundary conditions.

A practical implementation in the case of turbulent flow in a long channel induces the following geometry, the domain is a rectangle $\Omega = [0, L] \times [0, 1]$ as demonstrated in Fig. 1.

The rectangular domain was decomposed into L subdomains. First the particular solutions of either (4) or (5) were found in each subdomain. Then the singularity layers were introduced to match these solutions.

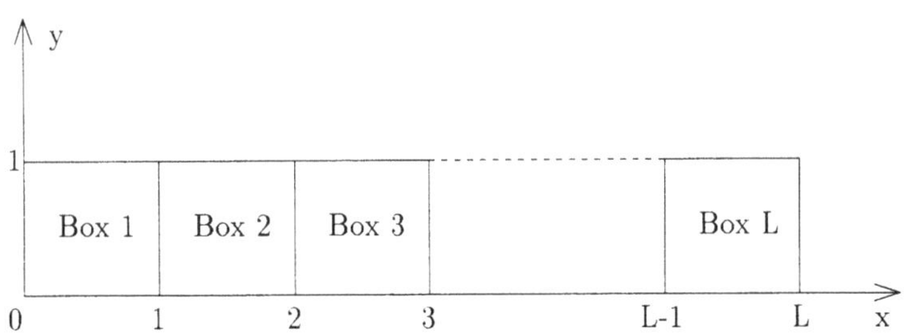

Figure 1: The rectangular domain is decomposed into L subdomains. Singularity layers are introduced on the $L - 1$ interfaces

2.2. Outline of the algorithm

The parallel algorithm consists of the following steps.

1. **Solution of the non-homogeneous equation in each subdomain.**
   Each subdomain is covered by a $N \times N$ grid, there are L subdomains. The right hand side of the Poisson or the non-homogeneous modified Helmholtz equation is defined on these grids and artificial local boundary conditions are determined subject to certain consistency requirements. If we use Dirichlet boundary conditions for each subdomain they have to match for adjacent subdomains and also they have to match the non-homogeneous right hand side at the knot points where the Laplacian can be computed from the boundary conditions. Now the solution of the non-homogeneous equation can be found in each subdomain using the modified Fourier spectral method of high accuracy [2,3]. The complexity of this step is $O(N^2 \log N)$.

2. **Matching of discontinuities.**
   In order to remove the discontinuities in the derivatives between the subdomains, the jump of the first derivative on the two sides of the interface is computed. This

requires only local communication between processors corresponding to adjacent subdomains. Before transmission, the difference function is compressed by projection on the multiwavelet basis [5] and thresholding. The remaining multiwavelet coefficients are transmitted from each processor to all other processors. A singularity layer based on these coefficients patches the discontinuity and modifies the solution everywhere else. Each processor can now compute the influence of all the layers on the interfaces which touch its own subdomains. Operators of the influences are also presented in wavelet bases, which leads to sparse matrices citeour. Sparse representations has the advantage of minimizing the operation count during the application of the operator. Dealing with sparse matrices also leads to a decrease of the computational time. The accumulation of all the influences determines the final local boundary conditions.

3. **Solution of the homogeneous equations in subdomains.**
The Laplace or the homogeneous modified Helmholtz equation is solved in each subdomain with the boundary conditions which were computed at the previous step. Combining the result with the solution of the non-homogeneous equation leads to a smooth global solution.

2.3. Numerical results

Assume that u is the exact solution and u' is the computed solution. In the examples we will use the following measures to estimate the errors:

$$\begin{aligned} \varepsilon_{MAX} &= \max \|u_i' - u_i\| \\ \varepsilon_{MSQ} &= \sqrt{\frac{\sum_{i=1}^{N}(u_i'-u_i)^2}{n}} \\ \varepsilon_{\mathcal{L}^2} &= \sqrt{\frac{\sum_{i=1}^{N}(u_i'-u_i)^2}{\sum_{i=1}^{N} u_i^2}} \end{aligned} \tag{7}$$

Example 1. We solve the Poisson equation with the boundary conditions corresponding to the exact solution which is a Gaussian function

$$u(x,y) = \exp\left\{-\alpha\left((x-x_0)^2 + (y-y_0)^2\right)\right\}, \tag{8}$$

with $x_0 = 1.5,\ y_0 = 0.5,\ \alpha = 2$ in the domain $[0,3] \times [0,1]$ divided into three equal boxes. Table 1 presents the errors obtained.

$N_x \times N_y$ in each box	ε_{MAX}	ε_{MSQ}	$\varepsilon_{\mathcal{L}^2}$
32×32	6.8e-6	2.0e-6	4.4e-6
64×64	3.9e-7	1.1e-7	2.4e-7
128×128	2.3e-8	6.6e-9	1.4e-8
256×256	1.4e-9	4.0e-10	8.5e-10
512×512	8.7e-11	2.4e-11	5.2e-11

Table 1: MAX, MSQ and $\mathcal{L}^2$ errors for the Poisson equation with the exact solution $u(x,y) = \exp\{-2(x-1.5)^2 - 2(y-0.5)^2\}$.

Example 2. We solve the modified Helmholtz equation (5) with $\lambda = 1$ and the same exact solution as in Example 1. The domain $[0,4] \times [0,1]$ is decomposed into four equal boxes. Table 2 presents the errors obtained.

$N_x \times N_y$ in each box	ε_{MAX}	ε_{MSQ}	$\varepsilon_{\mathcal{L}^2}$
32×32	8.5e-7	2.2e-7	3.5e-7
64×64	5.9e-8	1.6e-8	2.4e-8
128×128	3.9e-9	1.0e-9	1.6e-9
256×256	2.5e-10	6.6e-11	1.0e-10
512×512	1.6e-11	4.2e-12	6.5e-12

Table 2: MAX, MSQ and $\mathcal{L}^2$ errors for the modified Helmholtz equation with the exact solution $u(x,y) = \exp\{-2(x-1.5)^2 - 2(y-0.5)^2\}$ and $\lambda = 1$.

3. PARALLEL IMPLEMENTATION

3.1. Communication steps

The above algorithm incorporates two communication steps:

1. When the algorithm runs on a parallel computer, each subdomain is assigned to a distinct processor. Each processor sends to the adjacent processor either the values of the solution at the common boundary or the values for the first normal derivative at the same boundary. With these values every processor can evaluate the jumps in function (or normal derivative) at all its boundaries. This takes $O(N)$ operations.

2. In the second step the interface jump vectors are compressed by being represented in the multiwavelet bases. This information has to be transferred from each processor to all processors. Since we transfer between the processors only the information on the interfaces, the price is $O(NL)$, L is the number of layers. The representation by multiwavelets reduces the data to be transmitted to $O(L \log N)$, even to the nearest neighbors.

The algorithm was tested on the SGI Origin 2000 computer and IBM SP2 with the MPI interface. When the number of grid points in each of four subdomains is 256×256, on the SGI the communication time took 1% - 1.5% of the run time, on the IBM SP2 it took 0.1% - 0.3 %. The communication/computation time decreases as N grows. On the other hand, it increases with L. When the number of processors grows, the communication time can reach 10% - 15% of all the computational time. Communication time can be reduced by using efficient collective communication, since the same information (the density of singular layers) is to be distributed to all the processors. The timing results were presented in [7].

The algorithm was also adapted to shared memory systems, such as SGI parallel computer. It is to be emphasized that the algorithm is synchronized only twice. First, the densities of the singular layers are computed after each processor has finished the solution of the Poisson equation and the computation of either function values or normal derivatives at the interfaces. The second barrier is aimed to avoid the evaluation of the singular layers influence before all the layers are computed. Thus we obtain high efficiency when the parallel shared memory algorithm is compared to the same in its serial implementation.

3.2. Reduction of the communication time

Originally, each neighboring subdomain (processor) transferred to each other the jumps of the function or its first derivative represented in wavelet bases on the interface. Afterwards each processor computed all the correction functions at its interfaces. A more economical procedure utilizes the decay and smoothing out of the correction functions with distance. Here, each processor can compute the influence of its singulars layers on all the interfaces and transmits it (represented in multiwavelet bases). The number of coefficients decays with the distance from the source. Figure 2 illustrates the dependence of the total number of coefficients (negligible coefficients below a threshold of 10^{-12} are not transmitted) on L, the number of the interfaces where the influence of the singularity layer was computed.

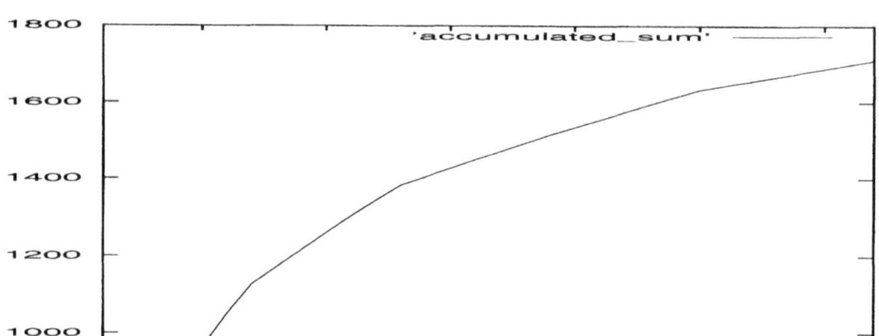

Figure 2: Total number of significant coefficients in multiwavelet representation (with 4 vanishing moments) when the influences of the double layer at all the parallel interfaces up to a certain distance are computed. The double layer was chosen near the peak of a steep Gaussian. The threshold is $\varepsilon = 10^{-12}$.

In the present simple configuration one does not have to use singularity layers. The singularity layers in effect add a discontinuous harmonic function to both sides of an interface in order to compensate for the discontinuity of the solution or its derivative. Instead of introducing singularity layers one can add a sum of certain harmonic functions (or functions satisfying the modified Helmholtz equation). Each of these functions has an exponential decay. The decay is faster for the modified Helmholtz equation. In Table 3 and table 4 we evaluate the number of coefficients above the threshold for the matching functions as a function of the distance from an interface for the case of the modified Helmholtz equation. The values are compressed in multiwavelet basis with four vanishing moments, the number of points on the interface is 512. The exact solution is a sum of 12 random Gaussians

$$\sum_{k=1}^{12} \exp\{-\alpha_k((x-x_k)^2+(y-y_k)^2)\}, \tag{9}$$

with $0.2 \leq \alpha_k \leq 7$. Table 3 presents the results for $\lambda = 1$ and table 4 for $\lambda = 10$.

Dist. / ε	10^{-5}	10^{-6}	10^{-7}	10^{-8}	10^{-9}	10^{-10}	10^{-11}
0	21	31	53	313	471	493	493
1	10	15	28	42	110	418	490
2	4	8	15	23	30	56	395
3	3	4	7	7	12	15	27
4	2	3	5	5	8	14	22
5	1	3	3	5	5	9	15
6	1	2	3	4	5	5	8
7	0	1	3	3	4	5	8
8	0	0	2	3	4	5	5
9	0	0	1	2	3	4	5
10	0	0	0	1	3	4	4
11	0	0	0	1	2	3	4
12	0	0	0	0	1	3	3
13	0	0	0	0	0	2	3
14	0	0	0	0	0	1	2

Table 3: The number of coefficients above the threshold for the matching functions as a function of the distance from an interface for the case of the modified Helmholtz equation with $\lambda = 1$. The multiwavelet basis is with four vanishing moments, the total number of coefficients is 512.

4. SUMMARY AND CONCLUSIONS

We developed a low communication parallel algorithm based on domain decomposition. The present algorithm enjoys the following properties:

Dist. / ε	10^{-5}	10^{-6}	10^{-7}	10^{-8}	10^{-9}	10^{-10}	10^{-11}
0	22	31	55	335	459	490	490
1	2	4	6	9	16	27	36
2	0	0	0	0	2	2	4
3	0	0	0	0	0	0	0

Table 4: The number of coefficients above the threshold for the matching functions as a function of the distance from an interface for the case of the modified Helmholtz equation, with $\lambda = 10$. The multiwavelet basis is with four vanishing moments, the total number of coefficients on the interface is 512.

1. The algorithm is adaptive (see [7]) since the resolution in a subdomain depends on the smoothness of the right hand. Moreover, the domain decomposition can be adaptive as in [6].

2. Low communication is achieved by an efficient representation of singularity layers or their influences (alternatively, the values of matching functions) at the other interfaces.

The method can be extended to a 3D case when a double layer is introduced on the interface between the cubic subdomains. This is a step in the algorithm for the fast solution of 3D Navier-Stokes equation using domain decomposition. The fast solution of 3D Poisson or Helmholtz equation in a cubic domain is described in [4] and [8].

The algorithm can be also developed such that the matching has a more economic (hierarchical) structure (at each step two adjacent boxes are matched, then we deal with larger adjacent boxes etc.). We presented these results on the 13th conference on Domain Decomposition in Lyon and they will appear in the proceedings of this conference.

REFERENCES

1. 1. E.G. Karniadakis, M. Israeli, S.A. Orszag, High-Order Splitting Methods for the Incompressible Navier-Stokes Equations, J. Comput. Phys. 97 (1991) 414.
2. A. Averbuch, M. Israeli, L. Vozovoi, On fast direct elliptic solver by modified Fourier method, Numerical Algorithms 15 (1997) 287.
3. A. Averbuch, M. Israeli, L. Vozovoi, A fast Poisson solver of arbitrary order accuracy in rectangular regions, SIAM J. Sci. Comput., 19 (1998) 933.
4. E. Braverman, M. Israeli, A. Averbuch, L. Vozovoi, A fast 3-D Poisson solver of arbitrary order accuracy, J. Comput. Phys. 144 (1998) 109.
5. B. Alpert, G. Beylkin, R. Coifman and V. Rokhlin, Wavelet-like bases for the fast solution of second-kind integral equations, SIAM J. Sci. Comput. 14 (1993) 159.
6. L. Greengard, J.-Y. Lee, A direct adaptive Poisson solver of arbitrary order accuracy, J. Comput. Phys. 125 (1996) 415.
7. A. Averbuch, E. Braverman, M. Israeli, Parallel adaptive solution of a Poisson equation with multiwavelets, SIAM J. Sci. Comput 22 (2000) 1053.
8. E. Braverman, M. Israeli and A. Averbuch,, A fast spectral solver for 3-D Helmholtz equation, SIAM J. Sci. Comput. 20 (1999) 2237.

Parallel Computational Fluid Dynamics – Trends and Applications
C.B. Jenssen et al. (Editors)

Parallel Multigrid on Cartesian Meshes with Complex Geometry

Marsha Berger[a*] and Michael Aftosmis[b] and Gedas Adomavicius[a*]

[a]Courant Institute, New York University,
251 Mercer St., New York, NY 10012

[b]NASA Ames Research Center,
Mail Stop T27B, Moffett Field, CA 94035

We describe progress towards an efficient and scalable inviscid flow solver for complex geometries using Cartesian grids. The Cartesian approach greatly simplifies the grid generation process. In the flow solver, Cartesian grids have regularity and locality properties that improve cache and parallel performance. Significant issues include domain partitioning for parallel computation and mesh coarsening for multigrid. We show that an on-the-fly space-filling curve approach leads to locally balanced partitions and to an effective multigrid coarsening algorithm. As a serial algorithm, our multigrid convergence rate is as good as any for realistic steady transonic flow simulation. On an SGI Origin 2000, we achieve nearly linear speedups with up to 64 processors.

1. INTRODUCTION

Fluid flow simulations about realistic aircraft (with pylons, nacelles, rudders, flaps, etc.) often require weeks of interactive grid generation followed by hours of computer time for the flow solver. Cartesian grids have become popular largely because they allow for robust, automatic, and fast grid generation [1,2,4,6,9]. In a Cartesian grid, all grid cells are regular except those cut by the body. The grid generation process for a Cartesian grid creates appropriately refined cells, determines which cells are inside or outside of the body, and computes geometric information for the cells that intersect the body surface. Cartesian grids also have advantages from the point of view of simplicity and accuracy of PDE discretization.

In view of the advantages of Cartesian grids discussed above, we have developed a new multigrid flow solver designed for Cartesian grids with embedded boundaries for modern cache-based parallel computers. The underlying discretization to the inviscid Euler equations uses an upwind second order finite volume spatial differencing. Runge-Kutta timestepping drives a multigrid acceleration scheme to steady state. A least squares procedure is used to reconstruct the gradient in each cell. The minmod limiter and van Leer flux function are used in all the computational examples. The remainder of this paper describes the parallel decomposition and the multigrid coarsening algorithm, both of

*This research was supported by AFOSR Grant 94-1-0132, and DOE Grant DE-FG02-88ER25053.

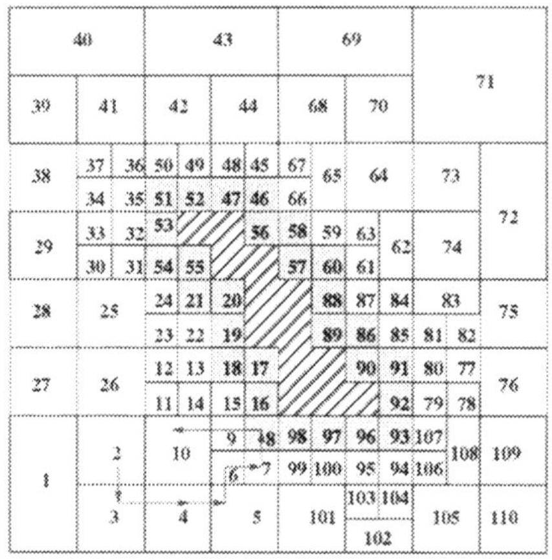

Figure 1. Illustration of Peano-Hilbert ordering for a two dimensional multi-level mesh with embedded geometry.

which make use of space-filling curves. We conclude with a representative computational example showing parallel scalability on a realistic three dimensional configuration.

2. SPACE-FILLING CURVES

Space-filling curves provide an inexpensive way to map an interval to a three-dimensional domain. In the last decade, these curves have found many applications, including multi-dimensional data mining, the solution of N-body problems and parallel domain decomposition [7,3,5,8]. Two common space-filling curves (henceforth SFC) are the Peano-Hilbert and Morton orderings.

Given our use of a Cartesian, multilevel mesh, either of these orderings can be easily generated, providing a linear ordering for each cell in the mesh. There are natural extensions for meshes with anisotropically refined cells and non-power-of-two meshes. Figure 1 illustrates in two dimensions the Peano-Hilbert ordering for a three level mesh with embedded geometry. Note that the so-called "solid" cells, (Cartesian cells removed from the mesh because they were inside the geometry) are simply skipped in our use of this ordering, since they are not part of the mesh.

Space-filling curves play two major roles in our approach. First, they are used for on-the-fly partitioning of the domain. Secondly, our multigrid mesh coarsening algorithm relies heavily on the SFC ordering in traversing the unstructured data structures of the fine mesh.

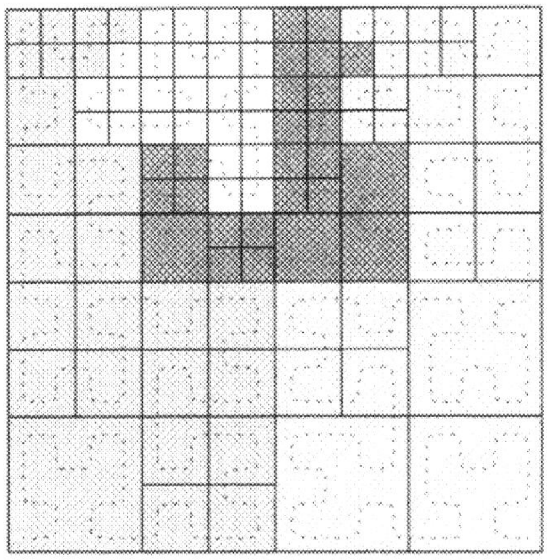

Figure 2. Illustration of two dimensional domain divided into four partitions using Peano-Hilbert space-filling curve ordering.

3. DOMAIN PARTITIONING

Space-filling curves are a natural choice for domain partitioning . At run-time, an SFC-ordered input mesh is partitioned so that the computational work in each partition is load balanced. This can be done dynamically for any number of processors, thus retaining maximum flexibility. The partitioning uses a two pass algorithm. The first pass scans the mesh, and computes both the total amount of work on the mesh and the ideal work per partition. For this work estimate, cut cells are weighted 2.7 times that of a full cell. The second pass actually reads the mesh, sequentially filling each partition until its work quota is reached. The partitioning is done on the cells; the faces in the mesh are assigned to the partitions that own the cells. One layer of overlap cells are duplicated around each partition boundary, so that one stage of the numerical scheme can be completed between communication steps.

Figure 2 illustrates the use of a space-filling curve to order a two dimensional mesh with 100 cells, divided into four partitions. We note that space-filling curves order the solution unknowns in a way that preserves locality *within* a partition as well, thus enhancing cache performance. In figure 3 this partitioning strategy is used to subdivide a mesh around a space shuttle into 16 partitions. As a measure of the quality of the partitions, table 1 compares the number of overlap cells for this mesh using 8 partitions and 64 partitions with the ideal case of a perfectly uniform Cartesian mesh with the same number of cells (N=1023277.) The SFC results are within 30 to 50% of the uniform case.

Figure 3. Domain partitioning example using Peano-Hilbert space-filling curve ordering for space shuttle.

4. MULTIGRID COARSENING ALGORITHM

Multilevel Cartesian meshes allow highly specialized and efficient mesh coarsening algorithms. The issues here concern generating the coarse meshes in a way that well approximates the coarsened solution, yet attains sufficient coarsening to get good speedups. The difficulties here are due to the unstructured multilevel starting mesh, and the fact that a Cartesian cell may be split by the geometry into several distinct regions.

The coarsening algorithm is based on the usual nesting of Cartesian refined cells. Note however that the finest mesh is global (covering the entire domain), and is composed of cells at many levels of refinement. This is in contrast to AMR-type algorithms, which usually group cells at the same level of refinement into a block-structured mesh level that covers only part of the domain. In our mesh, a cell can coarsen if all its siblings (i.e.

# Partitions	# Olap Cells (SFC)	# Olap Cells (Uniform)
8	97237	60928
64	242231	182783

Table 1
Comparison of the number of overlap cells for the space shuttle mesh using space-filling curve partitioning and a uniform rectangular decomposition of an isotropic mesh with the same number of cells.

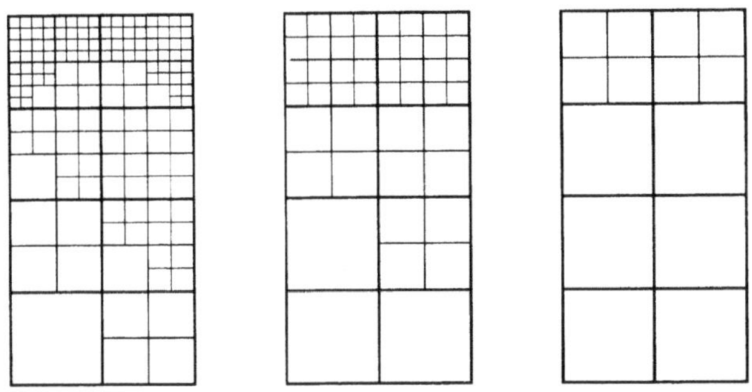

Figure 4. Two dimensional illustration of mesh coarsening algorithm, showing how a Cartesian cell cannot coarsen until all its sibling cells are at the same level of refinement.

cells with the same parent at the same level of refinement) can coarsen with it. If one of the sibling cells has been further refined, it cannot coarsen until its children have first coarsened. In addition, a cell can not coarsen if it would create an interface with more than 2 to 1 refinement. In the current implementation a cell cannot coarsen beyond the coarsest level specified when the mesh was generated. Thus, in general not all cells will be able to coarsen at each level, reducing the mesh coarsening ratio from its theoretical maximum of 8. Figure 4 illustrates the grid coarsening algorithm in two dimensions. Notice for example the coarsest cell in the top left corner has refined cells that can not coarsen on the first pass of the coarsening algorithm.

The space-filling curve ordering plays a crucial role in generating coarse meshes for multigrid. It allows us to generate a coarser mesh by looking only at neighboring cells. By traversing the cells in SFC order, those with the same parent cell are sequentially indexed and may be coarsened directly.

The enumeration of the cells in figure 1 demonstrates this point in two dimensions. When looking at cell 4, we can easily check the status of siblings 5 through 10 to decide if they all can coarsen. (They can't, since cells 6 through 9 are at a finer level.) If a cell can coarsen (e.g. cells 1–3 or 25–28) they get replaced by their parent in the coarser mesh. If a cell can not coarsen, it gets directly inserted into the coarser mesh as is. Thus, the coarsening is accomplished by a single pass (with recursive calls to examine finer levels) through the SFC-ordered mesh. Numerical results verify that this algorithm has linear asymptotic complexity.

Split cells present another challenge to the coarsening algorithm. When a fine cell is coarsened, the coarser cell may become split by the geometry into two or more control volumes. Conversely, the fine cell may be split, and may either remain split or become unsplit on the coarse mesh. Some of these possibilities are illustrated in figure 5. All such cases can be handled uniformly by using the face list on the fine grid to determine whether the fine cells comprising a coarse cell are fully connected (each fine cell in up the

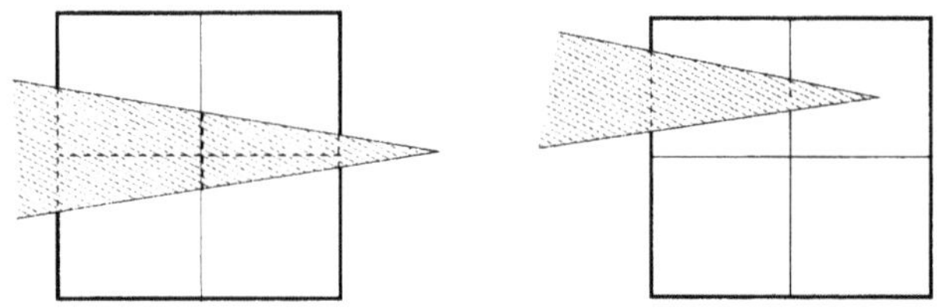

Figure 5. (Left) The four fine cells are cut but not split; the coarse cell is split. (Right) Only one of the fine cells is split, and the coarse cell is not split.

coarse cell is reachable from the others), or the cell is split.

Figure 6 shows a sequence of 4 meshes where this algorithm has been applied to generate grids around a space shuttle configuration. The finest grid has 1.02 million cells, the third coarser level grid has 6,060 cells. The coarsening ratios for this example are 6.63, 5.78, 4.41, 3.33, and 3.19. The finest mesh was generated in 2.8 minutes on a single CPU SGI R12000 Octane. It took 47 seconds to reorder the mesh in the Peano-Hilbert ordering, and 39 seconds to generate the 5 coarser meshes, the first three of which are shown in figure 6.

Note that the coarse meshes are generated in the same SFC ordering as the fine grid. Our parallel mesh partitioner can then be applied to each grid level separately. This is in contrast to an approach that first partitions the fine grid, and then generates the coarse meshes within each partition. This latter approach would be done to insure that the fine and coarse grids fall on the same partition. We do not have to specifically enforce this, since the SFC ordering automatically provides this. We can assess the SFC's success in this by measuring the amount of communication introduced by the multigrid restriction

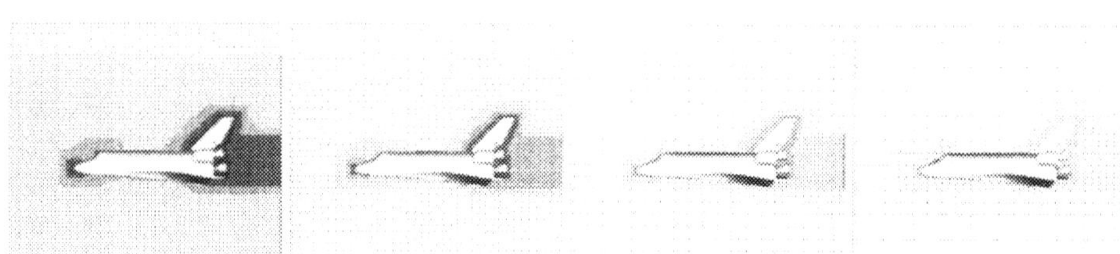

Figure 6. Cutting planes for a sequence of multigrid meshes around a space shuttle configuration. The coarsening ratios for this example are 6.63, 5.78, 4.41, 3.33, and 3.19 (last two meshes not shown).

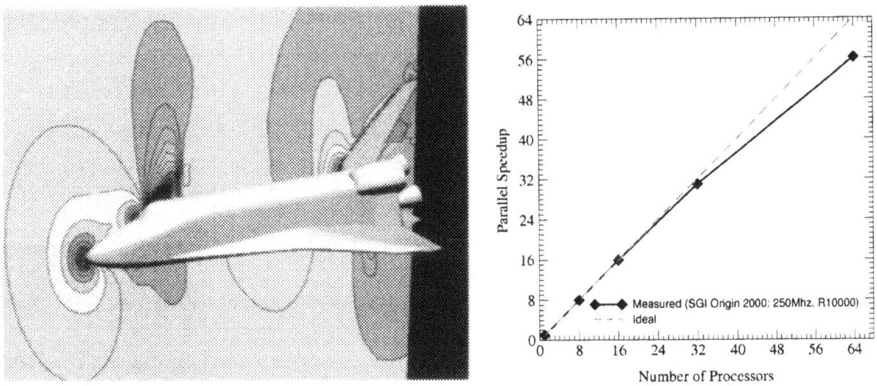

Figure 7. The figure on the left shows pressure contours for Mach .9 flow around a space shuttle configuration at 10 degrees angle of attack. Computational speedup using between 1 and 64 processors of an SGI Origin 2000 with 6 levels of multigrid are shown on the right.

and prolongation operations. If this is small, then the coarse and fine grid are well aligned. For the fine mesh shown in figure 6, on 8 partitions 91% of the fine cells are on the same partition as the corresponding coarse cells. On 16 partitions this number is 81%. As one would expect, this number remains relatively constant over several grid levels, until the grids are extremely coarse (on the order of 300 cells per processor).

5. PARALLEL SCALABILITY

Figure 7 shows the computed pressure contours for Mach 0.9 flow around a space shuttle configuration at 10 degrees angle of attack. The multigrid algorithm uses volume weighted restriction of the solution and piecewise constant prolongation, with a 5 stage Runge-Kutta scheme. The computational speedups using from 1 to 64 processors on an SGI Origin 2000 for the 6 level multigrid simulation are shown on the right. These scalability results show a speedup in excess of 56 on 64 processors. With 64 partitions, there are only 16,000 cells per processor on the finest mesh. The modest drop in scalability between 32 and 64 subdomains is most likely due to the small size of this mesh and the increased relative cost of communication.

6. CONCLUSIONS

This paper presents an effective, scalable, multigrid algorithm for Cartesian meshes with embedded boundaries. Although not the subject of this paper, there are still numerical improvements in the flow solver that can be made, particularly with respect to the handling of cut cells. We have demonstrated a flow solver turnaround time reduced from days to hours, bringing it more into line with the mesh generation time of minutes that is now possible with Cartesian mesh methods. Further improvement will come from the addition of preconditioning, which will be the subject of future work.

ACKNOWLEDGMENTS

The authors would like to thank Scott Baden and Rainold Löhner for introducing them to some uses of space-filling curves.

REFERENCES

1. M. Aftosmis, J. Melton, and M. Berger. Robust and efficient Cartesian mesh generation for component-based geometry. *AIAA J.* **36**(6), June, 1998.
2. S.A. Bayyuk, K.G. Powell, and B. van Leer. A simulation technique for 2-D unsteady inviscid flows around arbitrarily moving and deforming bodies of arbitrary geometry. *AIAA-93-3391*, July, 1993.
3. M. Griebel, N. Tilman and H. Regler. Algebraic multigrid methods for the solution of the Navier-Stokes equations in complicated geometries. *Intl. J. Numer. Methods for Heat and Fluid Flow* **26**, 1998.
4. S.L. Karman. SPLITFLOW: A 3D unstructured Cartesian/prismatic grid CFD code for complex geometries. *AIAA-95-0343*, Jan. 1995.
5. M. Parashar and J.C. Browne. Distributed Dynamic Data-Structures for Parallel Adaptive Mesh Refinement. *Proc. Intl. Conf. High Performance Computing*, (1995).
6. R.B. Pember, J.B. Bell, P. Colella, W.Y. Crutchfield and M.W. Welcome. Adaptive Cartesian Grid Methods for Representing Geometry in Inviscid Compressible Flow. *AIAA-93-3383*, July, 1993.
7. J.R. Pilkington and S.B. Baden. Dynamic partitioning of non-uniform structured workloads with spacefilling curves. *IEEE Trans. Parallel and Distrib. Systems* **7**(3), March, 1996.
8. J.K. Salmon and M.S. Warren and G.S. Winckelmans. Fast parallel tree codes for gravitational and fluid dynamical *N*-body problems. *Intl. J. Supercomp. Appl.* **8**(2), 1994.
9. Z.J. Wang. A Fast Nested Multi-Grid Viscous Flow Solver for Adaptive Cartesian/Quad Grids. *AIAA-96-2091*, June, 1996.

Parallel Computational Fluid Dynamics – Trends and Applications
C.B. Jenssen et al. (Editors)

Parallelisation of a CFD code: the use of Aztec library in the parallel numerical simulation of Extrusion of Aluminium

E. Celledoni[a], G. Johannessen[b], and T. Kvamsdal[c]

[a]IMF, NTNU 7049 Trondheim, Norway

[b]IMF, NTNU 7049 Trondheim, Norway

[c]SINTEF Applied Mathematics, N-7465 Trondheim, Norway

The parallelisation of a CFD (Computational Fluid Dynamics) code for simulating the extrusion of aluminium is considered by parallelising the linear algebra tasks in the code with the aid of Aztec library. In this paper we focus our attention on the parallel solution of the linear systems arising form the discretization of the fluid problem with preconditioned Krylov subspace methods. We present some preliminary theoretical remarks and some results on the convergence of Krylov subspace methods. We discuss then some numerical experiments obtained applying the methods and compare different preconditioning strategies.

1. INTRODUCTION

The problem addressed with this paper is how to achieve a well performing parallel version of the code *Extrud* developed for the simulation of extrusion of aluminium by Okstad and collaborators, [6]. Mainly for practical reasons, we choose to approach the problem from the linear algebra point of view. The availability of reliable packages like the Aztec library for solving numerically in parallel large linear systems, makes the parallelisation quickly achievable. In this way we have been able to experiment and compare different iterative methods and preconditioners for the linear systems arising from our problem and the outcome of these experiments is hereby reported. The initial part of this work has been more extensively reported in [5]. In the next section we give a description of the iterative methods and preconditioners used in our experiments in the third section we illustrate the numerical results.

2. ITERATIVE SOLUTION OF LINEAR SYSTEMS

Without intending to repeat the scopes of [9], we give here a brief introduction of those methods implemented in the Aztec library that showed the best performance in the numerical solution of the extrusion problem.

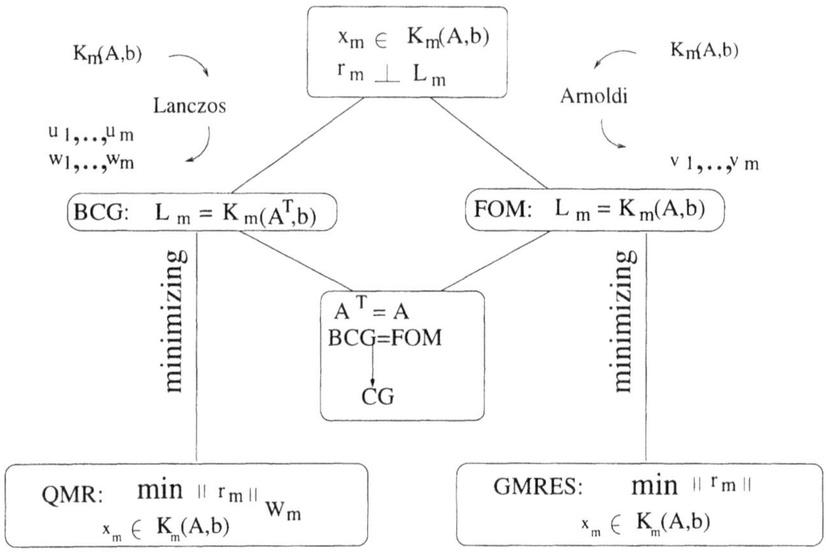

Figure 1. Convergence of GMRES.

2.1. Krylov subspace methods

Consider the solution of

$$A\boldsymbol{x} = \boldsymbol{b}, \quad A \in \mathbb{R}^{n\times n}, \boldsymbol{b} \in \mathbb{R}^n,$$

for constructing an iteration generating approximations of x we take

$$\boldsymbol{x}_m \in K_m(A, b) = \text{span}\{\boldsymbol{b}, \text{A}\boldsymbol{b}, \ldots, \text{A}^{m-1}\boldsymbol{b}\},$$

and assume for simplicity, but without loss of generality, that $\boldsymbol{x}_0 = 0$, $\boldsymbol{r}_0 = \boldsymbol{b} - A\boldsymbol{x}_0 = \boldsymbol{b}$. Given $V_m = [v_1, \ldots, v_m]$ an $n \times m$ matrix whose columns form a basis of the Krylov subspace $K_m(A, b)$, it is possible to write the approximation $\boldsymbol{x}_m = \sum_{i=1}^{m} \lambda_i \boldsymbol{v}_i = V_m \boldsymbol{\lambda}$ with $\boldsymbol{\lambda} = [\lambda_1, \ldots, \lambda_m]$; imposing different conditions on $\boldsymbol{r}_m = \boldsymbol{b} - A\boldsymbol{x}_m$ one obtains different iterative schemes.

In order to give a brief description of the various iterative methods implemented in the Aztec library in figure 1 we constructed a diagram classifying different Krylov subspace methods obtained following the procedure described above. Classically one distinguishes in between two different main strategies for constructing a basis of $K_m(A, b)$: the one corresponding to the Arnoldi algorithm, on the right part of the diagram, giving an orthonormal basis; the one corresponding to the Lanczos algorithm, on the left part of the diagram, constructing a basis for $K_m(A, b)$ biorthogonal to a basis of $K_m(A^{\text{T}}, b)$. Two different types of conditions are imposed on the residual r_m: the zero projection condition for the residual, for the methods on the upper part in the diagram, and the minimisation of some residual norm or seminorm, for the methods of the lower part of the diagram. In the numerical experiments we used a stabilised variant of the BCG method BICGstab and the GMRES method.

It is possible to show that

$$x_m^{GMRES} = \alpha x_m^{FOM} + (1-\alpha) x_{m-1}^{GMRES},$$

$$\alpha = \frac{\|r_{m-1}^{GMRES}\|^2}{\|r_m^{FOM}\|^2 + \|r_{m-1}^{GMRES}\|^2}$$

with $\|\cdot\|$ the Euclidean norm, [1], and

$$x_m^{QMR} = \alpha x_m^{BCG} + (1-\alpha) x_{m-1}^{QMR},$$

where the value of the parameter α is, in the latter case, computed using the seminorm $\|v\|_{W_n} := \|W_n v\|$, [3].

These relationship allow to derive estimates for the convergence of the projection methods in terms of the convergence of the minimisation methods. Moreover the propriety of minimisation of the Euclidean norm of the residual for the GMRES method, allows to derive estimates for virtually any other Krylov subspace method in terms of the GMRES residual.

In a Theorem by Campbell et al. [2] the convergence of GMRES method is described as follows:

Assume $A = I - K$, and consider the eigenvalues of K, s.t. $|\lambda_1,| \geq |\lambda_2| \ldots \geq |\lambda_n|$. If A is the matrix defining the linear system, dividing the eigenvalues of the matrix A in the complex plane in a cluster of eigenvalues internal to the disc D_ρ of radius $0 < \rho < 1$ around 1 and a set of external eigenvalues, after d initial iterations the norm of the residual of GMRES method is reduced by a factor ρ at each step. More precisely

$$\|r_{d+k}^{GMRES}\| \leq C\rho^k \|b\|,$$

with C not depending on k while the integer d depends essentially on the number of eigenvalues left outside the disc. Chosen a value of $0 < \rho < 1$ the value of d is determined by the problem to be solved. The goal of preconditioning is to maximally reduce d.

2.2. Preconditioners

Preconditioning is a sort of preprocessing of the original linear system that in theory is used for improving the convergence of the iterative methods, but in practice it is essential to the convergence and practical applicability of any Krylov subspace method in CFD problems.

Given an approximation B^{-1} of A^{-1} we can change our original problem into the equivalent preconditioned linear system: $B^{-1}Ax = B^{-1}b$. If we assume $A = B - E$ we can rewrite the previous as

$$(I - B^{-1}E)x = B^{-1}b,$$

corresponding to the format adopted in the previous Theorem and the convergence of the GMRES method depends now on the eigenvalues of $B^{-1}E$. Intuitively it is clear that if

$||E||$ is small we can expect fast convergence, or equivalently, from the observations of the previous section, a "big" number of eigenvalues of $I - B^{-1}E$ will be internal to the disc D_ρ.

When we construct a preconditioner B in a parallel environment the main requirements we need to fulfil are:

- B is such that $||E||$ is small;
- B is such that B^{-1} is easy to compute using parallel machines.

The strategies that have proved themselves successful in our problems correspond to the preconditioners of the Aztec library that implement the following philosophy: the use of "direct methods" for obtaining good approximations of A^{-1} combined with algebraic domain decomposition for the parallelisation.

In practice the original matrix is distributed among the processors and simplified subsystems are constructed locally on each processor. In the simplest case, the local subsystems are the diagonal blocks extracted from the set of subsequent rows belonging to each processor. When the overlapping options are activated, the local subsystems include some extra rows and columns obtained form other processors.

The local subsystems are inverted via direct methods, typically LU or approximate LU-factorizations, and the preconditioner is constructed and applied locally by means of these factorized local systems.

The main drawback in the use of a parallel linear algebra package in the parallelisation of our CFD code, is without any doubt the fact that for the moment the matrices of the linear systems are not constructed locally on the processors, but distributed assuming that the whole matrix is constructed in a sequential way. We believe anyway that there is an enormous potential for experimentation and improvement in this sense.

The distribution of the matrix among the processors is now done simply computing the number of rows per processor given the number of processors to be used, and starting to distribute the rows from the top of the matrix downwards.

A more sophisticated algebraic approach is based on the use of graph partitioning algorithms applied for subdividing the matrix among the processors, this strategy will involve the use of the Chaco 2.0 package [4]. This kind of algebraic approach turns out to be reasonably flexible. The matrix is subdivided among the processors by means of abstract algebraic considerations that aim to minimise the interaction among the processors, the partition can be reused for different linear systems in the same fluid problem or can be recalculated after a certain number of linear systems solves.

3. NUMERICAL EXPERIMENTS: PARALLEL PERFORMANCE

The test problem for the numerical experiments is a Jeffrey-Hamel flow in the domain described by figure 2,

$$\begin{aligned} &\Delta u + \nabla p = 0 \\ &\nabla \cdot u = 0, \end{aligned} \tag{3.1}$$

zero on the boundary. This problem has been used to test the sequential version of *Extrud* and we use it here to get some indications of the performances of the Krylov subspace preconditioned methods for the extrusion problem.

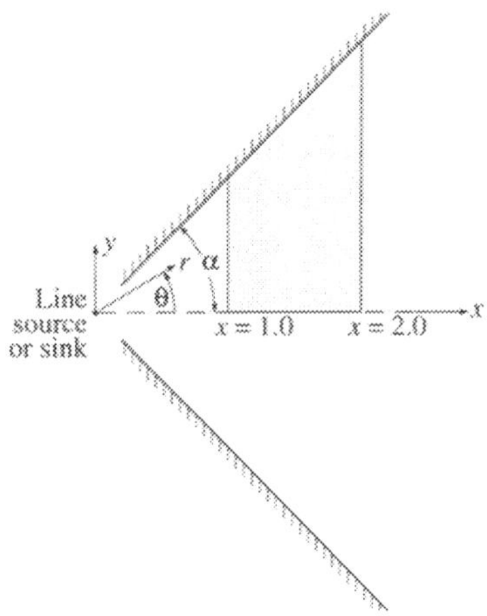

Figure 2. Jeffrey Hamel flow.

We here compare the performance of various iterative methods and preconditioners, the results are confined to the linear algebra parallel part of the code.

The iterative methods used are: GMRES (G) and BICGstab (B); the preconditioners are always obtained combining algebraic domain decomposition for the parallelisation, with exact or incomplete lu-decompositions to be performed on the local subsystems. The GMRES method is restarted every 30 iterations. The incomplete LU-decomposition [8] *ilut* yields different preconditioners depending on the number of subdomains considered and on the choice of two crucial input parameters: the dropping tolerance and the graph fill-in level.

The overlapping is established by the number of rows that each processor needs to obtain from other processors in order to construct the local subsystem [9].

In the tables containing the results we report execution times, and number of iterations, the speedup is reported in the figures 3 and 4.

In the first two tables we report the execution times and the number of iterations respectively for various methods. The capital letter indicates the type of iterative method and the dash is followed by the specifications for the preconditioners. On the subsystems we used *ilut* preconditioning with different dropping tolerances, with overlapping 2 (ov 2) or without overlapping. The graph fill-in parameter is set equal to 1.5.

The matrix of the linear system has $n = 21219$ rows. The maximum number of processors used is 16.

As we can observe the communication times increase with the use of the overlapping strategy even if in general the number of iterations is considerably reduced. In some cases the overlapping is crucial for getting the iteration to converge, for example for the case of BICGstab method with *ilut* preconditioner, with dropping tolerance $1e-3$, with 16

Table 1
Execution times

N–proc	B–1e-3	B–1e-3 ov 2	G–1e-3	G–1e-3 ov 2
1	30.82		26.99	
2	15.87	18.94	15.17	20.24
4	10.21	12.57	12.57	17.21
8	7.14	9.66	11.02	13.40
16	–	7.68	4.90	16.46

Table 2
Number of iterations

N-proc	B–1e-3	B–1e-3 ov 2	G–1e-3	G–1e-3 ov 2
1	54		76	
2	62	61	100	117
4	90	77	198	206
8	133	112	360	292
16	–	149	352	638

processors the method has a breakdown in the non-overlapped version and it converges successfully in the overlapped version. In general as the number of processors increases the preconditioners become easier to compute but also less effective.

In the next experiments we tried to keep the computational load per processor constant as number of processors increased. For achieving this in the first two columns of the table the graph fill-in parameter was increased form 1.5 (sequential case) to 1.9 (on 16 processors), mainly an increase of 0.1 each time we doubled the number of processors. In the experiments of third and fourth column we instead decreased the dropping tolerance starting form a value 0.016 for the sequential case and halving it each time we doubled the number of processors.

As one can see with this strategy the performance of the preconditioners does not deteriorate too much as we increase the number of processors and this fact can be also appreciated in the speedup plots. A robust solver for the linear algebra tasks of a CFD code based on the use of Aztec library should take into account this phenomenon and implement a dynamical choice of the option parameters depending on the number of processors used in the simulation. In figure 3 we plotted the speedup of a selection of the previous methods, computed taking the sequential version of each method as a reference sequential time. The best performing methods are BiCGstab and GMRES with *ilut*, dropping tolerance $1e-2$ and increasing graph fill-in parameter, (dashed and full line respectively), we reported for comparison also the times obtained by BiCGstab with *ilut* with dropping tolerance $1e-4$ and graph fill-in parameter fixed equal to 1.5 with and without overlapping (dashed dotted and dotted line).

In figure 4 we present the speedup obtained for a problem of much larger dimension ($n = 148739$ rows in the linear system), this problem is too large to be solved on just one processor therefore we considered the time obtained by solving the linear system on four

Table 3
Execution times

N-proc	B-1e-2	G-1e-2	B-1.9	G-1.7
1	28.66	33.11	26.72	26.41
2	14.95	16.23	15.48	16.06
4	8.82	11.00	9.88	11.00
8	6.03	7.28	6.93	9.57
16	4.31	5.34	4.37	5.00

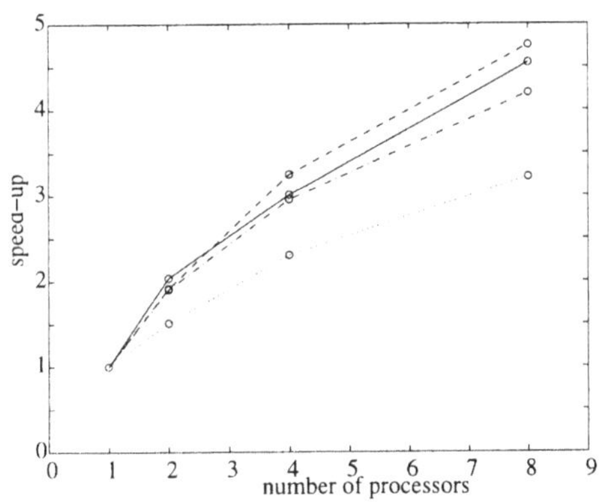

Figure 3. Speedup for the small matrix.

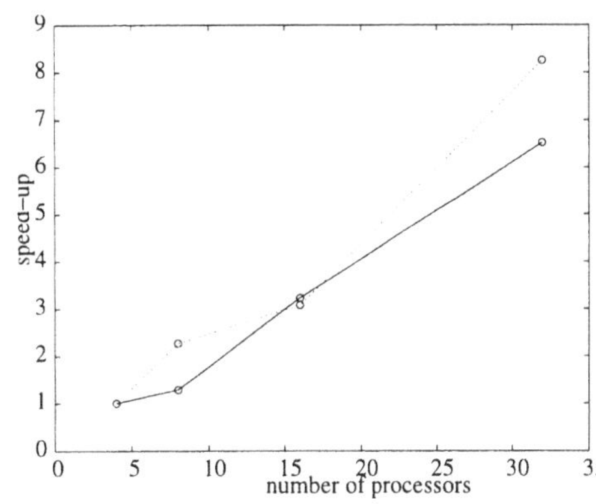

Figure 4. Speedup for the big matrix.

processors as reference time for our speedup. The methods considered are BiCGSTAB (full line) and GMRES (dotted line) with dropping tolerance $d = 1e - 2$ and graph fill-in parameter $t = 2$. The speedup is very good for this big matrix and for the GMRES method we obtain times that on 32 processors are 1/8 of the times obtained with 4 processors.

4. CONCLUSIONS

In this paper we have reported some preliminary studies on the use of parallel preconditioned Krylov subspace methods for the parallelisation of a code modelling the extrusion of aluminium. The use of Aztec library as a part of the *Extrud* program can be a robust strategy for the solution of the linear systems in the code when a fixed optimal computational load per processor is guaranteed. Our tests are for the moment limited to a simplified model problem (Jeffrey-Hamel flow), and the simple application of these methods to the full Navier-Stokes problem presented some extra difficulties that we are now trying to overcome. Inspired by some ideas of [7] we are at present experimenting a new preconditioner for the full Navier-Stokes problem based on the use of factorization methods and algebraic splitting techniques implemented using the Aztec library.

REFERENCES

1. P.N. Brown. A theoretical comparison of the Arnoldi and GMRES algorithms. *SIAM J. Sci. Stat. Comput.*, 12:58–78, 1991.
2. S.L. Campbell, I.C. Ipsen, C.T. Kelley, and C.D. Meyer. GMRES and the minimal polynomial. *BIT*, 36(4):664–675, 1996.
3. E. Celledoni. *Metodi di Krylov per sistemi lineari di equazioni differenziali ordinarie.* PhD thesis, Universita' degli Studi di Padova, 1997.
4. B. Hendrickson and R. Leland. The Chaco user's guide-version 2.0. Technical Report SAND95-2344, Sandia National Laboratories, Sandia National Laboratories, Albuquerque, NM 87185-1110, July 1995.
5. G. Johannessen. Parallel preconditioned Krylov subspace methods for large sparse linear systems. Master's thesis, Norwegian University of Science and Technology, March 2000.
6. K.M. Okstad, T. Kvamsdal, and S. Abtahi. A coupled heat and flow solver for extrusion simulation. In *ECOMAS 2000*, 2000.
7. A. Quarteroni, F. Saleri, and A. Veneziani. Factorization methods for the numerical approximation of Navier-Stokes equation. Technical report, EPFL, 2000.
8. Y. Saad. *Iterative methods for sparse linear systems.* PWS Publishing Company, 1996.
9. R.S. Tuminaro, M. Heroux, S.A. Hutchinson, and J.N. Shadid. Official Aztec user's guide-version 2.1. Technical Report SAND99-8801J, Massively Parallel Computing Research Laboratory, Sandia National Laboratories, Albuquerque, NM 87185, November 1999.

Parallel Computational Fluid Dynamics – Trends and Applications
C.B. Jenssen et al. (Editors)

An efficient highly parallel multigrid method for the advection operator *

Boris Diskin [a] , Ignacio M. Llorente [b] and Ruben S. Montero [c]

[a]Institute for Computer Applications in Science and Engineering, Mail Stop 132C, NASA Langley Research Center, Hampton, VA 23681-2199 (email: bdiskin@icase.edu)

[b]Departamento de Arquitectura de Computadores y Automática, Universidad Complutense, 28040 Madrid, Spain (email: llorente@dacya.ucm.es)

[c]Departamento de Arquitectura de Computadores y Automática, Universidad Complutense, 28040 Madrid, Spain (email: rubensm@dacya.ucm.es)

Standard multigrid algorithms fail to achieve optimal grid-independent convergence rates in solving non-elliptic problems. In many practical problems appearing in computational fluid dynamics, the non-elliptic part of a problem is represented by the convection operator. Downstream marching, when it is viable, is the simplest and most efficient way to solve this operator. However, in a parallel setting, the sequential nature of marching degrades the efficiency of the algorithm. The aim of this report is to present, evaluate and analyze an alternative highly parallel multigrid method for 3-D convection-dominated problems. This method employs semicoarsening, a four-color plane-implicit smoother, and discretization rules allowing the same cross-characteristic interactions on all the grids involved to be maintained.

1. INTRODUCTION

The convergence properties of multigrid algorithms are defined by two factors: (1) the smoothing rate which describes the reduction of high-frequency error components and (2) the quality of the coarse-grid correction which is responsible for dumping of smooth error components. In elliptic problems, all the fine-grid smooth components are well approximated on the coarse grid built by standard (full) coarsening. In nonelliptic problems, however, some fine-grid *characteristic* components that are much smoother in the characteristic direction than in other directions, cannot be approximated with standard multigrid methods (see [1–3,5]).

Several approaches aimed to cure the characteristic-component problem were studied in literature. These approaches fall into two categories: (1) development a comprehensive relaxation scheme eliminating not only high-frequency error components but the

*This research was supported by the National Aeronautics and Space Administration under NASA Contract No. NAS1-97046 while the authors were in residence at the Institute for Computer Applications in Science and Engineering (ICASE), NASA Langley Research Center, Hampton, VA 23681-2199. Ignacio M. Llorente and Ruben S. Montero were also supported in part by the Spanish research grant TIC 99/0474

characteristic error components as well; (2) devising an adjusted coarse-grid operator approximating well the fine-grid characteristic error components.

In many practical problems appearing in computational fluid dynamics (CFD), the nonelliptic part is represented by advection. For advection, the most efficient and comprehensive relaxation is downstream marching. If the target discretization is a stable upwind scheme, the downstream marching reduces all (high-frequency and smooth) error components, solving a *nonlinear* advection equation in just a few sweeps (a single downstream sweep provides the exact solution to a linearized problem). Incomplete LU (ILU) decomposition methods act similarly, given a suitable ordering [4]. The downstream marching technique was successfully applied for solution of many CFD problems associated with non-recirculating inviscid flows (see, e.g., [3]). However, if the discretization is not fully upwind (e.g., upwind biased) the downstream marching in its pure form is not viable. One of the most efficient (also marching type) alternatives often applied to the schemes that cannot be directly marched is a defect-correction method (see, e.g., [9,10]). Usually the efficiency of these methods is quite satisfactory. Sometimes, however, the convergence rate of defect-correction method is grid dependent (see [6]). Another, very important, drawback associated with all marching and ILU methods is a low parallel grade because the efficiency of these methods is essentially based on the correctness of the sequential marching order.

The efficiency of methods belonging to the second category is often order independent, i.e., most of the operations can be performed in parallel. These methods are much more attractive for massive parallel computing. The necessary requirements for coarse-grid operators used in second-category methods were formulated in [12]. Among the options available in conjunction with full coarsening are Galerkin coarsening [13], matrix-dependent operators [4], and corrected coarse-grid operators [7]. Analysis in [12] showed that all these methods have certain drawbacks. Another way to construct an appropriate coarse-grid operator is to employ semicoarsening [5]. The semicoarsening algorithm presented in this paper is an efficient highly parallel method solving the three-dimensional (3-D) advection operator defined on a cell-centered grid.

2. PROBLEM STATEMENT AND DISCRETIZATION

The model problem studied in this paper is the three-dimensional (3-D) constant-coefficient convection equation

$$LU \equiv \frac{1}{|\bar{a}|}\left(\bar{a} \cdot \nabla\right)U = f(x,y,z), \tag{1}$$

where $\bar{a} = (a_1, a_2, a_3)$ is a given vector and $|\bar{a}| = \sqrt{a_1^2 + a_2^2 + a_3^2}$. The solution $U(x,y,z)$ is a differentiable function defined on the unit square $(x,y,z) \in [0,1] \times [0,1] \times [0,1]$. Let ϕ_y and ϕ_z be the nonalignment angles and $t_y = \tan\phi_y = a_2/a_1$ and $t_z = \tan\phi_z = a_3/a_1$. For simplicity, we assume a horizontal inclination $a_1 \geq a_2, a_3 \geq 0$ and, therefore, $1 \geq t_y, t_z \geq 0$.

Equation (1) can be rewritten as

$$\partial_\xi U = f(x,y,z), \tag{2}$$

where $\xi = \frac{x + t_y y + t_z z}{\sqrt{1+t_y^2+t_z^2}}$ is a variable along the characteristic of (1). Equation (1) is subject to Dirichlet boundary conditions at the inflow boundary $x = 0$ and periodic conditions in the y and z directions.

The problem (2) is discretized on the 3-D Cartesian uniform grid with mesh sizes h in the three directions (target grid). Let u_{i_1,i_2,i_3} be a discrete approximation to the solution $U(x,y,z)$ at the point $(x,y) = (i_1h, i_2h, i_3h)$. To derive a proper discretization, we exploit the idea of a *low-dimensional prototype* introduced in [2]. For our studies, we choose the low-dimensional prototype to be the (one-dimensional) first order accurate two-point discretization of the first derivative, corresponding to the pure upwind scheme

$$\frac{1}{h\sqrt{1+t_y^2+t_z^2}}\left(u_{i_1,i_2,i_3} - u_{i_1-1,i_2-t_y,i_3-t_z}\right) = f_{i_1,i_2,i_3}. \tag{3}$$

The 3-D discretization is obtained from the low-dimensional prototype by replacing function values at the ghost points (points with fractional indexes) by weighted averages of the values at adjacent genuine grid points. The resulting *narrow* discretization scheme is defined by

$$\begin{aligned}
&L^h u_{i_1,i_2,i_3} \equiv \frac{1}{h\sqrt{1+t_y^2+t_z^2}}\Big(u_{i_1,i_2,i_3} \\
&-\big((1-t_z)((1-t_y)u_{i_1-1,i_2,i_3} + t_y u_{i_1-1,i_2-1,i_3}) \\
&+t_z((1-t_y)u_{i_1-1,i_2,i_3-1} + t_y u_{i_1-1,i_2-1,i_3-1})\big)\Big) = f_{i_1,i_2,i_3},
\end{aligned} \tag{4}$$

where the discretization of the right-hand-side function is $f_{i_1,i_2,i_3} = f(i_1h, i_2h, i_3h)$.

The coarse grids are 3-D Cartesian rectangular grids with aspect ratios $m_y = h_x/h_y$ and $m_z = h_x/h_z$, where h_x, h_y and h_z are the mesh sizes in the x, y and z directions respectively. The problem (2) is discretized on a coarse grid as

$$\begin{aligned}
&L^{(h_x,h_y,h_z)} u_{i_1,i_2,i_3} \equiv \frac{1}{h_x\sqrt{1+t_y^2+t_z^2}}\Big(u_{i_1,i_2,i_3} \\
&-\big((1-s_z)((1-s_y)u_{i_1-1,i_2+k_y,i_3+k_z} + s_y u_{i_1-1,i_2+k_y+1,i_3+k_z}) \\
&+s_z((1-s_y)u_{i_1-1,i_2+k_y,i_3+k_z+1} + s_y u_{i_1-1,i_2+k_y+1,i_3+k_z+1})\big)\Big) = f_{i_1,i_2,i_3};
\end{aligned} \tag{5}$$

where $k_y + s_y = -m_y t_y$ and $k_z + s_z = -m_z t_z$, k_y and k_z are integers, $0 \le s_y, s_z < 1$. See Figure 1 for a pictorial explanation of the discretization stencil.

The cross-characteristic interaction (CCI) introduced by a discrete operator (*inherent* CCI) can be quantitatively estimated by the coefficients of the lowest pure cross-characteristic derivatives appearing in the first differential approximation (FDA) (see [11]) to the discrete operator. In our model problem, the CCI appears only because of interpolation in the y-z plane. Therefore, the true CCI is actually determined by the FDA coefficients of ∂_{yy} and ∂_{zz}. The FDA to the discretization (4) taken for a characteristic component u ($\partial_x u \gg \partial_\xi u, \partial_y u \gg \partial_\xi u$) is given by

$$FDA\big(L^h\big) = \partial_\xi - T_y^h h^2 \partial_{yy} - T_z^h h^2 \partial_{zz}, \qquad T_y^h = \frac{t_y(1-t_y)}{2h\sqrt{1+t_y^2+t_z^2}}, \qquad T_z^h = \frac{t_z(1-t_z)}{2h\sqrt{1+t_y^2+t_z^2}}. \tag{6}$$

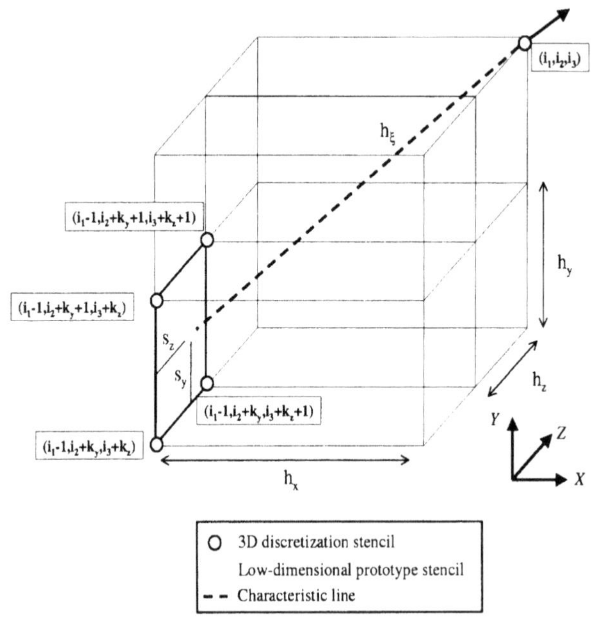

Figure 1. Discretization stencil.

The first differential approximation to the discretization (5) for rectangular grids is

$$
\begin{aligned}
&FDA\left(L^{(h_x,h_y,h_z)}\right) = \partial_\xi - T_y^{(h_x,h_y,h_z)} h_y^2 \partial_{yy} - T_z^{(h_x,h_y,h_z)} h_z^2 \partial_{zz}, \\
&T_y^{(h_x,h_y,h_z)} = \frac{s_y(1-s_y)}{2h_x\sqrt{1+t_y^2+t_z^2}}, \qquad T_z^{(h_x,h_y,h_z)} = \frac{s_z(1-s_z)}{2h_x\sqrt{1+t_y^2+t_z^2}}.
\end{aligned}
\tag{7}
$$

Previous studies on different types of nonelliptic equations (see [2] and [3]) have shown that the main difficulty in constructing an efficient multigrid solver is a poor coarse-grid approximation to the fine-grid characteristic error components. It was observed that a coarse-grid operator defined on a grid built by full coarsening unavoidably introduces a too strong cross-characteristic interaction. On the other hand, a narrow discretization (5) on a semicoarsened grid (only the x-directional mesh size is doubled) results in a coarse-grid CCI that is lower than required. However, operator (5) on the semicoarsened grid can be supplied with additional terms (*explicit* CCI), so that the *total coarse-grid* CCI would be exactly the same as on the fine grid. Thus, one could derive the final form of the coarse-grid operator by comparing the coarse-grid FDA operator (7) with the target

sol.	64^3 case 1	64^3 case 2	64^3 case 3	64^3 case 4	128^3 case 1	128^3 case 2	128^3 case 3	128^3 case 4
1	0.09/0.09	0.09/0.10	0.08/0.09	0.07/0.08	0.09/0.11	0.11/0.10	0.08/0.10	0.08/0.11
2	0.18/0.16	0.10/0.11	0.28/0.25	0.07/0.08	0.29/0.24	0.19/0.17	0.44/0.39	0.08/0.10

Table 1
Asymptotic/geometric-average convergence rate for the solver-case combinations.

grid FDA operator (6)

$$\begin{aligned}
L^{(h_x,h_y,h_z)}u_{i_1,i_2,i_3} \equiv \frac{1}{h_x\sqrt{1+t_y^2+t_z^2}}\Bigg(&u_{i_1,i_2,i_3} \\
-\Big(\big(1-s_z\big)\big((1-s_y)u_{i_1-1,i_2+k_y,i_3+k_z} &+ s_y u_{i_1-1,i_2+k_y+1,i_3+k_z}\big) \\
+ s_z\big((1-s_y)u_{i_1-1,i_2+k_y,i_3+k_z+1} &+ s_y u_{i_1-1,i_2+k_y+1,i_3+k_z+1}\big)\Big)\Bigg) \\
-(T_y^h - T_y^{(h_x,h_y,h_z)})(u_{i_1,i_2-1,i_3} &- 2u_{i_1,i_2,i_3} + u_{i_1,i_2+1,i_3}) \\
-(T_z^h - T_z^{(h_x,h_y,h_z)})(u_{i_1,i_2,i_3-1} &- 2u_{i_1,i_2,i_3} + u_{i_1,i_2,i_3+1}).
\end{aligned} \tag{8}$$

The proposed multigrid construction employs semicoarsening and narrow coarse-grid discretization schemes supplied with explicit terms (which are discrete approximations to $h_y\partial_{yy}$ and $h_z\partial zz$ with suitable coefficients) to maintain on the coarse grids the same CCI as on the target (fine) grid. This construction ensures that all the *characteristic error components* are well eliminated by the coarse-grid correction. The *noncharacteristic error components* must be reduced in relaxation. Successive semicoarsening implies a fast decrease in the *inherent* CCI on coarse grids and, hence, a fast increase in the weight of the explicit terms in the coarse-grid operators (since the total CCI remains fixed). Thus, plane smoothers should be applied to reduce the noncharacteristic error components.

3. THE MULTIGRID METHOD

The studied semicoarsening multilevel $V(1,1)$ cycle consists of a four-color plane-implicite relaxation scheme, an upwind-biased restriction operator for the residual transfer and a linear prolongation operator for the coarse-grid correction. (See detailed description in [8].) The discrete operators on all the grids are supplied with an additional dissipation term approximating $-\lambda h_x\sqrt{1+t_y^2+t_x^2}\partial_{\xi\xi}$. The parameter λ is chosen to provide good smoothing rate in the four-color relaxation. The value $\lambda = 0.25$ has been found numerically.

4. NUMERICAL RESULTS

The inflow boundary conditions for the test problems were chosen so that the function $U(x,y,z) = \cos(\omega(y+z-(t_y+t_z)x))$ is the exact continuous solution of the homogeneous $(f_{i_1,i_2,i_3} = 0)$ problem (2). The initial approximation was interpolated from the solution on the previous coarse grid. The frequencies $\omega = 8\pi$ for a 64^3 grid and $\omega = 16\pi$ for a 128^3 grid

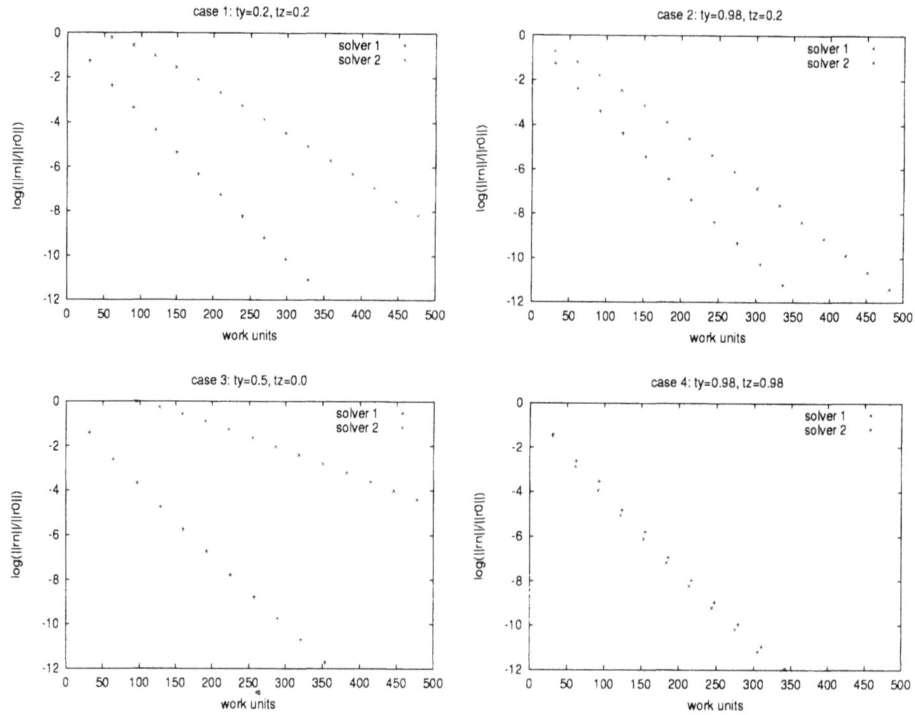

Figure 2. Residual versus work units for $V(1,1)$ cycles with semicoarsening: *Solver 1* (with explicit CCI terms) and *Solver 2* (without explicit CCI terms) for a 128^3 grid and $w = 16\pi$.

were chosen to reduce the total computational time exploiting periodicity and to provide a reasonable accuracy in approximating the true solution of the differential equation. Two cycles, with (*solver 1*) and without (*solver 2*) explicit CCI terms in coarse-grid operators, were compared on 64^3 and 128^3 uniform grids for the nonalignment parameters $t_y = 0.2$ $t_z = 0.2$ (*case 1*), $t_y = 0.98$ $t_z = 0.2$ (*case 2*), $t_y = 0.5$ $t_z = 0.0$ (*case 3*), and $t_y = 0.98$ $t_z = 0.98$ (*case 4*). Table 1 contains the asymptotic and geometric average convergence rates and Figure 2 shows the residual history versus work units; the work unit is the computer-operation count in the target-grid residual evaluation.

The plane solver used in the 3-D smoothing procedure is a robust two-dimensional (2-D) multigrid V(1,1)-cycle employing full coarsening and alternating-line smoothers; the approximate 2-D solution obtained after one 2-D cycle is sufficient to provide robustness and good convergence rates in the 3-D solvers. The combination of semicoarsening, the four-color plane-implicit smoother, and introduction of explicit CCI terms in coarse-grid discretizations (solver 1) yields a multigrid solver with fast grid-independent convergence rates for any angles of nonalignment and especially for $t = 0.5$ (case 3).

The plane smoothing without explicit CCI terms (solver 2) presents a worser and grid-dependent convergence rate for $t = 0.5$ (case 3) that roughly corresponds to the maximum inherent CCI. Notice that for $t = 0.98$ (case 4), which corresponds to a low inherent CCI, the convergence is optimal for both the solvers.

5. CONCLUSIONS AND FUTURE WORK

The combination of semicoarsening, a four-color plane-implicit smoother and the introduction of explicit CCI terms in the discretization of all grids yields an efficient highly parallel multigrid solver for the convection equation with fast grid-independent convergence rates for any angle of non-alignment. This solver permits the parallel solution of a convective process that is sequential in nature.

Although not included in the paper, experiments with pointwise smoothers were also performed. On finest grids, where the direction of the strongest coupling approximately coincides with the characteristic direction, a pointwise smoother can be as efficient as a plane-implicit smoother. Using pointwise smoothers on finest (most expensive) grids considerably reduces the work-unit count of the approach. The coupling analysis provides a criterion to switch between point and plane smoothers in the multigrid cycle. Such an approach was successfully applied in [5] for 2-D vertex-centered grids and its extension for 3-D cell-centered grids will be a subject for future studies.

We intend to continue the work on the effective solution of convection-dominated problems. In particular, we will apply the CCI correction technique to improve the parallel properties and convergence rate of the multigrid resolution of the Navier-Stokes equations by distributive smoothers.

The program code has been implemented in C language and has been parallelized with the standard OpenMP directives for shared-memory parallel computing to exploit the parallelism in the smoother. An efficiency equal to 0.7 has been obtained on 8 processors of a SGI Origin 2000 for a 128^3 grid.

6. ACKNOWLEDGMENTS

We would like to thank CSC (Centro de Supercomputación Complutense), ICASE and the Computational Modeling and Simulation Branch at the NASA Langley Research Center for providing access to the parallel computers that have been used in this research.

REFERENCES

1. A. Brandt. Multigrid solvers for non-elliptic and singular-perturbation steady-state problems. (unpublished). The Weizmann Institute of Science, Rehovot, Israel, December 1981.
2. A. Brandt and B. Diskin. Multigrid solvers for nonaligned sonic flows. *SIAM J. Sci. Comp.*, 21(2):473–501, September 1999.
3. A. Brandt and I. Yavneh. On multigrid solution of high-Reynolds incompressible entering flow. *J. Comput. Phys.*, 101:151–164, 1992.

4. P. M. de Zeeuw. Matrix-dependent prolongations and restrictions in a blackbox multigrid solver. *J. Comput. Appl. Math.*, 33:1–27, 1990.
5. B. Diskin. Solving upwind-biased discretizations II: Multigrid solver using semicoarsening. ICASE Report 99-25, July 1999.
6. B. Diskin and J. L. Thomas. Solving upwind-biased discretizations: Defect-correction iterations. ICASE Report 99-14, March 1999.
7. J. E. Dendy Jr. Black box multigrid for nonsymmetric problems. *Appl. Math. Comput.*, 13:261–283, 1983.
8. I. M. Llorente and B. Diskin. A highly parallel multigrid solver for 3-D upwind-biased discretizations of convection-dominated problems. ICASE Report in preparation, 1999.
9. C. W. Oosterlee, F. J. Gaspar, T. Washio, and R. Wienands. Multigrid line smoothers for higher order upwind discretizations of convection-dominated problems. *J. Comput. Phys.*, 139:274–307, 1998.
10. J. L. Thomas, B. Diskin, and A. Brandt. Distributed relaxation multigrid and defect correction applied to the compressible Navier-Stokes equations. AIAA Paper 99-3334, 14th Computational Fluid Dynamics Conference, Norfolk, VA, July 1999.
11. N.N. Yanenko and Y.I.Shokin. On the correctness of first differential approximations of difference schemes. *Dokl. Akad. Nauk SSSR*, 182:776–778, 1968.
12. I. Yavneh. Coarse-grid correction for non-elliptic and singular perturbation problems. *SIAM J. Sci. Comput.*, 19(5):1682–1699, 1998.
13. S. Zeng, C. Vink, and P. Wesseling. Multigrid solution of the incompressible Navier-Stokes equations in general coordinates. *SIAM J. of Num. Anal.*, 31:1764–1784, 1994.

Parallel Computational Fluid Dynamics – Trends and Applications
C.B. Jenssen et al. (Editors)

A parallel robust multigrid algorithm for 3-D boundary layer simulations.*

Ruben S. Montero [a] , Ignacio M. Llorente [b] and Manuel D. Salas [c]

[a]Departamento de Arquitectura de Computadores y Automática, Universidad Complutense, 28040 Madrid, Spain (email: rubensm@dacya.ucm.es)

[b]Departamento de Arquitectura de Computadores y Automática, Universidad Complutense, 28040 Madrid, Spain (email: llorente@dacya.ucm.es)

[c]Institute for Computer Applications in Science and Engineering, Mail Stop 132C, NASA Langley Research Center, Hampton, VA 23681-2199 (email: salas@icase.edu)

Anisotropies occur naturally in CFD where the simulation of small scale physical phenomena, such as boundary layers at high Reynolds numbers, causes the grid to be highly stretched leading to a slow down in convergence of multigrid methods. Several approaches aimed at making multigrid a robust solver have been proposed and analyzed in literature using the scalar diffusion equation. This paper contains numerical results of the behavior of a popular robust multigrid approach, plane implicit smoothers combined with semi-coarsening, for solving the 3-D incompressible Navier-Stokes equations in the simulation of the boundary layer over a flat plate on a stretched grid.

1. INTRODUCTION

Standard multigrid techniques are efficient methods for solving many types of partial differential equations (particularly elliptic problems) due to their optimal complexity (work linearly proportional to the number of unknowns), optimal memory requirement, and good parallel efficiency and scalability in parallel implementations [8]. In the last decades, an important effort has been made to adapt multigrid solvers to hyperbolic initial-boundary value problems. These problems are of particular interest in the field of Computational Fluid Dynamics (CFD), where ones deals with systems of non-linear coupled equations that contain hyperbolic factors [2].

The efficiency of standard multigrid methods degenerates on anisotropic problems. Several methods that combine implicit point, line or plane relaxation with partial and full coarsening have been proposed in the multigrid literature [7,4,10,11] to solve anisotropic operators and achieve robustness when the coefficients of a discrete operator vary throughout the computational domain. This anisotropic condition arises in a natural manner in CFD where the simulation of small scale physical phenomena, such as boundary layers at

*This research was supported by the National Aeronautics and Space Administration under NASA Contract No. NAS1-97046 while the authors were in residence at the Institute for Computer Applications in Science and Engineering (ICASE), NASA Langley Research Center, Hampton, VA 23681-2199. Ignacio M. Llorente and Ruben S. Montero were also supported in part by the Spanish research grant TIC 99/0474.

high Reynolds numbers, causes the grid to be highly stretched leading to a slow down in convergence.

The aim of this work is to present a highly parallel FAS (full approximation scheme) multigrid algorithm that combines plane-implicit smoothing and semicoarsening to solve the incompressible 3-D Navier-Stokes equations for the simulation of a boundary layer over a finite flat plate. Our approach is based on a second-order staggered finite volume scheme. For the convective terms a first order upwind scheme is used and second-order accuracy is achieved with a defect correction procedure based on a QUICK scheme. Textbook Multigrid Convergence (TMC) is attained for the model problem. We define TMC as the solution of the governing system of equations in a fixed amount of work units independent of the grid size, grid stretching factor and Reynolds number. The code has been parallelized with the standard OpenMP directives for share-memory parallel computing to exploit the parallelism in the smoother.

2. NUMERICAL PROBLEM AND DISCRETIZATION

The dimensionless steady-state incompressible Navier-Stokes equations in the absence of body forces may be written as:

$$\begin{aligned} (\mathbf{u}\cdot\nabla)\mathbf{u} &= -\nabla p + \frac{1}{Re}\Delta\mathbf{u}, \\ \nabla\cdot\mathbf{u} &= 0, \end{aligned} \tag{1}$$

where $\mathbf{u} \in \Re^3 = (\mathbf{u},\mathbf{v},\mathbf{w})$ is the non dimensional velocity field and p is the dimensionless pressure, Re being the Reynolds number defined as $Re = \frac{U_\infty \cdot L}{\nu}$ with U_∞ a characteristic velocity, L a characteristic length and ν the kinematic viscosity.

In order to obtain the discrete expression of the non-linear system (1) the solution domain is divided in a finite set of control-volumes (CV). In the present work we will use an orthogonal cell-centered structured grid where each control volume will be an hexahedron like in figure 1. Our discretization is based on a staggered location of the unknowns (i.e the velocities are evaluated at the faces of the CV and the pressure field at the center of each CV). The staggered arrangement was implemented due to its accuracy, stability and conservation properties [6].

The procedure to discretize the u-momentum equation will be carried out with some detail. A CV surrounding a generic node u_{ijk} is built, displaced from the CV of the continuity equation (see figure 1 left-hand chart). Integrating the convective terms of the u-momentum equation over this volume we get:

$$\int_\Omega \mathbf{u}\nabla \mathrm{u}\, \mathrm{dV} = \int_{\partial\Omega} \mathrm{u}(\mathbf{u}\cdot\mathbf{n})\, \mathrm{dS} \approx \sum_{\mathrm{k}} \mathrm{m_k u_k dS}, \;\; \mathrm{k} = \mathrm{e,w,s,n,t,b}, \tag{2}$$

where the mass fluxes m_k are easily evaluated using linear interpolation and the velocities at the faces are calculated using an upwind interpolation.

Second-order accuracy is obtained using a defect correction technique within the multigrid cycle as explained subsequently. The higher order operator is based on the QUICK scheme proposed by Hayase et al [5]. With these considerations the algebraic coefficients for the convective terms can be written as:

$$L_e^c = \min(0, \mathrm{m_e}), \;\; \mathrm{L_w^c} = \min(0, \mathrm{m_w}), \;\; \mathrm{L_n^c} = \min(0, \mathrm{m_n});$$

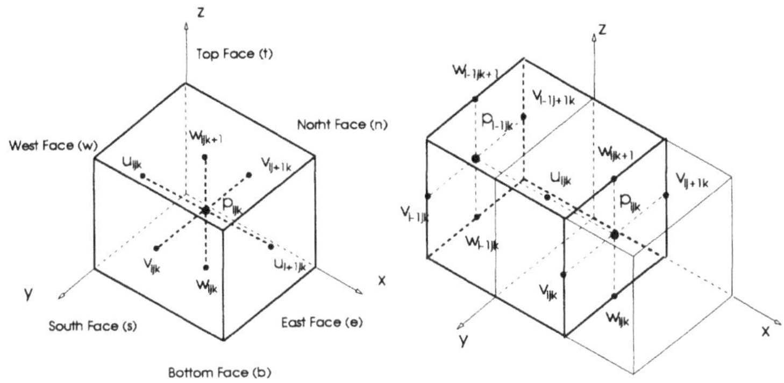

Figure 1. Placement of the unknowns in the CV (left-hand chart). Control Volume where the u-momentum equation is integrated (right-hand chart).

$$L^c_s = \min(0, \mathrm{m}_s),\ \mathrm{L}^c_t = \min(0, \mathrm{m}_t),\ \mathrm{L}^c_b = \min(0, \mathrm{m}_b);$$
$$L^c_p = -(L^c_e + L^c_n + L^c_s + L^c_b + L^c_t + L^c_w). \tag{3}$$

The expression for L^c_p has been obtained using the continuity equation. In a similar way the coefficients for the difussive part gives:

$$L^d_e = \frac{\Delta Y \Delta Z}{Re(x_{i+1jk} - x_{ijk})},\ L^d_w = \frac{\Delta Y \Delta Z}{Re(x_{ijk} - x_{i-1jk})},\ L^d_n = \frac{\Delta X \Delta Z}{Re(y_{ij+1k} - y_{ij-1k})};$$
$$L^d_s = \frac{\Delta X \Delta Z}{Re(y_{ij+2k} - y_{ijk})},\ L^d_t = \frac{\Delta Y \Delta X}{Re(z_{ijk+1} - z_{ijk-1})},\ L^d_b = \frac{\Delta Y \Delta X}{Re(z_{ijk+2} - z_{ijk})};$$
$$L^d_p = -(L^d_e + L^d_n + L^d_s + L^d_b + L^d_t + L^d_w). \tag{4}$$

The coefficients of the algebraic equation for a generic node u_{ijk} are the sum of the difussive and convective parts, i.e. $L^u_l = L^c_l + L^d_l$ with $l = e, w, n, s, b, t$. The resulting discrete system of equations is written as follows:

$$\begin{pmatrix} L^h_u & 0 & 0 & L^h_p \\ 0 & L^h_v & 0 & L^h_p \\ 0 & 0 & L^h_w & L^h_p \\ L^h_m & L^h_m & L^h_m & 0 \end{pmatrix} \begin{pmatrix} u \\ v \\ w \\ p \end{pmatrix} = \begin{pmatrix} f_u \\ f_v \\ f_w \\ f_p \end{pmatrix}, \tag{5}$$

where the source terms f_u, f_v, f_w and f_p in the right-hand side of the system (5) include the discretization of the boundary conditions and the contribution of the QUICK scheme.

3. THE MULTIGRID METHOD

Our approach uses a FAS full multigrid algorithm [1], where a certain number of F(2,1)-cycles (see figure 2) is applied in each level. The restriction and prolongation operators are dictated by the staggered arrangement of unknowns and for the coarsening strategy

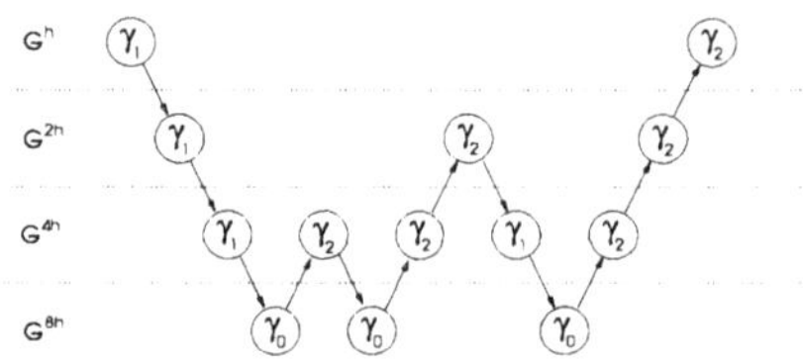

Figure 2. Scheme of an F-cycle, F(γ_1,γ_2), where γ_0 represents the number of iterations of the smoother performed to solve the coarsest level

used. One of the most important parts in a multigrid algorithm is the smoothing process. Several smoothers have been proposed for the Navier-Stokes equations and can be classified in two types: (1) coupled smoothing (where the momentum and continuity equation are satisfied simultaneously [14,12]), and (2) distributive smoothing [13,3] (where the momentum equation are solved in a first step, and then the continuity equation is satisfied). We have chosen a cell-implicit Symmetric Coupled Gauss Seidel (SCGS) method as smoother because of its higher stability and rapid convergence [14]. In the SCGS scheme all the variables involved in each CV are updated simultaneously by inverting a small matrix of equations, and so the operation count per multigrid cycle is greater than for the distributive relaxation.

In terms of the residuals and corrections at the control-volume ijk (see fig 1 right-hand chart) the equation can be written as follows:

$$\begin{pmatrix} L^u_{ijk} & 0 & 0 & 0 & 0 & 0 & \frac{1}{\Delta X} \\ 0 & L^u_{i+1jk} & 0 & 0 & 0 & 0 & -\frac{1}{\Delta X} \\ 0 & 0 & L^v_{ijk} & 0 & 0 & 0 & \frac{1}{\Delta Y} \\ 0 & 0 & 0 & L^v_{ij+1k} & 0 & 0 & -\frac{1}{\Delta Y} \\ 0 & 0 & 0 & 0 & L^w_{ijk} & 0 & \frac{1}{\Delta Z} \\ 0 & 0 & 0 & 0 & 0 & L^w_{ijk+1} & -\frac{1}{\Delta Z} \\ -\frac{1}{\Delta X} & \frac{1}{\Delta X} & -\frac{1}{\Delta Y} & \frac{1}{\Delta Y} & -\frac{1}{\Delta Z} & \frac{1}{\Delta Z} & 0 \end{pmatrix} \begin{pmatrix} \Delta u_{ijk} \\ \Delta u_{i+1jk} \\ \Delta v_{ijk} \\ \Delta v_{ij+1k} \\ \Delta w_{ijk} \\ \Delta w_{ijk+1} \\ \Delta p_{ijk} \end{pmatrix} = \begin{pmatrix} r^u_{ijk} \\ r^u_{i+1jk} \\ r^v_{ijk} \\ r^v_{ij+1k} \\ r^w_{ijk} \\ r^w_{ijk+1} \\ r^m_{ijk} \end{pmatrix}. \quad (6)$$

The system (6) is linearized by computing the fluxes in the volume using the actual solution. The linear system of equations can be easily solved using Gaussian elimination and does not need any further discussion. The velocity components and the pressure are updated using under-relaxation:

$$\begin{aligned} \mathbf{u}^{\text{new}} &= \mathbf{u}^{\text{old}} + \mathrm{r_u}\Delta\mathbf{u}, \\ \mathrm{p}^{\text{new}} &= \mathrm{p}^{\text{old}} + \mathrm{r_p}\Delta\mathrm{p}. \end{aligned} \quad (7)$$

The under-relaxation technique has the effect of adding a pseudo-time dependent term in the equations. In the following simulations the under-relaxation for the pressure has been fixed to 1.0, while the under-relaxation factor for the velocities is strongly problem dependent and has been set empirically.

A plane-implicit smoother combined with semicoarsening has been used to deal with the anisotropy due to the grid stretching (the cell-implicit smoother diverges). All the variables involved in the plane are updated simultaneously via under-relaxation. It was observed that this block implicit solver needs to be applied along the sub-characteristics for convection dominated problems. In fact, the alternating-plane smoother was found to exhibit very poor convergence rates. The plane solver used in the 3-D smoothing procedure is a robust two-dimensional (2-D) multigrid F(2,1)-cycle employing semi-coarsening and a line smoother; the approximate 2-D solution obtained after one 2-D cycle is sufficient to provide robustness and good convergence rates in the 3-D solvers.

As mentioned before second-order accuracy is obtained using a defect correction inside the multigrid cycle that can be summarize as follows: whenever the metrics are re-calculated the source terms from the QUICK scheme are computed based on the current approximation and added to the right-hand side of the system of equations.

4. RESULTS

We consider an square plate placed in the middle of the solution domain. In the west face ($x = 0$) we define the inflow boundary with no angle of attack, thus the east face will hold the outflow condition. On the plate a no-slip boundary condition is imposed, and symmetric condition elsewhere in the domain boundary. As the velocity gradient normal to the wall is very large only in the boundary layer, the thin-layer approximation which only retains those terms can be adopted. However in the following simulations the original form (1) of the Navier-Stokes equations is solved. In order to capture the viscous effects

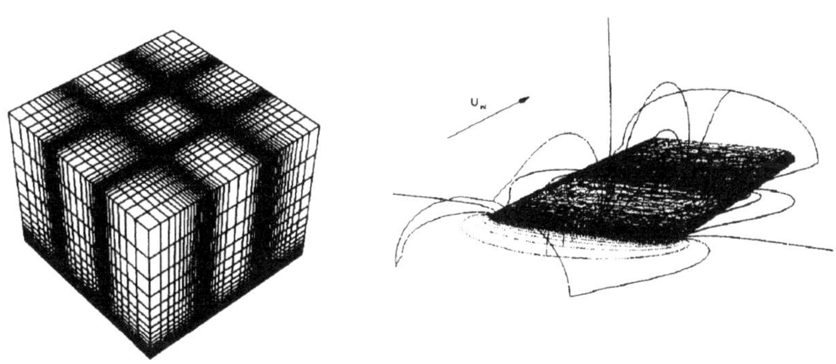

Figure 3. $48x48x32$ grid used for the flat-plate simulation (left-hand chart). Pressure contour lines and boundary layer for $Re = 10^4$ (right-hand side).

the grid is highly stretched near the plate, see figure 3. To ensure that sufficient number of grid points will lay inside the boundary-layer the mesh space for a uniform mesh will impose too high demands on the computation. For example, approximating the boundary-layer thickness with $\delta \sim \frac{1}{\sqrt{Re}}$ for $Re = 10000$ we have $\delta \sim 0.01$ that implies at least 100 grid points in a uniform grid which cannot be considered due to memory limitations.

Thus the grids are stretched in the z-direction using a geometric factor $h_k = \beta h_{k-1}$ with $\beta = 1.3$ for the $24x24x32$ grid and $\beta = 1.1$ for the $48x48x64$ grid.

The L_2-norm of the average residuals of the equations is shown in the figure 4 for four different grids with lexicographic and zebra ordering and Z semi-coarsening. The residual norm is reduced nearly five orders of magnitude in the first five cycles in all the cases (note that four orders of magnitude are reduced in the first two cycles for the $48x48x64$ grid). The asymptotic convergence rate is equal to 0.19 which is close to the one obtained for the Poisson equation with a semi-coarsened smoother [9]. In fact, the full multigrid algorithm converges the solution below truncation error with one F(2,1) cycle per level.

Results obtained for first-order accuracy are even better. The residual norm is reduced nearly five orders of magnitude in the first three cycles and the asympotityc convergence rate is equal to 0.1.

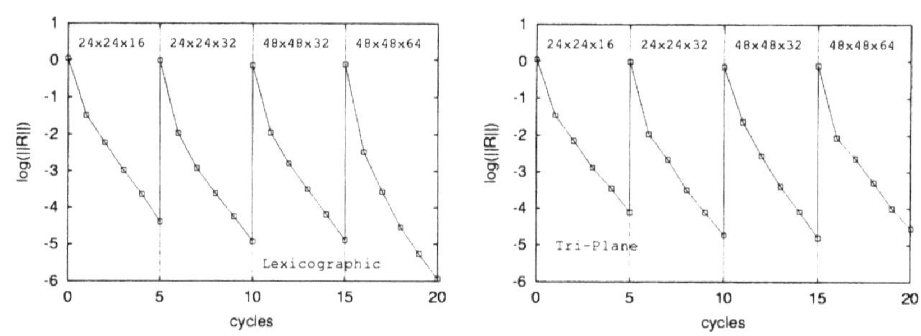

Figure 4. L_2-norm of the residual for the simulation of a 3-D flat plate for $Re = 10^4$ with lexicographic (left chart) and tri-plane (right chart) orderings.

In order to verify the solution, the u-velocity in the middle of the plate is compared with the Blasius analytical solution for a 2-D plate (figure 5). The little discrepancy near the layer edge is due to the highly stretched grid used in this simulation. The second-order accuracy has been verified using the solution in the three finest grids.

Convergence rates, independent of the Reynolds number, the grid size and the stretching factor, were achieved for the resolution of the boundary layer over a flat plate. The Tri-plane ordering of the planes exhibits similar properties to the lexicographic ordering (see figure 4) and allows the parallel implementation of the algorithm.

5. CONCLUSIONS AND FUTURE WORK

The robustness of plane smoothers combined with semi-coarsening has been investigated through the solution of the incompressible 3-D Navier-Stokes equations. Convergence results have been obtained for a common benchmark in CFD; the flow over a flat plate. Robustness has been defined as the ability of the multigrid method to solve the model problem with a convergence rate per work unit independent of grid size, stretching factor

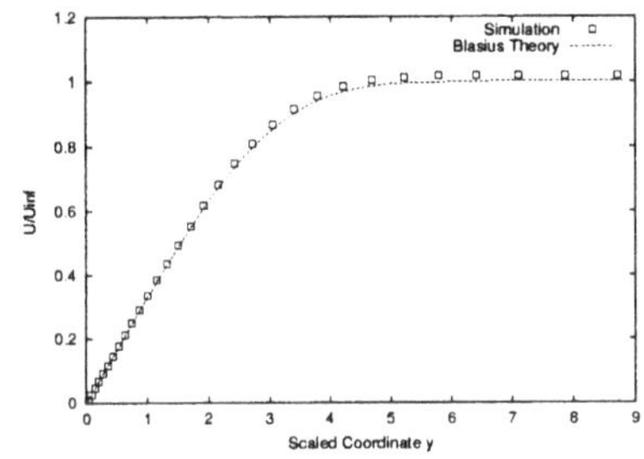

Figure 5. Simulation comparison with Blasius theory at the middle of the plate with $Re = 10^4$.

and Reynolds number (*textbook multigrid convergence rate*).

The combination of xy-plane smoothing and Z semi-coarsening has been found to be the best choice for the flat plate simulation. Its convergence rate is independent of grid size, stretching and Reynolds number, and the tri-plane variant exhibits similar properties to the lexicographic ordering and allows the parallel implementation of the algorithm. It has been parallelized with the standard OpenMP directives for share-memory parallel computing to exploit the parallelism in the smoother. Preliminary results show an efficiency of about 0.6 for 32 processors on a SGI Origin 2000 for a 48x48x64 grid.

6. ACKNOWLEDGMENTS

We would like to thank CSC (Centro de Supercomputación Complutense), ICASE and the Computational Modeling and Simulation Branch at the NASA Langley Research Center for providing access to the parallel computers that have been used in this research.

REFERENCES

1. A. Brandt. Multigrid techniques: 1984 guide with applications to fluid dynamics. Technical Report GMD-Studien 85, May 1984.
2. A. Brandt. Barriers to Achieving Textbook Multigrid Efficiency (TME) in CFD. *ICASE Interim Report No. 32*, 1998.
3. A. Brandt and I. Yavneh. On multigrid solution of high-Reynolds incompressible entering flows. *Computational Physics*, 101:151–164, 1992.
4. J. E. Dendy, S. F. McCormick, J.W. Ruge, T.F. Russell, and S. Schaffer. Multigrid methods for three-dimensional petroleum reservoir simulation. In *Tenth SPE Symposium on Reservoir Simulation*, February 1989.
5. T. Hayase, J.A.C. Humphrey, and R. Greif. A Consistently Formulated QUICK scheme for fast and stable convergence using finite-volume iterative calculation procedures. *Computational Physics*, 98:108–118, 1992.
6. J. Linden, B. Steckel, and K. Stuben. Parallel multigrid solution of the Navier-Stokes equations on general 2-D domains. *Parallel Computing*, 7:461–475, 1988.

7. I. M. Llorente and N. D. Melson. Robust multigrid smoothers for three dimensional elliptic equations with strong anisotropies. Technical Report 98-37, ICASE, 1998.
8. I. M. Llorente and F. Tirado. Relationships between efficiency and execution time of full multigrid methods on parallel computers. *IEEE Trans. on Parallel and Distributed Systems*, 8(6), 1997.
9. R. S. Montero, M. Prieto, I. M. Llorente, and F. Tirado. Robust multigrid algorithms for 3-D elliptic equations on structured grids. In *Proceedings of the 6th European Multigrid Conference. (EuroMG'99)*, September 1999.
10. N. H. Naik and J. V. Rosendale. The improved robustness of multigrid elliptic solvers based on multiple semicoarsened grids. *SIAM J. of Num. Anal.*, 30:215–229, February 1993.
11. C. W. Oosterlee. The convergence of parallel multiblock multigrid methods. *Applied Numerical Mathematics*, 19:115–128, 1995.
12. C. W. Oosterlee. A GMRES-Based plane smoother in multigrid to solve 3-D anisotropic fluid flow problems. *Computational Physics*, 130:41–53, 1997.
13. J.L. Thomas, B. Diskin, and A. Brandt. Textbook multigrid efficiency for the incompressible navier-stokes equations: High reynolds number wakes and boundary layers. Technical Report 99-51, ICASE, 1999.
14. S. P. Vanka. Block-Implicit Multigrid Solution of Navier-Stokes Equations in Primitive Variables. *Computational Physics*, 65:138–158, 1986.

Parallel Computational Fluid Dynamics – Trends and Applications
C.B. Jenssen et al. (Editors)

Parallel Computing Performance of an Implicit Gridless Type Solver

K. Morinishi

Department of Mechanical and System Engineering,
Kyoto Institute of Technology,
Matsugasaki, Sakyo-ku, Kyoto 606-8585, Japan

This paper describes an implicit gridless type solver and its performance on a parallel computer. The solver consists of the gridless evaluation of gradients, upwind evaluation of inviscid flux, central evaluation of viscous flux, and an implicit temporal discretization using LU-SGS method. The implicit solver is more than one order of magnitude efficient compared to an explicit counterpart in sequential computations. Domain decomposition method without overlapping is introduced for the parallel computing. In order to improve parallel computing speedup, a corrector stage is introduced. Numerical experiments are carried out on the Hitachi SR2201 parallel computer at Kyoto Institute of Technology up to 16 PUs. A typical speedup ratio of 14.5 is attained using 16 PUs.

1. INTRODUCTION

Points, instead of a grid, are first distributed all over the flow field considered in a gridless type flow simulation. As an example, Figure 1 shows the computational points distributed around a four-element airfoil of Suddhoo and Hall [1]. Figure 2 shows a cloud of points selected for each point, with which spatial derivatives of flow quantities are evaluated. The governing equations of fluid dynamics are numerically solved on each point. The solution can be applied on the points of structured grids, unstructured grids, and even on the points randomly distributed over the flow field [2].

The parallel computing performance of an explicit gridless type solver was demonstrated in the previous study [2]. The solver consists of a central gridless approximation and an artificial dissipation for the spatial terms and an explicit rational Runge-Kutta method for the temporal term. The solver generally works pretty well for inviscid and viscous flow simulations. For applications to viscous high Reynolds number flow simulations, however, uncertain reliability of the artificial dissipation term and slow convergence rate of the explicit method must be improved.

In this paper, the parallel computing performance of an implicit upwind gridless type solver is demonstrated. The inviscid flux is evaluated with Roe's approximate Riemann solver. The linear system derived from Euler implicit temporal approximation is solved using LU-SGS method. In order to improve the parallel computing efficiency, a corrector stage is introduced to the LU-SGS method. The reliability and efficiency of the solver are examined for typical inviscid and viscous flow simulations.

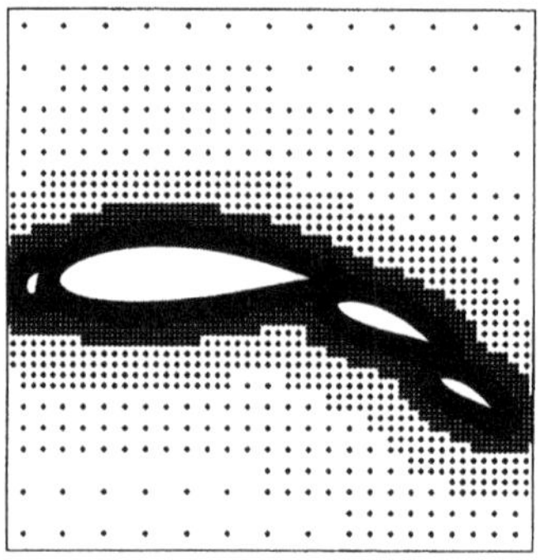

Figure 1. Points distributed around a four-element airfoil.

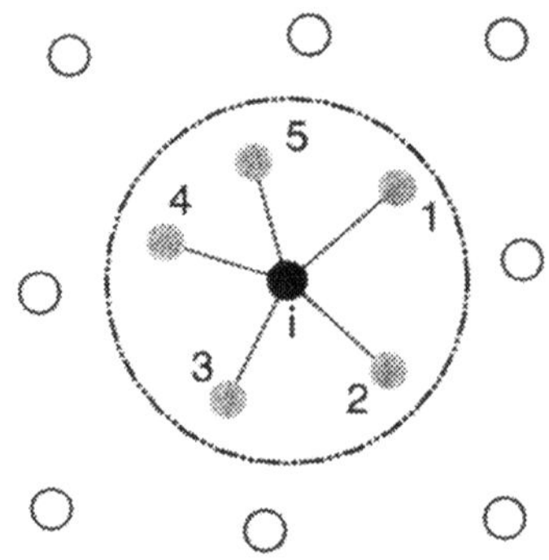

Figure 2. Cloud $C(i)$ for point i.

2. NUMERICAL PROCEDURE

The governing equations of this study are the compressible Euler and Navier-Stokes equations in Cartesian coordinates:

$$\frac{\partial \mathbf{q}}{\partial t} + \frac{\partial \mathbf{E}}{\partial x} + \frac{\partial \mathbf{F}}{\partial y} = \frac{1}{Re}\left(\frac{\partial \mathbf{R}}{\partial x} + \frac{\partial \mathbf{S}}{\partial y}\right) \tag{1}$$

where $\mathbf{q}$ is conservative vector, $\mathbf{E}$ and $\mathbf{F}$ the convective flux terms, and $\mathbf{R}$ and $\mathbf{S}$ the viscous flux terms. The gridless type solution for the equations are described in this section.

2.1. Evaluation of spatial derivatives

Spatial derivatives of any function f are evaluated at every point with the following linear combination forms:

$$\left.\frac{\partial f}{\partial x}\right|_i = \sum_{k \subset C(i)} a_{ik} f_{ik}\,, \quad \left.\frac{\partial f}{\partial y}\right|_i = \sum_{k \subset C(i)} b_{ik} f_{ik} \tag{2}$$

where the subscript k denotes the index of points which belong to the cloud $C(i)$ for the point i considered as shown in Figure 2 and f_{ik} are evaluated at the midpoint between the points i and k.

The coefficients a_{ik} and b_{ik} are once obtained at the beginning of computation using a weighted least-squares method and stored [2]. It should be noted that if the method is applied on a point of uniform Cartesian grids with usual five point stencil for its cloud, the coefficients are strictly identical to those of the conventional second-order central difference approximations.

2.2. Evaluation of inviscid flux

If Eqs.(2) are applied to the inviscid flux of the Navier-stokes equations, the following expression is obtained.

$$\begin{aligned}\frac{\partial \mathbf{E}}{\partial x} + \frac{\partial \mathbf{F}}{\partial y} &= \sum a_{ik}\mathbf{E}_{ik} + \sum b_{ik}\mathbf{F}_{ik} \\ &= \sum \mathbf{G}_{ik}\end{aligned} \tag{3}$$

The flux term $\mathbf{G}$ at the midpoint is expressed as:

$$\mathbf{G} = \begin{pmatrix} \rho U \\ \rho u U + ap \\ \rho v U + bp \\ U(e+p) \end{pmatrix} \tag{4}$$

where U are defined with:

$$U = au + bv \ . \tag{5}$$

An upwind method may be obtained if the numerical flux on the midpoint is obtained using Roe's approximate Riemann solver as:

$$\mathbf{G}_{ik} = \frac{1}{2}\left(\mathbf{G}(\tilde{\mathbf{q}}_{ik}^{+}) + \mathbf{G}(\tilde{\mathbf{q}}_{ik}^{-}) - |\tilde{\mathbf{A}}|(\tilde{\mathbf{q}}_{ik}^{+} - \tilde{\mathbf{q}}_{ik}^{-})\right) \tag{6}$$

where $\tilde{\mathbf{q}}$ are the primitive variables and $\tilde{\mathbf{A}}$ are the flux Jacobian matrices

The primitive variables at the midpoint are obtained from:

$$\tilde{\mathbf{q}}_{ik}^{-} = \tilde{\mathbf{q}}_i + \frac{1}{2}\phi^{-}\delta\tilde{\mathbf{q}}_{ik}^{-} \ , \quad \tilde{\mathbf{q}}_{ik}^{+} = \tilde{\mathbf{q}}_k - \frac{1}{2}\phi^{+}\delta\tilde{\mathbf{q}}_{ik}^{+} \tag{7}$$

where $\tilde{\mathbf{q}}_{ik}^{-}$ and $\tilde{\mathbf{q}}_{ik}^{+}$ are defined with:

$$\delta\tilde{\mathbf{q}}_{ik}^{-} = \nabla\tilde{\mathbf{q}}_i \cdot \mathbf{r}_{ik} \ , \quad \delta\tilde{\mathbf{q}}_{ik}^{+} = \nabla\tilde{\mathbf{q}}_k \cdot \mathbf{r}_{ik} \ . \tag{8}$$

The flux limiters ϕ^{-} and ϕ^{+} are defined as:

$$\phi^{-} = \frac{\delta\tilde{\mathbf{q}}_{ik}\delta\tilde{\mathbf{q}}_{ik}^{-} + |\delta\tilde{\mathbf{q}}_{ik}\delta\tilde{\mathbf{q}}_{ik}^{-}|}{\delta\tilde{\mathbf{q}}_{ik}^{2} + \delta\tilde{\mathbf{q}}_{ik}^{-2} + \epsilon} \ , \quad \phi^{+} = \frac{\delta\tilde{\mathbf{q}}_{ik}\delta\tilde{\mathbf{q}}_{ik}^{+} + |\delta\tilde{\mathbf{q}}_{ik}\delta\tilde{\mathbf{q}}_{ik}^{+}|}{\delta\tilde{\mathbf{q}}_{ik}^{2} + \delta\tilde{\mathbf{q}}_{ik}^{+2} + \epsilon} \tag{9}$$

where ϵ is very small number which prevents null division in smooth flow regions and $\delta\tilde{\mathbf{q}}_{ik}$ are defined as:

$$\delta\tilde{\mathbf{q}}_{ik} = \tilde{\mathbf{q}}_k - \tilde{\mathbf{q}}_i \ . \tag{10}$$

The gradients of primitive variables $\nabla\tilde{\mathbf{q}}$ are obtained using Eqs.(2) at each point. These gradients of primitive variables are also used for the evaluation of the viscous stress in the Navier-Stokes equations.

2.3. Evaluation of viscous flux

The viscous terms of the Navier-Stokes equations are evaluated using Eqs.(2) at each point as:

$$\frac{\partial}{\partial x}\left(\mu\frac{\partial u}{\partial x}\right)\bigg|_i = \sum_{k\subset C(i)} a_{ik}\left(\mu\frac{\partial u}{\partial x}\right)_{ik} . \tag{11}$$

Instead of a simple arithmetical average, the first derivative at the midpoint is obtained with:

$$\frac{\partial u}{\partial x}\bigg|_{ik} = \frac{\Delta x}{\Delta s^2}(u_k - u_i) + \frac{1}{2}\frac{\Delta y}{\Delta s^2}\left(\Delta y\left(\frac{\partial u}{\partial x}\bigg|_i + \frac{\partial u}{\partial x}\bigg|_k\right) - \Delta x\left(\frac{\partial u}{\partial y}\bigg|_i + \frac{\partial u}{\partial y}\bigg|_k\right)\right) \tag{12}$$

where

$$\Delta x = x_k - x_i\ , \quad \Delta y = y_k - y_i\ , \quad \Delta s^2 = \Delta x^2 + \Delta y^2\ . \tag{13}$$

2.4. Temporal discretization

A linearized implicit Euler method is used for the temporal discretization of the gridless type solver with the following linearizing assumption.

$$\mathbf{G}_{ik}^{n+1} = \mathbf{G}_{ik}^{n} + \mathbf{A}_{ik}^{+}(\mathbf{q}_i^n)\Delta\mathbf{q}_i + \mathbf{A}_{ik}^{-}(\mathbf{q}_k^n)\Delta\mathbf{q}_k \tag{14}$$

Here the superscript n denotes the time index and the split flux Jacobian matrices are defined as:

$$\mathbf{A}^{\pm} = \mathbf{X}\mathbf{\Lambda}^{\pm}\mathbf{X}^{-1}\ . \tag{15}$$

Then the linearized implicit Euler method is obtained as:

$$\left(\frac{1}{\Delta t_i}\mathbf{I} + \sum_{k\subset C(i)}\mathbf{A}_{ik}^{+}(\mathbf{q}_i^n)\right)\Delta\mathbf{q}_i + \sum_{k\subset C(i)}\mathbf{A}_{ik}^{-}(\mathbf{q}_k^n)\Delta\mathbf{q}_k = \mathbf{W}(\mathbf{q}_i^n) \tag{16}$$

where $\mathbf{W}$ are residuals obtained from the gridless evaluation of the inviscid and viscous terms of the Navier-Stokes equations and $\Delta\mathbf{q}$ are defined as:

$$\Delta\mathbf{q} = \mathbf{q}^{n+1} - \mathbf{q}^{n}\ . \tag{17}$$

The solution of this linear system of equation is obtained with LU-SGS procedure as:

$$\Delta\mathbf{q}_i^* = \mathbf{D}_i^{-1}\left[\mathbf{W}(\mathbf{q}_i^n) - \sum_{k\in L(i)}\mathbf{A}_{ik}^{-}(\mathbf{q}_k^n)\Delta\mathbf{q}_k^*\right] \tag{18}$$

$$\Delta\mathbf{q}_i = \Delta\mathbf{q}_i^* - \mathbf{D}_i^{-1}\sum_{k\in U(i)}\mathbf{A}_{ik}^{-}(\mathbf{q}_k^n)\Delta\mathbf{q}_k \tag{19}$$

where $\mathbf{D}_i$ are defined with

$$\mathbf{D}_i = \left(\frac{1}{\Delta t_i}\mathbf{I} + \sum_{k\subset C(i)}\mathbf{A}_{ik}^{+}(\mathbf{q}_i^n)\right)\ . \tag{20}$$

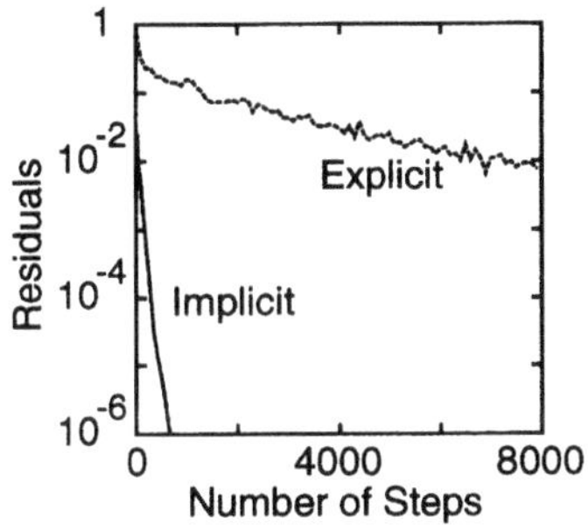

Figure 3. Comparison of convergence histories.

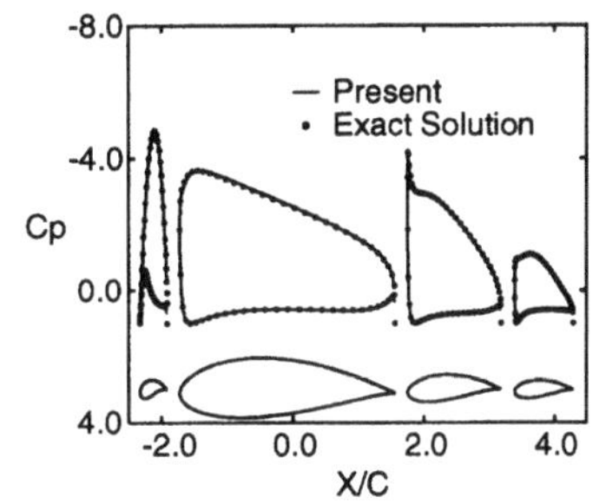

Figure 4. Comparison of surface Cp distributions.

This implicit solver is more than one order of magnitude efficient compared to an explicit counterpart in sequential computations. Figure 3 shows the comparison of residual histories of the implicit solver and a typical explicit solver. The results obtained for an inviscid flow about the four-element airfoil of Suddhoo and Hall [1] at a free stream Mach number of 0.16. The convergence rate of the explicit solver is intolerably slow at this low Mach number flow condition, whereas the convergence rate of the implicit one is quite satisfactory. Suddhoo and Hall obtained the exact pressure distributions along the airfoil for incompressible potential flows. Figure 4 shows the comparison of the pressure distributions. The pressure distributions predicted with the gridless solver are also quite satisfactory.

In parallel computation using the domain decomposition method without overlapping, the convergence rates of residual histories decrease as the number of small subdomain increases. In order to overcome the deficiency following corrector stage is introduced to the LU-SGS method.

$$\mathbf{W}^{(1)} = 2\mathbf{W}(\mathbf{q}_i^n) - \sum_{k\subset C(i)} \mathbf{A}_{ik}^-(\mathbf{q}_k^n)\Delta\mathbf{q}_k^{(1)} - \mathbf{D}_i\Delta\mathbf{q}_i^{(1)} \tag{21}$$

$$\Delta\mathbf{q}_i^* = \mathbf{D}_i^{-1}\left[\mathbf{W}^{(1)} - \sum_{k\in L(i)} \mathbf{A}_{ik}^-(\mathbf{q}_k^n)\Delta\mathbf{q}_k^*\right] \tag{22}$$

$$\Delta\mathbf{q}_i = \Delta\mathbf{q}_i^* - \mathbf{D}_i^{-1}\sum_{k\in U(i)} \mathbf{A}_{ik}^-(\mathbf{q}_k^n)\Delta\mathbf{q}_k \tag{23}$$

Here $\Delta\mathbf{q}^{(1)}$ is the prediction of $\Delta\mathbf{q}$ obtained with Eqs.(18) and (19).

3. PARALLEL COMPUTING RESULTS

Numerical experiments were carried out on the Hitachi SR2201 parallel computer at Kyoto Institute of Technology. The computer has 16 PUs which are connected by a crossbar network. Each PU consists of a 150MHz PA-RISC chip, a 256MB memory, and two cascade caches. The message passing is handled with express Parallelware.

For parallel computing of the implicit gridless type solver, the domain decomposition method without any overlapping is adopted. In every time step, the message passings are carried out three times between the corresponding PUs . In the explicit stage (evaluation of the inviscid and viscous fluxes), conservative quantities $\mathbf{q}$ are first transferred and later the gradients of primitive variables $\nabla\tilde{\mathbf{q}}$ are transferred. In the implicit stage, $\Delta\mathbf{q}_i^*$ obtained from Eq.(18) are transferred before the process of Eq.(19) in the standard LU-SGS method. For the LU-SGS method with the corrector stage, $\Delta\mathbf{q}^{(1)}$ in stead of $\Delta\mathbf{q}_i^*$ are transferred before the corrector stage.

3.1. Invisicd flow over a 4-element airfoil

An application to the Euler equations is first shown. The inviscid flow about the four-element airfoil of Suddhoo and Hall [1] is considered at a free stream Mach number of 0.16 and a null angle of attack. Figure 1 shows the partial view of computational points distributed over the flow field. There are 7890 points in total with 200 points on the main element, 80 on the slat, 160 and 128 on the two flaps, respectively. Figure 5 shows speedup ratios par step as a function of the number of processors. Since operational counts par step are almost equally divided, the linear speedup is attained par step up to 16 PUs. Figure 6 shows the number of steps for obtaining five orders of magnitude reduction of residuals as a function of the number of processors. The number of steps for the standard LU-SGS method increases as the number of processors increases. Since the converged solution using 8 PUs and 16 PUs are only obtained at small CFL numbers, the large numbers of steps are required for those cases. The total speedup ratios for obtaining the converged solution are shown in Figure 7. The speedup ratios for 8 PUs and 16 PUs are reduced to 6.4 and 10.3, respectively.

Introducing the corrector stage to the LU-SGS method, the converged solution are obtained at a large CFL number up to 16 PUs and the increase in number of steps at 16 PUs is less than 10 percent as shown in Figure 6. The total speedup ratios for the corrected LU-SGS method at 8 PUs and 16 PUs are 7.8 and 14.4, respectively. Figure 8 shows the normalized CPU times for obtaining the converged solution. The CPU times are normalized with that of the standard LU-SGS method using 1 PU. Though the corrector stage requires considerable operational counts, the number of steps for converged solution using the stage substantially decreases as shown in Figure 6. As a result, the normalized CPU times of the corrected LU-SGS method are 0.84 using 1 PU and 0.059 using 16 PUs.

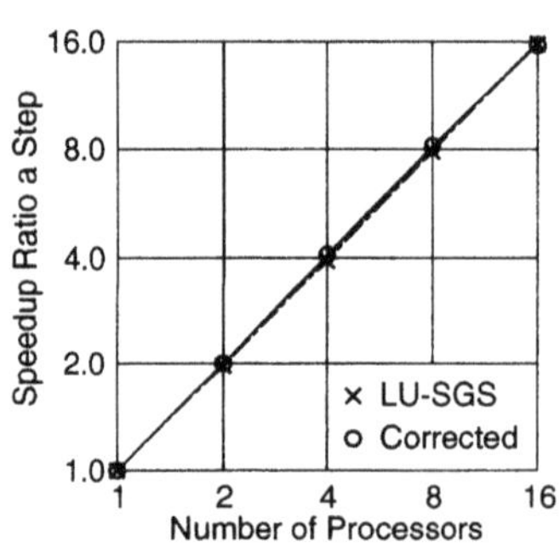

Figure 5. Comparison of speedup ratios par step.

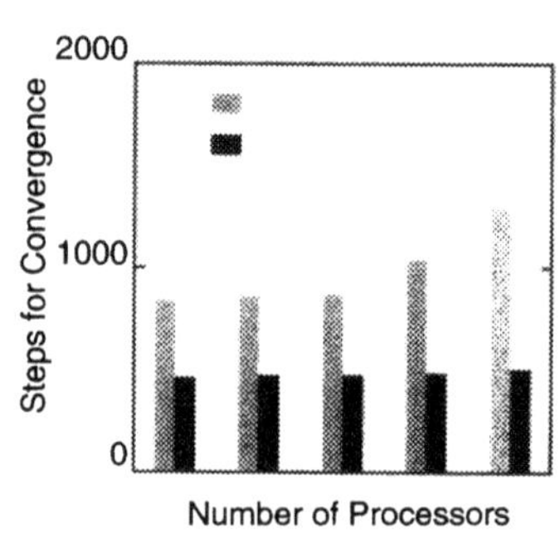

Figure 6. Number of steps for steady state solution.

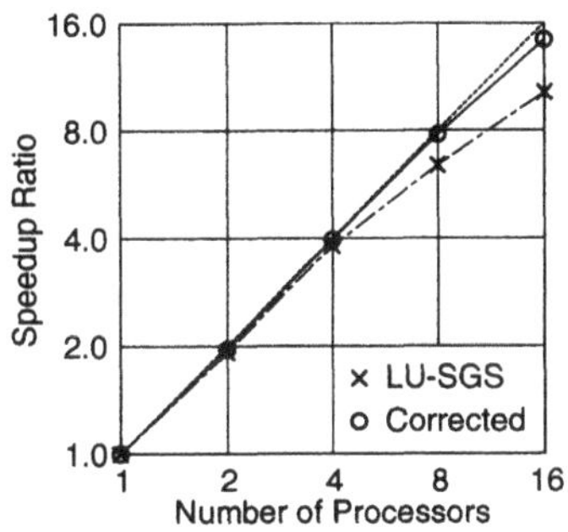

Figure 7. Comparison of final speedup ratios for steady state solution.

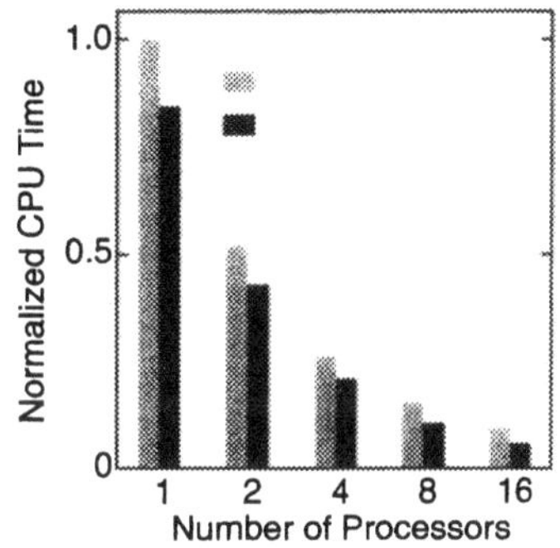

Figure 8. Comparison of normalized CPU times for steady state solution.

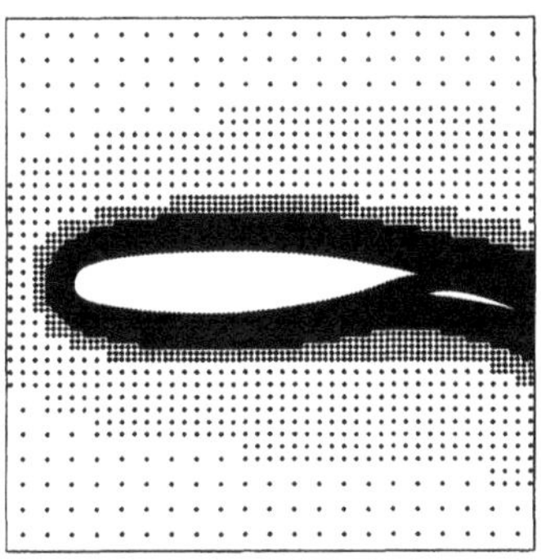

Figure 9. Points distributed around NLR7301 airfoil.

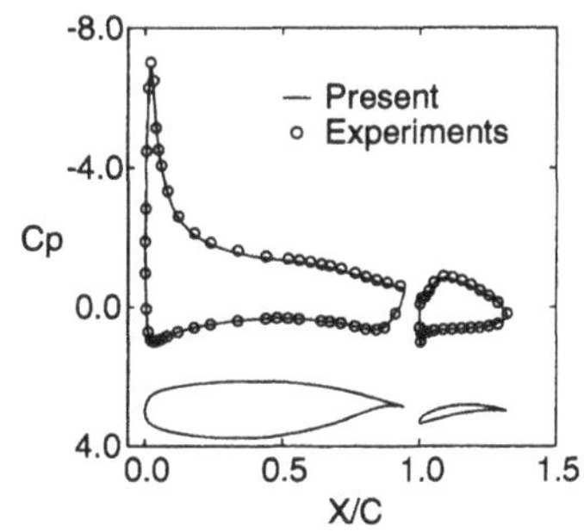

Figure 10. Comparison of surface Cp distributions.

3.2. Turbulent flow over a NLR7301 2-element airfoil

Figure 9 shows the partial view of computational points distributed around a NLR7301 two-element airfoil [3]. The total number of points distributed in the computational domain is 22152 with 193 points on the main element surface and 129 points on the flap. The pressure distributions, which are obtained along the airfoil at a free stream Mach number of 0.185, an attack angle of 6°, and a Reynolds number of 2.51×10^6, are compared with experimental data in Figure 10. Again the pressure distributions predicted with the implicit gridless solver are quite satisfactory.

Figure 11 shows the number of steps for obtaining the five orders of magnitude reduction of residuals as a function of the number of processors. Since the converged solution using the standard LU-SGS method with 16 PUs is only obtained at a small CFL number, the number of steps increases by 18 percent compared with one using 1 PU. Though the linear speedup is again attained par step up to 16 PUs for this turbulent flow case, the final speedup ratios for obtaining the converged solution are 7.6 and 13.0 for 8 PUs and 16 PUs, respectively.

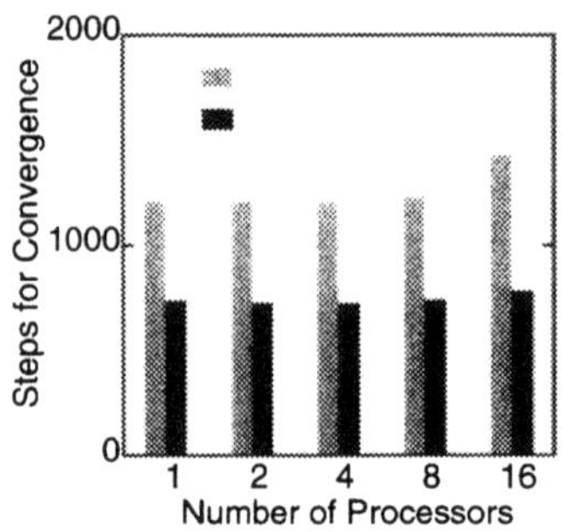

Figure 11. Number of steps for steady state solution.

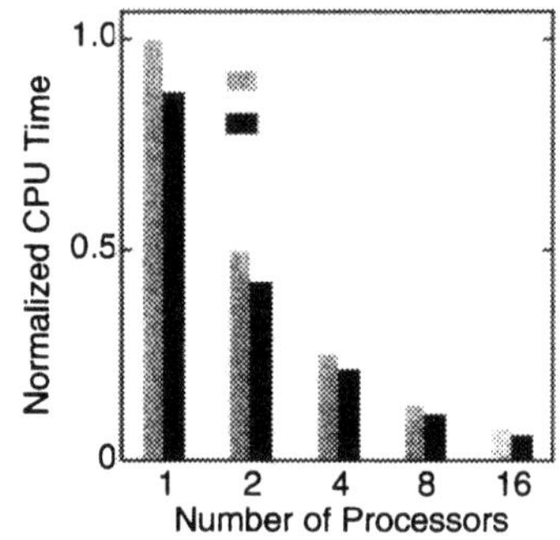

Figure 12. Comparison of normalized CPU times for steady state solution.

Using the LU-SGS method with the corrector stage, the converged solution is obtained at a large CFL number up to 16 PUs and the increase in number of steps at 16 PUs is only 6 percent. The total speedup ratios for the corrected LU-SGS method for 8 PUs and 16 Pus are 8.0 and 14.5, respectively. Figure 12 shows the normalized CPU times for obtaining the converged solution. The CPU times are normalized with that of the standard LU-SGS method using 1 PU. The normalized CPU times of the LU-SGS method with the corrector stage are 0.87 using 1 PU and 0.06 using 16 PUs. Generally the robustness and efficiency of LU-SGS method are improved with the corrector stage.

4. CONCLUSIONS

The parallel computing performance of the implicit upwind gridless type solver has been demonstrated. The robustness and efficiency of the method for decomposed computational subdomains are improved using the corrector stage. The typical speedup ratio of 14.5 is attained using 16 PUs.

This study was supported in part by the Research for the Future Program (97P01101) from Japan Society for the Promotion of Science and a Grant-in-Aid for Scientific Research (12650167) from the Ministry of Education, Science, Sports and Culture of the Japanese Government.

REFERENCES

1. Suddhoo, A. and Hall, I.M., Test Cases for the Plane Potential Flow past Multi-element aerofoils, Aeronautical Journal, December (1985), 403-4414.
2. Morinishi, K., A Gridless Type Solver for Parallel Simulation of Compressible Flow, Parallel Computational Fluid Dynamics, Development and Applications of Parallel Technology, Elsevier, pp. 293-300, (1999).
3. Van Den Berg, B., Boundary Layer Measurements on a Two-dimensional Wing with Flap, NLR TR 79009, (1979).

Parallel Computational Fluid Dynamics – Trends and Applications
C.B. Jenssen et al. (Editors)

Efficient Algorithms for Parallel Explicit Solvers

A. Ecer and I. Tarkan

Computational Fluid Dynamics Laboratory
Department of Mechanical Engineering
Indiana University-Purdue University Indianapolis
723 W. Michigan St., Indianapolis, IN 46202, USA

1. INTRODUCTION

For the solution of CFD problems, one has to choose between implicit and explicit schemes. Explicit schemes are more popular for the solution of unsteady problems. Implicit schemes are favored to solve steady flow problems since explicit schemes are restricted by stability limitations in the time step size. The time step is usually restricted by the largest eigenvalue of the operator, although the transient solutions corresponding to such large eigenvalues decay rapidly if the scheme is stable. One can assume that the slow converging waves produce residuals which vary smoothly both in time and in space. If one can filter the high frequency components of the residuals, one can integrate the equations by using a larger time step [1]. In this paper, time filtering and extrapolation in time are employed to reach a steady state solution. One can employ these techniques to unstructured grids and can treat discontinuities in space, such as shocks, efficiently. The proposed technique resembles multi-grid schemes where larger time-steps to accelerate the slow waves are achieved by using filtering the solution through several grids with different levels of grid refinement. In the present approach, a single grid is utilized and the solution is filtered in time by using digital filters.

Regarding parallelization, in the present paper, data parallelization is assumed. The flow domain is divided into sub-regions (blocks) which are connected to each other by smaller sub-regions (interfaces). An explicit algorithm is running for each block. Interfaces are required only for neighboring blocks, which involves only local communication between a pair of processors. No global message broadcasting is required. The objective is to reach a steady state at the shortest elapsed time using all available computers. This involves reducing the time of iterations required to reach a steady state (computation cost) and the number and size of the messages to be communicated between the processors (communication cost)[2].

2. BLOCK CORRECTION SCHEME:

The developed procedure can be described for the solution of the following one-dimensional heat equation:

$$\frac{\partial u}{\partial t} = \frac{\partial^2 u}{\partial x^2} \qquad (1)$$

We will consider the solution of this equation with two different boundary conditions:

Case a: $u(0,t) = u(1,t) = 0$;

$$u_{exact}(x,t) = \sum_{n=1}^{\infty} c_n \sin(n\pi x) \cdot e^{-n^2\pi^2 t}, \quad \text{where} \quad c_n = 2\int_0^1 f(\zeta)\sin(n\pi\zeta)\, d\zeta .$$

Case b: $u(0,t) = 0; u(1,t) = 1$;

$$u_{exact}(x,t) = x + \sum_{n=1}^{\infty} c_n \sin(n\pi x) \cdot e^{-n^2\pi^2 t}, \quad \text{where} \quad c_n = 2\int_0^1 \{f(\zeta) - \zeta\}\sin(n\pi\zeta)\, d\zeta$$

If we employ forward time and centered space differencing, the difference formula can be written as,

$$u_i^{t+1} = u_i^t + d \cdot (u_{i+1}^t - 2u_i^t + u_{i-1}^t) \tag{2}$$

where, ($d=\alpha\Delta t/\Delta x^2$) has a stability limit of $d < 0.5$ which ensures that all components of the residual vector are decaying in time

The integration time step is defined by the stability requirements of the highest frequency component of the residual vector at a given time. The residual curves for cases a, with $d = 0.5$ are shown in figure 1 for a grid with 100 points. The speed of convergence to a steady state is controlled by the lowest frequency component of the residual vector and is quite slow. On the other hand, if one considers the accuracy of different Fourier components of the residual vector either in time or in space, one realizes that only a small portion of these waves are accurate [1].

In the present approach we implement a time filter directly to the residual vector at each time step and filter the high frequency components. We then accelerate the remaining low frequency components by extrapolation.

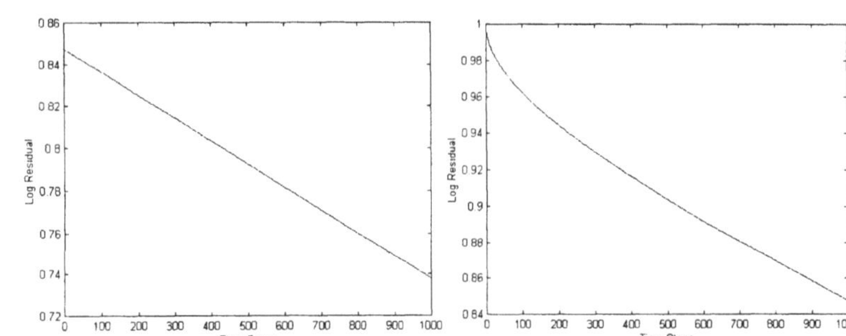

Figure 1. Residual curve a) with an initial condition of a sine wave, b) with an initial condition of a point source in the middle.

As a simple low pass filter, consider the time averaging of the residual vector over m time steps. We can then extrapolate to calculate a steady state solution by assuming that the low frequency components are varying slowly in time.

$$u_{ss} = u_{n+2m} - \frac{\left(\Delta t \sum_{i=0}^{m-1} R_{n+m+i}\right)^2}{\Delta t \sum_{i=0}^{m-1} R_{n+m+i} - \Delta t \sum_{i=0}^{m-1} R_{n+i}} \tag{4}$$

After each extrapolation, large frequency components develop within the residual vector. If one integrates the solution for n time steps, the high frequency components rapidly decay. The above filtering and extrapolation cycle is then repeated as summarized in figure 2.

By using the above scheme, the first example was again chosen to be case a, with an initial condition of a sine wave. The variation of residuals with the standard explicit scheme with d=0.5 and the proposed scheme are shown in figure 3.10. As can be seen from these results the low frequency components are accelerated while the high frequency components do not produce instability.

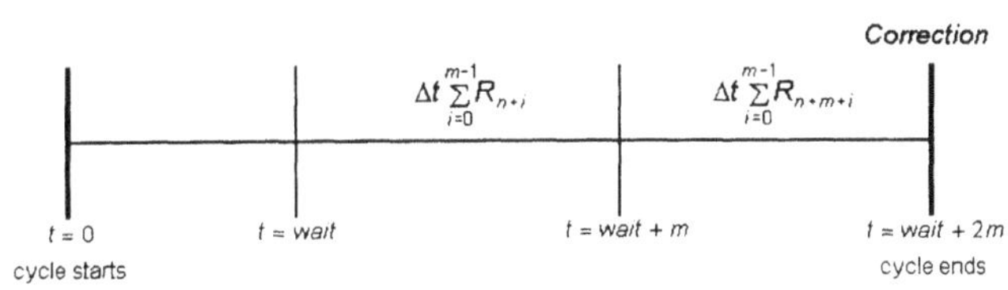

Figure 2. Block correction scheme.

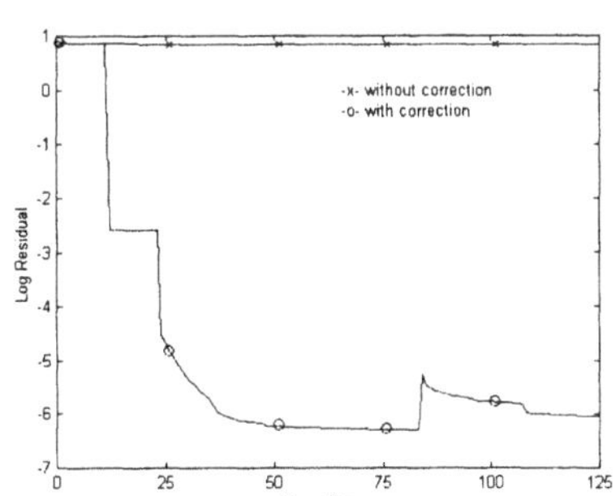

Figure 3. Block correction (residual curve for homogeneous boundary conditions, with an initial sinusoidal profile).

The second example was chosen to be case a, with an initial condition of one at the center node and zero at all other nodes. The variation of residuals with the standard explicit scheme with c=0.5 and the proposed scheme are shown in figure 4. As can be seen from these results the low frequency components are accelerated considerably.

The third example was chosen to be case b, with an initial condition of u = x +sin x/L. The variation of residuals with the standard explicit scheme with d=0.5 and the proposed scheme are shown in figure 5. As can be seen from these results the steady state is reached rapidly by accelerating the decay of the sine wave.

The fourth example is case b with the same initial condition of the second example: point source in the middle. The residual curves for the standard and corrected solutions are shown in figure 6. These results are not as impressive as the previous examples. This is due the fact that the steady state solution is not contained in the initial solution. In this case, the development of the steady state configuration can not be accelerated by simple extrapolation.

The fifth example involves case b with random initial conditions. In this case, the initial conditions are chosen randomly at each node between 0 and 1. Figure 7 demonstrates the improvement in convergence in comparison with the previous case by simply changing initial conditions.

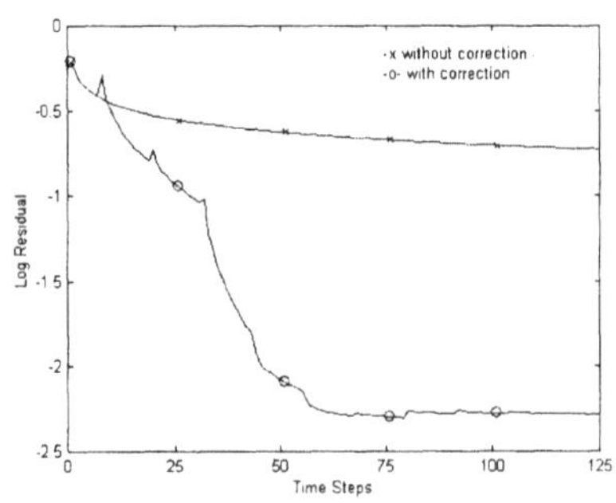

Figure 4. Block correction (residual curve for homogeneous boundary conditions, with an initial disturbance at the midpoint).

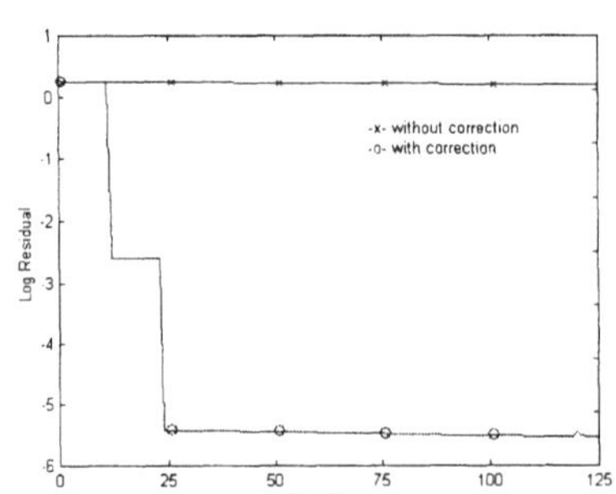

Figure 5. Block correction (residual curve for non-homogeneous boundary conditions, with a sinusoidal initial condition on top of exact solution).

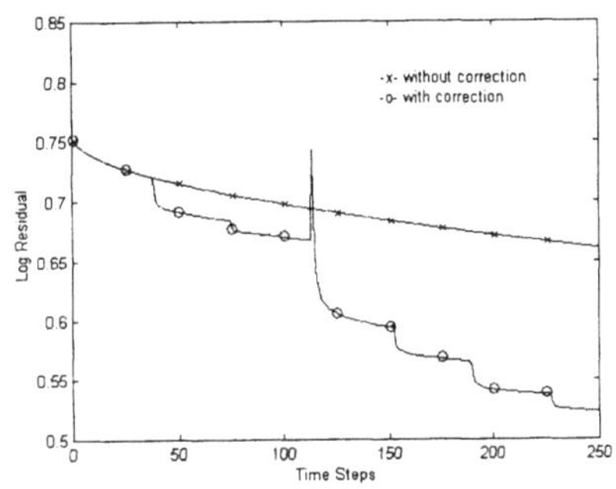

Figure 6. Block correction (residual curve for non-homogeneous boundary conditions, with an initial point source).

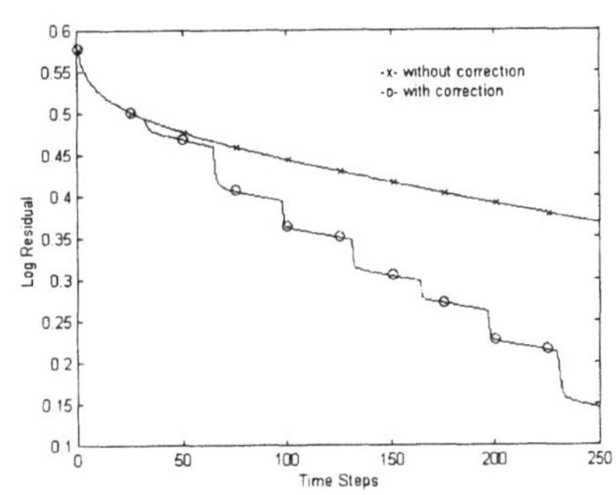

Figure 7. Block correction (residual curve for non-homogeneous boundary conditions, with an initial random profile).

3. INTERFACE CORRECTION SCHEME

Communication cost becomes significant for the parallel computation of explicit schemes where the ratio of communication to computation is high. Here, the basic concept to reduce communication cost is not to communicate between the blocks at each time step [2,3]. Again, we employ extrapolation and filtering. The procedure is summarized in figure 8.

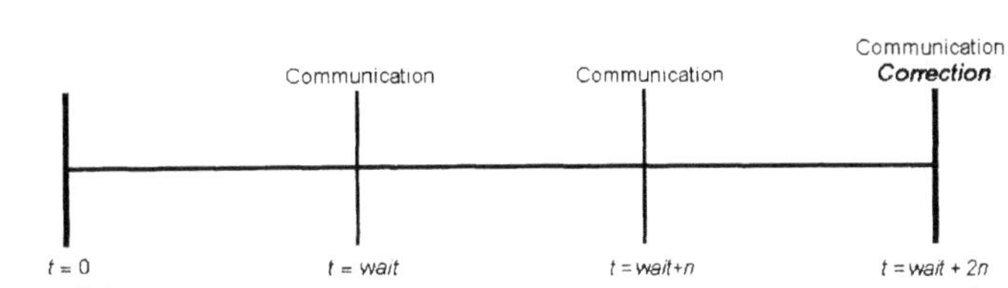

Figure 8. Interface correction scheme.

We wait for m time steps for each block without communicating, until the high frequency components due to the last corrections on the interface boundary conditions decay. We communicate between the neighboring blocks. We wait for n steps and communicate again. We utilize the interface solutions after two communications to estimate the correct boundary conditions by extrapolation. For the two blocks shown in fig 9, the scheme can be summarized as follows:

$$U_I = u_I^{2n} - \frac{c^2}{(1-c^2)}(u_I^{2n} - u_I^0) \tag{5}$$

$$U_{II} = u_{II}^{2n} - \frac{c^2}{(1-c^2)}(u_{II}^{2n} - u_{II}^0) \tag{6}$$

where

$$c = \frac{u_{II}^n - u_I^n - u_{II}^{2n} + u_I^{2n}}{u_{II}^n - u_I^n - u_{II}^0 + u_I^0}. \tag{7}$$

Two examples are presented: The first example is case a, with sinusoidal initial conditions and consists of 10 blocks each with 10 grid points. Only, interface corrections are applied. As can be seen from fig 10, with the correction and communicating every 50 steps, the convergence rate is only slightly reduced.

The second example is case b again with half sine wave as the initial condition as shown in fig. 11. In this case the convergence rate is in fact faster, probably due to time effects of decaying of the high frequency component of the residuals by waiting m steps before communicating.

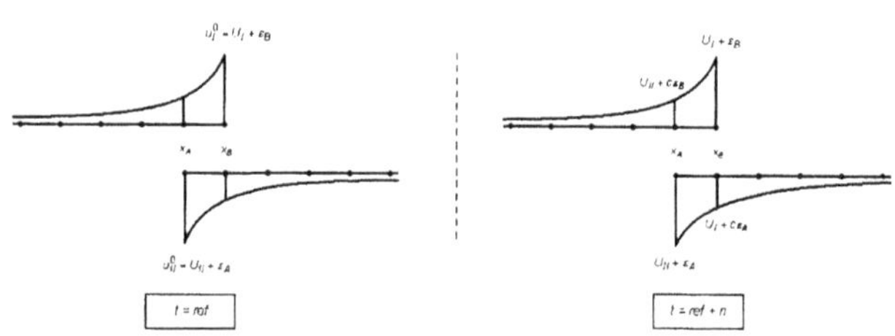

Figure 9. Propagation of the boundary error between neighboring blocks.

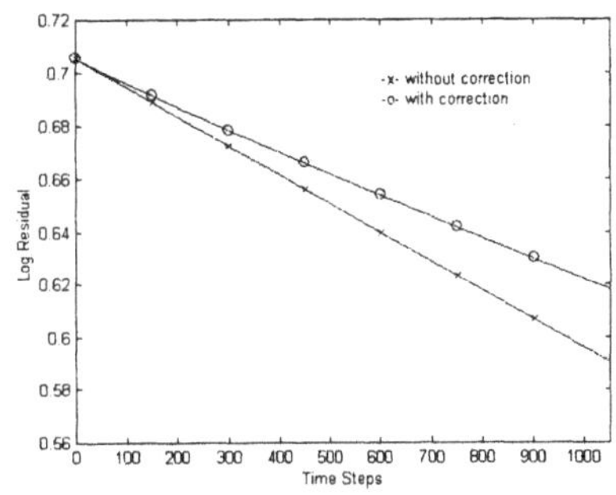

Figure 10.Interface correction (residual curve for homogeneous boundary conditions, with a sinusoidal initial solution).

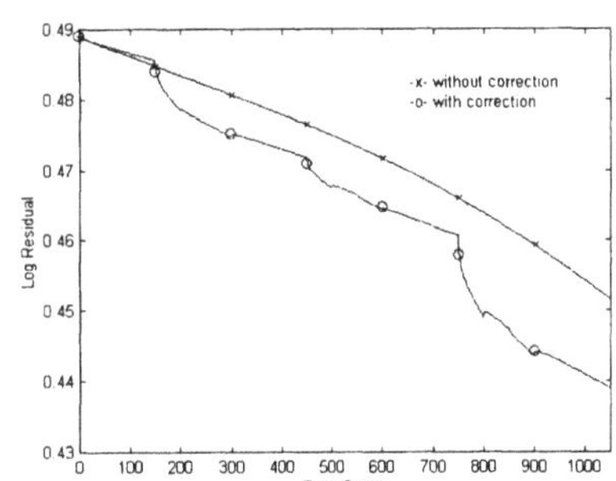

Figure 11. Interface correction (residual curve for non-homogeneous boundary conditions with a sinusoidal initial condition).

4. BLOCK CORRECTIONS IN THREE-DIMENSIONS:

The above schemes were tested for the solution of the following three-dimensional equation:

$$\frac{\partial u}{\partial t} = \alpha \nabla^2 u \tag{8}$$

over a cube with 100*100*100 = 1,000,000 grid points, by using a forward time centered space scheme. Figure 12 shows the solution for homogeneous boundary conditions with two different initial conditions with filtered and unfiltered schemes. As can be seen from the results the low frequency components slow down the convergence of the unfiltered solution. Figure 13 shows the results for the same problem with non-homogeneous boundary conditions. In this case, a randomly chosen initial condition improves the speed of convergence since as an initial condition contains low frequency components of the steady state solution.

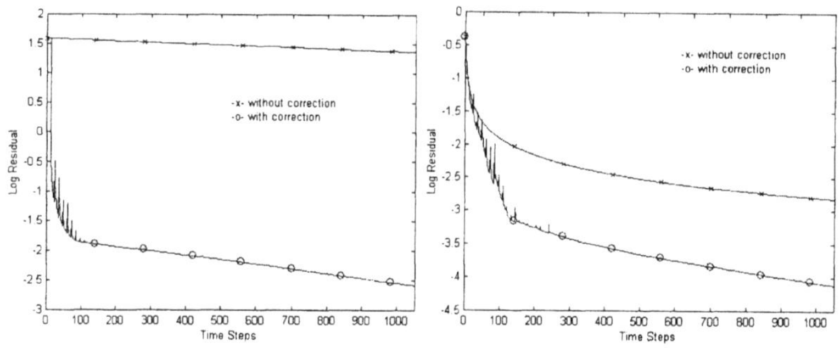

Figure 12. Block correction for the 3-D case(residual curve for homogeneous boundary conditions, with an initial sinusoidal profile and a point source at the middle).

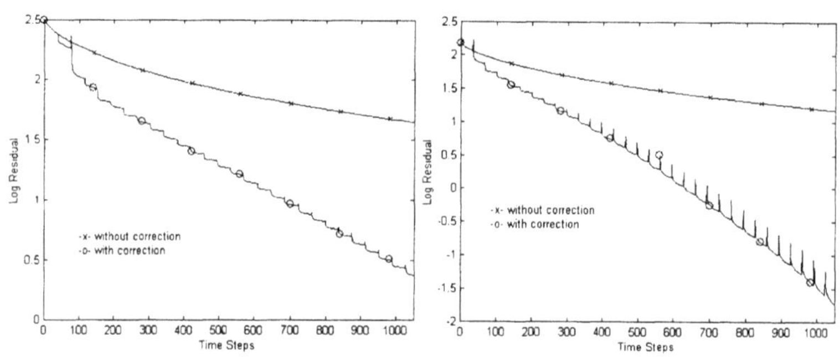

Figure 13 Block correction (residual curve for non-homogeneous boundary conditions, with an initial point source in the middle and with random initial conditions).

ACKNOWLEDGEMENTS

This research was supported by the NASA Glenn Research Center under Grant No.: NAG3-2260.

REFERENCES

1. A. Ecer, N. Gopalaswamy, H.U. Akay, and Y.P. Chien, "Digital Filtering Techniques for Parallel Computation of Explicit Schemes," International Journal of Computational Fluid Dynamics, pp. 211-222, Vol.13, 2000.
2. N. Gopalaswamy, A. Ecer, H.U. Akay and Y.P. Chien, "Efficient Parallel Communication Schemes for Computational Fluid Dynamics Codes," AIAA Journal vol. 36, No.6, pp.1-7, June 1998.
3. A.Ecer, I. Tarkan, and E. Lemoine, "Communication Cost Evaluations for the Parcfd Test Case", Proceedings of Parallel CFD '99, May 22-26, 1999, Williamsburg VA, USA.

Parallel Computational Fluid Dynamics – Trends and Applications
C.B. Jenssen et al. (Editors)

Parallel Spectral Element Atmospheric Model

S. J. Thomas and R. Loft [a]

[a]Scientific Computing Division, National Center for Atmospheric Research,
1850 Table Mesa Drive, Boulder CO 80303, USA

A semi-implicit formulation of the parallel spectral element atmospheric model (SEAM) is presented. Spectral element methods are *h-p* type finite element methods that combine the geometric flexibility of finite elements with the exponential convergence of pseudospectral methods. To avoid spurious 'pressure' modes and obtain a symmetric positive definite Helmholtz operator, a staggered $\mathcal{P}_p \times \mathcal{P}_{p-2}$ method is adopted. Parallel preconditioning strategies for an iterative conjugate-gradient solver are described and performance of the model on the NCAR SP supercomputer is summarized.

Keywords. Spectral elements, semi-implicit time-stepping.

1. INTRODUCTION

Semi-implicit time-stepping schemes for geophysical fluid dynamics were first applied in a spectral atmospheric model to remove the stability constraint associated with fast-moving gravity waves. The time step could then be increased by a factor of six without adversely affecting the meteorologically important Rossby waves. Solution of the resulting Helmholtz problem in a spectral transform model is trivial since the spherical harmonic basis functions are eigenfunctions of the Laplacian operator on the sphere and the computational overhead is minimal. Semi-implicit schemes have now been successfully implemented over a wide range of scales including both global and regional atmospheric models using low-order finite difference and finite element methods.

In this paper we describe the application of the semi-implicit scheme to the Spectral Element Atmospheric Model (SEAM) developed by Mark Taylor at NCAR (Taylor et al. 1997a,b) SEAM originally employed an explicit Eulerian leap-frog time-stepping scheme. Spectral elements combine the accuracy of conventional spectral methods and the geometric flexibility of finite element methods. In a spectral element discretisation, the computational domain is divided into rectangular elements, within which variables are approximated by a polynomial expansion of high degree. The discrete equations are derived using Gauss-Lobatto-Legendre quadrature together with the Lagrangian interpolants on the collocation grid (Legendre cardinal functions) as basis functions, resulting in diagonal mass matrices. There are several practical advantages to using a relatively new, high-accuracy numerical method such as the Spectral Element Method (SEM) over current methods. In particular, the local nature of the computations implies that the method is ideally suited to cache-based RISC microprocessor computer architectures. The explicit code has demonstrated excellent parallel scalability.

2. SHALLOW-WATER EQUATIONS

The shallow-water equations have been used as a vehicle for testing promising numerical methods for many years by the atmospheric modeling community. The equations contain the essential wave-propagation mechanisms found in more complete models. The governing equations for inviscid flow of a thin layer of fluid in two dimensions are the horizontal momentum and continuity equations for the wind $\mathbf{u}$ and geopotential height Φ.

$$\frac{\partial \mathbf{u}}{\partial t} + (\mathbf{u} \cdot \nabla)\, \mathbf{u} + f\, \hat{\mathbf{k}} \times \mathbf{u} + \nabla\Phi = 0 \tag{1}$$

$$\frac{\partial \Phi}{\partial t} + \mathbf{u} \cdot \nabla\, \Phi + \Phi\, \nabla \cdot \mathbf{u} = 0 \tag{2}$$

Given the initial basic state $u = 0$, $v = 0$ and $\Phi = \Phi^*$, linearization of the equations followed by a Von Neumann stability analysis reveals that solutions exist having phase speeds $c = 0$ and $c = \pm\{\Phi^* + f^2/k^2\}^{1/2}$. The former represent slow movements of the atmosphere, related to the propagation of Rossby modes, whereas the latter are the fast-moving gravity wave oscillations. For numerical integration by an Eulerian leapfrog scheme, the above analysis suggests that a time step would be limited by the speed of the fastest-moving gravity modes. In atmospheric models it is known that gravity-wave modes propagate at a speed many times larger than the Rossby modes, implying that time steps would be many times smaller than those required for an explicit treatment of passive advection. To overcome this limitation, Robert (1969) introduced a semi-implicit scheme in a primitive equations spectral model. A semi-implicit scheme applied to the shallow water equations combines an explicit leapfrog scheme for the diagonal advection terms with a Crank-Nicholson scheme for the off-diagonal geopotential gradient and divergence terms. So unless an accurate representation of gravity-wave oscillations is important, the time step can be increased significantly.

3. SPECTRAL ELEMENT DISCRETISATION

Spectral element methods are high-order weighted-residual techniques for the solution of partial differential equations that combine the geometric flexibility of h-type finite element methods with p-type pseudo-spectral techniques (Canuto et al 1988, Karniadakis and Sherwin 1999). In the spectral element discretisation introduced by Patera (1984), the computational domain is partitioned into subdomains (spectral elements) in which the dependent and independent variables are approximated by p-th order tensor-product polynomial expansions within each element. A variational form of the equations is then obtained whereby inner products are directly evaluated using Gaussian quadrature rules. Exponential (spectral) convergence to the exact solution is achieved by increasing the degree p of the polynomial expansions while keeping the minimum element size h fixed.

Consider the time-discretised form of the shallow water equations employing the semi-implicit scheme,

$$\mathbf{u}^{n+1} + \Delta t\, \nabla\, \Phi^{n+1} = \mathbf{F}_{\mathbf{u}} \tag{3}$$

$$\Phi^{n+1} + \Delta t\, \Phi_0\, \nabla \cdot \mathbf{u}^{n+1} = F_{\Phi} \tag{4}$$

$$\mathbf{F_u} \equiv \mathbf{u}^{n-1} - \Delta t\, \nabla\, \Phi^{n-1} + 2\Delta t\, \mathbf{f}_\mathbf{u}^n$$
$$F_\Phi \equiv \Phi^{n-1} - \Delta t\, \Phi_0\, \nabla \cdot \mathbf{u}^{n-1} + 2\Delta t\, f_\Phi^n$$

Φ_0 is the mean geopotential reference state, the tendencies $\mathbf{f_u}$ and f_Φ contain nonlinear advection and Coriolis terms. The spectral element discretisation of (3) and (4) is based on the equivalent variational formulation for the equations over a biperiodic domain Ω. Find $(\mathbf{u}, \Phi) \in \mathcal{X} \times \mathcal{M}$ such that for all $(\mathbf{w}, q) \in \mathcal{X} \times \mathcal{M}$,

$$\langle \mathbf{u}^{n+1}, \mathbf{w} \rangle - \Delta t \langle \Phi^{n+1}, \nabla \cdot \mathbf{w} \rangle = \mathbf{R_u}, \tag{5}$$
$$\langle \Phi^{n+1}, q \rangle + \Delta t\, \Phi_0 \langle q, \nabla \cdot \mathbf{u}^{n+1} \rangle = R_\Phi, \tag{6}$$

$$\mathbf{R_u} \equiv \langle \mathbf{u}^{n-1}, \mathbf{w} \rangle + \Delta t \langle \Phi^{n-1}, \nabla \cdot \mathbf{w} \rangle + 2\Delta t \langle \mathbf{f}_\mathbf{u}^n, \mathbf{w} \rangle$$
$$R_\Phi \equiv \langle \Phi^{n-1}, q \rangle - \Delta t\, \Phi_0 \langle q, \nabla \cdot \mathbf{u}^{n-1} \rangle + 2\Delta t \langle f_\phi^n, q \rangle$$

The inner products appearing above are defined by

$$f, g, \mathbf{v}, \mathbf{w} \in \mathcal{L}^2(\Omega), \quad \langle f, g \rangle \equiv \int_\Omega f(\mathbf{x}) g(\mathbf{x})\, d\mathbf{x},$$
$$\langle \mathbf{v}, \mathbf{w} \rangle \equiv \int_\Omega \mathbf{v}(\mathbf{x}) \cdot \mathbf{w}(\mathbf{x})\, d\mathbf{x}.$$

The equations have been formulated to solve for a perturbation about a mean state which nearly preserves the non-divergent flow. In particular, it is well known that the variational formulation of the Stokes problem can lead to spurious 'pressure' modes when the Ladyzhenskaya–Babuska–Brezzi (LBB) *inf-sup* condition is violated (see Brezzi and Fortin 1991). For spectral elements, solutions to this problem are summarized in Bernardi and Maday (1992). To avoid spurious modes, the discrete velocity $\mathcal{X}^{h,p}$ and geopotential $\mathcal{M}^{h,p}$ approximation spaces are chosen to be subspaces of polynomial degree p and $p-2$ over each spectral element. Thus a staggered grid is employed with Gauss-Lobatto-Legendre quadrature points for the velocity and Gauss-Legendre quadrature points for the geopotential. The spectral element model described in Taylor et al (1997a) does not employ the weak variational formulation and so the equations are discretised on a single collocation grid. However, a staggered grid was adopted for the shallow water ocean model described in Iskandarani et al (1995). The major advantage of a staggered mesh in the context of semi-implicit time-stepping is that the resulting Helmholtz operator is symmetric positive definite and thus a preconditioned conjugate gradient elliptic solver can be used to compute the geopotential perturbation.

To simplify the discussion we first describe a one dimensional decomposition, which is straightforward to extend to higher dimensions: Spectral elements are obtained by partitioning the domain Ω into N_h disjoint rectilinear elements of minimum size h.

$$\overline{\Omega} = \bigcup_{\ell=1}^{N_h} \overline{\Omega}_\ell, \qquad \Omega_\ell \equiv]\, a_\ell,\, a'_\ell [, \qquad a'_\ell \equiv a_{\ell+1}.$$

The variational statement (5) - (6) must be satisfied for the polynomial subspaces $\mathcal{X}^{h,p} \subset \mathcal{X}$ and $\mathcal{M}^{h,p} \subset \mathcal{M}$ defined on the Ω_ℓ,

$$\mathcal{X}^{h,p} \equiv \mathcal{X} \cap \mathcal{P}^{h,p}(\Omega), \quad \mathcal{M}^{h,p} \equiv \mathcal{M} \cap \mathcal{P}^{h,p-2}(\Omega),$$

$$\mathcal{P}^{h,p} \equiv \left\{ f \in \mathcal{L}^2(\Omega) : f_{|\Omega_\ell} \in \mathcal{P}_p(\Omega_\ell) \right\},$$

where $\mathcal{P}_p(\Omega_\ell)$ denotes the space of all polynomials of degree $\leq p$ with respect to each of the spatial variables. Note that the polynomial degree for the geopotential space is two less than for the velocity space. For a staggered mesh, two integration rules are defined by taking the tensor-product of Gauss and Gauss-Lobatto formulae over each spectral element. The local Gauss points and weights $(\tilde{\xi}_j, \tilde{w}_j)$ $j = 1, \ldots, p-1$ and the local Gauss-Lobatto nodes and weights (ξ_j, w_j), $j = 0, \ldots, p$ are mapped to the global quadrature points and weights as follows:

$$\begin{aligned} \tilde{x}_{j,\ell} &\equiv \vartheta_\ell(\tilde{\xi}_j), \quad x_{j,\ell} \equiv \vartheta_\ell(\xi_j), \\ \tilde{w}_{j,\ell} &\equiv \tilde{w}_j(a'_\ell - a_\ell)/2, \qquad w_{j,\ell} \equiv w_j(a'_\ell - a_\ell)/2, \end{aligned}$$

$$\vartheta_\ell(\xi) \equiv a_\ell + (a'_\ell - a_\ell)(\xi + 1)/2,$$

The two integration rules are defined according to:

$$\langle f, g \rangle_{\mathrm{G}} \equiv \sum_{\ell=1}^{N_h} \sum_{j=1}^{p-1} f(\tilde{x}_{j,\ell})\, g(\tilde{x}_{j,\ell})\, \tilde{w}_{j,\ell}$$

$$\langle f, g \rangle_{\mathrm{GL}} \equiv \sum_{\ell=1}^{N_h} \sum_{j=0}^{p} f(x_{j,\ell})\, g(x_{j,\ell})\, w_{j,\ell}$$

The discrete form of (5) - (6) can now be given as follows. Find $(\mathbf{u}^{h,p}, \Phi^{h,p}) \in \mathcal{X}^{h,p} \times \mathcal{M}^{h,p}$ such that for all $(\mathbf{w}, q) \in \mathcal{X}^{h,p} \times \mathcal{M}^{h,p}$,

$$\langle \mathbf{u}^{h,p}, \mathbf{w} \rangle_{\mathrm{GL}} - \Delta t \, \langle \Phi^{h,p}, \nabla \cdot \mathbf{w} \rangle_{\mathrm{G}} = \mathbf{R}_{\mathbf{u}}^{h,p}, \tag{7}$$

$$\langle \Phi^{h,p}, q \rangle_{\mathrm{G}} + \Delta t \, \Phi_0 \, \langle q, \nabla \cdot \mathbf{u}^{h,p} \rangle_{\mathrm{G}} = R_{\Phi}^{h,p} \tag{8}$$

To numerically implement these equations, a set of basis functions must be specified for $\mathcal{X}^{h,p} \times \mathcal{M}^{h,p}$. The velocity is expanded in terms of the p-th order Lagrangian interpolants ϕ_j^p (the Legendre cardinal functions):

$$\mathbf{u}^{h,p} \circ \vartheta_\ell(\xi) \equiv \sum_{j,j'=0}^{p} \mathbf{u}_{\ell,j,j'}\, \phi_j^p(\xi^1)\, \phi_{j'}^p(\xi^2)$$

where $\mathbf{u}_{\ell,j,j'} \equiv \mathbf{u} \circ \vartheta_\ell(\xi_j^1, \xi_{j'}^2)$ is the velocity at tensor-product Gauss-Lobatto-Legendre points in subdomain (spectral element) Ω_ℓ. A point $\mathbf{x} \in \Omega_\ell$ in this element is mapped by ϑ_ℓ from a point $\xi \in]-1, 1[^2$. The geopotential is represented using the $(p-2)$-th order Lagrangian interpolants $\tilde{\phi}_j^p$

$$\Phi^{h,p} \circ \vartheta_\ell(\xi) \equiv \sum_{j,j'=1}^{p-1} \Phi_{\ell,j,j'}\, \tilde{\phi}_j^p(\xi^1)\, \tilde{\phi}_{j'}^p(\xi^2)$$

at Gauss-Legendre points $(\tilde{\xi}_{j,\ell}^1, \tilde{\xi}_{j,\ell}^2)$ in the spectral element Ω_ℓ. The test functions $\mathbf{w} \in \mathcal{X}^{h,p}$ are chosen to be unity at a single $x_{j,\ell}$ and zero at all other Gauss-Lobatto-Legendre points. Similarly, the $q \in \mathcal{M}^{h,p}$ are unity at a single $\tilde{x}_{j,\ell}$ and zero at all other Gauss-Legendre points. The resulting discretised shallow water equations are given by

$$\mathbf{B}\,\mathbf{u}^{n+1} - \Delta t\,\mathbf{D}^T\,\Phi^{n+1} = \mathbf{R}_\mathbf{u} \quad (9)$$

$$\tilde{B}\,\Phi^{n+1} + \Delta t\,\Phi_0\,\mathbf{D}\,\mathbf{u}^{n+1} = R_\Phi \quad (10)$$

where $\mathbf{B} = (B^1, B^2)$ and $\tilde{B}$ are the diagonal velocity and geopotential mass matrices whose elements are products of the Gaussian weights (with the appropriate scaling for computation on the global domain). The derivative matrix $\mathbf{D} \equiv (D^1, D^2)$ is obtained by differentiating the Legendre cardinal functions ϕ_j^p in the velocity expansion at Gauss-Legendre points. In the spectral element method, $\mathcal{C}^0$ continuity of the velocity is enforced at inter-element boundaries which share Gauss-Lobatto-Legendre points. Thus, direct stiffness summation is applied to assemble the global matrices.

The Legendre cardinal functions $\phi_j^p(x)$ on a Gauss-Lobatto-Legendre grid are defined in Appendix A of Ronquist (1988) along with the pseudo-spectral differentiation matrix D defined by

$$D_{jj'} \equiv \frac{d\phi_{j'}^p}{d\xi}(\xi_j)$$

in the expansion of the derivative of a polynomial $u(\xi)$ at nodal points ξ_j

$$u'(\xi_j) = \sum_{j'=0}^{p} \frac{d\phi_{j'}^p(\xi_j)}{d\xi}\, u(\xi_{j'}) = \sum_{j'=0}^{p} D_{jj'}\, u(\xi_{j'})$$

(See also Chap 2 of Canuto et al 1987, and Karniadakis and Sherwin 1999).

4. HELMHOLTZ PROBLEM

A Helmholtz problem for the geopotential perturbation is obtained by solving for the velocity $\mathbf{u}^{n+1}$

$$\mathbf{u}^{n+1} = \mathbf{B}^{-1}\left(\Delta t\,\mathbf{D}^T\,\Phi^{n+1} + \mathbf{R}_\mathbf{u}\right) \quad (11)$$

and then applying back-substitution to obtain

$$\tilde{B}\,\Phi^{n+1} + \Delta t^2\,\Phi_0\,\mathbf{D}\,\mathbf{B}^{-1}\,\mathbf{D}^T\,\Phi^{n+1} = R'_\Phi \quad (12)$$

where

$$R'_\Phi \equiv R_\Phi - \Delta t\,\Phi_0\,\mathbf{D}\,\mathbf{B}^{-1}\,\mathbf{R}_\mathbf{u}$$

Once the updated geopotential Φ^{n+1} is computed, the velocity $\mathbf{u}^{n+1}$ is computed from (11). Quarteroni and Valli (1994) note that the LBB *inf-sup* condition is satisfied if and only if $\ker \mathbf{D}^T = 0$, i.e. the nullspace of $\mathbf{D}^T$ is empty.

The resulting Helmholtz operator is symmetric positive definite and thus preconditioned conjugate gradient iterative solvers can be applied. Ronquist (1991), notes that the pseudo-Laplacian operator $\mathbf{D}\,\mathbf{B}^{-1}\,\mathbf{D}^T$ is not as well conditioned as the spectral element Laplacian operator (in weak form). Initially, we have adopted the domain decomposition preconditioner of Ronquist (1991) which combines deflation with a local element direct solver. Timings for our parallel implementation on an IBM SP indicate that the semi-implicit code requires half the CPU time of the explicit shallow water model for typical T42 climate resolutions. The explicit time step is 120 sec whereas the semi-implicit step is 1200 sec. Both models exhibit close to linear scaling out to 30 processors. The coarse solver required in the deflation step is a potential serial bottleneck and a parallel sparse Cholesky solver is being implemented. If this strategy proves not to scale out to large numbers of processors then additive Schwarz methods are an alternative.

5. CONCLUSIONS

From a computational point of view, spectral elements have several attractive features which lead us to believe that exceptional performance for the dynamical core of a General Circulation Model (GCM), even at modest climate simulation resolutions (> 1 degree), is achievable on parallel computers composed of RISC microprocessors. The basic computational kernel of a spectral element model computes pseudo-spectral derivatives as matrix-vector products of relatively small (typically 8×8) matrices which are naturally cache-blocked. Spectral elements also have desirable boundary-exchange communication patterns that are reminiscent of finite difference models. We have demonstrated single processor performance figures of 30% of peak for the explicit model and 25% of peak for the semi-implicit code on the 375 MHz IBM Power-3 processors at NCAR.

Fast computation and linear scaling curves can ultimately result in a useful kernel for climate simulation only if the amount of model time computed per unit wall clock time is sufficiently large. An efficient semi-implicit solver is essential if this all important performance metric is to be substantially increased. Our current goal is to build a GCM dynamical core capable of 30 Gflops/sec sustained performance on clustered RISC/cache architectures using a hybrid MPI/OpenMP programming model.

ACKNOWLEDGMENT

The first author would like to thank Einar Ronquist of NTNU for helpful discussions regarding the theory and implementation of spectral element methods.

REFERENCES

1. Bernardi, C. and Y. Maday, 1992: *Approximations spectrales de problèmes aux limites elliptiques.* Mathématiques et Applications, vol. 10, Springer-Verlag, Paris, France, 242p.
2. Brezzi, F. and M. Fortin, 1991: *Mixed and Hybrid Finite Element Methods.* Springer-Verlag, New York, 350p.
3. Canuto, C., M. Y. Hussaini, A. Quarteroni, and T. A. Zang, 1988: *Spectral Methods in Fluid Dynamics.* Springer-Verlag, New York, 557p.
4. Iskandarani, M., D. B. Haidvogel, and J. P. Boyd, 1995: A staggered spectral element model with application to the oceanic shallow water equations. *Int. J. Numer. Meth. Fluids*, **20**, 394–414.
5. Karniadakis, G. M., and S. J. Sherwin, 1999: *Spectral/hp Element Methods for CFD.* Oxford University Press, Oxford, England, 390p.
6. Quarteroni, A., and A. Valli, 1994: *Numerical Approximation of Partial Differential Equations.* Springer-Verlag, New York, 543p.
7. Patera, A. T., 1984: A spectral element method for fluid dynamics: Laminar flow in a channel expansion. *J. Comp. Phys.*, **54**, 468.
8. Robert, A., 1969: The integration of a spectral model of the atmosphere by the implicit method. In *Proceedings of WMO/IUGG Symposium on NWP*, VII, pages 19–24, Tokyo, Japan, 1969.
9. Ronquist, E. M., 1988: *Optimal Spectral Element Methods for the Unsteady Three Dimensional Navier Stokes Equations*, Ph.D Thesis, Massachusetts Institute of Technology, 176p.
10. Ronquist, E. M., 1991: A domain decomposition method for elliptic boundary value problems: Application to unsteady incompressible fluid flow. *Proceedings of the Fifth International Symposium on Domain Decomposition Methods for Partial Differential Equations*, SIAM, Philadelphia, 545–557.
11. Taylor, M., J. Tribbia, and M. Iskandarani, 1997a: The spectral element method for the shallow water equations on the sphere. *J. Comp. Phys.*, **130**, 92–108.
12. Taylor, M., R. Loft, and J. Tribbia, 1997b: Performance of a spectral element atmospheric model (SEAM) on the HP Exemplar SPP2000. NCAR Technical Note 439+EDD.

7. Optimization Dominant CFD Problems

Parallel Computational Fluid Dynamics – Trends and Applications
C.B. Jenssen et al. (Editors)
2001 Elsevier Science B.V.

Domain Decomposition Methods Using GAs and Game Theory for the Parallel Solution of CFD Problems

H. Q. Chen[a], J. Periaux[b] and A. Ecer[c]

[a]Institute of Aerodynamics, Nanjing University of Aeronautics and Astronautics, Nanjing 210016, P. R. China

[b]Dassault Aviation, 78 quai Marcel Dassault, 92214 Saint-Cloud, France

[c]IUPUI Department of Mechanical Engineering, Indianapolis, IN 46202-5132

In this paper, a Nash game using binary encoded GA players is investigated for the simulation of flow problems using domain decomposition methods on a distributed parallel environment. In this new approach, a Nash strategy is used to match flow field characteristics at interfaces of overlapped regions, which are used as a measure for the fitness function of GAs. GA players have a decentralized passive or reactive role for optimizing local inverse problems associated with the global inverse one. A reactive role of players is introduced consisting of a message processing step depending of local available knowledge and implemented to increase the convergence speed of the game to its Nash equilibrium. Numerical experiments on 2D incompressible or transonic potential flows in convergent-divergent nozzles are first performed in parallel computer environment consisting of a cluster of 6 IBM RS6000 machines. Various comparisons of performances between roles of passive and reactive players are presented with encouraging results. Other numerical DDM experiments of simple incompressible potential flows in 3D nozzles are presented and illustrate both the potentiality and robustness of the method for large parallel simulations of complex nonlinear flows.

1. INTRODUCTION

Domain decomposition methods (DDM) are currently used for flow simulation in distributed parallel environments. Genetic algorithms (GAs) are robust and simple adaptive optimization tools mimicking natural evolution mechanisms with Darwin's survival of the fittest principle[1]. As a remainder GAs have been introduced by J. Holland who explained the adaptive process of natural system and laid down two main principles of GAs: the ability of simple bit-string representation to encode complex structures and the power of simple transformations to improve such structures. GAs process population of binary encoded candidate solutions or individuals. Each individual has a fitness function which measures how fit it is for the environment (i.e. problem). Then a set of GAs operators such as selection, crossover and mutation are defined to create a new population at each generation. At each generation or evolutionary step, individuals in the population

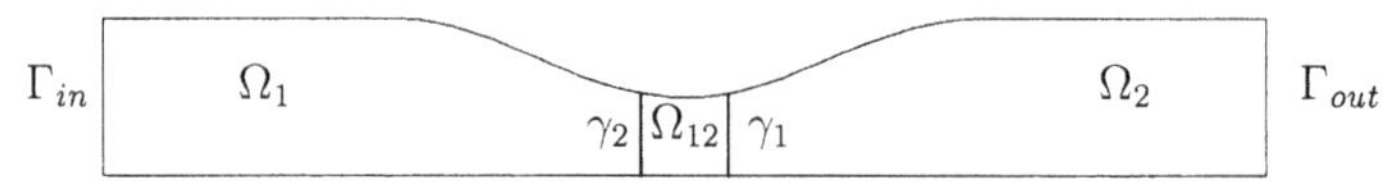

Figure 1. Description of a nozzle with two subdomains

strings are decoded, evaluated in order to measure their fitness value, and then the GA operators are applied in order to form the following generation. This process is iterated until convergence is achieved or a near optimal solution found. Based on derivative free information and requiring almost no assumptions about the physical space, nowadays it is quite often admitted that GAs is favored for capturing global solution of non convex optimization occuring namely in CFD problems. The Game Theory approach introduced here associates to a single criterion optimization problem a multi criteria Nash game [2] with several players solving local constrained sub-optimization tasks under conflict.

Using domain decomposition techniques, the global flow solutions are computed through matching of overlapped subdomain solutions. This matching process can be realized through searching an appropriated set of virtual control parameters on the interfaces of overlapped subdomains. In previous work [4], GAs have been used to realize the matching process through the minimization of appropriated fitness functions defined by the distance of local solutions on the overlapping regions. In this paper, a further extension via GAs and new parallelizable Game Theory ingredients is considered in order to realize a decentralized matching process [3] as follows: the global matching optimization problem is replaced by a non-cooperative game based on a Nash equilibrium with several GA players in charge of optimizing local inverse problems associated with the global inverse one. Moreover, reactive role of players consisting of a message processing step depending of local available knowledge is implemented to increase the convergence speed of the game to its Nash equilibrium.

2. DESCRIPTION OF THE FLOW SOLUTION USING DDM

The flow problems considered here using DDM are incompressible or transonic flows in a nozzle modeled by the full potential equation with Dirichlet boundary conditions at the entrance and exit and homogeneous Neumann conditions on the walls. For sake of simplicity, the computational domain Ω is decomposed into two subdomains Ω_1 and Ω_2 with overlapping Ω_{12} whose interfaces are denoted by γ_1 and γ_2 as shown in Fig.1. We shall prescribe potential values, g_1 on γ_1 and g_2 on γ_2, as extra Dirichlet boundary conditions in order to obtain potential solutions Φ_1 and Φ_2 in subdomain Ω_1 and Ω_2 respectively. Using domain decomposition techniques, the problem of the flow can be reduced to minimize the following functional:

$$J(g_1, g_2) = \frac{1}{2} \parallel \Phi_1(g_1) - \Phi_2(g_2) \parallel^2 \quad (1)$$

where Φ_1 and Φ_2 are solutions in the overlapping subdomain Ω_{12} , $\parallel \bullet \parallel$ denotes an appropriate norm , whose choice will be made precise in the following examples.

In references [4], the global fitness function used in GAs is the distance of local solutions on the overlapping domains(see (1)), defined in (2)

$$J(g_1, g_2) = \frac{1}{2}\int_{\Omega_{12}} |\Phi_1(g_1) - \Phi_2(g_2)|^2 d\Omega \tag{2}$$

Using boundary integrals instead of the domain integral, we choose for (2) the boundary criteria introduced in (3). The minimization problem (1) can be reduced to minimize the following function based on boundary integral:

$$JB(g_1, g_2) = \frac{1}{2}\int_{\gamma_1} |\Phi_1(g_1) - \Phi_2(g_2)|^2 d\gamma_1 + \frac{1}{2}\int_{\gamma_2} |\Phi_1(g_1) - \Phi_2(g_2)|^2 d\gamma_2 \tag{3}$$

Associated to the global fitness function $JB(g_1, g_2)$, the decentralized multi fitness functions $JB_1(g_1, \underline{g_2})$ and $JB_2(\underline{g_1}, g_2)$ are then defined with the following two minimizations:

$$\begin{aligned} &\inf_{g_1} JB_1(g_1, \underline{g_2}) \quad with \quad JB_1(g_1, \underline{g_2}) = \frac{1}{2}\int_{\gamma_1} |\Phi_1(g_1) - \Phi_2(\underline{g_2})|^2 d\gamma_1 \\ &\inf_{g_2} JB_2(\underline{g_1}, g_2) \quad with \quad JB_2(\underline{g_1}, g_2) = \frac{1}{2}\int_{\gamma_2} |\Phi_1(\underline{g_1}) - \Phi_2(g_2)|^2 d\gamma_2 \end{aligned} \tag{4}$$

In the following sections, implementations of Nash/GAs with decentralized passive or reactive role of players is addressed for the solution of the global DDM flow problem through the Nash equilibrium search of the multi criteria problem (4).

3. DESCRIPTION OF NEW NASH/GAs GAMES

Following the description of a traditional Nash/GAs game(see reference [3] for details) , we can apply Game Theory and simulate the DDM flow optimization problem as a game with two decentralized *passive* players, named *Flow-GA1* and *Flow-GA2* , with objective functions

$$J_{p1}(\phi_1(g_1), \Phi_2(\underline{g_2})) = JB_1(g_1, \underline{g_2}), \quad J_{p2}(\phi_1(\underline{g_1}), \Phi_2(g_2)) = JB_2(\underline{g_1}, g_2) \tag{5}$$

respectively. Note that each player optimizes the corresponding objective function with respect to non-underlined variables.

In this paper, we introduced the notion of *active* players with objective functions modified as follows:

$$J_{a1}(\phi_1(g_1), \Phi_2(\underline{g_2})) = \frac{1}{2}\int_{\gamma_1} |\Phi_1(g_1) - \Phi_1^T)|^2 d\gamma_1 \tag{6}$$

$$J_{a2}(\phi_1(\underline{g_1}), \Phi_2(g_2)) = \frac{1}{2}\int_{\gamma_2} |\Phi_2^T - \Phi_2(g_2)|^2 d\gamma_2 \tag{7}$$

where Φ_1^T and Φ_2^T are targets whose choices will be defined in *Remark 1* in the sequel.

After discretization of the computational domain occupied by the flow in the nozzle, we have $g_1 = g_{1i}$ and $g_2 = g_{2i}$, i=1,ny (ny is mesh size in y direction). For each interface, *only* one point is binary encoded (for instance, g_{11} for γ_1 and g_{21} for γ_2). Other values of g_{1i} and g_{2i} (i $\geq$ 2) are corrected by numerical values. Then the algorithm based on the information exchange between players can be found in the reference [3].

Remark 1: Based on the current existing or newly received information of each player, the targets Φ_1^T and Φ_2^T mentioned above are defined by

$$\Phi_1^T = \beta\phi_2(\underline{g_2})|_{\gamma=\gamma_1} + (1-\beta)\phi_1(\underline{g_1})|_{\gamma=\gamma_1} \quad (8)$$

$$\Phi_2^T = \beta\phi_1(\underline{g_1})|_{\gamma=\gamma_2} + (1-\beta)\phi_2(\underline{g_2})|_{\gamma=\gamma_2} \quad (9)$$

where β is a real positive parameter, which could be specified as an appropriate constant.
Remark 2: If $\beta = 1$, the present method reduces to a traditional Nash game using GAs. In this case players do their own optimization passively based on received information. In the present method, the players have the possibility to behave reactively by chosing a value of the auxiliary β parameter. In other words we have introduced an extra message processing step with aim at capturing Nash equilibrium faster. This approach is confirmed by the numerical experiments presented in the sequel.
Remark 3: This extra message processing step does not affect the existing parallel implementation of this Nash/GAs, since the processed information is still a current one.

4. PARALLEL IMPLEMENTATION WITH MULTI BLOCKS

The DDM approach described above can be generalized to the flow simulation using several blocks which associate one interface per GAs process (See ref. [3] for details). In this paper we consider one block per GAs process in order to achieve parallelization based on a domain decomposition database called GPAR. This library has been developed at the CFD laboratory at IUPUI specifically for parallel computing. Let us suppose block i consisting of $K(i)$ interfaces denoted by $\gamma_{ik}, (k = 1, K(i))$ and let $J_{\gamma_{ik}}$ be a boundary integral on the interface γ_{ik}, such that $J_{\gamma_{ik}} = \frac{1}{2}\int_{\gamma_{ik}} |\Phi_i - \Phi_{ni}|^2 d\gamma_{ik}$, where subscript ni is the index of a neighboring block related to the interface γ_{ik}. Then for a computational domain split into N blocks, the global fitness function based on boundary integrals can be written as

$$J_G = \sum_i^N \sum_k^{K(i)} J_{\gamma_{ik}} \quad (10)$$

In view of implementing the game described above, each $J_{\gamma_{ik}}$ or an appropriate neighboring combination associated to each block can be chosen as local fitness function. In this paper, one interface per block is chosen for the decomposition in one direction and a combination of two neighboring interfaces is selected for the two directional decomposition. Let $K'(i)$ and J_i be the total number of selected interfaces and the local decentralized fitness function associated to the i block, respectively, such that $J_i = \sum_k^{K'(i)} J_{\gamma_{ik}}, i = 1, 2, ..., N$, then interfaces with dotted points are selected and operated by GAs players described above using J_i as their fitness functions (see Figure 2). A key potential value as genetic parameter is chosen at each selected interface and binary encoded by the GAs (See section 3 for details). The potential values at the other interfaces are updated by the corresponding fittest solution of the neighboring blocks.

In this approach parallelization is realized at the level of one block associated with one GAs player per computer process. The exchange of data between block interfaces for matching process via the GPAR data structure is achieved using the Message-Passing Interface (MPI) library.

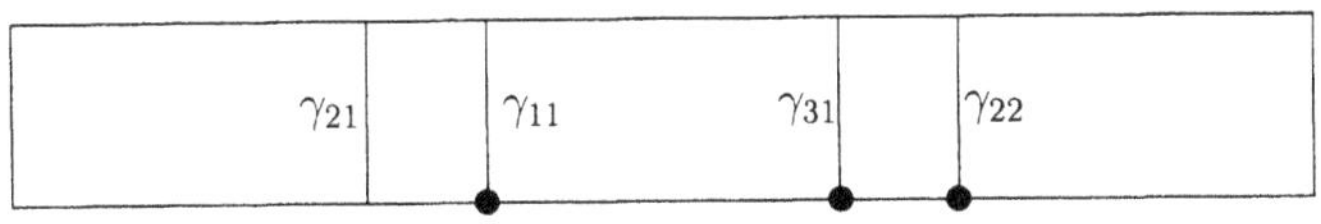

Figure 2. Description of selected interfaces (dotted) in a domainon split in one direction

5. RESULTS AND ANALYSIS

In order to illustrate the above new approach, numerical experiments on linear or non-linear transonic potential flows in a convergent-divergent nozzle are performed with a parallel environment consisting of a cluster of 6 IBM RS6000 machines available at the CFD laboratory of IUPUI. The global meshes of test cases 1-3 are shown in Figure 3. Meshes of cases 1,3 are used for simulating potential flow with a finite element Laplace's solver using a direct Choleski method. The mesh of case 2 is generated for simulating transonic flows with a finite difference AF3 solver [6]. Parameters used in GAs players are 0.85 for crossover rate and 0.09 for mutation rate. Exchange frequency number of the Nash game specified is 1. Convergence histories are measured by the J_G (see (10)).

5.1. Test case 1: Half nozzle split in one direction

As depicted in Figure 3, the computational domain is uniformly split into 3 or 6 blocks with one layer overlap. During the parallel computation, each GAs player with one computer process is assigned in different available machines. The preliminary results of tracing convergence histories using 3 or 6 blocks are represented on Figure 4. Potential solutions with continuity between blocks are successfully interfaced for both 3 and 6 block cases. In this test computation different values of β (see (8)) namely $\beta = 1$ denoting passive players and $\beta = 0.5$ reactive players as shown in Figure 4 have been chosen. It can be noticed that the convergence speed with reactive players is faster in comparison with the results using only passive players.

5.2. Test case 2: transonic flows over a nozzle split in one direction

As mentioned above, the finite difference AF3 solver is used for this test case. According to the method the nozzle is decomposed with a two layer overlap region in order to approximate the center difference operators. Figure 5 shows block assembled iso-Mach lines for both 3 and 6 block cases. For both cases, numerical results show a similar behaviour of convergence historie, the same location of the captured shock and indicate an evident speed up due to parallelization using 6 blocks.

5.3. Test case 3: a 3D nozzle split in one direction

A similar 3D parallel implementation is achieved as a straightforward extension of 2D approches mentioned above. Considered just as a preliminary test problem, the geometry used here is representative of a 3-D nozzle with square cross section. The global 3-D mesh is generated with $6 \times 61 \times 6 = 2196$ nodes and 9000 elements and then is almost uniformly split in y direction to form subdomains or blocks with one layer overlap(see Fig. 3). Results obtained with the same above methodology of block assembled iso-Mach lines are presented for both 3 and 6 block cases on Figure 6. Table 1 and Table 2 show the scalability effect between the two considered cases. The figures on tables indicate that the computing CPU time is reduced when the number of blocks increases, while the communication time increases moderately. Other computation with larger numbers of blocks and mesh points are presently under investigation in this parallel environment.

Table 1
3D Case 3: 3 blocks with 3 GA players

Block number	Block size (points)	Computation elapse time as second	Communication + waiting elapse time as second	Compucation CPU time as second	Communication CPU time as second
1	756	0.137	0.124	0.107	0.000
2	792	0.175	0.087	0.110	0.000
3	792	0.158	0.103	0.111	0.000

Table 2
3D Case 3: 6 blocks with 6 GA players

Block number	Block size (points)	Computation elapse time as second	Communication + waiting elapse time as second	Compucation CPU time as second	Communication CPU time as second
1	396	0.085	0.191	0.055	0.000
2	432	0.090	0.188	0.058	0.001
3	432	0.110	0.168	0.056	0.006
4	432	0.114	0.164	0.058	0.005
5	432	0.088	0.189	0.060	0.000
6	432	0.192	0.085	0.060	0.000

6. CONCLUSION AND FURTHER COMMENTS

A parallel version of the Nash/GAs procedure via GPAR data structure has been successfully implemented for both 2D and 3D test cases. For a more detailed description of Nash/GAs and associated numerical experiments see [7]. The convergence speed with reactive players for the considered test cases is encouraging in comparison with results using only passive players. An introductory notion of the reactive role of players consisting of a message processing step has been developed in this CFD application and can be extended to other ones depending of locally distributed available information. Further levels of parallelization could be considered using a special parallel GAs in each block or applying dynamic load balancing [5] namely in dimension 3.

7. ACKNOWLEDGEMENT

The authors are grateful to J.L. Lions for fruitful discussions on decentralized strategies for solution of PDEs, and their colleague B. Mantel for cooperative work in this research.

R. Payli is also acknowledged for friendly accessing the IUPUI CFD Lab and his help for using the GPAR data structure with MPI protocole in the parallel environment.

REFERENCES

1. D. Goldberg, *Genetic Algorithms in Search, Optimization, and Machine Learning*, Addison-Wesley, Reading, Mass., (1989).
2. J. F. Nash "Noncooperative games", *Annals of Mathematics*, pp 54–289, (1951).
3. J. Periaux, J.L. Lions and H. Q. Chen, "Decentralized Nash/GAs Optimization Strategies for the Solution of Multi-criteria Inverse Fluid Dynamics Problems", *Cedyap 99*, Las Palmas de Gran Ganaria, September 21-24, (1999).
4. J. Periaux and H.Q. Chen, "Domain Decomposition Method using GAs for Solving Transonic Aerodynamic Problems", *DDM-8*, Beijing, (1995)
5. N. Gopalaswamy, H.U. Akay, A. Ecer and Y.P. Chien, "Parallelization and Dynamic Load Balancing of NPARC Codes",*32nd AIAA/ASME/SAE/ASEE Joint Propulsion Conference*, Lake Buena Vista, FL, July 1-3, (1996).
6. H. Q. Chen and M. K. Huang, "An AF3 Algorithm for the Calculation of Transonic Nonconservative Full Potential Flow over Wings or Wing- Body Combinations", *Chinese J. of Aero.*, Vol.5, No.3, (1992).
7. H.Q. Chen, J. Periaux, M. Sefrioui, H.T. Sui, ," Evolutionary computating for solving complex design problems in Aerospace Engineering", *CMES*, to appear.

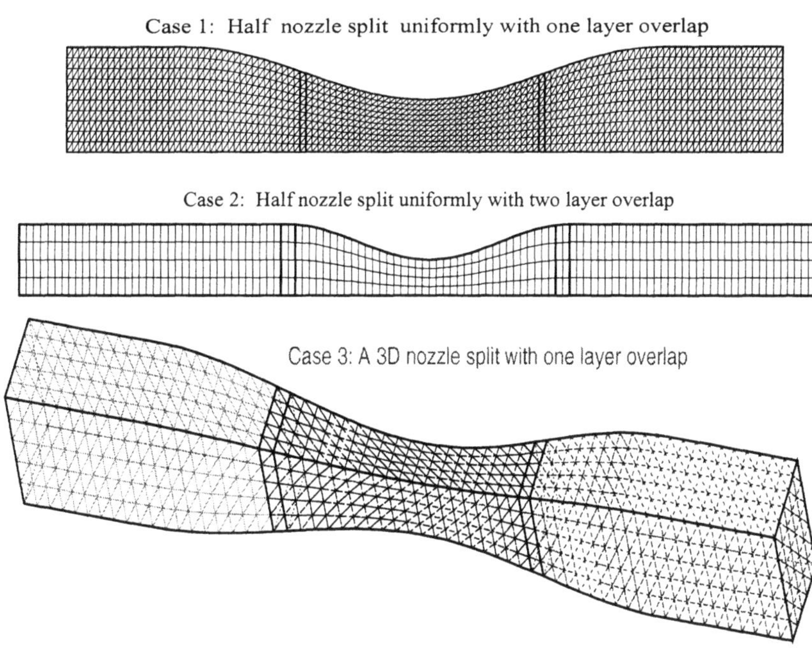

Figure 3. Different meshes for three test cases (3 blocks in x direction)

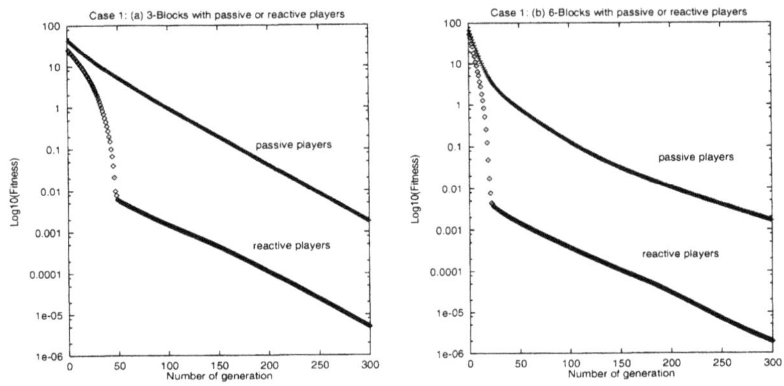

Figure 4. Tracing convergence histories of case 1: (a) 3 blocks and (b) 6 blocks

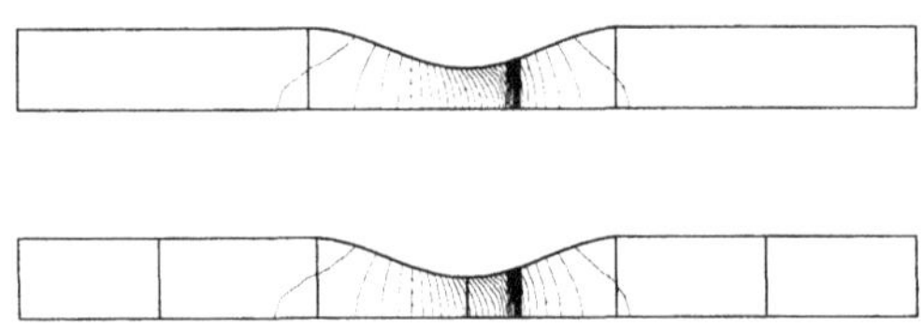

Figure 5. Assembly iso-Mach lines of case 2: (a) 3 blocks and (b) 6 blocks

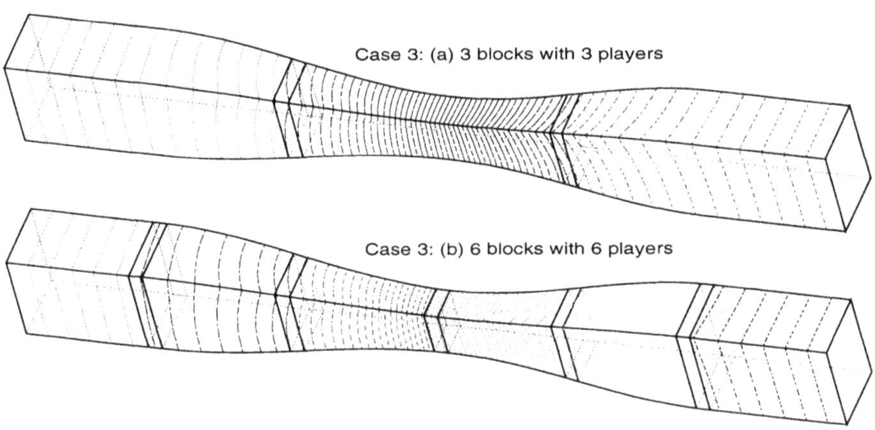

Figure 6. Assembly potential isolines of case 3 : (a) 3 blocks and (b) 6 blocks

Parallel Computational Fluid Dynamics – Trends and Applications
C.B. Jenssen et al. (Editors)

A Parallel *CFD* Method for Adaptive Unstructured Grids with Optimum Static Grid Repartitioning

A. P. Giotis, D. G. Koubogiannis and K. C. Giannakoglou [a]

[a]Lab. of Thermal Turbomachines, Mechanical Enineering Department,
National Technical University of Athens,
P.O. Box 64069, 15710 Athens, Greece.

A technique for redressing load-imbalancies which arise in parallel *CFD* methods using adapted unstructured grids is presented. This technique is applicable to any *CFD* code parallelized through the multidomain concept and message-exchange protocols, like the PVM. Data subsets, i.e. unstructured grid partitions, are defined using an efficient, Genetic Algorithm (*GA*) based tool and the adapted grid is repartitioned after each adaptation cycle. Despite its distinct sequential parts, the so-called Static Grid Repartitioning (*SGR*) algorithm, yields high parallel efficiencies and presents certain advantages over Dynamic Load-Balancing (*DLB*) algorithms which rely on the exchange of excess grid entities between processors.

1. INTRODUCTION

Load-balancing between processors is of primary concern in all parallel *CFD* methods. According to the multidomain concept, grids and the relevant data should first be partitioned off, defining thus subdomains or grid/data subsets; then, each processor will be associated with one subdomain at a time. Evident prerequisites for the high parallel efficiency are (a) load-balancing between processors and (b) minimum communication overhead. Aiming at load-balanced processors, each subdomain should possess the same number of grid entities. To reduce the communication overhead, the exchange of data between processors during the iterative part of the algorithm should be kept as low as possible. Practically, this can be taken into account by building subdomains with minimum interface "length". The generation of optimum subsets for unstructured grids is by no means trivial and several relevant algorithms can be found in the literature [1], [2], [3].

Nowadays, grid adaptation to the evolving solution is built in all modern *CFD* codes. However, in *CFD* codes parallelized using the multidomain technique this is a source of imbalance between the initially evenly loaded processors which may degrade their parallel efficiency. For each subdomain/processor, the degree of imbalance is measured in terms of the deviation of its population from the mean population of all subdomains. In order to combine the advantages of both grid adaptation and parallelization, without their side-effects, several load-balancing algorithms have been proposed [4], [5], [6]. These can be classified in *SGR* (grid subsets are redefined from scratch after each adaptation) and *DLB* (excess grid entities migrate to less loaded processors) algorithms.

In the past, the present authors presented a fully parallel *DLB* algorithm for the solution of the Navier-Stokes equations based on adaptive unstructured grids [7]. Here, a *SGR* algorithm is introduced, where the grid adaptation and the partitioner parts of the algorithm are sequential. However, these are very fast and yield satisfactory parallel efficiencies. A *GA*-based method carries out the grid partitioning, at negligible CPU cost.

2. THE PARALLEL FLOW SOLVER

The flow solver operates on unstructured grids with triangular elements. The discretization of the governing equations relies on the finite-volume technique with a vertex-centered storage system. At any node P, the control volume C_P is defined as in fig. 1, by connecting the barycenters of the surrounding triangles and the mid-points of any edge emanating from P. The integration of the flow equations over C_P results to the balance of fluxes crossing its boundary ∂C_P and a pseudo-time derivative, namely

$$(\frac{\partial \vec{U}}{\partial t})_P \iint_{C_P} dxdy + \int_{\partial C_P} \vec{f}\, ds = \vec{0} \tag{1}$$

where $\vec{U} = (\rho,\ \rho u,\ \rho v,\ E)^T$ is the conservative variables array. The fluxes $\vec{f}$ that cross ∂C_P are computed using the Roe flux-difference splitting scheme [8]. Their computation involves pairs of nodes (like P and Q) linked with a grid edge. To achieve second-order accuracy in space, the primitive flow variables array $\vec{V} = (\rho,\ u,\ ,v,\ p)^T$ at a left (L) and a right (R) state defined in the middle M of PQ should be computed as follows

$$\vec{V_L} = \vec{V_P} + \frac{1}{2}\vec{PQ} \cdot (\nabla \vec{V})_P \quad , \quad \vec{V_R} = \vec{V_Q} - \frac{1}{2}\vec{PQ} \cdot (\nabla \vec{V})_Q \tag{2}$$

Practically, the grid edges are swept in order to compute the inviscid fluxes or to employ the iterative Jacobi solution scheme, whereas triangles are swept to compute the (constant) $\nabla \vec{V}$ in each one of them before scatter-adding to the nodes, using an area-weighted scheme. A similar sweep over triangles is used to compute the local time-steps for the pseudo-time integration.

It is a matter of linearization of the discretized equations to get the final equation per grid node P. The latter is numerically solved by the Jacobi scheme, involving a number of sub-iterations (index k) per pseudo-time-step (index n), namely

$$\delta \vec{U_P}^{n+1,k+1} = D_P{}^{-1} \cdot (\vec{R_P}^{n} - \sum_{Q \in K_N(P)} O_{PQ} \cdot \delta \vec{U_Q}^{n+1,k}) \tag{3}$$

where $K_N(P)$ is the set of nodes that are linked to P through an edge, D_P and O_{PQ} are the diagonal and off-diagonal matrix coefficients, respectively.

The flow solver is parallelized using the PVM [9] message passing system on a dedicated (Beowulf) cluster of 4 identical dual-processor PCs (Intel Pentium III/ 500Mhz, 512MB RAM) under Linux (Redhat 6.1 distribution) and connected via a fast ethernet switch (100 Based-TX) has been used along with the Pentium-optimized GNU compiler. From the programming point of view the master-slave model has been employed where the slave processes stand for 'copies' of the flow solver operating on different subdomains. The parallel and the sequential algorithms are much alike, though a number of communication tasks distinguish the former from the latter, as discussed in [10].

The grid adaptation to the evolving flow solution is based on the grid edges' marking for refinement (sometimes for derefinement, too) according to some sensor functions and relevant criteria. Each triangle is allowed to yield either two or four offspring triangles depending on the number of its marked edges. Several rules have been laid down in order to avoid modifying the initial grid, creating hanging nodes or going through an endless refinement-derefinement procedure, etc.

3. UNSTRUCTURED GRID PARTITIONING METHOD

Recently, an unstructured grid partitioning method has been developed by the authors, [3], [11]. The definition of non-overlapping subdomains takes the form of a minimization problem with the interface length between subdomains considered as the cost function. The optimization problem is solved using *GAs*. As all of the operations are carried out using the equivalent graph of the grid, 2-D or 3-D, homogeneous or hybrid grids can be readily partitioned. This method is based on the concept of recursive bisections, giving rise to 2^n subdomains; thus, only a single bisection problem will be analyzed below.

The concept of the partitioner is simple, though innovative; it uses two points (A and B, located at (x_A, y_A) and (x_B, y_B)) which are free to move in the graph space (x, y), their roles being to create scalar fields around them. Any graph node located at (x, y) is affected by both A and B, according to laws that mimic the static electric field generated around point charges, namely $F(x, y) = +(k_A/r_A^2) - (k_B/r_B^2)$, where $r_A^2 = (x_A - x)^2 + (y_A - y)^2$ and $r_B^2 = (x_B - x)^2 + (y_B - y)^2$.

By giving arbitrary values to the six quantities $(x_A, y_A, k_A, x_B, y_B, k_B)$, a unique scalar field over the graph space is created. Half of the graph nodes with the lower $F(x, y)$ values are assigned to the first subdomain and the rest form the second one. By construction, these subdomains are evenly loaded so that the optimum partition is the one with the minimum interface length. One of the k coefficients can be given an arbitrary value $(k_A = -1)$. The remaining five parameters $(x_A, y_A, x_B, y_B, k_B)$ are controlled by the *GA* that carries out the search for the optimum partition.

A basic feature of this partitioner is that it relies on the multilevel scheme. So, at each bisection, the initial graph G^0 becomes coarser by collapsing neighboring graph nodes into groups. Through M coarsening steps, the final or coarser graph (highest level, G^M) is formed. Using *GA*, the G^M graph is first partitioned in two subsets yielding thus good starting solutions for the G^{M-1} graph bisection. The G^{M-1} graph is further improved through fast heuristics and this is repeated up to the finest level.

4. THE LOAD-BALANCING ALGORITHM

4.1. The Dynamic Load-Balancing Algorithm

In a previous work by the authors [7], a parallel *DLB* algorithm was designed and implemented on an Intel/Paragon computer. The algorithm was fully parallelized, in the sense that not only the solver but the grid adaptation and load-balancing tasks as well, were carried out concurrently for all subdomains. Load-balancing was achieved through regular load-redistributions among the processors. This involved migrations of grid entities between neighbouring subdomains, designed to perform in a multi-pass, tree-like manner and to operate on groups of processors, in conformity with the tree-like grid partition. In order to retain the capability of grid derefinement, only triangles of the initial (coarse) grid were allowed to migrate, carrying with them the topological data for their offspring. Each such migration was realized by exchanging packed messages. In [7], it has been shown that, besides the fact that *DLB* was theoretically fully parallelized, the results were sensitive to the coarseness of the initial grid, the partitioning requirements were often poorly satisfied whereas a variety of complicated migration scenaria should be taken into account.

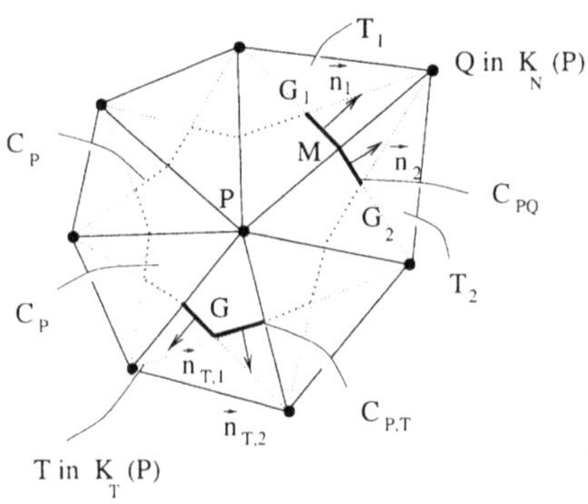

Figure 1. The finite-volume at node P.

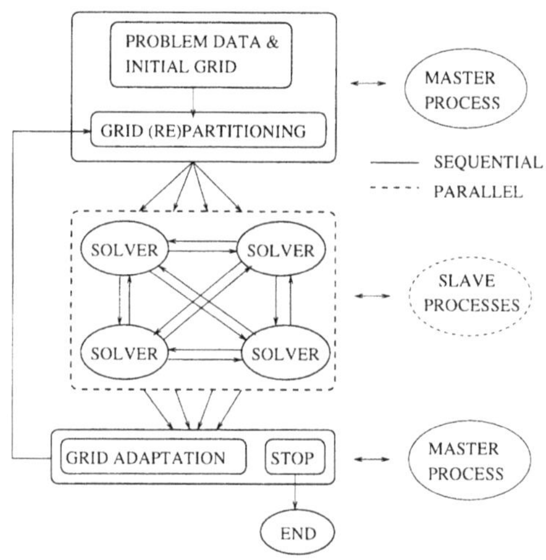

Figure 2. The *SGR* algorithm.

4.2. The Static Grid Repartitioning Algorithm

In contrast to *DLB*, the repartitioning algorithm described below will be refered to as "static" in the sense that the parallel execution is suspended and the load redistribution is performed from scratch through regular calls to the GA-based partitioner (fig. 2).

A few comments on the proposed method are necessary. There are two distinct subtasks, namely the grid adaptation and (re)-partitioning, which are carried out sequentially, by the master process. It is evident that sequential parts which interfere in the parallel ones are expected to badly affect the parallel efficiency. However, we should point out that both sequential parts are very fast. Over and above, the sequential grid adaptation circumvents a considerable amount of inter-processor communication for the regular exchange of information relevant to the marking of interfacial nodes and to the numbering of new grid entities. Instead of them, the new algorithm requires a single, though major communication task, during which the master process recovers data from the slave ones, and after adaptation and redistribution, a second similar communication in the reverse direction is due.

The repartitioning of the grid by the master process ensures optimality and creates subdomains which are both perfectly balanced and with minimum interface length. *SGR* is a very general procedure that overcomes the uncontrollable patching up of the damaged optimum partition.

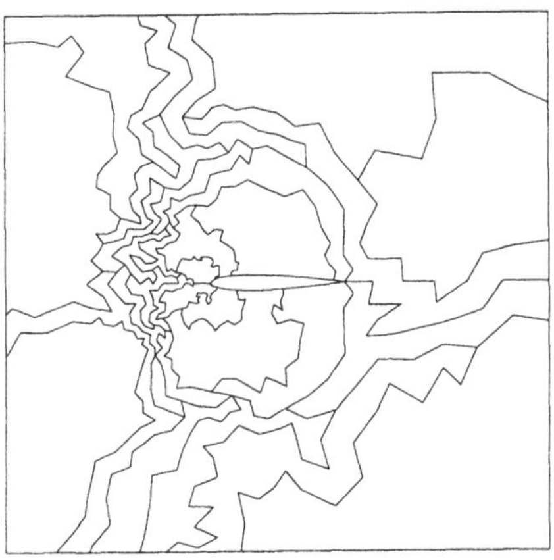

Figure 3. The 32 subdomains created through the *DLB* algorithm, after four adaptation cycles (from [7]).

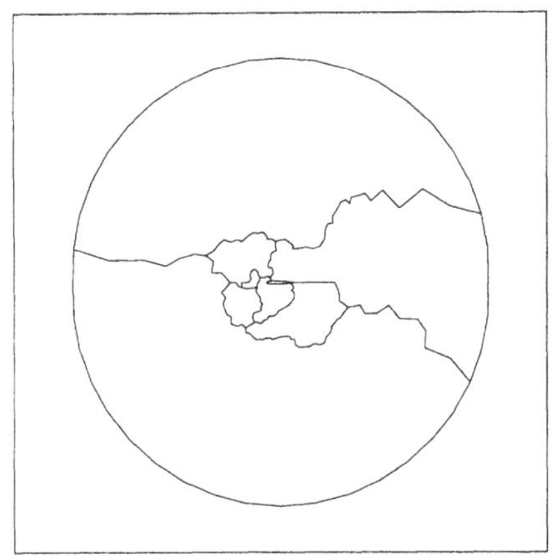

Figure 4. The 8 subdomains created through the *SGR* algorithm, after four adaptation cycles.

5. RESULTS AND DISCUSSION

The proposed parallel solver with grid adaptation and load-balancing via *SGR* will be tested in two cases of inviscid, supersonic flow. The first case concerns the study of the flow which develops around an isolated NACA12 airfoil at 1.2 freestream Mach number and zero incidence. The second one refers to the study of the flow in a 2-D compressor cascade with wedge-shaped blades at 1.2 inlet Mach number and 60 deg. inlet flow angle. Supersonic flows, like those examined herein, exhibit strong shock waves where grid adaptation is mandatory.

In the isolated airfoil case, a bow shock close to the leading edge and two oblique shocks emanating symmetrically from its trailing edge define the areas where grid adaptation mainly occurs. Fig. 3 illustrates the shapes of the 32 subdomains formed through *DLB* after four grid adaptation steps. The shapes of these domains deviate considerably from one of our objectives, that of minimum inter-processor communication and are affected by the coarseness of the initial grid. The same study is repeated with 8 processors using

SGR. The so-obtained subdomains are illustrated in fig. 4 which, in contrast to fig. 3, illustrates well-shaped interface boundaries.

With both *DLB* and *SGR*, four adaptation cycles have been performed with the density difference along each edge as adaptation criterion. Blown-up views of the initial, second, third and fourth (final) grids along with their partitions (obtained through the newly proposed *SGR* method) close to the airfoil are shown in figures 5, 6, 7 and 8 respectively.

For the supersonic cascade, shock waves traverse the entire passage and reflect on the adjacent blades. Periodicity had to be included in all the components of this method (solver, grid partition, adaptation) to account for this case. The initial grid partitions as well as the final ones, obtained through *SGR* are shown in figures 9 and 10 respectively. The subdomains remain equally balanced and with minimum interface length.

The isolated airfoil case was further used to investigate the parallel efficiency of *SGR*. Fig 11 summarizes results from five runs along with two theoretical curves. The first of them corresponds to the diagonal (speed-up=number of processors) whereas the second takes into account the serial fraction of the computational algorithm (stands for grid adaptation and SGR) but ignores any communication overhead. These two curves indicate the unattainable theoretical limits for the cases without and with adaptation, respectively. Four more curves are plotted in the same figure using four different grids, without adaptation. These grids are the four adapted grids around the NACA12 profile (the outcome of the first up to the fourth adaptation cycles). The diagram exhibits a good scaling of the speed-up by increasing the load per processor. A seventh curve is included in this figure and this corresponds to the speed-up of the Euler solver run with the four adaptation cycles included. The latter resembles the behaviour of the third finer grid. The parallel efficiency of the code with adaptation is excellent, at least up to four processors.

Fig. 12 shows the results of a similar investigation carried out using the cascade problem. The ideal speed-up curve (diagonal) in the absence of grid adaptation and the maximum theoretical speed-up with adaptation (based on the measured serial fraction of the parallel code) have been computed as in the previous case. Two curves with measured speed-up values are also shown. The first corresponds to runs with a single process assigned to each two-processor networked personal computer, whereas in the second both processors are in use in each computer. The comparison of these curves shows that the parallel runs are more efficient when the use of both processors in each computer is avoided.

6. CONCLUSIONS

A flow solver for adaptive unstructured grids was ported to a low-cost, distributed memory parallel platform using the multidomain technique and the PVM communication protocol. In order to redress load imbalancies during the parallel execution, the so-called *SGR* algorithm, where one processor undertakes the repartitioning of the flow domain after each adaptation cycle, is proved to perform better than *DLB*, which is based on the migration of grid cells between processors. Despite its sequential parts, *SGR* is capable of maintaining optimality in load-balancing and ensures reduced communication cost during the iterative part of the algorithm. The advantages of the *SGR* are mainly due to the *GA*-based grid partitioner.

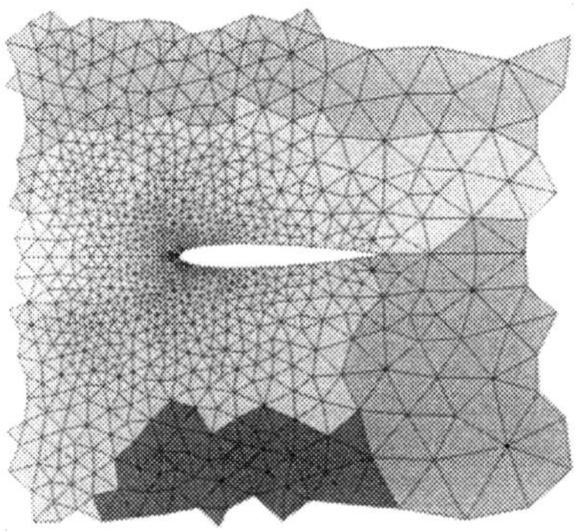

Figure 5. Initial grid (1858 nodes / 3616 triangles).

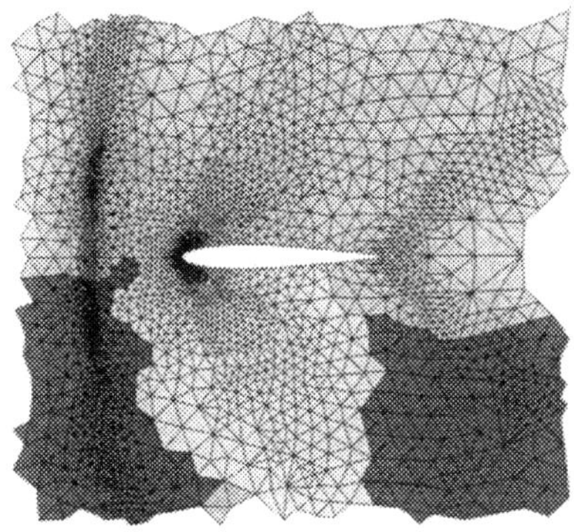

Figure 6. Grid after two adaptation cycles (5120 nodes / 10102 triangles).

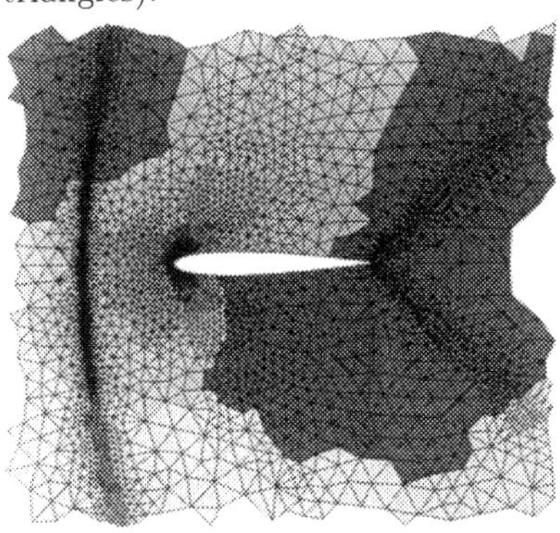

Figure 7. Grid after three adaptation cycles (8013 nodes / 15870 triangles).

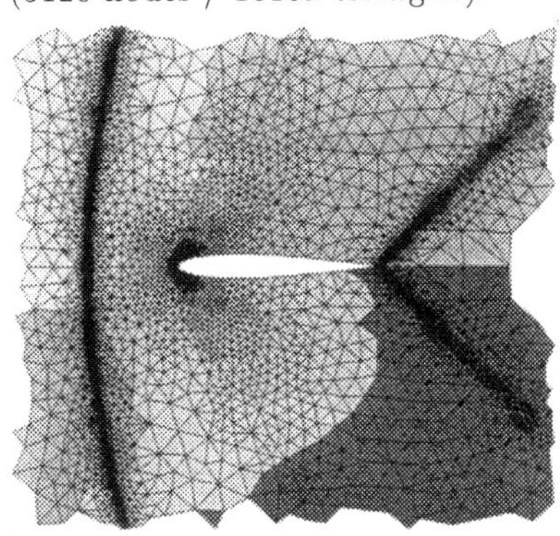

Figure 8. Grid after four adaptation cycles (13483 nodes / 26804 triangles).

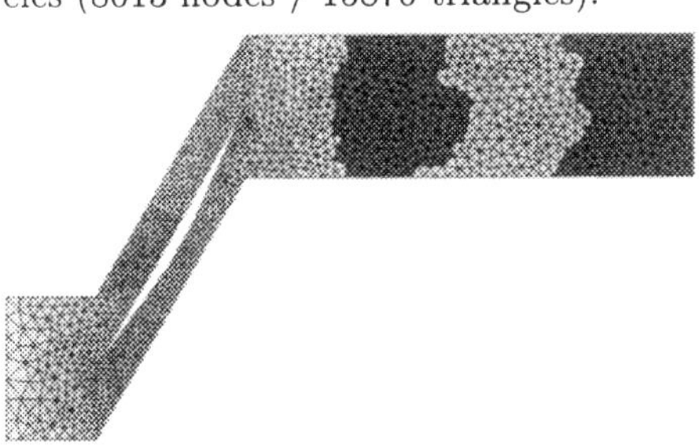

Figure 9. Supersonic cascade. Initial grid (1580 nodes / 2851 triangles), 8 processors.

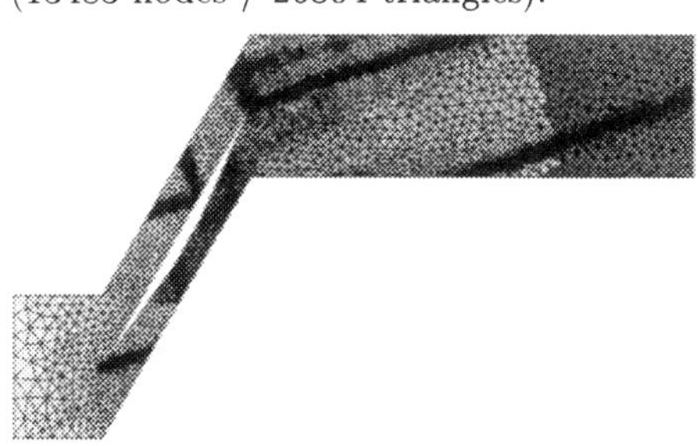

Figure 10. Grid after five adaptation cycles (12689 nodes / 24791 triangles), 8 processors.

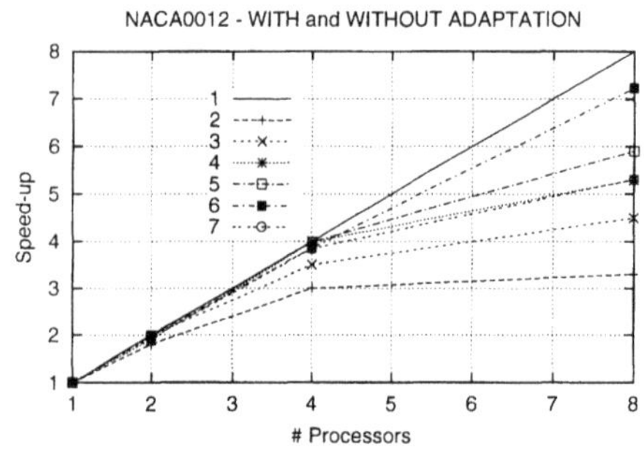

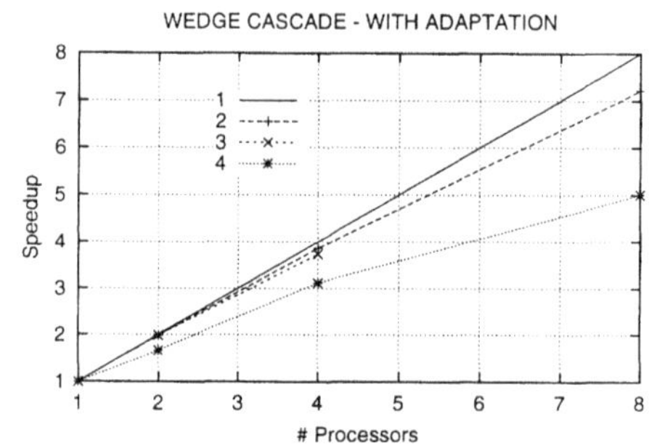

Figure 11. Speed-up (Sp) without adaptation: 1,2,3,4,5 are the ideal and measured curves for the 3380, 5120, 8013, 13483-node grids respectively, Sp with adaptation: 6,7 are theoretical and measured ones.

Figure 12. 1 is the ideal speed-up, speed-up with adaptation: 2 is the theoretical, 3 is the measured with 1 processor per computer, 4 is the measured with 2 processors per computer.

Acknowledgments. Parts of the present research have been separately funded by Dassault Aviation, France and the General Secretariat for Research and Technology, Greece.

REFERENCES

1. H. D. Simon, Computing Systems in Engineering 2 (1991) 135.
2. D. Roose, R. van Driessche, AGARD-R-807 2-1 (1995).
3. A.P. Giotis and K.C. Giannakoglou, Advances in Engineering Software 29 No. 2 (1998) 128.
4. C. Ozturan, H.L. deCoughy, M.S. Shephard and J.E. Flaherty, Comp. Meth. in Appl. Mech. and Eng. 119 (1994) 123.
5. A. Vidwans, Y. Kallinderis and V. Venkatakrishnan, AIAA J. 32 No. 3 (1994) 497.
6. T. Minyard, Y. Kallinderis and K. Schulz, AIAA Paper 96-0295, Reno Nevada, January (1996).
7. D.G. Koubogiannis, K.C. Giannakoglou and K.D. Papailiou, ECCOMAS 98, Proccedings of the European Computational Fluid Dynamics Conference, John Wiley & Sons, 2 (1998) 171.
8. P. Roe, J. of Comp. Phys. 43 (1981) 357.
9. A. Geist, A. Beguelin, J. Dongarra W. Jiang, R. Manchek and V. Sunderam, PVM: Parallel Virtual Machine. A User's Guide and Tutorial for Networked Parallel Computing, The MIT Press, 1994.
10. D.G. Koubogiannis, L.C. Poussoulidis, D.V. Rovas and Giannakoglou K.C., Comp. Meth. in Appl. Mech. and Eng. 160 (1998) 89.
11. K.C. Giannakoglou and A.P Giotis, ECCOMAS 98, Proccedings of the European Computational Fluid Dynamics Conference, John Wiley & Sons, 2 (1998) 186.

Parallel Computational Fluid Dynamics – Trends and Applications
C.B. Jenssen et al. (Editors)

Parallel implementation of genetic algorithms to the solution for the space vehicle reentry trajectory problem

S. Peigin[a] and J.-A. Désidéri[b]

[a] Faculty of Aerospace Engineering, TECHNION,
Technion city, Haifa, 32000, Israel

[b] INRIA Sophia Antipolis,
2004, route des Lucioles - B.P.93, 06902 Sophia Antipolis Cedex, France

A solution of a space-vehicle atmospheric-reentry trajectory optimization problem is considered. The objective consists in minimizing the maximum wall heat flux, located at the stagnation point of a 3D blunt body, and integrated with respect to time along the trajectory path, subject to physical constraints on the body deceleration and on the maximum value of the equilibrium temperature at the body surface. This problem is solved by means of the parallel version of floating-point Genetic Algorithms (GA) using the MPI package via calculation of the heat flux distribution along the reentry trajectory in the framework of the non-equilibrium multicomponent thin viscous shock layer model (TVSL). The influence of the GA parameters as well as number of processors on the optimum solution and on the algorithm arithmetical cost efficiency is assessed.

1. INTRODUCTION

Evidently, constructing a cost-efficient parallel algorithm to implement a classical conjugate gradient type optimization method is very difficult if the cost-function evaluations are made independently. In such a case, the uniform load-balancing of processors cannot be achieved, since at moments, certain processors wait to execute their tasks for the synchronisation with the others. Contrasting with this, in the case of a Genetic Algorithm, greater parallel efficiency is achieved because information coming from other processors can be exchanged at each generation of the optimization process. In fact, when separate calculations of the objective function necessitate different CPU times, these calculations can be performed at independent times. In this sense, the proposed algorithm is *asynchronous.*

One crucial question raised by the parallel implementation of an algorithm is the efficiency in the exchange of data between the different processors. From a widely accepted point of view, the main goal is to minimize the CPU time related with this process of data transfer. Using this point of view, we take into account this CPU time as a loss only. Hence, in this approach, the efficiency of parallelization cannot greater than 100%.

From a more general, or more " philosophical " point of view, the incidence of the data transfer between processors is twofold : from one hand, it results in a loss, by increasing the CPU time of the amount necessary to it, but, from the other hand, it also results in a gain, since individual processes gain additional information, thus improving the rate of convergence to the optimum, or the convergence to a better optimum. Consequently, at least theoretically, a parallelization efficiency greater than 100 % is conceivable. Of course, the most difficult question is to recognize which precise information should be transferred in order that the gain surpass the loss.

But for GAs, the situation is favorable to giving a positive constructive answer to this general question : by exchanging information related to the best invididuals of subpopulations, one can expect the convergence rate of the GA evolution process to improve.

2. PROBLEM STATEMENT

To formule the optimization problem, let us suppose that t_m is the total flight time of the vehicle moving along this high part of the trajectory. In this case, the total convective heat flux at the stagnation point of the 3D blunt body surface Q along trajectory can be calculated using the following relation:

$$Q(V, H, t_m) = \int_0^{t_m} q_w(H(t), V(t), R^*, k, k_{wi})dt. \tag{1}$$

Here $q_w(H(t), V(t), R^*, k, k_{wi})$ is the heat flux at the stagnation point of the space vehicle 3D blunt body surface, $V(t)$ is the flight velocity, $H(t)$ is the flight altitude, k is equal to the ratio of the body surface main curvatures at the stagnation point, R^* is the character linear size of the problem, k_{wi} are the known parameters of the heterogeneous chemical reactions occuring at the body surface.

Let us consider the following variational problem: find the smooth functions $V(t)$ (velocity) and $H(t)$ (altitude) $(0 \le t \le t_m)$ which have the following two properties:

1. The total integral of the convective heat flux $Q(V, H, t_m)$ at the stagnation point of the 3D space vehicle body surface along the high part of the trajectory is minimum;

2. The equilibrium temperature of the body surface at the stagnation point $T_w(t)$ does not exceed the preassigned limit value T_w^{max}.

This problem is solved with the following constraints:

$$V(0) = V_0, \quad H(0) = H_0; \qquad V(t_m) = V^*, \quad H(t_m) = H^* \tag{2}$$

$$|\dot{V}(t)| < ag, \qquad |\dot{V}| \le \frac{S^*}{m}\frac{\rho_\infty(t)V^2(t)}{2} \tag{3}$$

Here t_m is the total trajectory flight time, S^* is a characteristic body surface, m is the mass of the vehicle.

We assume that the vehicle mass is constant, the beginning acceleration is absent, the Earth is a sphere, its swirling is absent and the gravity force does not depend on altitude [1].

The relations in (2) mean that the trajectory begins from the point (V_0, H_0) and finishes at the point (V^*, H^*). The first relation in (3) is connected with the medical restrictions to the flight conditions. The second condition in (3) is rather obvious from physical point of view and in fact it requires, that the deceleration along trajectory do not exceed the maximum drag deceleration depending on a maximum drag force for the given body, for the given altitude and for the given flight velocity.

As initial mathematical model for calculating the integral in (1) we will use the thin (hypersonic) multicomponent nonequilibrium viscous shock layer equations. These equations are the asymptotic form of the Navier-Stokes equations and permit us to correctly describe the flow structure between the body surface and the shock wave if the following conditions are realized:

$$\varepsilon = \frac{\gamma - 1}{\gamma + 1} \to 0, \;\; Re \to \infty, \;\; K = \varepsilon Re \geq O(1).$$

The analysis demonstrates that this model has a good accuracy for our range of V and H for all possible trajectories [2].

It is rather obvious from physical point of view, that because T_w strongly depends on the body geometry the initial variational problem has no solution for low T_w^{max} value, because for all trajectories the maximum of the body surface temperature T_w will exceed T_w^{max} and the problem constraint on T_w will be not satisfied.

For this reason we have also considered the following minimax problem: find optimum trajectory with minimum integral heat flux and having minimum possible T_w^{max} value (named as minimax optimum trajectory).

3. METHOD OF SOLUTION

For the solution of the above optimization problem a variant of the floating-point Genetic Algorithms[3] was applied. We used ordinary single point, uniform or arithmetical crossover operator and the nonuniform mutation operator defined by Michalewicz [4] with distance-dependent mutation approach suggested by Periaux[5]. For approximation of the trajectory we used Bezier curve of order N. Our string $S=(a_1, a_2, ..., a_{N-1}, a_N, ..., a_{2N-2})$ contained $2N-2$ values of control points. These values were varied from Min_i to Max_i which are lower and upper bounds of the variable a_i.

Because the constraints on the optimization problem are inequalities during realization of the algorithm we modified the objective function. We used the following modified objective function Q^*:

$$Q^* = \begin{cases} q_1 + q_2(|\dot{V}(t)|/g - a), & \text{if } |\dot{V}(t)| > ag, \\ q_3 + q_4(T_w^{\max} - T_w)/T_w^{\max}, & \text{if } T_w > T_w^{\max}, \\ q_5 + q_6(|\dot{V}(t)|/g - A(t)/g) & \text{if } |\dot{V}(t)| > A(t), \\ Q, & \text{in other cases} \end{cases} \quad (4)$$

where $A(t) = S^* \rho(t) V^2(t)/(2m)$ and q_i are the problem depended parameters.

In fact, this approach enables one to extend the search space and to evaluate, in terms fitness, the individuals, that do not satisfy the constraints on the optimization problem. It is very important to note that differing from conjugate gradient optimization methods the GA methods are not restricted the smoothness of this extension.

The initial thin (hypersonic) multicomponent nonequilibrium viscous shock layer equations were solved with boundary conditions at the shock wave and at the body surface. At the shock wave the hypersonic approximation of generalized Rankine - Hugoniot conditions was used. At the body surface heterogeneous chemical reactions were taken into account and heat discharge inward of body was neglected. The numerical solution of this boundary value problem was obtained using high accuracy computational algorithm[6].

It is understood that the calculation of the integral (1) via the direct numerical solution of the TVSL equations system at each trajectory point requires a very large computer resource. For this reason, we have used the following approximate approach[7]. At the first step the initial boundary value problem was solved for given values R^*, k, k_{wi} on grid 21×22 with step $\Delta V_\infty = 0.25km/s$ (for velocity range from 7.8 km/s to 2.3 km/s) and with step $\Delta H = 2.5km$ (for altitude range from 100 km to 50 km). At the second step we calculated the objective function (1) via interpolation of the obtained results on the velocity an on the altitude using B-spline representation.

The comparisons demonstrated that the difference between integral heat flux along different trajectories, calculated on the basis exact and the above listed two-step approximate approach did not exceed 0, 5%, which permits us to use this method for serial calculations.

For parallel implementation of the GA algorithm, an asynchronous parallel solevr has been implemented. As a base software tools the wellknown MPI package was used. Our parallel optimization method involved the following algorithmic steps:

1. The initial population using random search is independently obtained on each processor.

2. The cost function Q^* for each individual is computed.

3. Non-blocking data receiving from other processors. If the requested message has not arrived this step is skipped. Otherwise the obtained individuals are included in the population, fitnes function values are compared and the worst individuals are eliminated.

4. The crossover and the mutation operators are applied to the population and the new generation is computed.

5. With a preassigned value of the generation step the information about the best individual in the new generation is broadcast to other processors.

6. If the convergence accuracy is achieved then stop. Otherwise, the iterative process is repeated from step **3**.

4. RESULTS AND DISCUSSIONS

The solution of this variational problem was obtained for the following range of problem and GA parameters:

$$S^*/m = 2.5 \cdot 10^{-3}\text{m}^2/\text{kg}; \quad a = 3.0; \quad k = 0.2, 1.0; \quad R^* = 1.0m \quad V_0 = 7.8\text{km/s},$$

$$V^* = 2.3\text{km/s}, \quad H_0 = 100\text{km}, \quad H^* = 50\text{km}, \quad 0.4 \leq P_c \leq 1.0; \quad 0.1 \leq M_m \leq 0.9;$$

$$S = 20 - 200; \quad b = 10, 20; \quad G = 10^3 - 10^4; \quad 8 \leq N \leq 13$$

where P_c is a crossover probability, M_m is a maximum allowed mutation probability, S is a population size.

As a whole the results of the systematic calculations demonstrated that the suggested algorithm is quite robust, has a good accuracy and a high level of arithmetical cost efficiency.

The results of these calculations for 8 processors, for non-catalytic surface and for k = 0.2 are presented in fig. 1, where the distributions of Q for various exchange steps L (fig.

1a) and the comparison of the solution for $L = 20$ (line 1. fig. 1b)) with the solutions without exchange information between processors (lines 2-5, fig. 1b) are shown.

As it can be seen from computations the exchange information about the best individuals on each evolution (each processor) essentially improved the global algorithm convergence and enables one to obtain the parallel efficiency of the algorithm more then 100%. For example, for 8 processors a parallel efficiency of the algorithm was equal about 200%.

It is interesting to question the influence of the exchange information step L. It is not optimal to use a too small value of L for which subpopulations exchange information before having had sufficient time to evolve significantly towards optimality. In an opposite case, when L is too large, little profit is made from exchange information between separate evolutions of simular type. For these reasons, the calculations demonstrated that best parallel efficiency is achieved when L assumes an intermediate value.

The analysis of calculations demonstrated the following: the distributions of $T_w(t)$ along optimum trajectories have two segments. Along the first one the equilibrium body surface temperature T_w reaches the maximum allowed value and has a local 'plateau'. Along the second segment, this value has a minimum. The length of this "plateau", the position and the value of this minimum both depend from total flight time t_m: if for $t_m = 30min$ the "plateau" length is equal to 5% of the total flight time and T_w has a local minimum then for $t_m = 15min$ the "plateau" length is increased up 30% of the total flight time and the minimum of T_w is reached at the trajectory end point.

It is necessary to note that this character of the heat flux distributions along optimal trajectories is rather clear from physical point of view. Because the heat flux to the body surface is decreased if a flight velocity V is decreased and (or) a flihgt altitude H is increased and taking into account boundary conditions for V and H the possible optimum strategy is to reach a lower velocity level at a higher altitude and then to move with smaller velocity.

In fact this strategy is realized in our optimum solution: taking into account the constraints on T_w and on a vehicle acceleration (depending from the aerodynamic characteristics of the space vehicle) at the first segment of trajectory the decrease of the velocity at the upper part of the trajectory is reached. The optimal strategy at the second flight segment depends on t_m. For long total flight time the altitude is not a monotonic decreasing function. Instead after first phase of flight the altitude is gained to reduce the heat flux and then again altitude decrease to match trajectory essentually target point. But for short t_m there is not enough time to realize an altitude non-monotonic part of trajectory and after first trajectory stage the altitude and the velocity are monotonically decreased to satisfy boundary conditions at the end point of trajectory.

It was also obtained that for minimax optimum trajectories a second non-monotonic flight altitude segment is absent. The comparisions also demonstrated that minimax optimum trajectory has a good correlation with the "BURAN" and "SPACE SHUTTLE" trajectories at the maximum heat flux domain.

It is very important to note that, as it is seen from these obtained solutions, physical constraints are very essential during most part of the trajectory. It means that the optimal point in our search domain is very close to the boundary of this domain and an application of the conjugate gradient optimization methods to the solution of this

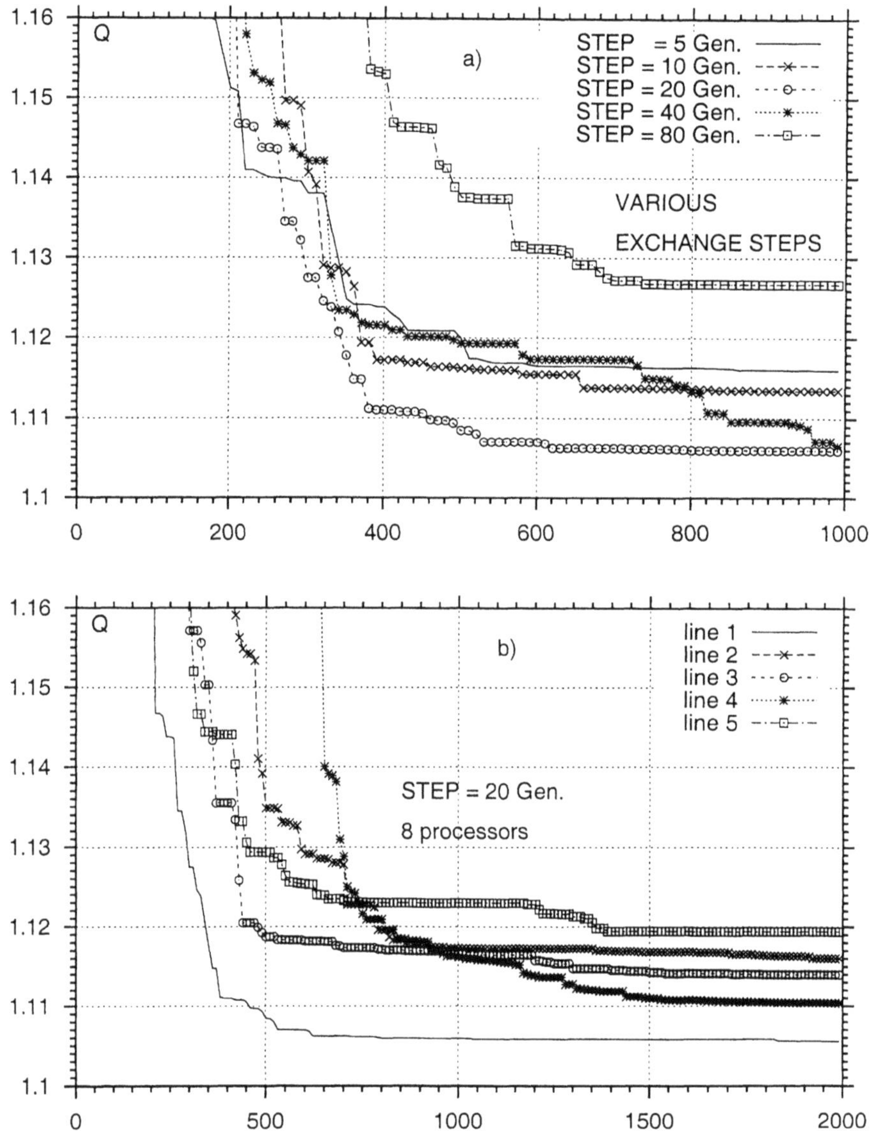

Figure 1. Comparisons of the objective function Q distributions versus generations number for various exchange steps (a) and with solutions without exchange between processors (b) (lines 2 - 5)

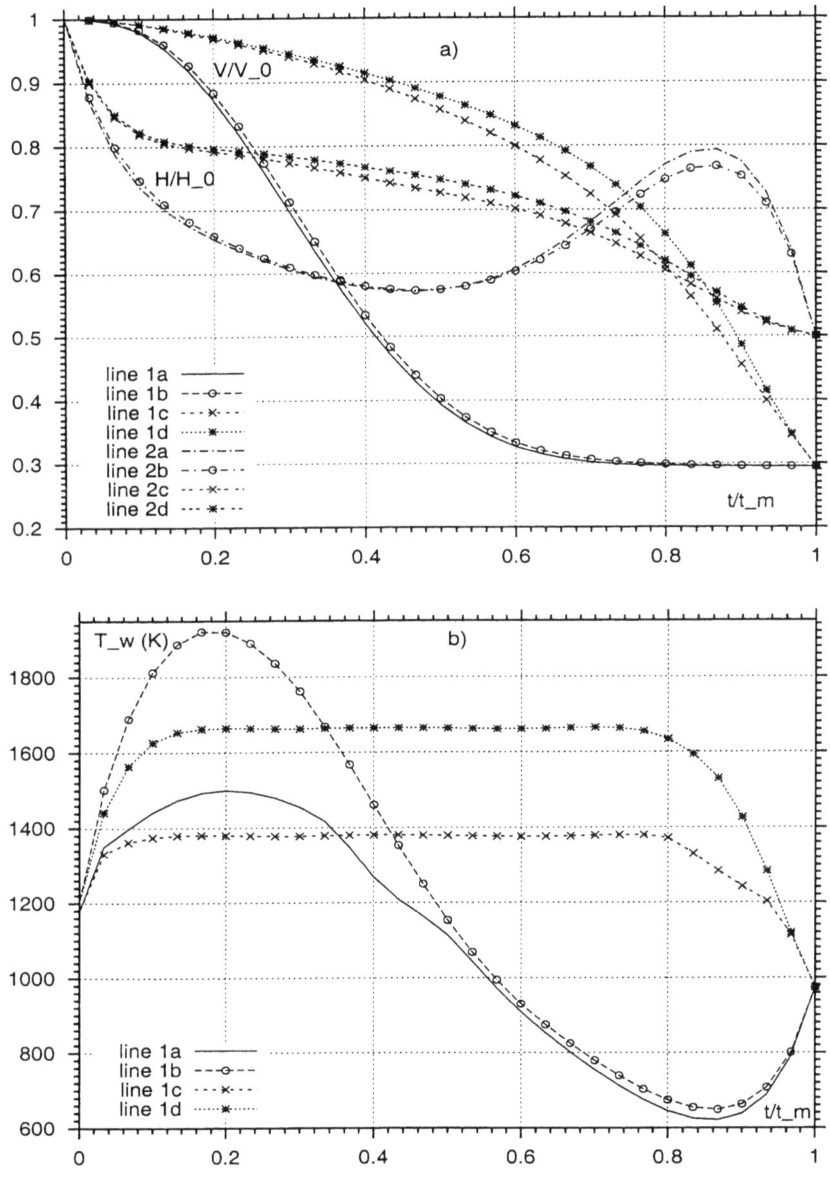

Figure 2. Influence of the catalytical activity of the body surface on optimal solutions for various T_w^{max} and for $t_m = 30min$. Lines a, c : non-catalytic surface, lines b, d : ideal catalytic one, lines a, b : $T_w^{max} = 2500K$, lines c, d : minimax optimum trajectories

optimization problem would experience a number of difficulties to calculate with high accuracy derivatives of the objective function near space search boundary.

From practical applications point of view it is very important to know how the optimal trajectory depends on the body geometry and on the rate of the heterogeneous chemical reactions proceeding at the body surface. Thus a number of calculations for various values of parameter k - ratio of the main curvatures of the body surface at the stagnation point (characterising body geometry) and for various catalytic activity of the body surface were carried out.

Some results of these calculations are shown in fig. 2 where the the distributions of altitude H, of velocity V versus τ (fig. 2a) and of temperature T_w versus τ (fig. 2b) along various optimal trajectories for $k = 1.0$ and for non-catalytic and ideal catalytic surface sre presented.

As a whole, the computations demonstrated that the influence of the body geometry and of the surface catalytic activity is rather low: for high T_w^{max} value the optimum trajectories practically coincide and for the opposite case the difference between minimax optimum trajectories is rather small also.

Finaly, we observe that the present approach of parallel implementation of a genetic algorithm provides a examplary case of a situation in which the gain realized by exchanging information among processors improves the convergence rate of the GA and surpasses the loss in CPU time necessary to the data transfer itself.

REFERENCES

1. V.V. Andrievskii, Dynamics of the space vehicle entry to Earth atmosphere, Mashinostroenie, Moscow, 1970.
2. S.V. Peigin, G.A. Tirskii et al., Super- and hypersonic aerodynamics and heat transfer, CRC Press, New-York, 1993.
3. F. Hoffmeister, T. Back, Genetic algorithms and evolution strategies: similarities and differences. Parallel Problem Solving from Nature - Proceedings of 1st Workshop. Dortmund (1991) 455.
4. Z. Michalewicz, Genetic algorithms + data structures = evolution programs, Springer-Verlag, Artificial Intelligence, New York, 1992.
5. M. Sefioui, J. Periaux J and J.-G. Ganascia, Fast convergence thanks to diversity. Evolutionary Programming V. Proc. of the 5th Annual Conference on Evolutionary Programming. MIT Press (1996).
6. S.V. Peigin, Numerical simulation of 3D hypersonic reacting flows over blunt bodies with catalytic surface. Computational method in Applied Sciences. Elsevier (1992) 127.
7. J.-A. Desideri, S.V. Peigin and S.V. Timchenko, Application of genetic algorithms to the solution of the space vehicle reentry trajectory optimization problem. INRIA. Rapport de recherche. No. 3843 (1999).

8. Lattice Boltzmann Methods

Parallel Computational Fluid Dynamics – Trends and Applications
C.B. Jenssen et al. (Editors)

Perspectives of the Lattice Boltzmann Method for Industrial Applications

J. Bernsdorf[a], T. Zeiser[b], P. Lammers[b], G. Brenner[b], F. Durst[b]

[a]C&C Research Laboratories (CCRLE), NEC Europe Ltd.,
Rathausallee 10, D-53757 St.Augustin, Germany

[b]Institute of Fluid Mechanics (LSTM), University of Erlangen-Nuremberg,
Cauerstr. 4, D-91058 Erlangen, Germany

The lattice Boltzmann (LB) method is based on the numerical evaluation of a Boltzmann type equation, treating the fluid from a statistical point of view. In recent years, this new method proved to be highly efficient for high performance computing (HPC) simulations in several research areas like porous media flow, multi phase flow, or the simulation of reaction-diffusion processes. In this paper, we discuss the perspectives of the LB method for industrial applications, and we present some successful examples from the automobile industry, chemical engineering and civil engineering.

1. Introduction

The reduction of turn around times in the design cycle is an important aspect of product development in many different industries. New tools for an efficient numerical simulation on HPC systems therefore play an increasingly important role in the simulation of fluid- and aerodynamics, e.g., in the areas of automobile design and chemical engineering.

The LB method [1–5] is a relatively new tool in computational fluid dynamics (CFD), with two outstanding features when compared to other conventional CFD methods:

- Very high performance (usually more than 50 % of the peak performance) and nearly ideal scalability on high performance vector-parallel computers due to the underlying cellular automata like algorithm.
- Very efficient handling of the equidistant cartesian meshes, which are generated by the semi-automatic discretization of arbitrarily complex geometries (conversion of CAD data or three dimensional computer tomography (3D CT)). [6].

While during the first ten years of its development the LB method was almost exclusively applied for academic research purposes, a significant tendency towards industrial applications can be observed nowadays.

After a short introduction into the method, we will discuss the advantages of an industrial application of the LB method for the two parties potentially involved: In industry, the LB methods are not yet widly known as a potential alternative to classical CFD methods, whereas some LB researchers appearently do not yet see the advantages of an industrial application of their research work.

A collection of industry-related flow simulations carried out with the LB method shall demonstrate the state of the art and give an evidence for the applicability of this new method.

2. The lattice Boltzmann method

The lattice Boltzmann method [1–5] treats the fluid on a statistical level, simulating the movement and interaction of ensemble-averaged particle density distributions by solving a velocity discrete Boltzmann type equation.

It has proved to be a very efficient tool for flow simulation in highly complex geometries (discretized by up to several million grid points) [7]. Among the different ways of approximating the collision integrals of the Boltzmann equation, the BGK approach [8,5] is nowadays the most common.

2.1. The lattice – BGK – approach

On every node $\vec{r_*}$ of an equidistant orthogonal lattice a set of i real numbers, the particle density distributions N_i, is stored. The updating of the lattice consists of basically two steps: a streaming process, where the particle densities are shifted in discrete time steps t_* through the lattice along the connection lines in direction $\vec{c_i}$ to their next neighboring nodes $\vec{r_*} + \vec{c_i}$ (left part of Eqn. 1), and a relaxation step, where locally a new particle distribution is computed by evaluating an equivalent to the Boltzmann collision integrals ($\triangle_i^{Boltz}$, Eqn. 2).

The complete lattice Boltzmann equation can be written as:

$$N_i(t_* + 1, \vec{r_*} + \vec{c_i}) = N_i(t_*, \vec{r_*}) + \triangle_i^{Boltz} , \tag{1}$$

$$\triangle_i^{Boltz} = \omega \left(N_i^{eq} - N_i \right) , \tag{2}$$

with a local equilibrium distribution function N_i^{eq}

$$N_i^{eq} = t_p \, \varrho \left\{ 1 + \frac{c_{i\alpha} \, u_\alpha}{c_s^2} + \frac{u_\alpha \, u_\beta}{2 c_s^2} \left(\frac{c_{i\alpha} \, c_{i\beta}}{c_s^2} - \delta_{\alpha\beta} \right) \right\} . \tag{3}$$

For every time step, all quantities appearing in the Navier-Stokes equations (velocity, density, pressure and viscosity) can be locally computed in terms of moments of this density distribution and (for the viscosity) of the relaxation parameter ω.

The local equilibrium distribution function N_i^{eq} has to be computed every time step for every node from the components of the local flow velocity u_α, the fluid density ϱ, a lattice geometry weighting factor t_p and the speed of sound c_s. It is chosen to recover the incompressible time-dependent Navier-Stokes equations [5].

2.2. Introducing complex geometries

Single lattice nodes are either *occupied* by an elementary obstacle, or they are *free* (*marker and cell* approach). Particle densities N_i, which are shifted to an occupied node owing to the streaming process, are simply bounced back to their original location during the next iteration, but with opposite velocity (indicated by the index i). This results in the desired no-slip (zero velocity) wall boundary condition.

2.3. Application strategy I: Automobile aerodynamics

In modern car design, a complete model of the automobile is already available as CAD data. Using some dedicated software, the voxel-data for the LB simulation can be generated automatically. Therefore, commercial LB packages usually come along with integrated modules for pre-processing (and sometimes also post-processing).

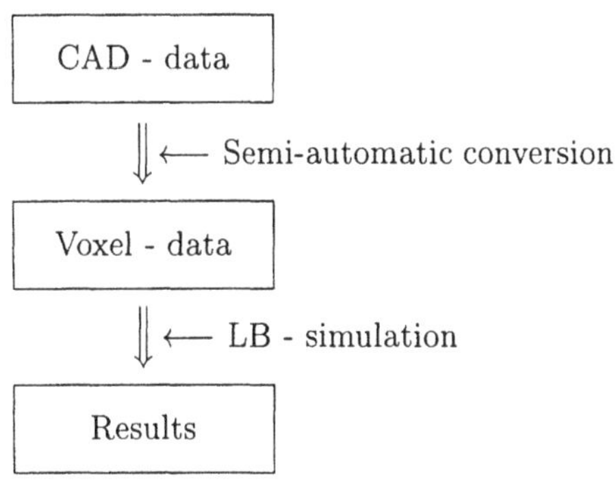

2.4. Application strategy II: Chemical engineering

For the development of new devices in the area of chemical engineering, a detailed knowledge of flow properties inside highly complex geometries is helpful (e.g., heterogeneous catalytic reactors) [9]. Usually, it is impossible to carry out the mesh generation for such geometries with conventional methods (either by hand or automatically for unstructured meshes). Using 3D computer tomography, arbitrary complex structures can be digitized and the CT data can easily be converted to LB-voxel-data.

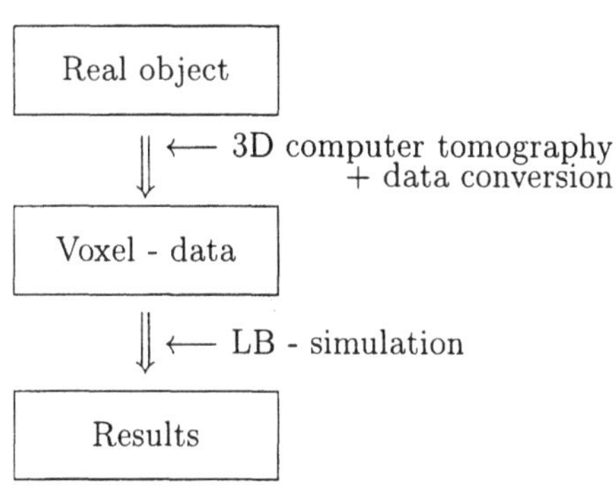

3. Arguments for industrial LB application

In this chapter, we will discuss different aspects of industrial applications of the LB method. At first, for the LB researchers, secondly for the industry.

3.1. Importance of industrial simulations for the LB community ...

A fact which is sometimes doubted by pure academic researchers, is the possible improvement of the numerical method during the process of practical application. The feedback on the quality of the simulation results, based on the extensive experience and/or databases compiled by industrial engineers is a good indicator of the reliability of the method.

Driven by the demand of continuous improvement, good suggestions for further developments can be given by the industrial engineers. This can lead to a fruitful cooperation with a mutual exchange of experience and finally result in a fast and efficient development of the method or its implementation.

Usually, in industry several different numerical codes are in use or under investigation. To participate in such an evaluation program is a good chance to find out the advantages and disadvantages of the LB method when compared with other commercial Navier-Stokes based codes.

Last but not least, a successful research project with an industry partner usually comes along with some financial support for the academic institutions involved. So, an increasing number of researchers can be employed, which definitely results in an improvement of the method.

3.2. ... and for the industry

LB codes are typically designed for easy applicability. The underlying scheme for the geometry discretization allows an (almost) automatic integration of arbitrary complex geometries, which can be either derived from CAD data by special software, or by 3D CT. It is no longer necessary to have a highly specialized CFD expert generating the mesh, a procedure which might easily take several weeks for complicated geometries. This can lead to a significant cost reduction during the industrial design process, and the simulation results are available usually *on the next day.*

In companies where already HPC platforms are installed, large scale simulations need to be carried out with software making optimally use of these expensive and powerful machines. Due to its cellular automata (CA) based algorithmic structure, LB codes can be implemented almost optimal for high end vector-parallel platforms.

Areas where CFD normally fails due to the impossibility of efficient mesh generation for complex geometries (e.g., simulation of heterogeneous catalytic reactions in chemical engineering) are also potential users of the LB method. The simple marker and cell approach in combination with 3D CT allows the discretization of almost every geometry and the calculation of several 10^7 lattice nodes on big HPC platforms.

4. Examples of successful industrial LB applications

In this section, we show examples of successful industrial applications of commercial and research LB codes. The applications are in the area of automobile aerodynamics (PowerFLOW and BEST), chemical engineering (BEST) and civil engineering (FLASH).

4.1. Automobile engineering (PowerFLOW / EXA Corp.)

Method: Geometry discretization from CAD data.

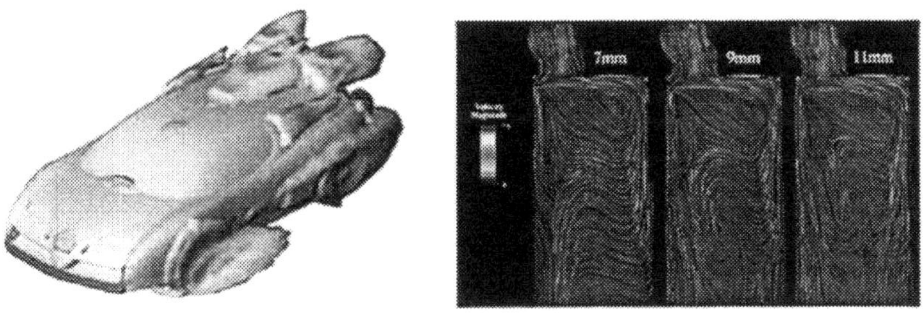

Figure 1. **Left:** external flow visualization of an Alfa Romeo Scighera (Courtesy of ItalDesign). **Right:** Streamlines through valve centerline for all three intake ports (Courtesy of EXA GmbH).

4.2. Automobile engineering (BEST / LSTM Erlangen, Invent Computing GmbH, NEC)

Method: Geometry discretization from CAD data.

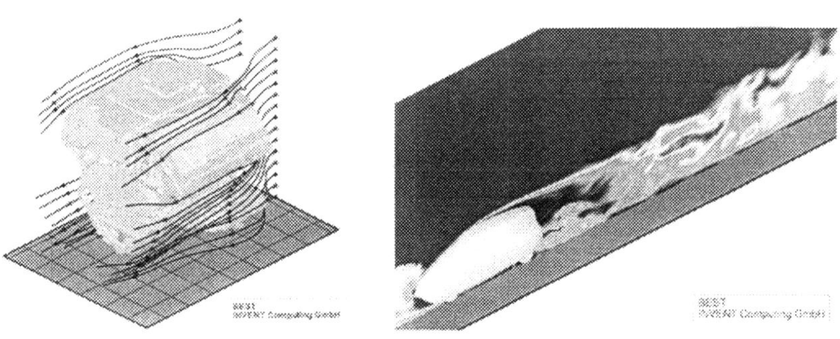

Figure 2. **Left:** Streamlines around an IC engine. **Right:** Turbulent flow field around an ASMO shape (both courtesy of INVENT Computing GmbH).

4.3. Chemical engineering (BEST / LSTM Erlangen, Invent Computing GmbH, NEC)

Method: Geometry discretization using 3D CT.

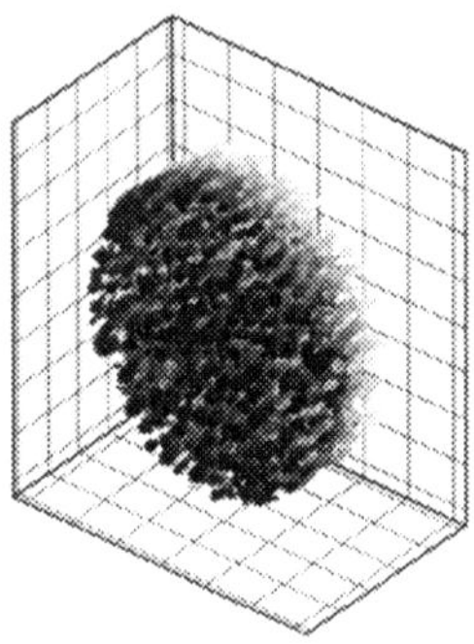

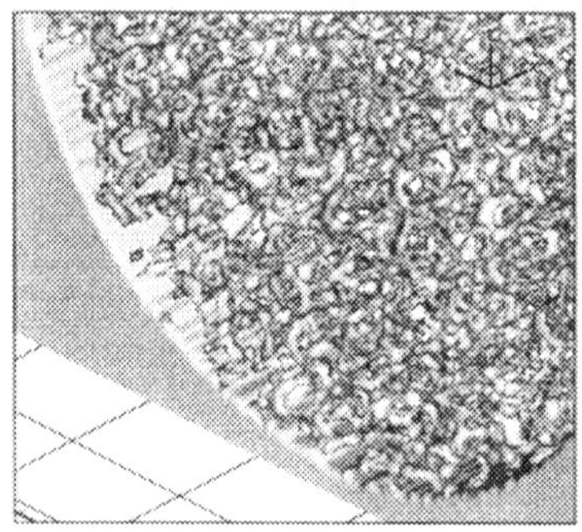

Figure 3. **Left:** Velocity iso-surface, colored with pressure. **Right:** 3D CT data (SiC-matrix of a porous burner, section).

4.4. Civil engeneering (FLASH / Lehrstuhl für Bauinformatik, TU München)

Method: Geometry discretization from CAD data.

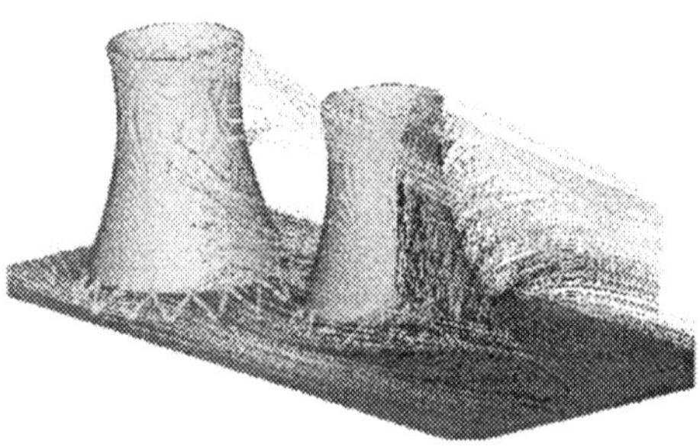

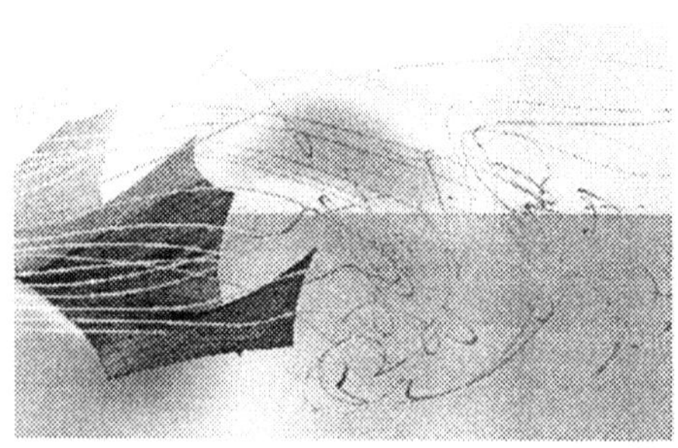

Figure 4. **Left:** Wind flow past cooling towers. **Right:** Wind flow over a tent roof (both courtesy of Lehrstuhl für Bauinformatik, TU München.)

5. Conclusion

In this paper, we presented some major arguments for the industrial application of the LB method. Two ways of almost automatically discretizing geometries (from CAD data and 3D CT) were pointed out, and several examples of successful simulations from different groups with different codes were presented.

From these promising results we believe that the LB method will become a major CFD application in the future which will help to save the industry time and money for carrying out highly accurate numerical simulations.

REFERENCES

1. G. R. McNamara, G. Zanetti, Use of the Boltzmann equation to simulate lattice gas automata, Phys. Rev. Lett. 61.
2. F. J. Higuera, J. Jiménez, Boltzmann approach to lattice gas simulations, Europhys. Lett. 9 (7) (1989) 663–668.
3. F. J. Higuera, S. Succi, R. Benzi, Lattice gas dynamics with enhanced collisions, Europhys. Lett. 9 (4) (1989) 345–349.
4. S. Succi, R. Benzi, F. Higuera, The lattice Boltzmann equation: A new tool for computational fluid-dynamics, Physica D 47 (1991) 219–230.
5. Y. H. Qian, D. d'Humières, P. Lallemand, Lattice BGK models for Navier-Stokes equation, Europhys. Lett. 17 (6) (1992) 479–484.
6. J. Bernsdorf, O. Günnewig, W. Hamm, M. Münker, Strömungsberechnung in porösen Medien, GIT Labor-Fachzeitschrift 4 (1999) 389.
7. J. Bernsdorf, F. Durst, M. Schäfer, Comparison of cellular automata and finite volume techniques for simulation of incompressible flows in complex geometries, Int. J. Numer. Met. Fluids 29 (1999) 251–264.
8. P. Bhatnagar, E. P. Gross, M. K. Krook, A model for collision processes in gases. I. small amplitude processes in charged and neutral one-component systems, Phys. Rev. 94 (3) (1954) 511–525.
9. T. Zeiser, P. Lammers, E. Klemm, Y. Li, J. Bernsdorf, G. Brenner, CFD-calculation of flow, dispersion and reaction in a catalyst filled tube by the lattice Boltzmann method, submitted to Chem. Engng. Sci .

Parallel Computational Fluid Dynamics – Trends and Applications
C.B. Jenssen et al. (Editors)

Parallel Efficiency of the Lattice Boltzmann Method for Compressible Flows

Andrew.T. Hsu, Chenghai Sun, and Akin Ecer
Department of Mechanical Engineering
Indiana University - Purdue University, Indianapolis
Indianapolis, IN 46202, USA

The lattice-Boltzmann method is believed to be particularly suitable for parallel computing. In the present work, a newly developed LB model for high Mach number compressible flows is evaluated for parallel computing efficiencies. The results show that up to 32 computer nodes, the new compressible LB model still show ideal or superlinear speedup.

1. INTRODUCTION

Significant progress has been made during the past few years in the development of lattice-Boltzmann method for computational fluid dynamics [1,2]. LB models have been successfully applied to various physical problems, such as single component hydrodynamics, multiphase and multi-component fluid flows, magneto-hydrodynamics, reaction-diffusion systems, flows through porous media, and other complex systems [3,4]. The LB method has demonstrated significant potentials and broad applicability with numerous computational advantages, such as, the simplicity of programming, the ability to incorporate microscopic interactions, and easy parallelization of algorithms.

The general LB method developed in the past suffered from the constraint of small Mach number because the particle velocities are limited to a finite set, and the macroscopic velocity and the Mach number are thus limited. Recently, we proposed a locally adaptive LB model [5], where particle velocities may have a large range of values. The support set of equilibrium distributions is determined by the mean velocity and internal energy. The fluid velocity is no longer limited by the particle-velocity set. Consequently, the model is suitable for a wide range of Mach numbers. The simulations of Sod shock-tube problem and two-dimensional shock reflection showed the model's capability of solving compressible Euler flows with shocks. This model has been extended to viscous flows that included heat transfer [6,7].

Many attempts have been made to use the LB method for parallel computing [8-10]. There are indications that the LB method is particularly suited for parallel algorithms. The computation of LB method consists of two alternant steps: particle collision and convection. The collision takes place at each node and is purely local, and is independent of information on the other nodes. The convection is a step in which particles move from a node to its neighbors according to their velocities. In terms of floating point operation, most of the computation for the LB method resides in the collision step and therefore is local. This feature makes the LB method particularly suited for parallel computations. Because of its local nature, the LB method is likely to be fully scalable on parallel machines [8-10].

The primary objective of the present paper is to develop an efficient parallel algorithm for the new high Mach number LB model. In the traditional LB models, the information is transferred to adjacent nodes, and the interface between two blocks is well defined, usually containing one or two columns. In the present model, however, due to the adaptive nature of the particle velocities, particles may be sent to nodes far away when the mean velocity is high. Information on only the block boundaries is not sufficient, and one must consider the interactions between interior points of the blocks. Therefore, information from interior points needs to be passed to the adjoining blocks. Because of these features of the compressible flow LB models, new parallel structures are designed in the present work.

2. LB MODEL FOR COMPRESSIBLE FLOWS

Conventionally, the LB method solves a discretized BGK model of the Boltzmann equation, where the unknown variable is the particle density distribution function f($\boldsymbol{x}$, $\boldsymbol{c}_j$, t), where $\boldsymbol{x}$ is the location of the lattice node, and $\boldsymbol{c}_j$ is the particle velocity. In conventional LB models, the particle velocity magnitude is restricted to $c_j=l/\Delta t$, where l is the length of the side of the lattice. The macroscopic velocity obtained from this model can only be less than c_j. On the other hand the speed of sound is in general the order of $l/\Delta t$. Thus the Mach number of the solution is severely limited and high-speed compressible flows cannot be solved.

In order to overcome this limitation on the macroscopic velocity, we introduce a larger particle velocity set, S={$\boldsymbol{r}$}, into the present model, where $\boldsymbol{r}$ is the migrating velocity of the particles. The set S={$\boldsymbol{r}$} is discrete because the nodes of lattice are discrete. The migrating velocity $\boldsymbol{r}$, unlike $\boldsymbol{c}_j$, is unrestricted so that the particles are allowed travel any number of lattice lengths. Once the velocity is defined, the momentum of the particle is determined.

The migrating velocity $\boldsymbol{r}$, can only have discrete values; this causes errors in the macroscopic solution. In order to minimize the discretization error, we introduce a continuous particle velocity, ξ, for the evaluation of particle momentum, $m\xi$, and particle energy, $m\zeta$, where m is particle mass. The difference between ξ and $\boldsymbol{r}$, as will be explained in detail later, is of the order of lattice size. The migrating velocity, $\boldsymbol{r}$, is used to calculate the location of the particle, and the continuous particle velocity, ξ, is used to calculate the exact particle momentum.

We have ξ and $\boldsymbol{r} \in D_1$, where D_1 is a bounded domain in $\Re^2$ for two-dimensional flow; $\zeta \in D_0$, where D_0 is a bounded domain in $\Re$. In the standard LB model, space, time and the particle velocity are all discrete. If we let ξ and ζ take the discrete values $\boldsymbol{r}$ and $\boldsymbol{r}^2/2$, respectively, then the present model will be consistent with the conventional LB model; however, the velocity set, S($\boldsymbol{r}$), is still larger than that of the conventional model.

With the above definition of velocities, momentum, and energy, we now define f($\boldsymbol{x}$, $\boldsymbol{r}$, ξ, ζ, t) as the particle density distribution function for particles located at $\boldsymbol{x}$, with a continuous velocity ξ, a discrete migrating velocity $\boldsymbol{r}$, and specific energy ζ. These particles will move to $\boldsymbol{x} + \boldsymbol{r}\Delta \boldsymbol{t}$ after Δt, and transporting with them a momentum $m\xi$ and energy $m\zeta$.

The macroscopic quantities, i.e., mass ρ, momentum $\rho \boldsymbol{v}$ and energy ρE, are defined as

$$\boldsymbol{Y} \equiv \sum_{\boldsymbol{r}} \int_{D} \eta f(x, \boldsymbol{r}, \xi, \zeta, t) d\xi d\zeta , \tag{1}$$

where $D=D_1\times D_0$, $\boldsymbol{Y} \equiv (\rho, \rho\boldsymbol{v}, \rho E)$, and $\eta \equiv (m, m\xi, m\zeta)$.

In a GBK type of LB model, the Boltzmann equation is written as

$$f(\boldsymbol{x}+\boldsymbol{r}\Delta t, \boldsymbol{r}, \xi, \zeta, t+\Delta t) - f(\boldsymbol{x}, \boldsymbol{r}, \xi, \zeta, t) = \Omega . \tag{2}$$

The collision operator is given as

$$\Omega = -\frac{1}{\tau}[f(\boldsymbol{x}, \boldsymbol{r}, \xi, \zeta, t) - f^{eq}(\boldsymbol{x}, \boldsymbol{r}, \xi, \zeta, t)],$$

where $f^{eq}(\boldsymbol{x}, \boldsymbol{r}, \xi, \zeta, t)$ is the equilibrium distribution, which is completely determined by the macroscopic variables such as the fluid density, momentum and energy. Details of the compressible flow solution procedure used in this work is given in Reference [18].

Using the Chapman-Enskog expansion, the following set of Navier-Stokes equations can be recovered from Eq. (2) [5,11]:

$$\frac{\partial \rho}{\partial t} + \mathrm{div}(\rho \boldsymbol{v}) = 0, \quad (3)$$

$$\frac{\partial \rho \boldsymbol{v}}{\partial t} + \mathrm{div}(\rho \boldsymbol{v}\boldsymbol{v}) + \nabla p = \mathrm{div}\{\mu[\nabla \boldsymbol{v} + (\nabla \boldsymbol{v})^{\mathrm{T}} - (\gamma-1)\mathrm{div}\boldsymbol{v}\boldsymbol{I}] + \mathrm{O}(\boldsymbol{v}\boldsymbol{v}'_k\boldsymbol{v}'_k)\}, \quad (4)$$

$$\frac{\partial \rho E}{\partial t} + \mathrm{div}(p\boldsymbol{v} + \rho E\boldsymbol{v}) = \mathrm{div}\{\mu\boldsymbol{v}\cdot[\nabla \boldsymbol{v} + (\nabla \boldsymbol{v})^{\mathrm{T}} - (\gamma-1)\mathrm{div}\boldsymbol{v}\boldsymbol{I}]\} +$$
$$\mathrm{div}\{\kappa\nabla e - (\gamma-1)e\nabla\kappa + \mathrm{O}(\boldsymbol{v}'_k\boldsymbol{v}'_k)\}, \quad (5)$$

where

$$\mu = \kappa = \Delta t\,[\tau - (1/2)]\sum_{v} m b_v d_v (1/D) c'^2_v, \quad (6)$$

$\boldsymbol{I}$ is a second-order unit tensor; D is space dimension; μ and κ are respectively viscosity and heat diffusivity. In Eq. (5) the first term and the second term of right-hand side correspond respectively to the dissipation and the heat diffusion.

3. NUMERICAL SIMULATION

If we regard the viscous terms and the diffusion terms of the right hand sides of Eqs. (4) and (5) as the discretization error the equations (3-5) become an inviscid Euler system. In fact, the viscosity and diffusivity are of order $[\tau - (1/2)]\, l^2/\Delta t$, where l is the length of the lattice and Δt is the unit time. In two previous papers [5,6] we simulated Sod shock-tube problem and two-dimensional shock reflection. The numerical results agree well with exact solutions. In the following, we simulate two more complex flows with strong shocks under the condition $\tau = 1$ and $\gamma = 1.4$.

A double Mach reflection of a strong shock (an normal shock passing a wedge) [14] is calculated on a 360×140 hexagonal lattice. Figure 1 shows the density contours at $t = 200^{\text{th}}$ iteration. Complex features, such as oblique shocks, contact discontinuity and triple points are well captured and are in good agreement with results obtained by a upwind method [14] and a gas kinetic method [13].

4. PARALLELIZATION

A parallel version of the compressible flow LB solver is developed using domain decomposition, which is illustrated in Fig. 3. In a standard LB model the particle velocities are constant, independent of lattice nodes and time, for example, the two-level LB models in Refs. [15-17] shown in Fig. 4. An interface of two columns is sufficient for data exchanges between two adjoining blocks. In the present model, due to the adaptive nature of the particle velocities, the velocities of the particles moving from one block to another vary from node to node and from time to time, depending on the mean velocity; therefore the interface is irregular (see Fig. 5), and in fact include interior block lattice points.

In order to transfer data from the interior points of one block to interior points of another block, six one-dimensional buffers are created to store the information to be passed to the adjoining blocks: two buffers for the coordinates (in x and y directions) of the node in the adjoining block to which particles move, one for the mass, two for the momentum, and one for the energy to be added to this node. The buffers are passed to the destination block by PVM commands after all the calculations have been done in the block in consideration.

We computed the double Mach reflection case on a PC cluster of 32 Pentium II 600 processors on 480×140 and 960×280 lattices. The density, pressure, and entropy distributions on the 960×280 lattice at $t = 500^{th}$ iteration are shown in Fig. 2. Because of the increased grid resolution, the shocks and the contact surface in Fig. 2 (a) are much finer compared to Fig. 1.

Table 1 shows the CPU time per iteration and the parallel efficiency versus the number of processors for lattice I (480×140) and lattice II (960×280). The parallel efficiency is calculated based on the CPU time of the virtual machine of one processor. For lattice I the efficiencies are nearly 100 percent. For lattice II the efficiency goes up and then slightly goes dawn as the number of processor increases. The efficiencies are more than 100 percent. The possible reason is that on a large lattice the paging efficiency on a single processor may drop. And if the memory is not sufficient it takes CPU time to swape.

Figure 6 shows the speed-up. With the high efficiencies reported above, the speed-up is nearly ideal. The number of nodes of the lattice II is four times that of lattice I. The CPU time ratios shown in the Table 2 are also about four. The total time is proportional to the total number of nodes.

4. CONCLUSIONS

An adaptive LB model for high-speed compressible flows has been implemented on parallel computers. In contrast to standard LB models, this model can handle flows over a wide range of Mach numbers and capture strong shock waves. The present results agree well with those of other computational methods. This compressible flow LB model is of same efficiency as standard LB models but consumes less computer memory. The total computation time is proportional to the total number of nodes. For the cases tested, high parallel efficiency is achieved. The model appears to be a promising scheme for large-scale parallel computational fluid dynamics.

ACKNOWLEDGEMENT

This work was funded by NASA Glenn Research Center's HPCC Program under grant NAG3-2399

REFERENCES

[1] H. Chen, S. Chen, and W. Matthaeus, Phys. Rev. A, **45** (1992), R5339.
[2] Y. H. Qian, D. d'Humières, and P. Lallemand, Europhys. Lett., **17** (1992), 479.
[3] S. Chen and G.D. Doolen, Annu. Rev. Fluid Mech., **30** (1998), 329.
[4] Y. H. Qian, S. Succi, and S. A. Orszag, Annu. Rev. of Comput. Phys. III, ed. Dietrich, W. S. (1995), 195.
[5] C. H. Sun, Phys. Rev. E, **58**(6) (1998), 7283.
[6] C. H. Sun, Phys. Rev. E, **61**(3) (2000), 2645.
[7] C. H. Sun, Chin. Phys. Lett., **17**(3) (2000), 209.
[8] G. Amati, S. Succi, and R. Piva, Inter. J. of Modern Phys. C, **8**(4) (1997), 869.
[9] N. Satofuka, T. Nisihioka, and M. Obata, in: Parallel Computational Fluid Dynamics, Recent Development and Advences Using Parallel computers, D. R. Emerson, A. Ecer, J. Peraux, N. Satofuka and P. Fox, editors, 1998 Elsevier Science, 601.
[10] N. Satofuka and T. Nisihioka, in: Parallel Computational Dynamics, Development and Applications of Parallel Technology, C. A. Lin, A. Ecer, J. Peraux, N. Satofuka and P. Fox, editors, 1999 Elsevier Science, 171.
[11] D. Bernardin, O. Sero-Guillaume, and C. H. Sun, Physica D, **47** (1991), 169.
[12] P. Woodward and P. Colella, J. Comput. Phys., **54** (1984), 115.
[13] K. Xu, L. Martinelli, and A. Jameson, J. Comput. Phys., **120** (1995), 48.
[14] P. Colella, J. Comput. Phys., **87** (1990), 171.
[15] C. H. Sun, Z. J. Xu, B. G. Wang, and M. Y. Shen, Commun. Nonlinear Sci. & Numer. Simul., **2** (1997), 212.
[16] C. H. Sun, Acta Mechanica Sinica, **30** (1998), 20. (in Chinese)
[17] C. H. Sun, Chinese J. Numer. Math. and Appl. **21** (1999), 9.
[18] A. Hsu and C.H. Sun, "Parallel Computation of Compressible Flows Using Lattice Boltzmann Method," To appear in *International Journal of Computational Fluid Dynamics*.

Table 1.
Parallel performance on PC600.

Number of lattice nodes	n: Number of processors	T_n: CPU time (sec/per iteration)	Parallel efficiency (T_1/nT_n, %)
I 480X140	1	2.4108	100.00
	4	0.59708	100.94
	8	0.29736	101.34
	16	0.14952	100.77
	32	0.07600	99.13
II 960X280	1	10.821	100.00
	4	2.3788	113.72
	8	1.1863	114.02
	16	0.59614	113.45
	32	0.30018	112.65

Table 2.
CPU time ratio (960X280 lattice to 480X140 lattice).

n	1	4	8	16	32
$T_n(II)/T_n(I)$	4.4886	3.9841	3.9804	3.9870	3.9497

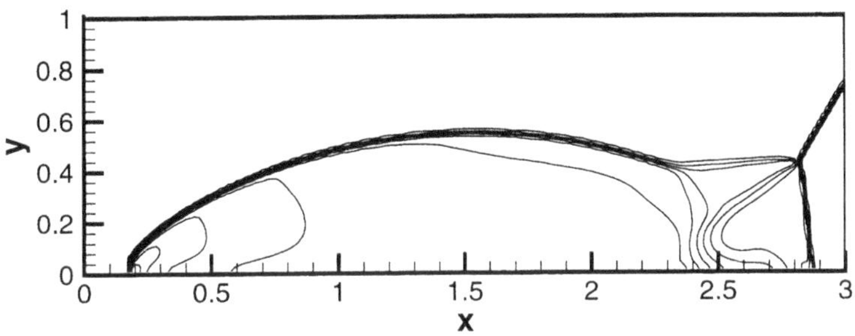

Figure 1. Density distribution for Mach 10 reflection shock on lattice 360×140 at 200th iteration.

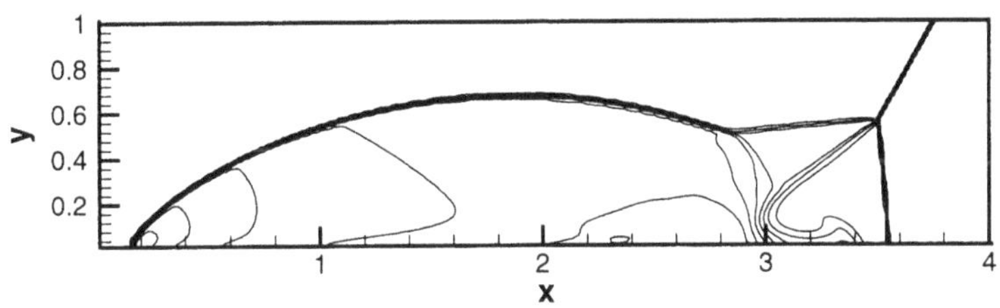

Figure 2. (a) Density distribution for Mach 10 reflection shock on lattice 960×280 at 500th iteration.

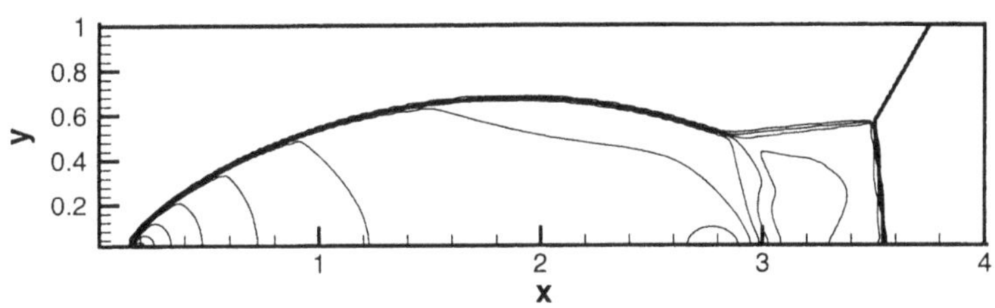

Figure 2. (b) Pressure distribution for Mach 10 reflection shock on lattice 960×280 at 500th iteration.

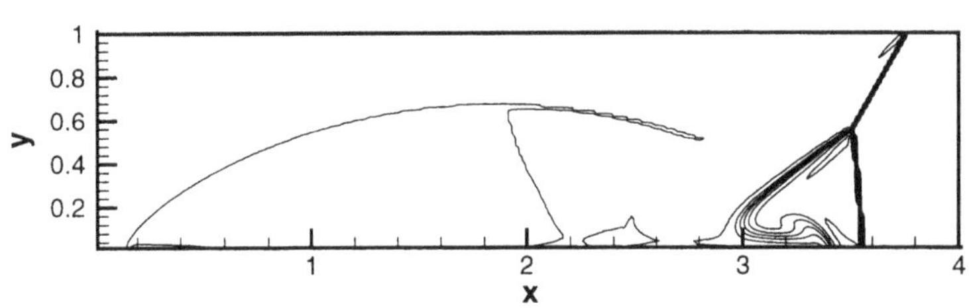

Figure 2. (c) Entropy distribution for Mach 10 reflection shock on lattice 960×280 at 500th iteration.

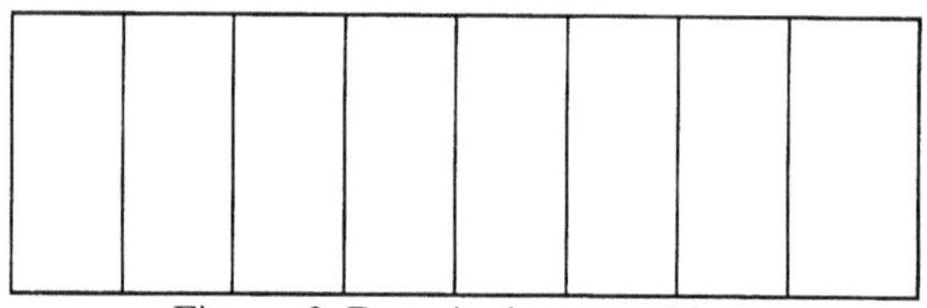

Figure. 3. Domain decomposition.

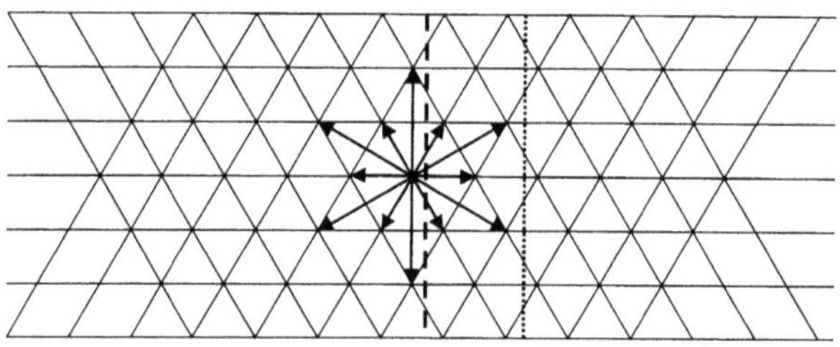

Figure 4. Particle velocities of a standard two-level LB model.
An interface of two columns is sufficient for parallel computing.

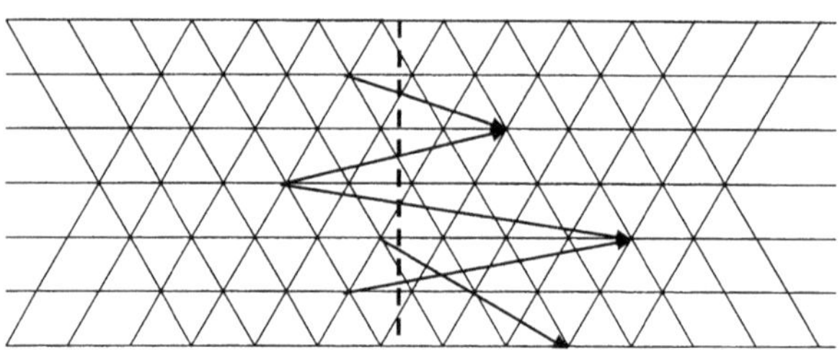

Figure 5. Present model: particles moving from domain i to domain ii.
The interface is irregular.

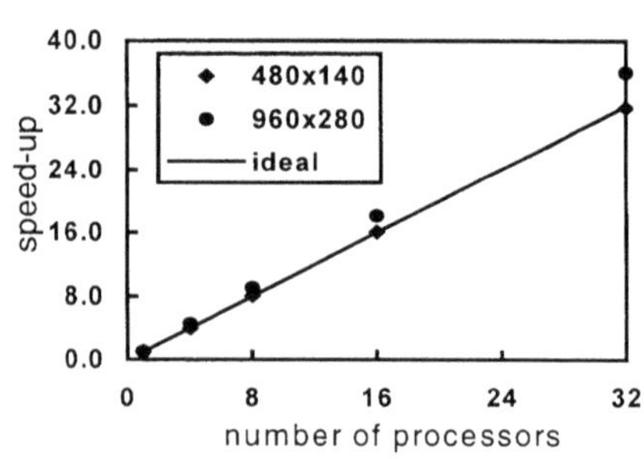

Figure 6. Speed-up

Parallel Computational Fluid Dynamics - Trends and Applications
C.B. Jenssen et al. (Editors)

Turbomachine flow simulations with a multiscale Lattice Boltzmann Method

F. Mazzocco[a], C. Arrighetti[a], G. Amati[b], G. Bella[a], O. Filippova[a] and S. Succi[a]

[a] Dip. di Meccanica e Aeronautica, Universitá di Roma "La Sapienza", I-00184 Rome, Italy;
[b] CASPUR, p.le A. Moro 5, I-00185 Roma, Italy;
[c] Dip. di Ingegneria Meccanica, Universitá di Roma "Tor Vergata", I-00133 Rome, Italy;
[d] Institute of combustion and gas dynamics, University of Duisburg, D-47048, Duisburg, Germany;
[e] Istituto Applicazioni Calcolo "M. Picone", Viale Policlinico 137, I-00161 Rome, Italy.

Multiscale Lattice Boltzmann schemes for fluid dynamic applications are described in connection with a novel application to axial compressor flows. Preliminary results in two-dimensions are presented, which indicate that these schemes hold good potential for the simulation of turbomachines.

1 INTRODUCTION

Lattice kinetic theory, and most notably the Lattice Boltzmann (LB) method, have received considerable interest in the last decade as an efficient method to compute a variety of fluid flows, ranging from low-Reynolds flows in porous media to highly turbulent flows [1, 3]. Until recently, LB applications to flows of engineering interest have been held back by a certain lack of flexibility to accomodate non-uniform grids.

To circumvent this limitation, a number of variants have been proposed. Most of these variants are based on a combination of LB with consolidated finite volume or finite difference techniques. This merge has considerably extended the range of application of the LB method at a fairly reasonable cost in terms of computational complexity. Nonetheless, they can accomodate only relatively smooth variations of the flow field because large deformations of the non-uniform mesh may result in numerical instabilities.

Many phenomena of physical and engineering interest exhibit violent excursions over highly localized regions (boundary layers, shock fronts) which require a correspondingly highly clustered mesh. A popular response to this kind of needs in modern computational fluid dynamics is provided by *unstructured* meshes, namely discrete grids where the number of neighbours of a given node may change from place to place. This allows much stronger distorsions of the computational grid, but only at the expenses of a significantly more complex data-structure. Another popular possibility is provided by *locally embedded*

grids, namely grids in which the local connectivity (number of neighbours) is unchanged but the lattice spacing is refined or coarsened locally, typically in steps of two for practical purposes. Local embedding is a specific instance of a more general framework known as *multiscale algorithms*.
Multiscale LBE schemes have been first developed by Filippova et al [4], who tested and validated it for moderate-Reynolds flows around cylinders. In this paper we shall demonstrate the viability of the multiscale LBE scheme for a new type of application, namely two-dimensional flows in a axial compressor cascade.

2 BASICS OF THE MULTISCALE LB METHOD

Our starting point is the lattice BGK formulation of fluid dynamics [5]:

$$f_i(\vec{x}+\vec{c}_i,t+1)-f_i(\vec{x},t)=-\omega[f_i-f_i^e](\vec{x},t) \tag{1}$$

Since this has been described at length in many papers, only essential information shall be provided below. Here f_i is a set of discrete populations representing the probability of finding a particle at position $\vec{x}$ at time t moving along the direction identified by the discrete speed $\vec{c}_i$. The right hand side of (1) represents the relaxation to a local equilibrium f_i^e in a time lapse ω^{-1}. Once the discrete populations are known, fluid density and speed are obtained by (weighted) sums all over the set of discrete speeds:

$$\rho=\sum_i f_i \tag{2}$$

$$\rho\vec{u}=\sum_i f_i\vec{c}_i \tag{3}$$

For future purpose, it proves convenient to recast the LBGK dynamics as a two-step (Collide+Stream) process:

$$f_i'(\vec{x},t)=(1-\omega)f_i(\vec{x},t)+\omega f_i^e(\vec{x},t) \tag{4}$$

$$f_i(\vec{x}+\vec{c}_i,t+1)=f_i'(\vec{x},t) \tag{5}$$

where prime means 'after-collision'.
The multiscale implementation of the LB method is based on the following steps. First, we introduce a coarse and fine-grained distribution F_i and f_i respectively. For simplicity, we shall consider the simplest instance in which only one level of refinement is used, so that the fine-grain distribution lives in a mesh with spacing $1/n$ as compared to the coarse-grained one. The simple case $n=2$ is illustrated in fig.1.

We identify three types of nodes:

1. Coarse-nodes (common to coarse and fine grids)
2. Fine-nodes (fine grid only)
3. Coarse-to-fine boundary nodes (at the interface between the two grids).

First, we perform a coarse-grain evolution (Collision+Streaming) step:

$$F_i'(\vec{X},t)=(1-\Omega)F_i(\vec{X},t)+\Omega F_i^e(\vec{X},t) \tag{6}$$

$$F_i(\vec{X}+\vec{C}_i,t+1) = F_i'(\vec{X},t) \tag{7}$$

where $\vec{X}$ denotes a generic coarse node, $\vec{C}_i = n\vec{c}_i$ is the associated discrete speed (connectivity) and Ω is the relaxation parameter in the coarse lattice.
This provides all the populations in the coarse-grain nodes. Next we need to perform the fine-grain dynamics, actually n steps of size $\delta t = 1/n$ each. To this end, we need the values at the boundary nodes at times $t = 0, 1/n, 2/n, \ldots 1 - 1/n$.
Let us begin by considering the first sub-step. The boundary values at time t are obtained by a simple space interpolation from the coarse nodes also at time t:

$$\tilde{f}_i(t) = \mathcal{I}F_i(t) \tag{8}$$

where $\mathcal{I}$ denotes some coarse-to-fine interpolator (typically first order).
These boundary values serve as an effective boundary initial condition to perform completely the first sub-step of the fine-grain dynamics:

$$f_i'(\vec{x},t) = (1-\omega)f_i(\vec{x},t) + \omega f_i^e(\vec{x},t) \tag{9}$$

$$f_i(\vec{x}+\vec{c}_i,t+1/n) = f_i'(\vec{x},t) \tag{10}$$

with the proviso that whenever a boundary node is involved, the corresponding interpolated value $\tilde{f}_i$ is used.
In order to proceed at the same Reynolds number on both coarse and fine grids, the corresponding relaxation parameters must relate as follows:

$$\omega = \frac{\Omega}{n + \Omega(1-n)/2} \tag{11}$$

Manifestly, $\Omega = \omega$ in the limit $n = 1$.
At completion of (9) and (10) we dispose of the fine-grain distribution everywhere, including the coarse and boundary sites. Since the latter serve as a boundary initial condition, they are overwritten by a second interpolation at $t + 1/n$ by means of the previously calculated coarse values at time t and $t + 1$ (space-time interpolation).
With this second interpolate, we are ready to complete the second fine-grain sub-step:

$$f_i'(\vec{x},t+1/n) = (1-\omega)f_i(\vec{x},t+1/n) + \omega f_i^e(\vec{x},t+1/n) \tag{12}$$

$$f_i(\vec{x}+\vec{c}_i,t+2/n) = f_i'(\vec{x},t+1/n) \tag{13}$$

Apparently, n of these steps complete the task, since we now dispose of updated values on both grids at time $t + 1$.
Such a procedure would however miss a crucial ingredient, namely fluxes continuity across the coarse-fine boundary. This crucial condition is secured by imposing the continuity of the first order non-equilibrium term $f_i^{ne} = f_i - f_i^e$ which, by virtue of (1), contains both space and time first order derivatives of the equilibrium distribution to first order in the Knudsen number:

$$f_i^{ne} \sim -\omega^{-1}[\partial_t + c_{ia}\partial_a]f_i^e \tag{14}$$

As shown by Filippova et al, this leads to the following two scale transformations between the post-collisional coarse and fine-grain populations:

$$F_i' = F_i^e + (f_i' - f_i^e)\omega' \tag{15}$$

and

$$f_i' = f_i^e + (\tilde{f}_i' - \tilde{f}_i^e)\omega'^{-1} \tag{16}$$

where tilde means interpolation from the coarse grid and

$$\omega' = \frac{\omega(1-\Omega)n}{\Omega(1-\omega)} \tag{17}$$

is a rescaling factor.
Obviously, $f_i^e = F_i^e$ at the same node because the local equilibrium depends on the macroscopic properties of the fluid. Note that the transformations (15) and (16) reduce to identities in the limit $n = 1$.
The final one-step algorithm reads as follows:

1. Move and Collide F
2. Scale F to f
3. For all subcycles k=0,1,...,$n-1$ do:
   I. Interpolate coarse-to-fine
   II. Move and Collide f
4. End do
5. Rescale f to F.

Besides allowing selective grid refinements, typically around solid bodies, the previous procedure provides a potential operational candidate for kinetic-based renormalisation-group formulations of fluid turbulence, as recently developed by H. Chen et al [2].

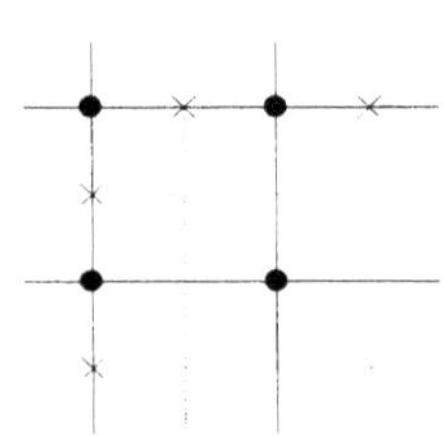

Figure 1: Coarse and fine grid interfaces for a refinement factor $n = 2$.

3 VALIDATION OF THE CODE

In order to validate the code and to estimate its computational performance, the flow around a NACA 4412 profile with small angles of attack is considered. Two different $k-\epsilon$ models, the $k-\epsilon$ RNG and the Standard $k-\epsilon$, are used and comparison with experiments is based upon data from [6].
The computational experiment is as follows: for the angle $\alpha = -0.5^o$ the flow is injected at the inlet and upper section of the computational domain with a LBGK Mach number of $U_{in}/C = 0.08$. For the angle $\alpha = 2.9^o$ the flow is injected at the inlet and lower section of the computational domain at a LBGK Mach number $U_{in}/C = 0.08$. Pressure at the sections of injection is extrapolated along the normal to the section from the outer flow. Pressure at outlet is assumed to be constant. At outlet the velocity is extrapolated along the normal from the outer flow, at the other "outlet" section the $x-$component of velocity is also extrapolated along the normal from the outer flow whereas the normal component of velocity is obtained under the condition of constant inclination α of the stream. Pressure at this section is defined according to the Bernoulli equation. Two values of the Reynolds numbers related to the length of the chord were considered, $Re = 10^6$ and $Re = 3 \cdot 10^6$, as far as in [6] the value of Reynolds number was not defined precisely ("approximately 3000000"). The computational domain consists of 160 $\times$ 111 nodes.
Grid refinement defined by the parameter $n = \delta_x^c/\delta_x^f$ is applied to a box $(65\delta_x^c, 17\delta_x^c)$ surrounding the profile with the left-down corner at (40,49). Convergence criteria is prescribed as: $max|u(t,\vec{r}) - u(t-1,\vec{r})| + |v(t,\vec{r}) - v(t-1,\vec{r})| < 10^{-6}$ in the nodes of the coarse grid. The accelerated scheme using 1 time-step on the fine grid versus 1 time-step on the coarse grid is used. The value of ω_{coarse} in the absence of turbulent viscosity is 1.95 for the simulations with the $k-\epsilon$ RNG model and 1.92 for the simulations with the Standard $k-\epsilon$ model.
The main results for the $k-\epsilon$ RNG model are collected in figures 2a, 2b and for the Standard $k-\epsilon$ model in figure 3a, 3b where the numerically and experimentally obtained pressure coefficients along a chord of profile are shown for different angles of attack. Parameter of refinement in the numerical simulations was taken as $n = 5$. In addition the numerical results for $Re = 10^6$ and $n = 3$ are shown in figure 2c with dotted lines.
As one can see from figures 2a, 2b and figures 3a, 3b both C_P curves for $Re = 10^6$ and $Re = 3 \cdot 10^6$ coincide closely. The agreement between experimental and numerical results is found to be good for both turbulence models.

4 COMPUTATIONAL EFFICIENCY AND PARALLEL COMPUTING

We wish to emphasize that grid refinement together with boundary-fitting formulation proves instrumental to ensure good agreement between experimental results and numerical results for turbulent flow simulation around blunt bodies at high Re numbers. The higher quality provided by grid refinement comes at a reasonable cost in CPU and especially in memory requirements. For instance, a case with $n = 5$ takes about 30 Mbytes of storage and about 40 minutes CPU time on a a PC AMD-K7, (500 Mhz).

Depending on the complexity of the physical problem, this corresponds to a factor 2-4 memory savings with respect to a standard uniform LB. Since the cost of updating a single LBE grid-point is independent of the size of the problem, there is a corresponding gain in CPU time too. The latter is however partly reduced by overheads due to the various interpolations between coarse and fine grids. Of course, the full potential of grid-refinement is best capitalized in three-dimensional applications. To date, no parallel version of the present multiscale LBE code has been developed. The multiscale features add significantly to the complexity of the code but in principle they should not hamper the basic locality of LBE. Therefore, the multiscale algorithm is expected to perform pretty well on parallel machines. However, as is well known, a number of issues have to be addressed, primarily load-balancing, since it is clear that highly refined regions require more computational work, an issue that simply does not exist in the uniform case. This is to be acknowledged by the domain-decomposition procedure which must secure an evenly distributed load among the various processors. This is left for future work.

References

[1] R. Benzi, S. Succi, M. Vergassola, *The lattice Boltzmann equation - theory and applications*, Phys. Reports 222, 145-197, 1992.

[2] H. Chen, S. Succi, S. Orszag, *Analysis of subgrid scale turbulence using the Boltzmann BGK kinetic equation*, Phys. Rev. E, Rap. Comm., 59(3), R2527, 1999.

[3] S. Chen, G.D. Doolen, Annual Rev. of Fluid Mechanics 30, 329, 1998.

[4] O. Filippova, D. Hanel, *Grid refinement for lattice-BGK models*, Journal of Computational Physics, 147, 219-238, 1998.

[5] Y.H. Qian, D. d'Humieres, P. Lallemand, *Lattice BGK models for the Navier-Stokes equation*, Europhys. Lett. 17(6), 479-484, 1992.

[6] R. M. Pinkerton, Calculated and measured pressure distributions over the midspan sections of the NACA 4412 airfoil, NACA Rept. N. 563, 1936.

[7] E. Baskharone, and A. Hamed, A new approach in cascade flow analysis using the finite element method, AIAA 19, N. 1, 1981.

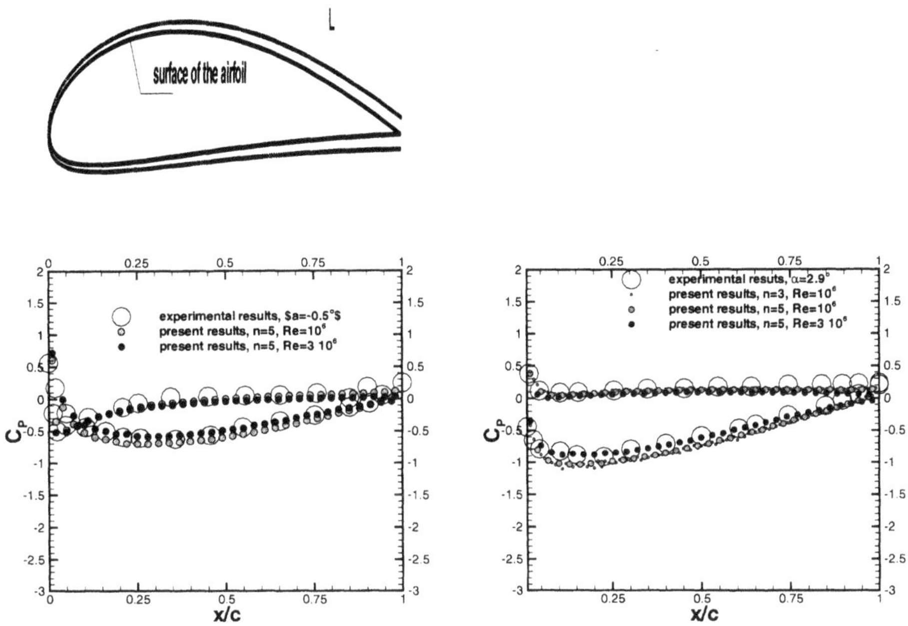

Figure 2: A. Relative positions of the surface L and the surface of the airfoil in the numerical simulation of the turbulent flow around NACA 4412 profile at $Re = 10^6$, $\alpha = 2.9^o$. B. Pressure coefficient C_P along the chord of the NACA 4412 profile, $\alpha = -0.5^o$. Solid circles - present solution, $k - \epsilon$ RNG model, open circles - experimental results. C. Pressure coefficient C_P along the chord of the NACA 4412 profile, $\alpha = 2.9^o$. Solid circles - present solution, $k - \epsilon$ RNG model, open circles - experimental results.

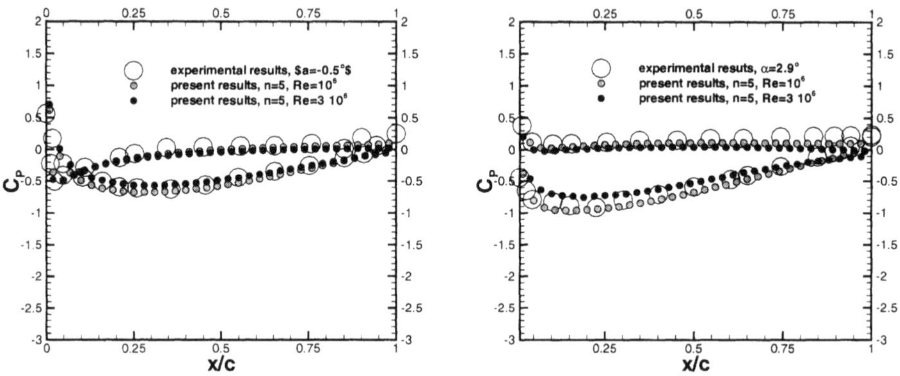

Figure 3: A. Pressure coefficient C_P along the chord of the NACA 4412 profile, $\alpha = -0.5^o$. Solid circles - present solution, Standard $k - \epsilon$ model, open circles - experimental results. B. Pressure coefficient C_P along the chord of the NACA 4412 profile, $\alpha = 2.9^o$. Solid circles - present solution, Standard $k - \epsilon$, open circles - experimental results.

Parallel Computational Fluid Dynamics – Trends and Applications
C.B. Jenssen et al. (Editors)
2001 Elsevier Science B.V.

Parallel Simulation of Three-dimensional Duct Flows using Lattice Boltzmann Method

N. SATOFUKA and M. ISHIKURA [a]

[a]Kyoto Institute of Technology,
Matsugasaki,Sakyo-ku,Kyoto 606-8585, Japan.

Numerical simulation using the lattice Boltzmann Method is presented for 3-dimensional square duct flows with/without sudden expansion. Fifteen-velocity cubic lattice model is used for the simulation. The LBGK method is able to reproduce the transition from uniform inlet flow to fully developed laminar velocity profile in the square duct and could be an alternative for solving the Navier-Stokes equations. The lattice Boltzmann method is parallelized by using domain decomposition and implemented on a distributed memory computer, HP Exemplar V - class with 32 CPU. A speedup of 11.8 was obtained on a 16 processors. Further investigation is needed on the accuracy and efficiency of cubic LBGK method.

1. INTRODUCTION

There are two approaches in CFD. Conventional one is to solve the Navier-Stokes equations based on the continuum assumption. In the case of incompressible computations one has to solve the momentum equation together with the Poisson equation to satisfy divergence free condition. The second approach starts from the Boltzmann equation using the Lattice Gas Method. The Boltzmann equation can recover the Navier-Stokes equations by using the Chapman-Enskog expansion. The scheme is ideal for massively parallel computing because the updating of a node only involves its nearest neighbors. Boundary conditions are easy to implement. Therefore the code is simple and can be easily written in the form suitable for parallel processing.

Among the lattice gas methods there exists the Lattice Gas Automata (LGA) [1]. LGA has fundamental difficulty in simulating realistic fluid flows obeying the Navier-Stokes equations. Beside its intrinsic noisy character which makes the computational accuracy difficult to achieve, it contains certain properties even in the fluid limit. The lattice gas fluid momentum equations cannot be reduced to the Navier-Stokes equations because of two fundamental problems. The first is the non-Galilean invariance property due to the density dependence of the convection coefficient. This limits the validity of the LGA method only to a strict incompressible region. Second the pressure has an explicit and unphysical velocity dependence. To avoid some of these problems, several lattice Boltzmann (LB) models have been proposed. [2]–[4] The main feature of the LB method is to replace the particle occupation variables n_i (Boolean variables) by the single-particle distribution function (real variables) $f_i = \langle n_i \rangle$, where $\langle \ \rangle$ denotes a local ensemble average, in the evolution equation, i.e., the lattice Boltzmann equation. The LB model proposed by Chen et al [5] and Qian et al [6] applies the single relaxation time approximation first introduced by Bhatnager, Gross, and Krook in 1954 [7], to greatly simplify the collision operator. This model is called the lattice BGK (LBGK) model.

In the present work, the LBGK method is used to simulate three-dimensional square duct flow with/without sudden expansion on a massively parallel computer, a HP Exemplar V - class. The accuracy, physical fidelity, and efficiency of the LBGK method are investigated. Speedup and CPU time are presented for a one-dimensional domain decomposition in three-dimensional computation.

2. LATTICE BOLTZMANN METHOD FOR THREE-DIMENSION

A cubic lattice [3] with unit spacing is used on which each node has fourteen nearest neighbors connected by fourteen links. Particles can only reside on the nodes and move to their nearest neighbors along these links in the unit time. Hence, there are two types of moving particles. Particles of type 1 move along the axes with speed $|\mathbf{e}_{1i}| = 1$ and particles of type 2 move along the links to the corners with speed $|\mathbf{e}_{2i}| = \sqrt{3}$. Rest particles with speed zero are also allowed at each node. The occupation of the three types of particles is represented by the single-particle distribution function, $f_{\sigma i}(\mathbf{x}, t)$, where subscripts σ and i indicate the type of particle and the velocity direction, respectively. When $\sigma = 0$, there is only f_{01}. The distribution function, $f_{\sigma i}$, is the probability of finding a particle at node $\mathbf{x}$ and time t with velocity $\mathbf{e}_{\sigma i}$. The particle distribution function satisfies the lattice Boltzmann equation

$$f_{\sigma i}(\mathbf{x} + \mathbf{e}_{\sigma i}, t + 1) - f_{\sigma i}(\mathbf{x}, t) = \Omega_{\sigma i} \tag{1}$$

where $\Omega_{\sigma i}$ is the collision operator representing the rate of change of the particle distribution due to collisions. According to Bhatnagar, Gross, and Krook (BGK) , the collision operator is simplified using the single time relaxation approximation. Hence, the lattice Boltzmann BGK (LBGK) equation (in lattice unit) is

$$f_{\sigma i}(\mathbf{x} + \mathbf{e}_{\sigma i}, t + 1) - f_{\sigma i}(\mathbf{x}, t) = -\frac{1}{\tau}\left[f_{\sigma i}(\mathbf{x}, t) - f_{\sigma i}^{(0)}(\mathbf{x}, t)\right] \tag{2}$$

where $f_{\sigma i}^{(0)}(\mathbf{x}, t)$ is the equilibrium distribution at $\mathbf{x}, t$ and τ is the single relaxation time which controls the rate of approach to equilibrium. The density per node, ρ, and the macroscopic velocity, $\mathbf{u}$, are defined in terms of the particle distribution functions by

$$\rho = \sum_{\sigma}\sum_{i} f_{\sigma i}, \qquad \rho\mathbf{u} = \sum_{\sigma}\sum_{i} f_{\sigma i}\mathbf{e}_{\sigma i} \tag{3}$$

A suitable equilibrium distribution can be chosen in the following form for particles of each type

$$f_{01}^{(0)} = \rho\alpha - \frac{3}{2}\rho\mathbf{u}^2$$

$$f_{1i}^{(0)} = \rho\beta + \frac{1}{3}\rho(\mathbf{e}_{1i}\cdot\mathbf{u}) + \frac{1}{2}\rho(\mathbf{e}_{1i}\cdot\mathbf{u})^2 - \frac{1}{6}\rho\mathbf{u}^2 \tag{4}$$

$$f_{2i}^{(0)} = \rho\frac{(1-4\beta-\alpha)}{4} + \frac{1}{12}\rho(\mathbf{e}_{2i}\cdot\mathbf{u}) + \frac{1}{8}\rho(\mathbf{e}_{2i}\cdot\mathbf{u})^2 - \frac{1}{24}\rho\mathbf{u}^2$$

Values of $\alpha = 2/9$ and $\beta = 1/9$ are used. The relaxation time is related to the viscosity by

$$\tau = \frac{6\nu+1}{2} \tag{5}$$

where ν is the kinematic viscosity measured in lattice units.

To impose pressure boundary conditions explicitly, it is better to transform the density distribution function $f_{\sigma i}$ to the pressure distribution function $p_{\sigma i}$ using $p_{\sigma i} = \rho c_s^2 f_{\sigma i}$, then Eq.(2) becomes

$$p_{\sigma i}(\mathbf{x}+\mathbf{e}_{\sigma i}, t+1) - p_{\sigma i}(\mathbf{x}, t) = -\frac{1}{\tau}\left[p_{\sigma i}(\mathbf{x}, t) - p_{\sigma i}^{(0)}(\mathbf{x}, t)\right]. \tag{6}$$

3. THREE-DIMENSIONAL DUCT FLOW

3.1. Description of the problem and boundary condition

Numerical simulations of the laminar flow development are carried out in a duct as shown in Figure 1 and also in a square duct which undergo a sudden expansion with uniform step height equal to 0.5 times the width of the inlet duct as shown in Figure 2. The noslip wall boundary conditions, $u = v = w = 0$, are used as in the three-dimensional simulations. At the inlet we assume uniform axial velocity $u = 0.1$, and the other velocity components are set to zero. The pressure at the outflow boundary is kept constant value of $p = 1.0$.

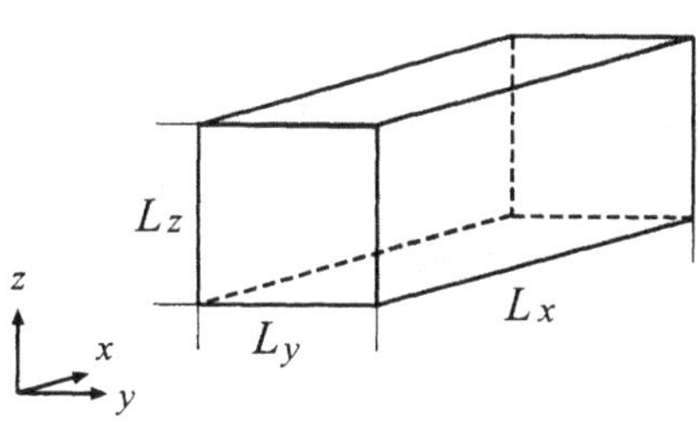

Figure 1. Computational domain of duct without sudden expansion.

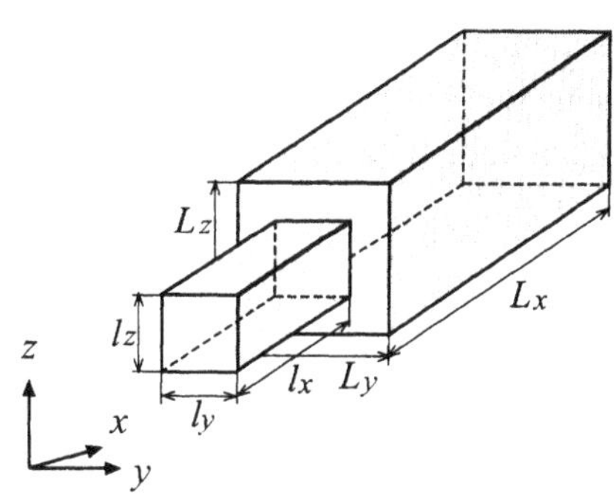

Figure 2. Computational domain of duct with sudden expansion.

Table 1
Computational domain

Re number	Computational domain	lattice points
! ! ! !! !	$L_x : L_y : L_z$	
100	10 : 1 : 1	321 !_33 !_33
200	10 : 1 : 1	641 !_65 !_65
400	40 : 1 : 1	5121 !_129 !_129

3.2. Results of simulations

In the case of a square duct without sudden expansion simulations were carried out by using a uniform cubic lattice tabulated in Table 1. Figures 3 and 4 show velocity distributions at x-z and y-z cross sections of the square duct flow, and center-line velocity profiles at several longitudinal positions for $Re = 200$ computed on a $641 \times 65 \times 65$ lattice. The transitional process from uniform to fully developed velocity profile is well captured. Figures 5 and 6 show velocity and pressure distributions at x-z and y-z cross sections for $Re = 400$. Figure 7 shows velocity distributions at x-z and y-z cross sections of sudden expansion flow for the case with $Re = 100$ computed on a $160 \times 33 \times 33$ lattice for upstream duct and $161 \times 65 \times 65$ lattice for downstream. Figure 8 shows center-line velocity profiles at $x = 5$ and $x = 10$. In this case the velocity profile is not yet fully developed due to a shortage of duct length.

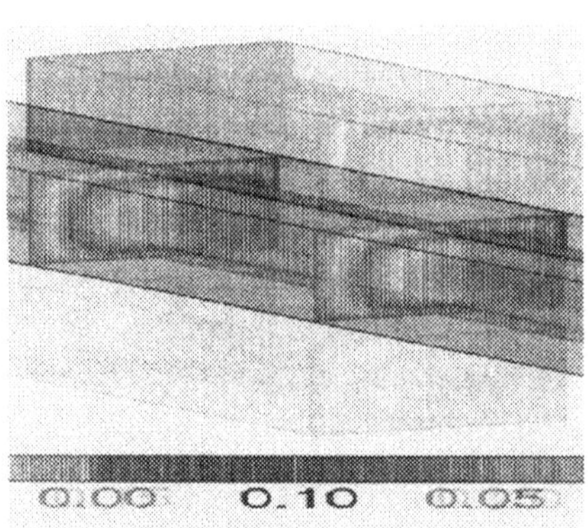

Figure 3. Velocity profiles for the mainstream direction.

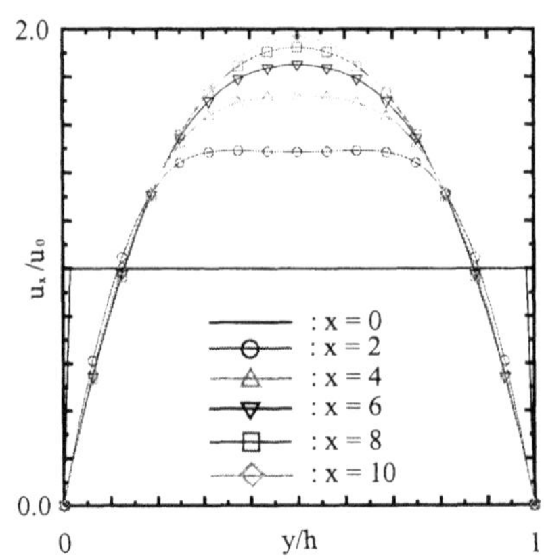

Figure 4. Development of velocity profiles with x.

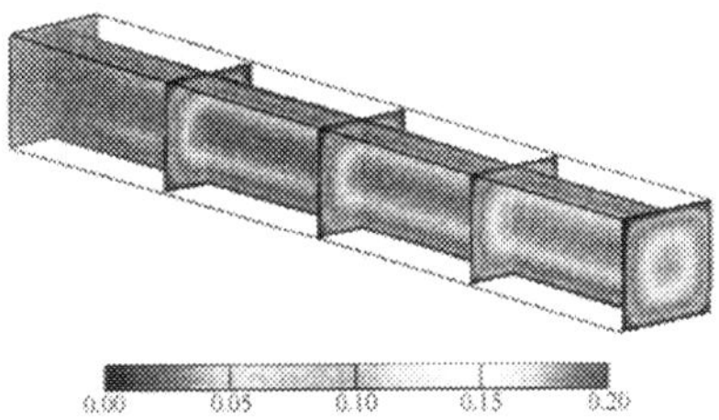

Figure 5. Velocity profiles for the main-stream direction.

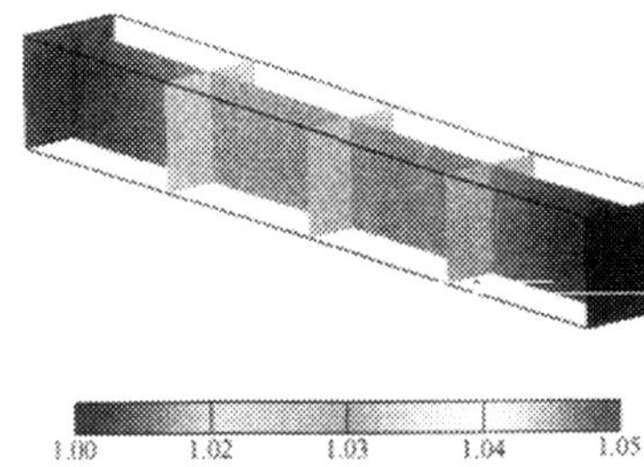

Figure 6. Pressure distribution.

4. PARALLELIZATION

4.1. Domain decomposition for three-dimension

Two-dimensional simulation shows that the longer the size of sub domain in horizontal direction, the shorter the CPU time. In other words, the longer the outer loop dimension of data, the shorter the CPU time [9]. For simulation of duct flow it is natural to decompose the computational domain in longitudinal direction (x-axis) as shown in Figure 9. Although fifteen distribution functions are placed on each node for the cubic lattice model, only five variables have to be sent across each domain boundary.

4.2. Speedup and CPU time

The speedup and CPU time are shown in Figure 10 and Table 2. For a $641 \times 65 \times 65$ lattice speedup of 11.8 was obtained on a 16 processor HP V - class. In spite of the fact that the number of lattice nodes is small we can get higher speedup as compared with two-dimensional simulation.

5. CONCLUSION

Numerical simulations of three-dimensional square duct with/without sudden expansion are carried out by using the lattice Boltzmann BGK method. The LBGK method is able to

Table 2
CPU time and speedup

Number of CPU	1	2	4	8	16
CPU time	18336	8921	4585	2660	1522
Speedup	1.0	2.0	4.0	6.9	11.8

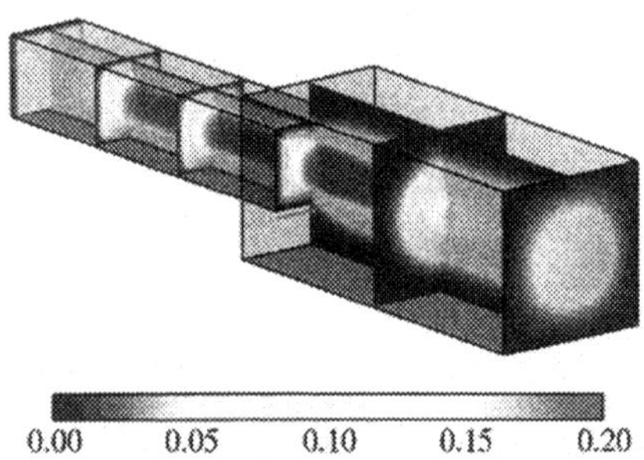

Figure 7. Velocity profiles for the main-stream direction.

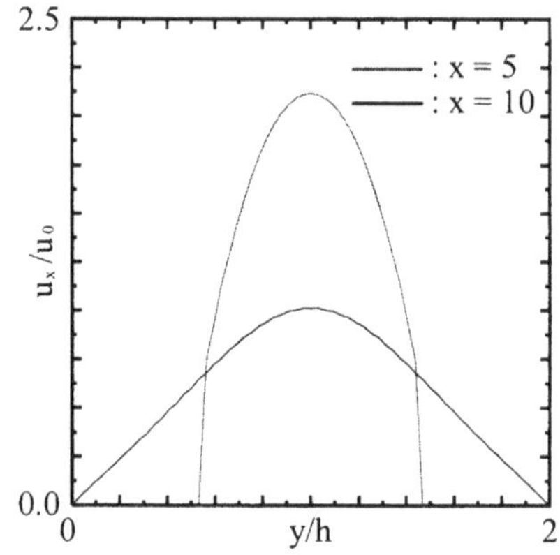

Figure 8. Development of velocity profiles with x.

reproduce the transition from uniform inlet flow to fully developed laminar velocity profile and could be an alternative for solving the Navier-Stokes equations. Parallel computations of three-dimensional duct flows were carried out on a HP Exemplar V - class with 32CPU. A parallel speed up of 11.8 was obtained with 16CPU.

Further investigation is needed on the accuracy and efficiency of cubic lattice BGK model.

ACKNOWLEDGEMENT

This study was supported in part by the Research for Future Program (97P01101) from Japan Society for the Promotion of Science.

REFERENCES

1. U. Frisch and B. Hasslacher and Y. Pomeau, Lattice-Gas Automata for the Navier-Stokes Equation, *Phys. Rev. Lett. 56*, (1986) 1505.
2. F. Heiguere and J. Jimenez, Simulating the Flow around a Cylinder with Lattice Boltzmann Equation, *Europhys, Lett. 9*, (1989) 663.
3. F. Heiguere and S. Succi, Simulating the Flow around a Cylinder with a Lattice Boltzmann Equation, *Europhys, Lett. 8*, (1989) 517.
4. G. McNamara and G. Zanetti, Use of the Boltzmann Equation to Simulate Lattice-Gas Automata, *Phys. Rev. Lett. 61*, (1988) 2332.
5. H. Chen and W.H. Matthaeus, Recovery of the Navier-Stokes Equation Using a Lattice-Gas Boltzmann Method, *Phys. Rev. A. 45*, (1992) 5539.
6. Y.H. Qian and D. D'Humieres and P. Lallemand, Lattice BGK Models for Navier-Stokes Equation, *Europhys. Lett. 17*, (1992) 479.

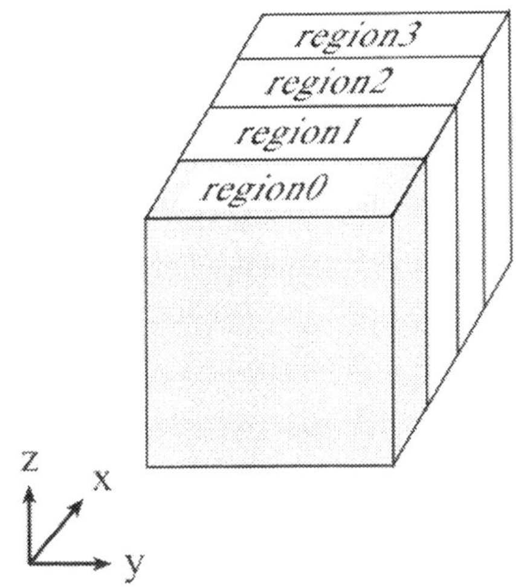

Figure 9. Domain decomposition.

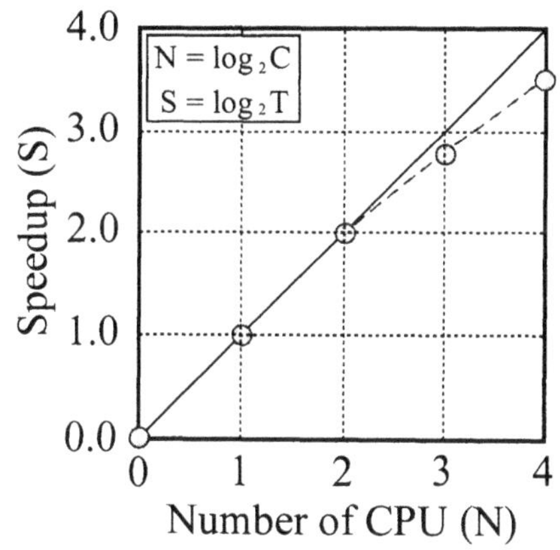

Figure 10. Number of CPU and speedup ratio.

7. P.L. Bhatnagar and E.P. Gross and M. Krook, A model for Collision Processes in Gases. 1. Small amplitude Processes in Charged and Neutral one-component Systems, *Phys. Rev. 94*, (1954) 511.
8. S. Hou and Q. Zou and S. Chen and G. Doolen and A.C. Cogley, Simulation of Cavity Flow by the Lattice Boltzmann Method, *J. Comput. Phys. 118*, (1995) 329.
9. N. Satofuka and T. Nishioka and M. Obata, Parallel Computation of Lattice Boltzmann Equations for Incompressible Flows, *Proc. Parallel CFD '97 Conference*, May 19-21, (1997) 601.

Parallel Computational Fluid Dynamics – Trends and Applications
C.B. Jenssen et al. (Editors)

Parallel computation of rising bubbles using the lattice Boltzmann method on workstation cluster

Tadashi Watanabe and Kenichi Ebihara

Research and Development Group for Numerical Experiments,
Center for Promotion of Computational Science and Engineering,
Japan Atomic Energy Research Institute,
Tokai-mura, Naka-gun, Ibaraki-ken, 319-1195, Japan

The three-dimensional two-phase flow simulation code based on the two-component two-phase lattice Boltzmann method, in which two distribution functions are used to represent two phases, is developed and parallelized using the MPI library. Efficient parallel computations are performed on a workstation cluster composed of different types of workstations. Rising bubbles and their coalescence in a static fluid are simulated as one of the fundamental two-phase flow phenomena.

1. INTRODUCTION

Multi-phase flow phenomena are seen in many engineering fields including nuclear reactor engineering. It is, however, very difficult to predict multi-phase flow phenomena accurately, since the interface between phases changes its shape and area. The interface plays an important role in multi-phase flows, because the phase change, heat transfer, and momentum exchange occur through the interface. In reactor safety analyses, for instance, computer codes based on the two-fluid model are used most commonly to predict two-phase flow phenomena under abnormal or accidental conditions. In these codes, the exchanges between phases through the interfaces are represented by the interfacial transfer terms in the conservation equations. Empirical correlations are often used for evaluation of the interfacial transfer terms including the interfacial area.

In order to simulate two-phase flows with interfaces, numerical techniques based on discrete particle approach have recently progressed a great deal. Motions of fluid particles or molecules are calculated instead of solving continuous fluid equations, and macroscopic flow variables are obtained from the particle motion. The interface or surface is calculated to be the edge or the boundary of the particle region. Among the particle simulation methods, the LGA (lattice gas automata) is one of the simple techniques for simulating phase separation and free surface phenomena as well as macroscopic flow fields. In the LGA introduced by Frisch et al. [1], space and time are discrete, and identical particles of equal mass populate a triangular lattice. The particles travel to neighboring sites at each time step, and obey simple collision rules that conserve mass and momentum.

Macroscopic flow fields are obtained by coarse-grain averaging in space and time. Since the algorithm and programming are simple and complex boundary geometries are easy to represent, the LGA has been applied to numerical simulations of hydrodynamic flows including multiphase flows [2]. The LGA has, however, some inherent drawbacks such as: velocity dependence of the equation of state, lack of Galilean invariance, and statistical noise in the results. These drawbacks are overcome by using the Boltzmann equation [3].

The main feature of the LBM (lattice Boltzmann method) is to replace the particle occupation variables, which are Boolean variables in the LGA, by single-particle distribution functions, which are the ensemble average of particle occupation and real variables, and neglect individual particle motion [4]. In the LBM which is used widely, the collision process of particle is simplified to be a relaxation process of distribution function toward the local equilibrium [5], and the local equilibrium distribution is chosen to recover the Navier-Stokes equations [6]. It was reviewed by Benzi et al. [7] and Chen and Doolen [3] that various kinds of fluid flows involving interfacial dynamics and complex boundaries were simulated using the LBM. The two-component two-phase LBM has been developed by Gunstensen and Rothman [8] based on the two-species variant of the LGA [9]. This model was extended by Grunau et al. [10] to include two-phase fluid flows that have variable densities and viscosities. Two distribution functions are used to represent two phases in these models, and the evolution of each distribution function is calculated. The additional collision operator related to the interfacial dynamics is taken into account besides the collision operator representing the relaxation process. Although a two-phase flow in a porous medium [8] and a Hele-Shaw flow [10] were simulated, few studies of the application of the two-component two-phase LBM to practical two-phase flow problems have been reported.

In this study, the three-dimensional two-phase flow simulation code based on the two-component two-phase lattice Boltzmann method, in which two distribution functions are used to represent two phases, is developed and parallelized using the MPI library. Parallel computations are performed on a workstation cluster composed of different types of workstations. Rising bubbles and their coalescence in a static fluid are simulated as one of the fundamental two-phase flow phenomena. The efficiency of parallel computations is discussed.

2. TWO-COMPONENT TWO-PHASE LATTICE BOLTZMANN METHOD

The two-component two-phase flow model proposed by Grunau et al. [10] is reviewed here. In this model, red and blue particle distribution functions $f_i^r(\boldsymbol{x},t)$ and $f_i^b(\boldsymbol{x},t)$ at space $\boldsymbol{x}$ and time t are introduced to represent two different fluids. The subscript i indicates the direction of moving particle on a lattice. The 3-D 15-direction model shown in Fig. 1 is used in this paper, while the 2-D hexagonal lattice was used by Grunau et al. [10]. Solid and broken arrows in Fig. 1 indicate the velocity vectors with the size of unity and $3^{1/2}$, respectively.

The total particle distribution function is defined as

$$f_i = f_i^r + f_i^b. \tag{1}$$

The lattice Boltzmann equation for both red and blue fluids is written as

$$f_i^k(\boldsymbol{x}+\boldsymbol{e}_i, t+1) = f_i^k(\boldsymbol{x}, t) + \Omega_i^k(\boldsymbol{x}, t), \tag{2}$$

where k denotes either the red or blue fluid, and

$$\Omega_i^k = (\Omega_i^k)^1 + (\Omega_i^k)^2 \tag{3}$$

is the collision operator. The first term of the collision operator, $(\Omega_i^k)^1$, represents the process of relaxation to local equilibrium. The linearized collision operator with a single time relaxation parameter τ_k is written as

$$(\Omega_i^k)^1 = \frac{-1}{\tau_k}(f_i^k - f_i^{k(eq)}). \tag{4}$$

Here $f_i^{k(eq)}$ is the local equilibrium state depending on the local density and velocity, and τ_k is a spatially dependent characteristic relaxation time for species k. Conservation of mass and momentum must be satisfied:

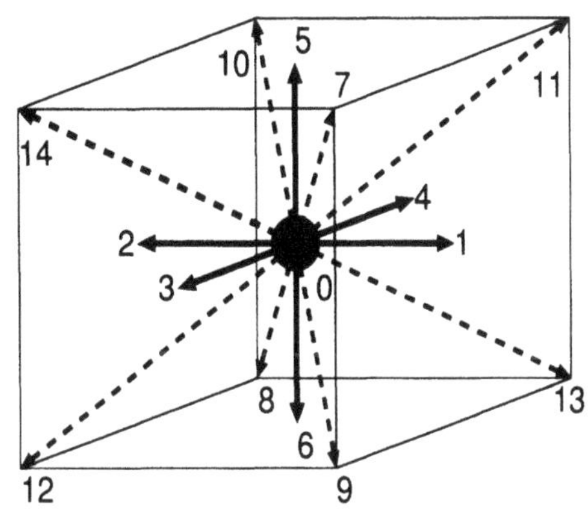

Figure 1: 3-D 15-direction lattice.

$$\rho_r = \sum_i f_i^r = \sum_i f_i^{r(eq)}, \tag{5}$$

$$\rho_b = \sum_i f_i^b = \sum_i f_i^{b(eq)}, \tag{6}$$

$$\rho \boldsymbol{u} = \sum_{i,k} f_i^k \boldsymbol{e}_i = \sum_{i,k} f_i^{k(eq)} \boldsymbol{e}_i, \tag{7}$$

where ρ_r and ρ_b are densities of the red and blue fluids, respectively,

$$\rho = \rho_r + \rho_b, \tag{8}$$

is the total density, and $\boldsymbol{u}$ is the local velocity. Using the Chapman-Enskog multiscale expansion, the second part of the collision operator for the 3-D 15-direction model is obtained as

$$(\Omega_i^k)^2 = A_k|\boldsymbol{F}|[\frac{(\boldsymbol{e}_i \cdot \boldsymbol{F})^2}{|\boldsymbol{e}_i|^2|\boldsymbol{F}|^2} - \frac{1}{3}], \tag{9}$$

where A_k is a parameter controlling the surface tension, and $\boldsymbol{F}$ is the local color gradient, defined as

$$\boldsymbol{F}(\boldsymbol{x}) = \sum_i \boldsymbol{e}_i[\rho_r(\boldsymbol{x}+\boldsymbol{e}_i) - \rho_b(\boldsymbol{x}+\boldsymbol{e}_i)]. \tag{10}$$

Note that $\boldsymbol{F}=0$ in a single-phase region, and the second term of the collision operator, $(\Omega_i^k)^2$, only has a contribution at two-phase interfaces. To maintain interfaces, the method of Rothman and Keller [9] is applied to force the local color momentum,

$$\boldsymbol{j} = \sum_i (f_i^r - f_i^b)\boldsymbol{e}_i, \tag{11}$$

to align with the direction of the local color gradient. In other words, the colored distribution functions at interfaces are redistributed to maximize $-\boldsymbol{j}\cdot\boldsymbol{F}$. The equilibrium distribution of the species k is obtained as

$$\begin{aligned} f_i^{k(eq)} &= \rho_k\left(\frac{n_k}{7+n_k} - \frac{1}{3}\boldsymbol{u}^2\right), && \text{for } |\boldsymbol{e}_i| = 0, \\ &= \rho_k\left[\frac{1}{7+n_k} + \frac{1}{3}(\boldsymbol{e}_i\cdot\boldsymbol{u}) + \frac{1}{2}(\boldsymbol{e}_i\cdot\boldsymbol{u})^2 - \frac{1}{6}\boldsymbol{u}^2)\right], && \text{for } |\boldsymbol{e}_i| = 1, \\ &= \frac{\rho_k}{8}\left[\frac{1}{7+n_k} + \frac{1}{3}(\boldsymbol{e}_i\cdot\boldsymbol{u}) + \frac{1}{2}(\boldsymbol{e}_i\cdot\boldsymbol{u})^2 - \frac{1}{6}\boldsymbol{u}^2)\right], && \text{for } |\boldsymbol{e}_i| = \sqrt{3}, \end{aligned} \tag{12}$$

so that the macroscopic mass and momentum conservation equations for continuous fluid are recovered. In the above equation, n is a parameter representing the ratio of rest and $|\boldsymbol{e}_i| = 1$ particles when velocity is zero. The equilibrium distribution functions obtained here are slightly different from those calculated for 2-D hexagonal lattice [10].

In order to simulate rising bubbles, the gravitational effect is taken into account for the two-component two-phase LBM. An external force field, $\boldsymbol{G}$, denoted by

$$-\,m_i\boldsymbol{G}\cdot\boldsymbol{e}_i, \tag{13}$$

is introduced in the right-hand side of Eq. (2), where m_i is a constant in the lattice direction i. The constant m_i is determined so that the macroscopic momentum conservation equation has an appropriate gravitational term: $-\rho g$ in the vertical direction, where g is the gravitational acceleration. Using the Chapman-Enskog multiscale expansion again, m_i is calculated to be 1/2 for $|\boldsymbol{e}_i| = 1$ and 1/8 for $|\boldsymbol{e}_i| = \sqrt{3}$. These values are different from those calculated for the 2-D square lattice [11].

3. SIMULATIONS OF RISING BUBBLES

The LBM code is developed and parallelized, and parallel computations are performed on a workstation cluster. The domain decomposition method is applied using the MPI library, and the three-dimensional simulation region is divided into small regions in the vertical direction. The workstation cluster consists of four COMPAQ au600 (Alpha 21164, 600MHz) workstations and four COMPAQ XP1000 (Alpha 21264, 500MHz) workstations. These workstations are connected through 100BaseTx (100 Mb/s) network and a switching hub.

As an example of parallel computations, two rising bubbles and their coalescence in a static fluid is shown in Fig. 2. The simulation region of this sample problem is $80\times80\times480$ lattice nodes, with periodic boundary conditions at the side boundaries. The bounce back

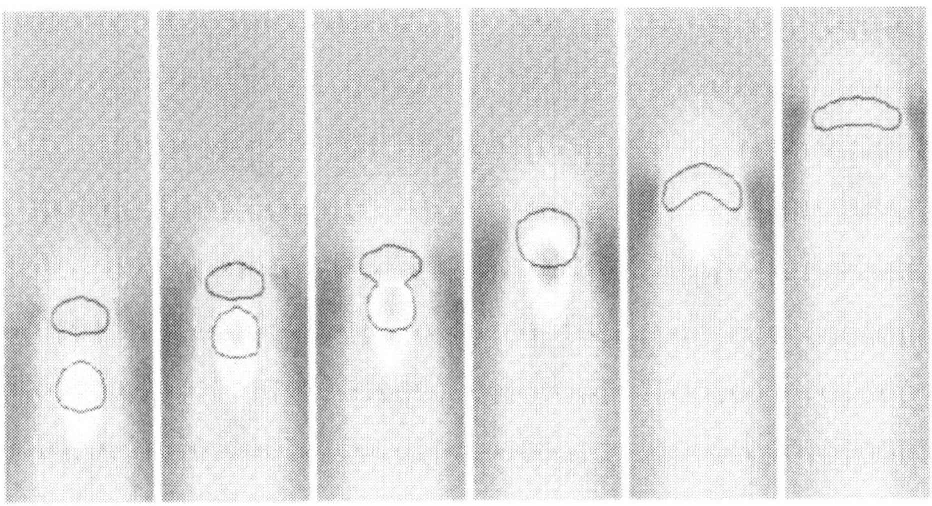

Figure 2: Coalescence of two rising bubbles.

condition is used as the no-slip velocity condition at the bottom boundary, and the first order extrapolation scheme is applied as the outflow condition at the top. Two red bubbles with the radius of 10 lattice nodes are initially placed in the blue fluid at the height of 20 and 70 lattice nodes along the vertical center axis. The shape of the interface is depicted by the contour line in the vertical cross section along the center axis. It is shown that the sharp and stable interface is obtained in our simulations. This is an important advantage to simulate two-phase flows in comparison with the LGA [12].

The flow velocity in the vertical direction is also indicated in Fig. 2: dark regions at the side of the bubbles are downward flow regions and light regions at the top and bottom of the bubbles are upward flow regions. Two bubbles go up due to the gravity and the vortex is formed around the bubbles. The lower bubble goes up faster since the upward flow is established in the wake of the upper bubble. The shape of the lower bubble is not much deformed in comparison with the upper bubble, though the rising velocity is larger. The change in the bubble shape shown in Fig. 2 is observed experimentally [13], and our simulations are found to be reasonable.

4. PARALLEL COMPUTATION ON A WORKSTATION CLUSTER

In our simulations, different types of workstations are used: COMPAQ au600 and XP1000. The processor of au600 is Alpha 21164 and the clock frequency is 600 MHz, while the processor of XP1000 is Alpha 21264 and the clock frequency is 500 MHz. The values of SPECfp 95 are 21.3 and 50, respectively. The simulation region is, thus, divided into two types of small regions as shown in Fig. 3: $80 \times 80 \times (nz)$ lattice nodes for au600 and $80 \times 80 \times (nz \times f)$ lattice nodes for XP1000. Each small region has boundary nodes at the top and bottom for data transfer as shown in Fig. 3. The factor f satisfies $(1+f) \times 4 = 480$. In order to see the effect of the factor, test simulations are performed with small simulation regions ($20 \times 20 \times 120 \sim 40 \times 40 \times 480$). The relative calculation time obtained by changing the value of the factor is shown in Fig. 4.

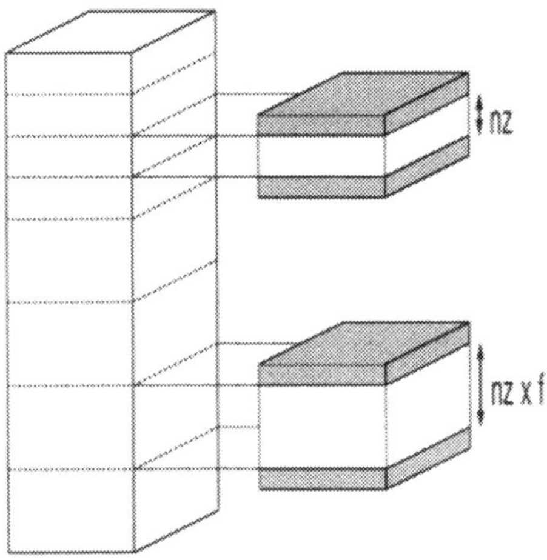

Figure 3: Domain decomposition.

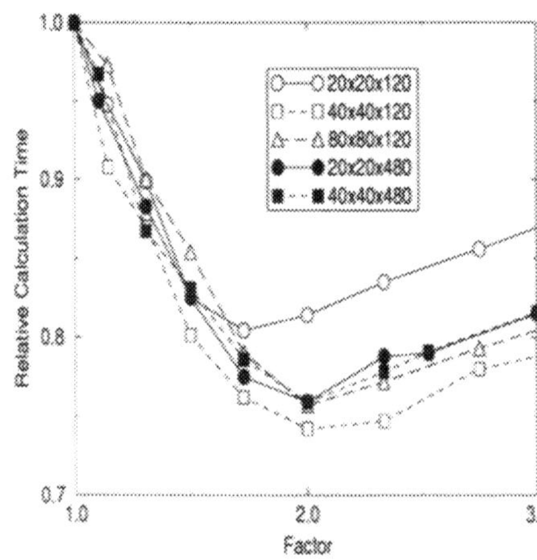

Figure 4: Calculation time for 100 steps.

The calculation time is normalized by the result of the base case with $f = 1$, in which each small region has the same size. In the base case, the amount of calculations is the same for au600 and XP1000, and XP 1000 waits the end of the calculations of au600 before data exchange. The efficiency of parallel computation is the worst for the base case as shown inf Fig. 4. The calculation time is reduced as the value of the factor increases. The calculation time is, however, increased when the factor is larger than the appropriate value. The appropriate value of the factor depends on the size of the simulation region, though the calculation time becomes minimum at around $f = 2$ as shown in Fig. 4.

In our parallel computations, the appropriate value of the factor is automatically determined at the beginning of the calculations according to the size of the simulation region. The trial calculation is performed first for some time steps with $f = 1$ before the LBM simulations, and the calculation time on each workstation is measured. The calculation time is then compared and the appropriate value of the factor is determined. The simulation region is divided into small regions using the appropriate value of the factor and the LBM simulations are performed.

The calculation time is shown in Figs. 5 and 6 for two cases with different size of the simulation region. The size of the simulation region is $20 \times 20 \times 120$ lattice nodes for Fig. 5 and $80 \times 80 \times 120$ for Fig. 6. The calculation time with the appropriate value of the factor is depicted by the black diamond in these figures. The results with $f = 1$ using 2 (au600 and XP1000), 4 (two au600 and two XP1000), 6 (three au600 and three XP1000), and

8 workstations are also shown by the broken lines. It is shown by using the appropriate value of the factor that 18 % and 28 % of the calculation time is reduced in Figs. 5 and 6, respectively, in comparison with the case with $f = 1$. These reduction of calculation time almost corresponds to the minimum value of the calculation time shown in Fig. 4. The appropriate value of the factor is, thus, found to be obtained automatically. In Figs. 5 and 6, the calculation time obtained by using the same type of workstations are also shown. The number of workstations is 1, 2, 3, and 4, for the calculations by au600 or by XP1000. The simulation region is divided into small regions with the same size in these cases. It is shown that the results with $f = 1$ using $2 \sim 8$ workstations are equivalent to the results using $1 \sim 4$ au600 workstations. It is thus confirmed that the calculation speed of a workstation cluster depends on the slowest workstation, and the load balance should be taken into consideration. The calculation time of XP1000 is much smaller than that of au600, and the results with the appropriate value of the factor are in between the results by au600 and XP1000 as shown in Figs. 5 and 6.

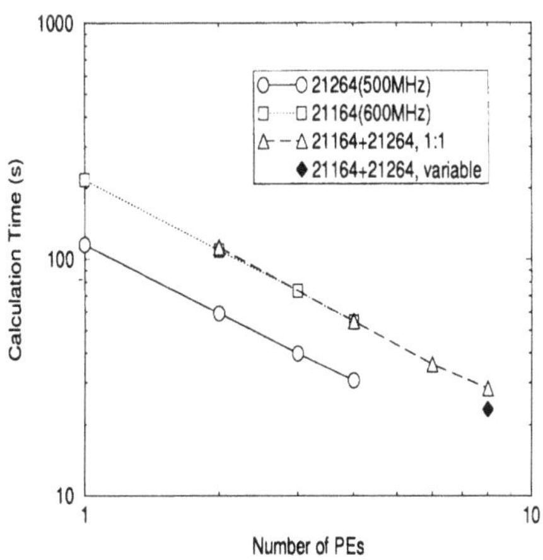

Figure 5: Calculation time for 20x20x120 lattice nodes.

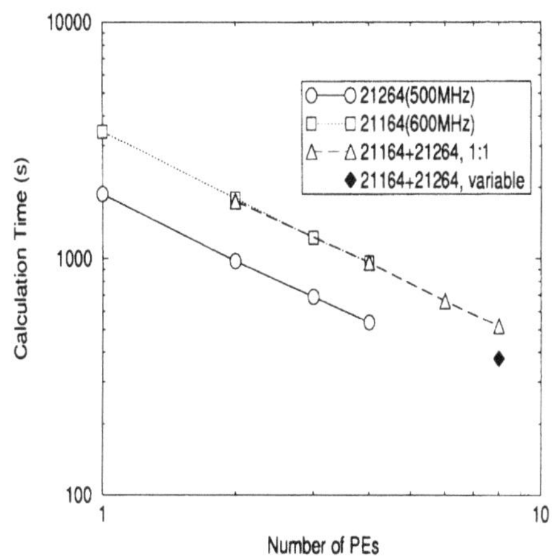

Figure 6: Calculation time for 80x80x120 lattice nodes.

5. SUMMARY

In this study, the three-dimensional two-phase flow simulation code based on the two-component two-phase lattice Boltzmann method has been developed and parallelized using the MPI library. Two rising bubbles and thier coalescence in a static fluid were simulated as one of the fundamental two-phase flow phenomena. Parallel computations were performed on a workstation cluster, which was composed of different types of workstations. The simulation region was divided into small regions by the domain decomposition method. The appropriate region size for each workstation was automatically obtained at the beginning of the parallel computations, and efficient computations were performed.

REFERENCES

[1] U. Frisch, B. Hasslacher, and Y. Pomeau, Phys. Rev. Lett. 56, 1505(1986).
[2] D. H. Rothman and S. Zaleski, Ref. Mod. Phys. 66, 1417(1994).
[3] S. Chen and G. D. Doolen,. Annu. Rev. Fluid Mech. 30, 329(1998).
[4] G. G. McNamara and G. Zanetti, Phys. Rev. Lett. 61, (1988).
[5] P. L. Bhatnagar, E. P. Gross, and M. Krook, Phys. Rev. 94, 511(1954).
[6] H. Chen, S. Chen, and W. H. Matthaeus, Phys. Rev. A, 45, R5339(1992).
[7] R. Benzi, S. Succi, and M. Vergassola, Phys. Rep. 222, 145(1992).
[8] A. K. Gunstensen and D. H. Rothman, Phys. Rev. A 43, 4320(1991).
[9] D. H. Rothman and J. M. Keller, J. Stat. Phys. 52, 1119(1988).
[10] D. Grunau, S. Chen, and K. Eggert, Phys. Fluids A 5, 2557(1993).
[11] X. Nie, Y. H. Qian, G. D. Doolen, and S. Chen, Phys. Rev. E 58, 6861(1998).
[12] T. Watanabe and K. Ebihara, Nucl. Eng. Des. 188, 111(1999).
[13] J. R. Crabtree and J. Bridgwater, Chem. Eng. Sci. 26, 839(1971).

Parallel Computational Fluid Dynamics – Trends and Applications
C.B. Jenssen et al. (Editors)

Performance Aspects of Lattice Boltzmann Methods for Applications in Chemical Engineering

T. Zeiser [a*], G. Brenner [a], P. Lammers [a], J. Bernsdorf [b], F. Durst [a]

[a] Institute of Fluid Mechanics, University of Erlangen-Nuremberg
Cauerstraße 4, D-91058 Erlangen, Germany

[b] C & C Research Laboratories, NEC Europe Ltd.
Rathausallee 10, D-53757 Sankt Augustin, Germany

In recent years, lattice Boltzmann automata (LBA) methods have emerged as an interesting alternative to the classical continuum mechanical methods in computational fluid dynamics. In particular, in very complex geometries, LBA methods have proved to be more efficient and robust. This is the reason why they are frequently used to analyze flows in porous media, catalyst-filled tubes and similar devices used in chemical engineering. In these situations, interest has to be focused not only on the flow field but also on the simulation of mass transport, homogeneous and heterogeneous chemical reactions and energy transport.

The goal of the present work was to evaluate different lattice Boltzmann approaches for the simultaneous simulation of mass transport and their applicability to engineering problems. In continuum mechanical methods based on the Navier-Stokes equations, these models are usually based on the solution of additional conservation equations for the chemical species together with Fick's law of diffusion. Within the frame of the LBA method, significantly more effort has to be spent on the simulation of multi-species systems. This causes greater computational effort with respect to CPU time and memory requirements.

Within the framework of lattice Boltzmann methods, parallelization is fairly easy owing to the mainly local algorithm. Especially for mid-range engineering applications, shared memory parallelization approaches are very promising. Little effort has to be spent on the shared memory parallelization but nevertheless very good parallel efficiencies and speedups can be achieved.

1. Introduction

In chemical engineering processes, packed beds are frequently used as catalytic reactors, filters or separation towers. In the design of these devices, fluid dynamics plays an important role, since the transport of chemical species, mixing or contacting of catalytic surfaces is determined by the fluid dynamic conservation laws.

In the present work, a lattice Boltzmann automata (LBA) based program developed by the authors was used to simulate directly the flow field in sphere-filled packed beds with different ratios of tube to particle diameters. In particular, hydrodynamic effects (e.g. the channeling effect) is discussed. Realistic geometric structures of the randomly packed beds used for the simulations are produced by a Monte Carlo technique [1].

*E-mail: thzeiser@lstm.uni-erlangen.de

The dispersion effect in a simple regular packed structure is analyzed by using LBA simulations to investigate the relation between dispersion and molecular transport. The extension to transport phenomena in packed bed reactors is the subject of ongoing work.

The typically way to shorten turn-around times is parallelization, which means dividing the work among several processors. This is typically done with message-passing methods for efficiency reasons. We show that for lattice Boltzmann methods the much easier shared memory parallelization approach is promising, at least for mid-range engineering applications. Owing to the intrinsic locality of the lattice Boltzmann algorithm, very good parallel efficiencies and speedups can be achieved.

2. Lattice Boltzmann automata (LBA)

An alternative approach to the well-known continuum mechanical techniques (i.e. solving the Navier-Stokes equations) is the lattice Boltzmann method. The fluid is treated on a statistical level, simulating the movement and interaction of particle density distributions representing the ensemble-averaged particle distribution approximating a discrete Boltzmann-type equation [2–4]. The lattice Boltzmann method has been shown to be a very efficient tool for flow simulations in highly complex geometric structures discretized by up to several million grid points [5–9].

Below we first give a short description of the general numerical method, then the formulation of passive-scalar tracer diffusion models is explained. Application aspects are presented to show the range of validity and applicability of the different diffusion models under realistic conditions in chemical engineering science.

2.1. General description of the lattice Boltzmann flow model

Our present 3-D lattice Boltzmann implementation is based on a 19-speed (D3Q19) model with a single-time BGK collision operator proposed by Qian et al.[3].

The computational domain is defined by an equidistant cartesian mesh. This can be done without a significant loss of computational performance since LBA require much less memory and CPU time than conventional finite volume methods. At each lattice node $\vec{r}_*$, the state of the fluid is represented by a set of i real numbers, the particle density distributions N_i. A time evolution of this state is done basically in two steps. First there is a streaming process, where the particle densities are shifted in discrete time steps t_* through the lattice along the connection lines in direction $\vec{c}_i$ to their vicinal nodes (left-hand side of Eq. 1). In the second step, the relaxation, new particle density distributions are computed at each node by evaluating an equivalent of the Boltzmann collision integrals Δ_i^{Boltz} (Eq. 2) with the single-time BGK collision operator [10]. For every time step, the macroscopic quantities (velocity, density and pressure) can locally be computed in terms of functions of these density distributions (Eqs. 4 and 5). The viscosity ν of the fluid is determined by the choice of the relaxation parameter ω (Eq. 6).

$$N_i(t_* + 1, \vec{r}_* + \vec{c}_i) = N_i(t_*, \vec{r}_*) + \Delta_i^{Boltz}, \tag{1}$$

$$\Delta_i^{Boltz} = \omega(N_i^{eq} - N_i), \tag{2}$$

$$N_i^{eq} = t_p \varrho \left\{ 1 + \frac{c_{i\alpha} u_\alpha}{c_s^2} + \frac{u_\alpha u_\beta}{2c_s^2} \left(\frac{c_{i\alpha} c_{i\beta}}{c_s^2} - \delta_{\alpha\beta} \right) \right\}, \tag{3}$$

$$\varrho(t_*, \vec{r}_*) = \sum_i N_i(t_*, \vec{r}_*), \tag{4}$$

$$\varrho(t_*, \vec{r}_*)\, u_i(t_*, \vec{r}_*) = \sum_i c_I\, N_i(t_*, \vec{r}_*), \tag{5}$$

$$\nu = \frac{1}{6}\left(\frac{2}{\omega} - 1\right). \tag{6}$$

The local equilibrium distribution function N_i^{eq} (Eq. 3, with implied summation over repeated Greek indices) has to be computed at each time step for each node from the components of the local flow velocities $\vec{u}$, the fluid density ϱ, the speed of sound c_s and a lattice geometry weighting factor t_p, which is chosen to recover the incompressible timedependent Navier-Stokes equations [3].

2.2. Simultaneous transport of passive scalar species

In the case of multi-component flows, the concentration ϱ_s of each chemical species s is treated as a passive scalar tracer which is governed by an advection-diffusion equation:

$$\frac{\partial \varrho_s}{dt} + u_i \frac{\partial \varrho_s}{\partial x_i} - D_s \frac{\partial^2 \varrho_s}{\partial x_i^2} = \mathcal{R} \tag{7}$$

with the molecular diffusion coefficient D and a source term $\mathcal{R}$ which is zero if no reactions occur.

The velocity field is still calculated as described in the previous section and can be seen as an external variable inserted into the advection-diffusion equation. Owing to the passive scalar approach, the concentrations have no influence on the fluid flow. This is a reasonable simplification for low solute concentrations.

To solve Eq. 7 within the lattice Boltzmann method, it can be shown [11] that exactly the same equations can be used as described for the fluid flow in Eqs. 1–4 with independent relaxation parameters ω_s for each species. Instead of the viscosity, we now obtain the molecular diffusion coefficient D_s from Eq. 6.

Since the advection-diffusion equation (Eq. 7) is only linear with respect to the velocity, a substantial simplification of the equilibrium distribution function and the number of necessary directions is possible for the tracer transport [12, 13]. For 3-D calculations, with a slight loss of accuracy, it is sufficient to use only the six vectors along the cartesian axes instead of 19 as for recovering the Navier-Stokes equations, together with an equilibrium distribution function of the form

$$N_{si}^{eq} = A + B(c_{i\alpha} u_\alpha). \tag{8}$$

The constants A and B are fixed by the requirements to conserve the mass of the species, namely $\varrho_s = \sum N_{si} = \sum N_{si}^{eq}$, and to predict the correct advective transport ($\varrho_s u_\alpha = \sum N_{si} c_{i\alpha} = \sum N_{si}^{eq} c_{i\alpha}$). It should be emphasized once more that this simplification can be applied only to the calculation of the concentration transport and not to the velocity field of the carrier fluid.

The great advantage of the simplified description of the concentration field is the saving of memory, memory operations and some calculations. The drawback is that the diffusion coefficient becomes velocity dependent once again as was known in the early days for lattice gas models. In Fig. 1, this effect is quantified from simulations of a decaying sinusoidal concentration field [14]. The deviation between the theoretical and the measured diffusion coefficients increases quadratically with the fluid velocity. For engineering applications this error normally is of only minor importance. Usually the fluid velocity is limited in the calculations to low mach numbers for stability reasons and accuracy of the flow calculation. For low velocities the error is fairly small and the typically remaining deviation is smaller than the experimental accuracies.

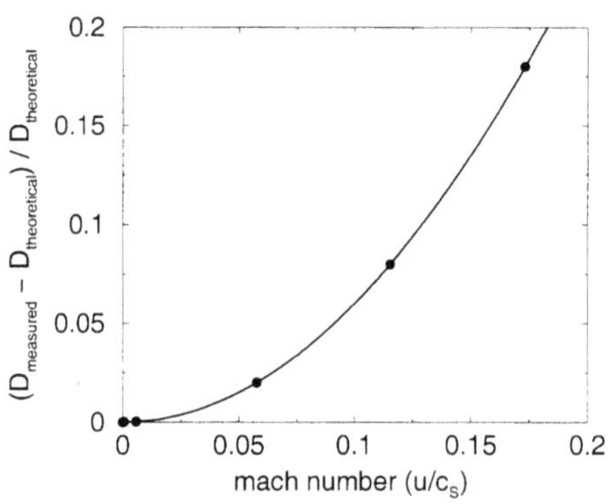

Figure 1. Deviation between the theoretical and the measured diffusion coefficients for the simplified advection-diffusion model as function of the velocity of the carrier fluid.

3. Results and discussion

3.1. Flow fields in sphere-packed beds

Experimental measurements of the void fraction of packed beds have provided the basis for the understanding of integral information, such as the radial distribution of the void fraction [15–19]. During the recent decades, the understanding of the processes has improved but detailed information is still rare because of the difficulties in measuring inside the bed.

Detailed CFD simulations can help in understanding the operations inside packed beds. The marker-and-cell representation of realistic packed bed geometries for CFD simulations can be generated either by computer tomography [7] or by a Monte Carlo simulation mimicking the packing process [1]. The latter method has the advantage, that it is easy to use and structures with arbitrary resolutions can be produced.

In the present simulations, we investigated the flow field in beds with low ratios of tube to particle diameters, since under these conditions the channeling effects, i.e. the formation of local stream tubes with high fluid velocity especially near the channel wall, become more severe. Fig. 2 shows the detailed flow structure in two cross-sections and the typical radial velocity distribution for the case of a tube-to-particle ratio of 5. One important feature of this odd packing ratio is that the mean flow velocity in the center of the tube (radial position = 0) is almost zero because the voidage approaches zero in the center.

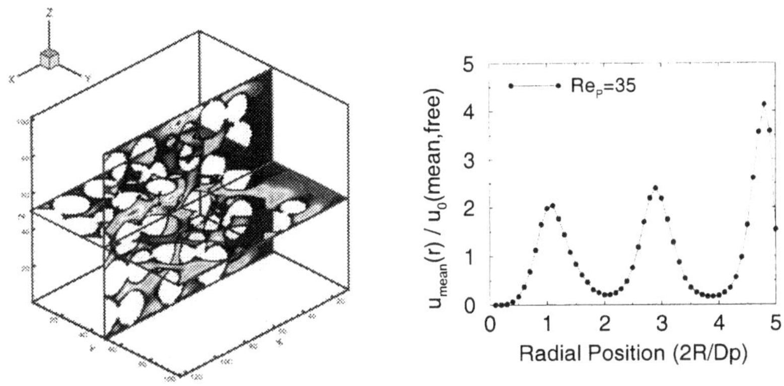

Figure 2. Detailed flow field in two cross-sections of the tube and mean radial velocity distribution for a tube-to-particle ratio of 5.

3.2. Dispersion

Dispersion in a packed bed is one of the important factors in the design of a fixed bed reactor and has been investigated recently by using LBA simulation [9, 13, 14]. As a first and simple, but nevertheless interesting, test case, the dispersion in an infinite array of a simple cubic packed domain is studied. The results provide basic information for the foregoing simulation of randomly packed beds.

The computer simulation is realized just as in a real experiment, i.e. after the flow field calculation has reached a converged solution, a pulse of a passive-scalar tracer is injected in a plane normal to the main flow direction. As was shown by Aris [20], after a certain characteristic time, the projection of the tracer distribution on the x-axis (flow direction) is of a nearly Gaussian type. The dispersion coefficient D^* can then be calculated by the momentum method according to

$$D^* = \frac{d\sigma^2}{2dt}, \tag{9}$$

where σ^2 is the variance of the tracer concentration distribution projected on the x-axis. In the simple cubic packed domain with $\varepsilon = 0.476$, the dispersion coefficients extracted from the simulations are scaled by the molecular diffusion coefficients of tracer. This ratio describes the increase of mass transfer due to the inhomogeneous flow field. Fig. 3 shows for a fixed Reynolds number of about 4 the simulation results of the variation of the axial dispersion coefficient normalized by the molecular diffusion coefficient with the product of Reynolds number Re and Schmidt number Sc, which is the molecular Peclet number

$$Pe_m = \frac{u_0 D_p}{D_m(1-\varepsilon)} \tag{10}$$

From the data of the simulation, a simple correlation equation can be extracted:

$$\frac{D^*}{D_m} = 0.758 + \frac{Pe_m^{1.83}}{39} \tag{11}$$

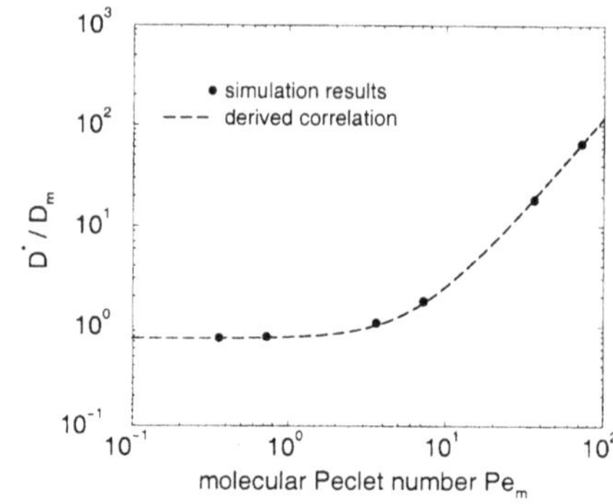

Figure 3. Dependence of the simulated dispersivity D^*/D_m in a simple cubic packing on the molecular Peclet number for a fixed Reynolds number of about 4.

This result shows less dependence on molecular Peclet number, as is known for the Taylor-Aris dispersion between parallel plates and in a capillary where the exponent of the molecular Peclet number is 2. This agrees with the results of similar configurations of simple cubic packings but for other flow conditions and other porous media [21–23]. It suggests that the LBA simulation can reveal the mixing nature of the fluids in the packed structure, and therefore is a potential tool for simulating related cases in a real packed reactor.

4. Performance

The LBA algorithm consists mainly of purely local data dependences and results in just a few subroutines with coarse granular loops. Therefore it is expected that a semi-automatic parallelization on shared memory parallel architectures by micro-tasking can lead to good parallel performance. The advantage for the programmer of a micro-tasking parallelization is the ease of a step-by-step change of a sequential program code to a parallel executable. Starting from the sequential code, either the fully automatic auto-tasking of common compilers on SMP machines is used, or directives (either vendor specific or OpenMP), instructing the compiler to parallelize certain loops, are inserted by the programmer before time-critical loops. Normally, this simplicity in programming must be paid for by lower parallel performance compared with an explicit programmed parallelization, e.g. in the context of a message-passing scheme.

The general speedup gain with semi-automatic parallelization by micro-tasking was investigated on several shared memory machines for an empty channel with one or 15×10^6 grid points. The results (Fig. 4, left) show good parallel efficiencies on all machines tested (HP V-Class, NEC SX-5/5E) in the case of up to 15 processors.

With all domain decomposition approaches, it is necessary to use advanced partitioning algorithms to achieve equal load balance among all processors for complex domains with regions of different computational effort. With the micro-tasking SMP approach it is expected that the load balancing will be done automatically as each processor takes small

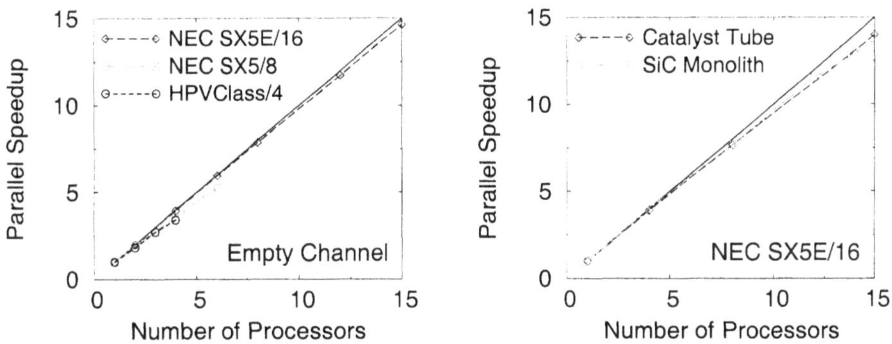

Figure 4. Parallel speedup for shared memory parallelisation on different parallel computers and different geometries.

slices of the work. The right part of Fig. 4 shows the speedup for two examples of complex geometries. The geometry is centered in a tube and makes up about 50% of the whole domain. The speedup results are almost as good as those for the empty channel.

For the LBA algorithm, the micro-tasking SMP approach is a very promising method. Current hardware developments (NEC SX-5/5E, Hitachi SR8000) emphasize this approach which consist of an SMP architecture within a node of a limited number of CPUs. For larger configurations involving more than one node, explicit message passing is preferred. Therefore, for grand challenge application a good mixture of both parallelization strategies has to be applied: a local micro-tasking SMP approach is combined with message passing on a much coarser area.

Acknowledgments

Financial support from the German Research Foundation (DFG) and the Bavarian Consortium for High Performance Computing in Science and Engineering (FORTWIHR) is gratefully acknowledged. The calculations were mainly carried out on machines of the Leibniz Rechenzentrum in Munich and the High-Performance Computing-Center Stuttgart, Germany.

REFERENCES

1. Y.-W. Li, T. Zeiser, P. Lammers, E. Klemm, G. Brenner, G. Emig, F. Durst, Structure features of sphere-packed beds and this consequence to the flow field: A computer simulation, in preparation for submission to AIChE J. .
2. U. Frisch, D. d'Humières, B. Hasslacher, P. Lallemand, Y. Pomeau, J.-P. Rivert, Lattice gas hydrodynamics in two and three dimensions, Complex Systems 1 (1987) 649–707.
3. Y. Qian, D. d'Humières, P. Lallemand, Lattice BGK models for Navier-Stokes equation, Europhys. Lett. 17 (6) (1992) 479–484.

4. R. Benzi, S. Succi, M. Vergassola, The lattice Boltzman equation: Theory and applications, Physics Reports (Review Section of Physics Letters) 222 (3) (1992) 145–197.
5. J. Bernsdorf, T. Zeiser, G. Brenner, F. Durst, Simulation of channel flow around a square obstacle with lattice-Boltzmann (BGK)-automata, Int. Journal of Modern Physics C 9 (8) (1998) 1129–1141.
6. J. Bernsdorf, F. Durst, M. Schäfer, Comparison of cellular automata and finite volume techniques for simulation of incompressible flows in complex geometries, Int. J. Numer. Met. Fluids 29 (1999) 251–264.
7. J. Bernsdorf, O. Günnewig, W. Hamm, M. Münker, Strömungsberechnung in porösen Medien, GIT Labor-Fachzeitschrift 4 (1999) 389.
8. J. Bernsdorf, F. Delhopital, G. Brenner, F. Durst, Prediction of pressure losses in porous media using the lattice Boltzmann method, in: Keil, Mackens, Voß, Werther (Eds.), Scientific Computing in Chemical Engineering II, Computational Fluid Dynamics, Reaction Engineering, and Molecular Properties, Vol. 1, Springer, Berlin, 1999, pp. 336–343.
9. B. Manz, L. Gladden, P. Warren, Flow and dispersion in porous media: Lattice-Boltzmann and NMR studies, AIChE Journal 45 (9) (1999) 1845–1854.
10. P. Bhatnagar, E. Gross, M. Krook, A model for collision processes in gases. I. small amplitude processes in charged and neutral one-component systems, Phys. Rev. 94 (3) (1954) 511–525.
11. X. He, N. Li, B. Goldstein, Lattice Boltzmann simulation of diffusion-convection systems with surface chemical reaction, Molecular Simulation, Internet Conference (Apr. 1999).
12. E. Flekkøy, Lattice Bhatnagar-Gross-Krook models for miscible fluids, Phys. Rev. E 47 (6) (1993) 4247–4257.
13. H. Stockman, R. Glass, Accuracy and computational efficiency in 3D dispersion via lattice-Boltzmann: Models for dispersion in rough fractures and double-diffusive fingering, Int. Journal of Modern Physics C 9 (8) (1998) 1545–1557.
14. T. Zeiser, Untersuchung von Diffusionsvorgängen in porösen Medien mit dem Lattice-Boltzmann Verfahren, Diplomarbeit, Lehrstuhl für Strömungsmechanik, Universität Erlangen-Nürnberg (2000).
15. L. Roblee, R. Baird, J. Tierney, Radial porosity variations in packed beds, AIChE J. 4 (1958) 460–464.
16. R. Benenati, C. Brosilow, Void fraction distribution in beds of spheres, AIChE J. 8 (3) (1962) 359–361.
17. M. Thadani, F. Peeles, Variation of local void fraction in randomly packed beds of equal spheres, Ind. Eng. Chem. Proc. Des. Dev. 5 (1966) 265–268.
18. K. Ridgway, K. Tarbuck, Randomly packed beds of spheres adjacent to a containing wall, J. Pharm. Pharmacol. 18 (1966) 1586–1595.
19. K. Ridgway, K. Tarbuck, Voidage fluctuations in randomly packed beds of spheres adjacent to a containing wall, Chem. Engng. Sci 23 (1968) 1147–1155.
20. R. Aris, On the dispersion of a solute in a fluid flowing through a tube, Proc. Royal Soc. 235A (1956) 67–77.
21. A. Eidsath, R. Carbonell, S. Whitaker, L. Herrmann, Dispersion in pulsed systems – III: Comparison between theory and experiments for packed beds, Chem. Engng. Sci 38 (11) (1983) 1803–1816.
22. D. Gunn, Theory of axial and radial disperion in packed beds, Trans. Inst Chem. Engrs 47 (1969) T351–359.
23. M. Sahimi, Flow and Transport in Porous Media and Fractured Rock, VCH, Weinheim, 1995.

9. Large Eddy Simulation

Parallel Computational Fluid Dynamics – Trends and Applications
C.B. Jenssen et al. (Editors)

PRICELES: a Parallel CFD 3-Dimensional Code for Industrial Large Eddy Simulations

U. Bieder[a], C. Calvin[a] and Ph. Emonot[a]

[a]CEA/DRN/DTP/SMTH/LDTA
CEA Grenoble, 17 Av. des Martyrs, 38054 Grenoble cedex 9, France

The CFD code PRICELES is presented. The code is based on an object oriented, intrinsically parallel approach and is especially designed for industrial Large Eddy Simulations on non-structured meshes. The parallelism conception is discussed in detail, and the basic numerical and physical models are presented. As an illustration, a typical industrial application (mixing of two flows of different temperature in a tube bend) is presented.

1. Introduction

The PRICELES project is a co-operation of the « Commissariat à l'Energie Atomique » and « Electricité de France ». Within this project, a new thermal-hydraulic code is under development which is especially designed for industrial Large Eddy Simulations (LES) [1] of several tens of millions of nodes. The code is based on an object oriented [2], intrinsically parallel approach, coded in C++ [3] and allows the user to define the discretisation on structured and non-structured grids. Appropriate physical models, several convection schemes and time marching schemes are available and variable boundary conditions can be defined. The parallelism is based on a domain decomposition method where either PVM [4] or MPI [5] libraries are used for the message passing between processors. The flexibility of the code, which is necessary for industrial applications, is implemented without a reduction of the overall performance.

The three main goals of the project are the following:

1. The solving of huge 3D flow problems on structured and unstructured meshes: The modeling of refined physical phenomena, like the study of turbulent flow in large complex geometry is a typical representative of such a problem.
2. This project has to be an open structure for the development, the coupling and the integration of other applications: such a structure is needed for the integration of different physical models (compressible, incompressible, turbulence ...), the development and integration of new numerical methods and discretisation methods (finite element, finite

volume, finite volume elements, iterative solvers, multigrid methods ...) and the coupling between various physical equations (thermal, hydraulic, mechanic, ...).
3. The portability: The code have to run on many different machines: workstation, parallel shared memory machines, parallel distributed memory machines.

In order to achieve the first objective the code is intrinsically parallel. The simulation of refined 3D physical phenomena requires meshing of several million elements, and the use of parallel distributed memory machines is compulsory. Nevertheless, the code could run also on other architectures, like parallel shared memory machines, network of workstations, or even in a sequential way on a single workstation. So the structure of the code has to take into account all these parameters in a way that does not burden the developer.
In order to achieve the portability and the open structure objectives, an object-oriented design, UML, is used. The implementation language chosen is C++. All the parallelism management and the communication routines have been encapsulated. Parallel I/O and communication classes over standard I/O streams of C++ have been defined, which allow the developer an easy use of the different modules of the application without dealing with basic parallel process management and communications. Moreover, the encapsulation of the communication routines, guarantee the portability of the application and allows us an efficient tuning of basic communication methods in order to achieve the best performances of the target architecture. At the present time, the message passing libraries encapsulated are PVM or MPI.
In the first section, the different models of parallelism used for the conception of the code are presented. In a second part, the parallel structure of the application and the different classes designed to manage the parallelism and the communication are described. We focus on the domain decomposition methods used. Finally, some experimental and performance results of large 3D simulation are presented.

2. Models of parallelism

The size of the considered problems can not be solved on standard machines. Moreover, even for the problems that can be solved on such machines, CPU times are prohibitive. Thus the mentioned massively parallel computers with distributed memory must be used in order to solve large 3D problems. However, to achieve the goals of generality of the code, standard machines, like sequential or shared memory computers, are considered too. One of the originality of the project is to design and intrinsically parallel application in order to hide the parallelism inside the code. This allows the development of new modules without taking care about their parallelization. The following designing choices have be done:

- *Parallelization model*: Data parallelism [6]. The initial domain is split into smaller ones and each of these sub-domains are distributed among the available processors. A domain decomposition method is used which is quite different form the standard ones, as one can see in a next section.
- *Programming model*: SPMD [6]. The same code is executed by all the processors but using different data.

- *Communication model*: Message passing [6]. Since irregular data structures are used, the management of communication between processors is carried out explicitly in order to achieve good performances. At the present time, either PVM or MPI library is used.

3. Design and implementation choices

3.1. Model of the parallel code

A model of a parallel code has been designed: the class `Process` is the model of a computational process. Each object of the hierarchy inherits from the class `Process`. Thus, each object can communicate with the other ones in a natural way.
One of the goal of the project is the portability. This goal can be achieved using the encapsulation mechanism of object oriented languages in order to hide the communications inside the application. A set of classes which manage all the communication between processors and parallel I/O has been designed.
The main idea which has driven the conception of these classes, is the transparency. Thus, a message is sent to another processor in the same way as data is written in a file. The same principle is used for the classes which redefine the I/O. According to the file type, the behavior is different, although the code expression is identical
Due to the polymorphism property of C++, only 4 routines are used to achieve all communication patterns [7]. Thus the user interface is greatly simplified. The same routines can be used for sending any kind of data (integer, double, strings or objects). According to the parameters of the routine any kind of communication scheme can be achieved (point-to-point, broadcast, distribution, synchronisation, ...). For instance, the call to: `send(source,target,tag,data)` realizes a send of the message data from processor "source" to processor "target" if both "source" and "target" are not equal to -1. On the contrary if "target" is equal to -1, then the processor "source" will send the same data to all the other processors, which corresponds to a broadcast.

3.2. Description of the domain decomposition technique used

PRICELES is based on a domain decomposition technique quite different from the standard ones [8,9,10]. The domain is distributed in load-balanced way among the different processors. Then each processor solves the problem on its own sub-domain more or less independently from the others.
The standard methods used in domain decomposition consist in solving in parallel the problems of each sub-domain, and then to deal with the frontiers problems in order to insure the continuity of the solution. Since the application is intrinsically parallel, there is only one problem which is solved by several processors. The elements which are located on a frontier are treated like internal ones, and a particular problem on the frontier has not to be solved.
Moreover, for the development of new numerical methods or physical models, the developer will not have to deal with the parallelism, since the data structure is parallel. To achieve this, the objects which represent the geometry and its discretization carry some information about the interface between sub-domains. So the connectivity between sub-domains is contained in the data structure, and thus each object of the geometry knows how to get an information

from a neighbor sub-domain. In order to optimize the communications, an overlap sub-domain method has been implemented, so the values of frontiers are only exchanged at each time step of the simulation.

We present the principle on figure 1. In this example, we consider two sub domains, and we detail, what we call real elements and virtual elements.

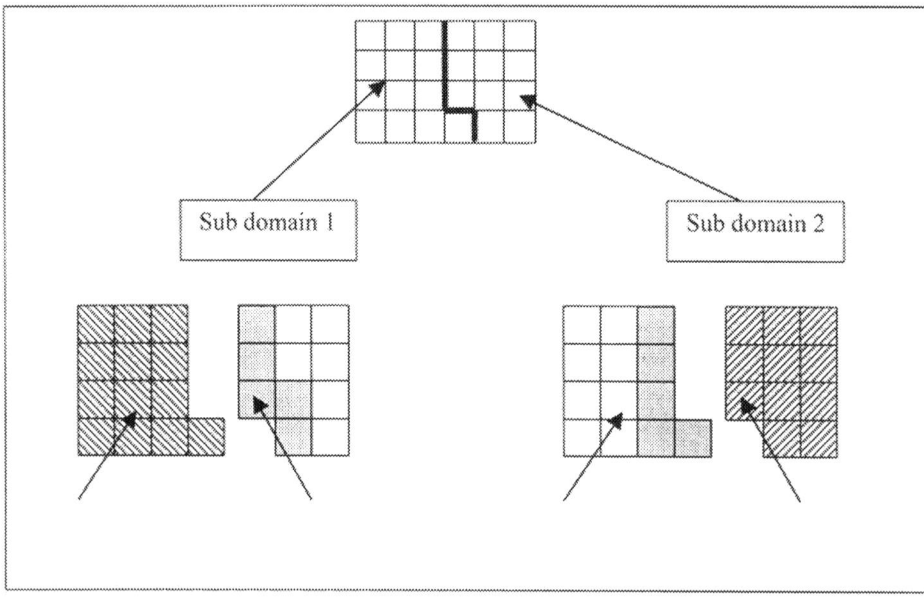

Figure 1. Principle of domain decomposition technique

This method is based on the notion of distributed arrays. Each distributed array owns a personal part which contains the values of its sub-domain and a virtual part which contains a copy of the values of the next sub-domains (see figure 2). On this figure, we describe the structure of a distributed array of processor 0 (PE 0), which contains values localized on element meshes. In this example, we suppose that, sub domain of processor 0 has two neighbors.

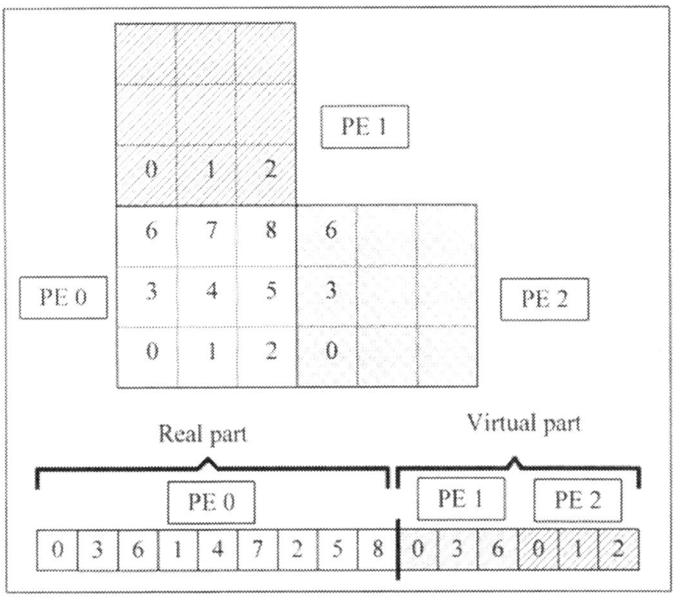

Figure 2. Distributed array notion

4. Parallel Linear Solvers

Different solvers and preconditionners are available [11,12]. All preconditionners are blocked algorithm: the precontionner is applied only on the local block owned by each processor. Thus no extra-communication are necessary for this operation. The code offers two main method as basic solvers:

- Conjugate gradient algorithm
- Multigrid methods

As preconditionners for the CG algorithm, many methods can be used, for instance:

- SSOR
- Direct Solver (Cholesky)
- Multigrid method

5. Physical and numerical models available for LES

For complex geometry, a finite volume element method based on a non-conforming discretisation tetrahedral meshes is used which is mathematically, numerically and physically very well adapted to the needs of LES [13,14]. In this discretisation scheme on overlapping co-volumes, the velocity nodes as well as the principal scalar unknowns are discretised in the center of the faces whereas the pressure unknowns are located in both the center of gravity and the vertices of a tetrahedral mesh.
Various sub-grid models are available for both structured and unstructured meshes, namely:

- Smagorinsky's model
- Structure function model
- Selective structure function models
- Dynamic sub-grid model using Smagorinsky's model

6. Some results of experiments

The performance of the code and the advantage of LES will be demonstrated on the example of a typical PWR application. Upstream of a tube bend, cold water is injected into the flow of hot water (Re about 700 000). This simulation has been achieved with 3 and 10 Million degrees of freedom using sequential and parallel calculations on 5 and 10 processors. We present, on figures 3, the results obtained using standard k-ε turbulence modeling, and the ones obtained using LES on figure 4. A typical instantaneous temperature distribution of the wall is shown below for the same instant. On one hand a standard k-ε model is used and on the other hand a LES is performed.

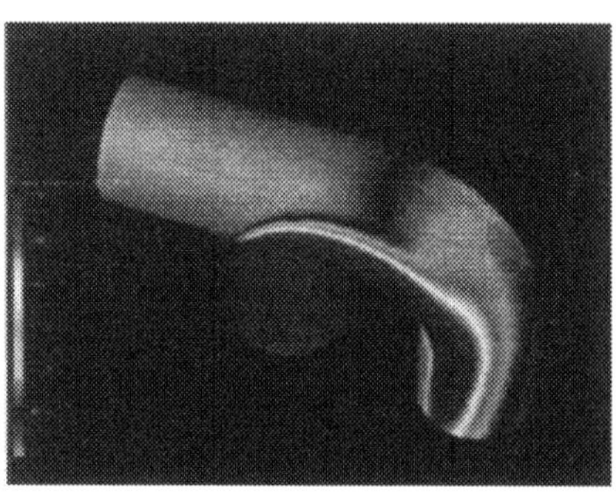

Figure 3. k-ε simulation on 10 processors

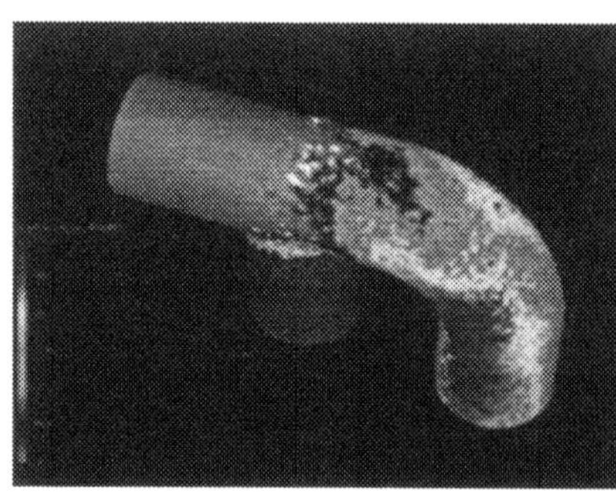

Figure 4. LES simulation on 10 processors

The mean CPU-times per time step for sequential calculation with 3 Million and 10 Million degrees of liberty were measured. In table 2 these times are compared to the corresponding parallel calculation on 5 and 10 processors, respectively. The calculations were performed on a COMPAQ SC232.

Table 2: Comparison of CPU-times

Calculation	CPU/timestep	SpeedUp
sequ. 3 Mio degree of liberty	5.45s	
paral. 3 Mio degree of liberty (5P)	1.2s	4.5
sequ. 10 Mio degree of liberty	23s	
paral. 10 Mio degree of liberty (10P)	2.8s	8.2

The speedup of the parallel calculation is linearly dependent on the used processors, i.e. the speedup by using 10 processors is in the order of a factor 8.2!

The curve of speedup from 1 to 15 processors on the calculation using 10 M degrees of freedom is presented in figure 5 and is compared to ideal one.

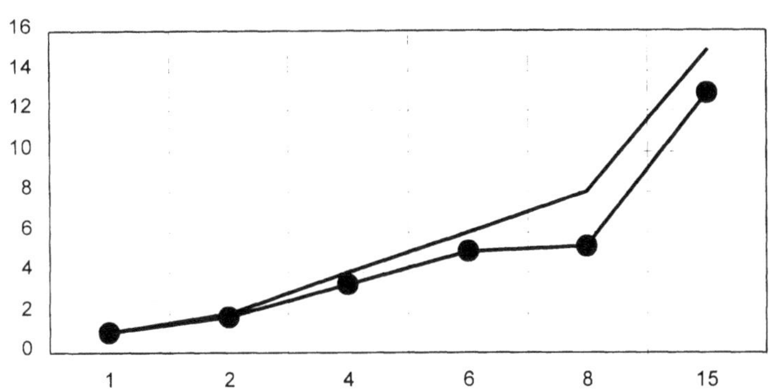

Figure 5. Speedup obtained for 10M degrees of freedom on COMPAQ SC232

7. Conclusion

The new thermal-hydraulic code PRICELES is presented which is based on a parallel, object oriented approach. In this approach, only one problem is treated for the whole calculation domain which is solved in parallel on several processors. The conservation equations are discretised on tetrahedral meshes using the Finite Volume Element method. Regarding large industrial applications, the speed-up of the parallelism allows to analyse problems with more than 100 Million degrees of liberty. As a consequence of the large number of nodes, on one hand more accurate results are expected by using the philosophy of mesh refinement. On the other hand, larger and more complex industrial problems can be treated.

References

[1] Lesieur M. : "Turbulence in Fluids.", Kluwers - Third edition – 1998
[2] J. Rumbaugh, M. Blaha, W. Premerlani, F. Eddy and W. Lorensen, "Object Oriented Modeling and Design", Prentice Hall, 1991.
[3] B. Stroustrup, "The C++ Programming Language, 2nd edition", Addison Wesley Publishers, 1992.
[4] A. Beguelin, J. Dongarra, A. Geist, R. Manchek, and V. Sunderam, "A User's Guide to PVM: Parallel Virtual Machine (version 3)", tech. Rep., Oak Ridge National Laboratory, May 1994.
[5] The MPI Forum, "Document for a Standard Message-Passing Interface", 1994.
[6] K. Hwang, "Advanced Computer Architecture - Parallelism, Scalability, Programmability", Mc Graw-Hill and MIT Press, 1993.
[7] F.T. Leighton, "Introduction to Parallel Algorithms and Architectures : Arrays - Trees - Hypercubes", Morgan-Kaufman Publishers, 1992.
[8] K. Hoffmann and J. Zou, "Parallel Efficiency of Domain Decomposition Methods", Parallel Computing, pp. 1375-1391, 1993.
[9] G. Meurant, "Domain Decomposition Methods for PDEs on Parallel Computers", International Journal of Supercomputing Applications, pp. 5-12, 1988.
[10] P. Le Tallec, "Domain Decomposition Methods in Computational Mechanics", Computational mechanics advances, vol. 1, pp. 121-220, 1994.
[11] G. Golub and C.V. Loan, "Matrix Computations", The Johns Hopkins University Press, Baltimore, second edition, 1993.
[12] Y. Saad, "Iterative Methods for Sparse Linear Systems.", PWS Publishing company, 1996.
[13] Bieder U, P. Emonot, D. Laurence, "PRICELES :Summary of the Numerical Scheme." CEA-Grenoble/DRN/DTP/SMTH/LATA/98-50, 1998
[14] U. Bieder U, C. Calvin and Ph. Emonot,"Industrial Applications of Large Eddy Simulations: Validation of a New Numerical Scheme", "8th International Symposium on Computational Fluid Dynamics", Bremen-Germany, 1999.

Parallel Computational Fluid Dynamics – Trends and Applications
C.B. Jenssen et al. (Editors)

Large eddy simulations of agitated flow systems based on lattice-Boltzmann discretization

Jos Derksen

Kramers Laboratorium voor Fysische Technologie, Department of Applied Physics, Delft University of Technology, Prins Bernhardlaan 6, 2628 BW Delft, The Netherlands,
e-mail: jos@klft.tn.tudelft.nl

Large eddy simulations in two stirred tank flow systems (*viz.* the flow driven by a Rushton turbine at Re=29,000, and the flow driven by a pitched blade turbine at Re=7,300) were performed. Lattice-Boltzmann discretization of the Navier-Stokes equations, along with a Smagorinsky subgrid-scale model was employed. The code was implemented on a Beowulf cluster. The results were extensively validated against experimental data. The phase-resolved average flow was represented accurately by the simulations. However, too high levels of the turbulent kinetic energy close to the impeller were predicted.

1. INTRODUCTION

Flows in stirred tanks are encountered in many industrial applications, especially in the chemical process industries. Applications include blending, crystallization, gas-liquid contacting, and emulsion polymerization. The mixing capacity of the tank mainly relies on the turbulent flow field that is generated by the revolving impeller. Therefore, realistic modeling of this type of flow systems has relevance for process optimization and scale-up.

At the same time, this type of flow system is quite difficultly accessible to numerical modeling. In the first place, it has time dependent boundaries (i.e. the rotating impeller), which make the flow inherently transient. In the second place, for their vast majority of applications, agitated tanks are operated in the turbulent regime. Finally, from the application side there is the demand to resolve as good as possible the flow phenomena that are influential to the rate determining steps in the process. Often these phenomena take place at the micro-scales, as they involve e.g. chemical reactions, or the behavior of particles or droplets or bubbles.

Obviously, a direct numerical simulation is out of the question due to the high Reynolds numbers (typically exceeding 10^5) that are encountered in industrial practice. Large eddy simulations (LES), however, are feasible. In relation to the time-dependent boundary conditions, they are even favorable, as they should be able to explicitly resolve the coherent motion induced by the impeller. The micro-scales are not resolved by the LES. This leaves the modeling of physical and chemical phenomena at these scales a speculative issue, albeit possibly less speculative than in simulations based on the Reynolds averaged Navier-Stokes (RANS) equations.

In this paper, we will discuss a large eddy approach to agitated flow systems based on a lattice-Boltzmann discretization of the Navier-Stokes equations. In the simulations, the agitator is represented by an adaptive force field, acting on the fluid, and a standard Smagorinsky model is used for subgrid-scale modeling. This approach has been introduced by Eggels [1]. The focus will be on the experimental validation of the simulation results, and on the parallel implementation of the numerical scheme on a relatively low-cost, Pentium III based Beowulf cluster.

2. FLOW CONFIGURATIONS AND SIMULATION PROCEDURE

Two flow configurations will be discussed, *viz.* the flow driven by a Rushton turbine in a baffled tank at Re=29,000 (see figure 1A) and the flow driven by a pitched blade turbine at Re=7,300 (see figure 1B). In stirred tank flow research, the Reynolds number is traditionally based on the impeller diameter (D), and the the angular frequency of the impeller (N in rev/s): $Re= ND^2/\nu$. The Rushton turbine system is the (*de facto*) standard case for stirred tank flow research. As a result, the flow is experimentally well documented. The definition of its geometry can e.g. be found in [2]. The pitched blade turbine case has been the subject of a high-quality experimental study ([3], see this article for the precise geometry definition) that allows for critical validation of the numerical results.

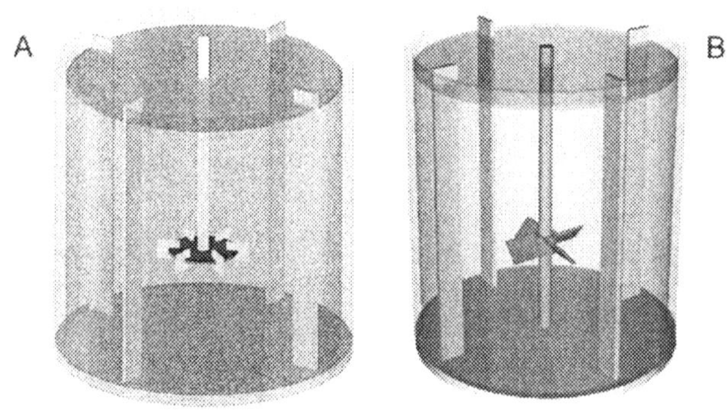

Figure 1. Flow configurations. **A**: baffled tank with a disk turbine with six flat blades (Rushton turbine). **B**: baffled tank with a pitched blade turbine (four flat blades under a 45° angle).

Lattice-Boltzmann discretization of the Navier-Stokes equations was chosen as the basis of our simulation procedure. Starting point for lattice-Boltzmann schemes is a many particle system, residing on a lattice. Every time step, the particles move to neighboring lattice sites where they collide, i.e. exchange momentum with all other particles involved in the collision. This simplified kinetic model for a fluid can be constructed in such a way that the macroscopic behavior of the system resembles a real fluid ([4], [5]). The specific scheme that is used in the present work is due to Somers and Eggels ([6], [7]). It has 18 velocity directions in three dimensions (D3Q18).

The boundary conditions can be stated in terms of reflection rules at the domain boundaries. For instance, a no-slip wall is a wall on which particles bounce back. To represent the revolving impeller, however, we need a way to smoothly model a moving object. For this, the impeller is viewed as a force-field acting on the fluid. The distribution of forces is then (iteratively) calculated in such a way that, at points on the impeller surface, the fluid velocity closely approximates the (prescribed) velocity of the impeller. This can be achieved by means of a control algorithm, in which a force opposes the mismatch between the actual velocity and the imposed velocity at a point (for details, see [2]).

For subgrid-scale modeling, a standard Smagorinsky model [8] with c_s=0.1 was applied. No wall-damping functions were used. As the Smagorinsky model is an eddy viscosity model, its incorporation in the lattice-Boltzmann scheme is straightforward. Rather than the molecular viscosity, the total viscosity (i.e. the sum of the eddy viscosity and the molecular viscosity) is used in the collision rule. The eddy viscosity is calculated from the resolved deformation rate, which is directly available because the stresses are part of the solution vector of the lattice-Boltzmann scheme.

3. IMPLEMENTATION ON THE BEOWULF CLUSTER

The memory requirements of the lattice-Boltzmann code are perfectly linear with the number of lattice-sites. Per lattice-site, 21 four-byte real values need to be stored (one real for each velocity direction, and three force components). Our largest simulations had grids of 360^3 (47 million) nodes. They used some 4 Gbyte of memory. The computer code was implemented on a Beowulf cluster consisting of 6 dual Pentium III machines running at 500 MHz, and connected through 100 BaseTX (100 Mbit/s) switched Ethernet. It was explicitly decided to develop and run the simulations on relatively cheap, pc-based computer hardware. This way, a potential obstacle for industrial application of the methodology can be overcome, as running the code does not require large costs in hardware and maintenance. The local nature of the operations involved in the lattice-Boltzmann scheme makes it well suited for distributed memory computing.

The code was written in Fortran77. MPI was used for message passing. The wall-clock time to simulate a single impeller revolution on the largest grid with the code running on 9 processing elements (PE's) amounted to 44 hours. The wall-clock time scaled almost perfectly with the number of processing elements (a speed-up of 8.9 on 9 PE's), and with the number of lattice sites used in the simulation.

4. RESULTS

4.1. Rushton turbine

An impression of the flow field is shown in Fig. 2. The average field is characterized by a strong, radial impeller outflow, which is redirected at the tank wall. Above and below the impeller disk plane, two circulation loops transport the fluid back into the impeller region. The single realization of the flow, also shown in Fig. 2, clearly indicates a strong turbulent activity in the impeller outflow, and a quiescent flow in the bulk of the tank. The time series clearly show the coherent fluctuations in the flow near the impeller tip due to the regular blade passage. Further away from the impeller, fluctuations become less intense and more random.

The trailing vortex system, emanating from the wake behind the impeller blades is clearly visible in the snapshot in Fig. 2. This vortex system has been studied experimentally to large extent (e.g. [9]-[11]). In Fig. 3 we present a comparison between experiments and the LES of the

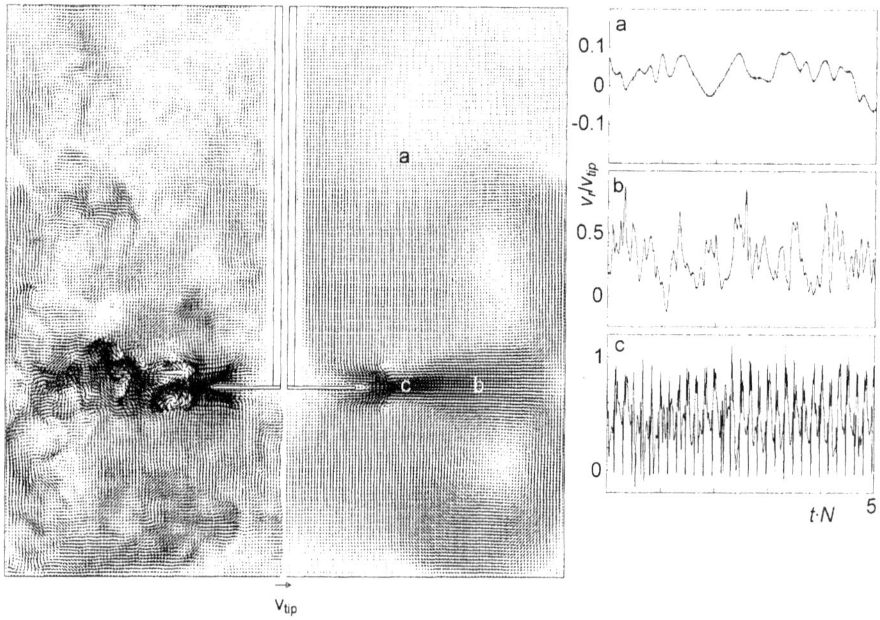

Figure 2. Left: a snapshot of the flow driven by the Rushton turbine in a vertical plane midway between two baffles. Center: the average flow field (averaging time: 24 impeller revolutions) in the same plane. Right: three velocity time series at different positions in the tank (as indicated in the center figure). The grid size was 180^3, the Reynolds number amounted to 29,000.

mean, phase-resolved flow in the vicinity of an impeller blade. Except for the near wake (i.e. within 10° behind the blade) there is excellent agreement between the high-resolution LES (the 360^3 grid) and the experimental values. There are some subtle differences between the high and low-resolution LES. For instance, the flow at disk level is too much vertically directed in the low-resolution case. The vortex core position is predicted well by both simulations.

4.1. Pitched blade turbine

The pitched blade turbine pumps in the axial direction, as becomes clear from the average flow fields, depicted in Fig. 4. In this figure, the phase-averaged flow throughout the tank, as measured by Schäfer et al. [3] is compared to simulation results on variously sized grids. Clear deviations between the low-resolved LES (the simulations on the 120^3 grid), and the experimental flow field can be observed. In the first place, the effective volume in the tank of the LES is smaller than it is in reality: at the tank wall, fluid is insufficiently pumped upwards. In the second place, the core of the big recirculation loop is located too high. In the third place, below the impeller, close to the bottom of the tank, the experiments show a small recirculation region. In the predictions on the 120^3 grid, the size of this region is overpredicted (in axial as well as in radial direction).

All three deviations become significantly smaller when enhancing the resolution of the LES (see the results of the more resolved simulations in Fig. 4). A problem that remains, however, is the too small size of the active volume in the tank.

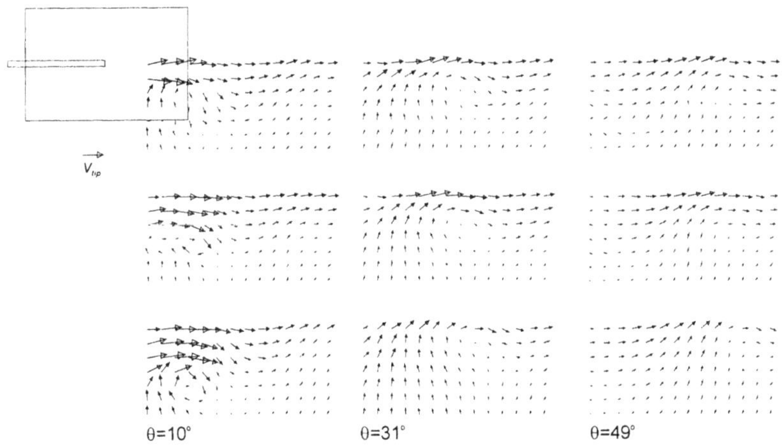

Figure 3. Phase-resolved average velocity fields at three different angles with respect to an impeller blade in the vicinity of the Rushton turbine, in the vertical plane, midway between two baffles. Comparison between experiments [11] and LES. Top row: LDA experiments, middle row: LES on a 360^3 mesh, bottom row: LES on a 180^3 mesh. In the top left graph, the θ=0° position of the impeller blade is indicated. The LES results have been linearly interpolated to the experimental grid. The reference vector serves all vector fields.

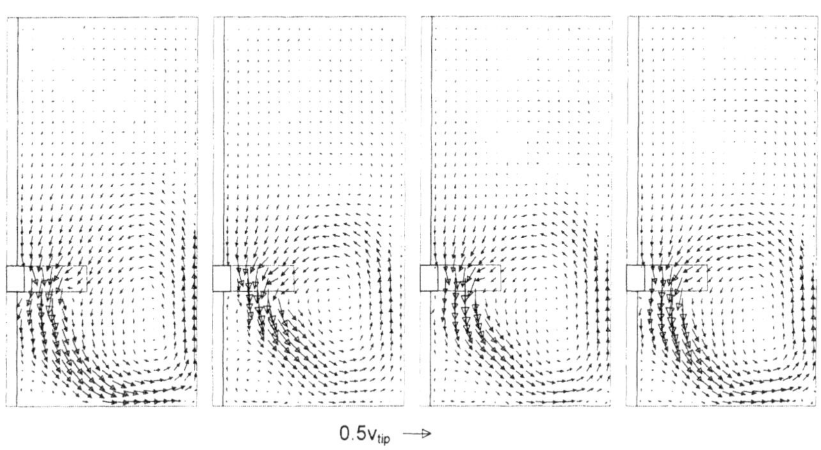

Figure 4. Phase averaged flow field in a plane midway between two baffles, induced by the pitched blade turbine at Re=7,300. From left to right: experiments [3]; LES on a 120^3 mesh; LES on a 240^3 mesh; LES on a 360^3 mesh. The LES results were linearly interpolated to the experimental grid. The reference vector serves all three plots.

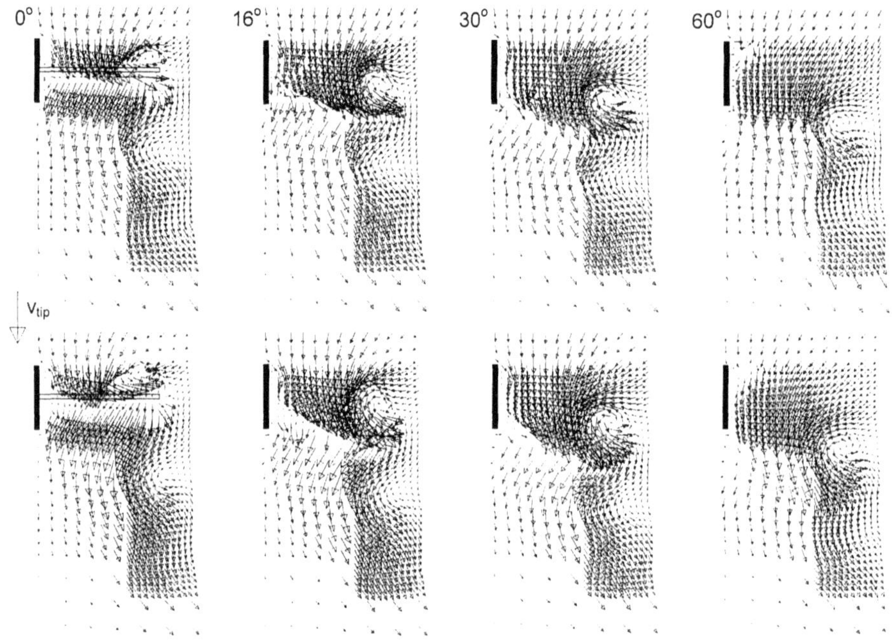

Figure 5. Phase-resolved velocity field at four different impeller angles in the vicinity of the pitched blade turbine (the field of view is indicated in the right diagram) at *Re*=7,300. Top row: experiments [3]. Bottom row: LES on a 240^3 grid. The reference vector serves all vector plots.

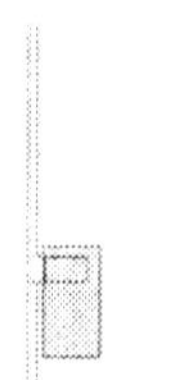

As there was no significant improvement in the phase-averaged flow predictions when going to the finest (i.e. 360^3) mesh, the rest of the results presented will be those of the 240^3 grid. In Fig. 5, phase-resolved velocity fields close to the impeller are depicted. The most eye-catching flow structure is the vortex that is formed at the upper part of the trailing side of the blade tip (Fig. 5, $\theta=0^{\circ}$). The vortex moves in the downward direction. Good correspondence between experiment and simulation can be observed. For instance, the squeezed shape of the vortex, just after it has been formed is represented well by the LES.

With a view to applications, the turbulence generated in the tank is at least as important as the average flow. It obviously contributes strongly to homogenization of the tank's content. Predictions on the turbulent kinetic energy in the impeller region by the LES are depicted in Fig. 6, along with their experimental counterparts. The LES overpredicts turbulence levels. In the lower-left plot ($\theta=20^{\circ}$), three high-energy regions can be distinguished: the upper one is related to the freshly formed vortex, the middle one is induced by the blade tip, and the lower one is the vortex from the preceding blade passage. These three high-energy regions can also be observed in the experiment (upper left graph of Fig. 6). Their energy levels are, however, significantly lower. This deviation remains for larger impeller angles.

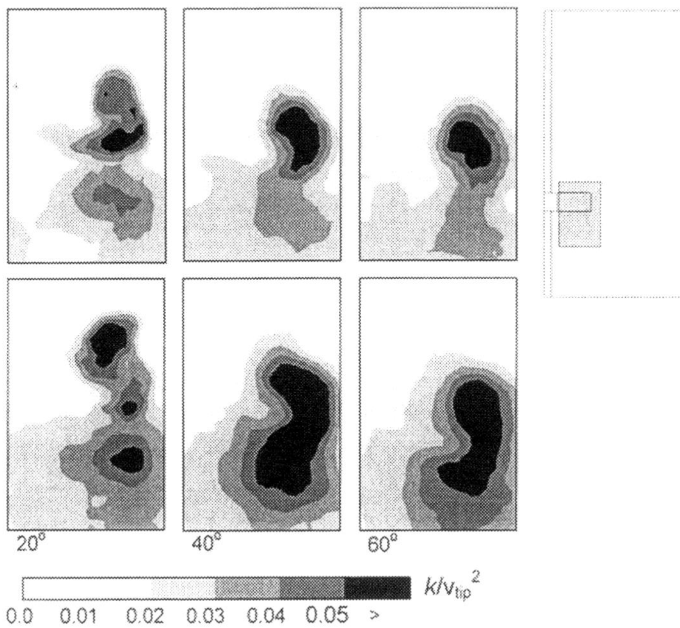

Figure 6. Phase-resolved contour plots at three different impeller angles of the turbulent kinetic energy in the vicinity of the pitched blade turbine (the field of view is indicated in the right diagram) at *Re*=7,300. Top row: experiments [3]. Bottom row: LES on a 240^3 grid.

5. CONCLUSIONS

Large eddy simulations (LES) of practically relevant flow systems (stirred tanks) were presented. The simulation algorithm was implemented on a Beowulf cluster, based on Pentium III processors. The mean flow fields corresponded well with experimental data. It has to be noted, however, that there was a significant effect of the spatial resolution of the simulations on the quality of their predictions. Turbulent kinetic energy in the vicinity of the impeller was overestimated by the LES. We speculate that this might be related to the Smagorinsky subgrid-scale model that was chosen in the present study.

REFERENCES

1. J.G.M. Eggels, Direct and large-eddy simulations of turbulent fluid flow using the lattice-Boltzmann scheme, Int. J. Heat Fluid Flow, **17**, 307 (1996).
2. J.J. Derksen, and H.E.A. Van den Akker, Large-eddy simulations on the flow driven by a Rushton turbine, AIChE Journal, **45**, 209 (1999).

3. M. Schäfer, M. Yianneskis, P. Wächter, and F. Durst, Trailing vortices around a 45° pitched-blade impeller, AIChE Journal, **44**, 1233 (1998).
4. U. Frisch, B. Hasslacher, and Y. Pomeau, Lattice-gas automata for the Navier-Stokes equation, Phys. Rev. Lett., **56**(14), 1505 (1986).
5. S. Chen, and G.D. Doolen, Lattice Boltzmann method for fluid flows, Annu. Rev. Fluid Mech., **30**, 329 (1998).
6. J.A. Somers, Direct simulation of fluid flow with cellular automata and the lattice-Boltzmann equation, Appl. Sci. Res., **51**, 127 (1993).
7. J.G.M. Eggels, and J.A. Somers, Numerical simulation of free convective flow using the lattice-Boltzmann scheme, Int. J. Heat and Fluid Flow, **16**, 357 (1995).
8. J. Smagorinsky, General circulation experiments with the primitive equations: 1. The basic experiment, Mon. Weather Rev., **91**, 99 (1963).
9. M. Yianneskis, Z. Popiolek, and J.H. Whitelaw, An experimental study of the steady and unsteady flow characteristics of stirred reactors, J. Fluid Mech., **175**, 537 (1987).
10. C.M. Stoots, and R.V. Calabrese, Mean velocity field relative to a Rushton turbine blade, AIChE Journal, **41**, 1 (1995).
11. J.J. Derksen, M.S. Doelman, and H.E.A. Van den Akker, Three-dimensional LDA measurements in the impeller region of a turbulently stirred tank, Exp. in Fluids, **27**, 522 (1999).

Parallel Computational Fluid Dynamics – Trends and Applications
C.B. Jenssen et al. (Editors)

Direct Numerical Simulation of three-dimensional transition to turbulence in the incompressible flow around a wing by a parallel implicit Navier-Stokes solver

Y. HOARAU, P. RODES, M. BRAZA[a] and
A. MANGO, G. URBACH, P. FALANDRY, M. BATLLE[b]

[a]Institut de Mécanique des Fluides de Toulouse,
Unité Mixte de Recherche CNRS/INPT UMR 5502,
Allée du Prof. Camille Soula, 31400 Toulouse, France

[b]Centre Informatique National de l'Enseignement Supérieur
950, rue de Saint Priest
34097 Montpellier CEDEX 5, France

The onset of secondary instability and successive steps of 3D transition to turbulence are examined in the flow around a wing of constant NACA0012 section along the spanwise direction. The wing is placed in uniform flow at 20 degrees of incidence, at the Reynolds numbers of 800 and 1200. The spanwise length is of order of 5 to 6 chord lengths according to different tests.

1. The numerical algorithm

1.1. The general algorithm

The numerical method is based on the three-dimensional full Navier-Stokes equations for an incompressible fluid.The pressure-velocity formulation is used as well as a predictor-corrector pressure scheme of the kind reported by Amsden & Harlow [1]. This formulation has been extended to an implicit one by Braza [2] and successfully used by Braza [3], Persillon [4], Nogues [5], Rodes [6] and Allain [7] at "l'Institut de Mécanique des Fluides de Toulouse".

As the values of the velocity and pressure are known at the time step n, the momentum equations are solved at the $(n+1)$ time step by using an approximate pressure field $P^* = P^n$. Therefore, these equations are solved for a corresponding velocity field V^*. The vector form of the exact momentum equation at the time step $(n+1)$ is

$$\frac{V^{n+1} - V^n}{\Delta t} + \nabla \cdot (V^n V^{n+1}) - \frac{1}{Re}\nabla \cdot (\nabla V^{n+1}) = -\nabla P^{n+1} \tag{1}$$

whereas the momentum equation for the approximate velocity field V^* is

$$\frac{V^* - V^n}{\Delta t} + \nabla \cdot (V^n V^*) - \frac{1}{Re}\nabla \cdot (\nabla V^*) = -\nabla P^*. \tag{2}$$

The velocity field V^* carries the exact vorticity but does not necessarily satisfy the mass conservation equation, as the case for the true velocity V^{n+1} at the step time $(n+1)$. As both fields V^* and V^{n+1} carry the same vorticity, they can be related by an auxiliary potential function Φ, such as: $V^{n+1} - V^* = -\nabla\Phi$. As $\nabla \cdot V^{n+1} = 0$, the potential Φ can be calculated by taking the divergence of the previous equation. A Poisson equation for Φ is then obtained :$\nabla \cdot V^* = \Phi$

The true velocity field can be now evaluated. The corresponding pressure is deduced by combining the exact momentum equation 1, the approximated one 2 and the relation between V^* and V^{n+1}. When the momentum equation is approximated by a fully explicit scheme, the pressure equation is reduced to $P^{n+1} = P^n + \frac{\Phi}{\Delta t}$

In this case of a semi-implicit scheme, the exact form for the pressure gradient is derived by subtracting equation 1 and 2 and by replacing $(V^{n+1} - V^*)$ by $(-\nabla\Phi)$. The relation finally obtained is

$$\nabla P^{n+1} = \nabla(P^n + \frac{\Phi}{\Delta t}) + \nabla \cdot (V^n \cdot \nabla\Omega) - \frac{1}{Re}\nabla^2(\nabla\Omega). \tag{3}$$

Braza [?] demonstrated that there were no major differences in term of solution accuracy between the two formulations. So, the above mentioned simplified pressure correction is used.

1.2. The space transformation

The method used to solve the set of equations is of the classical finite volume type. The equations are integrated over a control volume, and then the Gauss theorem is used to transform the volume integrals into surface integrals. The method uses a staggered mesh; the pressure is treated in the center of the control volume and the velocities are computed in the center of the faces.

The Navier-Stokes equations are written for any curvilinear grid. To solve these equations we make a transformation from the original space (x,y,z) to a new one (ξ, η, ζ) which is a cube. In this new space the momentum equation becomes :

$$\begin{aligned}
&\iiint_{\Omega_v} \frac{\partial v}{\partial t} J\, d\xi\, d\eta\, d\zeta + \iint_{\Gamma_1} [(Uv)\, J_1 + (Vv)\, J_2]\, d\eta d\zeta \\
&+ \iint_{\Gamma_2} \left[(Uv)\, J_1^{'} + (Vv)\, J_2^{'}\right] d\xi d\zeta + \iint_{\Gamma_3} (Wv)\, J_3^{''} d\xi d\eta \\
&= \nu \iint_{\Gamma_1} [v_\xi \alpha_1 + v_\eta \beta_1 + v_\zeta \gamma_1]\, d\eta d\zeta + \nu \iint_{\Gamma_2} [v_\xi \beta_1 + v_\eta \beta_2 + v_\zeta \gamma_2]\, d\xi d\zeta \\
&+\nu \iint_{\Gamma_3} v_\zeta \gamma_3 d\xi d\eta + S_v
\end{aligned} \tag{4}$$

For each component the pressure terms are: $S_U = -\iint_{\Gamma_1} P J_1 d\eta d\zeta - \iint_{\Gamma_2} P J_1^{'} d\xi d\zeta$, $S_V = -\iint_{\Gamma_1} P J_2 d\eta d\zeta - \iint_{\Gamma_2} P J_2^{'} d\xi d\zeta$, $S_W = -\iint_{\Gamma_3} P J_3^{''} d\xi d\eta$

1.3. The numerical scheme

To solve these equations we use the Douglas Alternating Direction Implicit method [8] which is second order both in time and space. Furthermore this method leads to tridiagonal systems, which can be efficiently solved by a Choleski algorithm. With this method each velocity momentum equation integrated over its corresponding control volume is written for three fractional steps ($^\bullet$,$^{\bullet\bullet}$ and *) as follows :

First ADI step :

$$
\begin{aligned}
2\frac{\upsilon^\bullet}{\Delta t}J + \iint_{\Gamma_1}\left(\upsilon^\bullet G_1^n - \nu\upsilon_\xi^\bullet\alpha_1\right)d\eta d\zeta \;=\; & \\
& 2\frac{\upsilon^n}{\Delta t}J + \iint_{\Gamma_1}\left(\upsilon^n G_1^n - \nu\upsilon_\xi^n\alpha_1\right)d\eta d\zeta \\
& +2\iint_{\Gamma_1}\upsilon_\eta^n\beta_1 d\eta d\zeta \\
& -2\iint_{\Gamma_2}\left[\upsilon^n G_2^n - \nu\left(\upsilon_\xi^n\beta_1 + \upsilon_\eta^n\beta_2\right)\right]d\zeta d\xi \\
& -2\iint_{\Gamma_3}\left[\upsilon^n G_3^n - \nu\upsilon_\zeta^n\gamma_3\right]d\xi d\eta \\
& +S_\upsilon^n
\end{aligned}
\qquad (5)
$$

Second ADI step :

$$
\begin{aligned}
2\frac{\upsilon^{\bullet\bullet}}{\Delta t}J + \iint_{\Gamma_2}\left(\upsilon^{\bullet\bullet} G_2^n - \nu\upsilon_\eta^{\bullet\bullet}\beta_2\right)d\xi d\zeta \;=\; & \\
& 2\frac{\upsilon^\bullet}{\Delta t}J + \iint_{\Gamma_2}\left(\upsilon^n G_2^n - \nu\upsilon_\eta^n\beta_2\right)d\xi d\zeta
\end{aligned}
\qquad (6)
$$

Third ADI step :

$$
\begin{aligned}
2\frac{\upsilon^*}{\Delta t}J + \iint_{\Gamma_3}\left(\upsilon^{n+1} G_3^n - \nu\upsilon_\zeta^{n+1}\gamma_3\right)d\xi d\eta \;=\; & \\
& 2\frac{\upsilon^{\bullet\bullet}}{\Delta t}J + \iint_{\Gamma_3}\left(\upsilon^n G_3^n - \nu\upsilon_\zeta^n\gamma_3\right)d\xi d\eta
\end{aligned}
\qquad (7)
$$

where $G_1^n = (UJ_1 + VJ_2 + WJ_3)^n$, $G_2^n = \left(UJ_1^{'} + VJ_2^{'} + WJ_3{'}\right)^n$, $G_3^n = \left(UJ_1^{''} + VJ_2^{''} + WJ_3{''}\right)^n$.
The Poisson equation for the potential Φ is also solved using ADI method : we introduce in the Poisson equation a pseudo-temporal term and within each time step we look for the convergence of this equation.

2. The parallelisation

The parallelisation has been made in the spanwise direction by domain decomposition.There is a memory independence of each processor which means that all the matrices of the code are dimensioned relatively to the domain number of points each processor is in charge with. This has been done so that the code is able to compute big domains on distributed memory super-computers. The principles of the parallelisation are explained on Figure 1.

Fisrt each processor reads the input data. As said in the previous section, the ADI scheme leads to solving an implicit tridiagonal system in each direction by using the

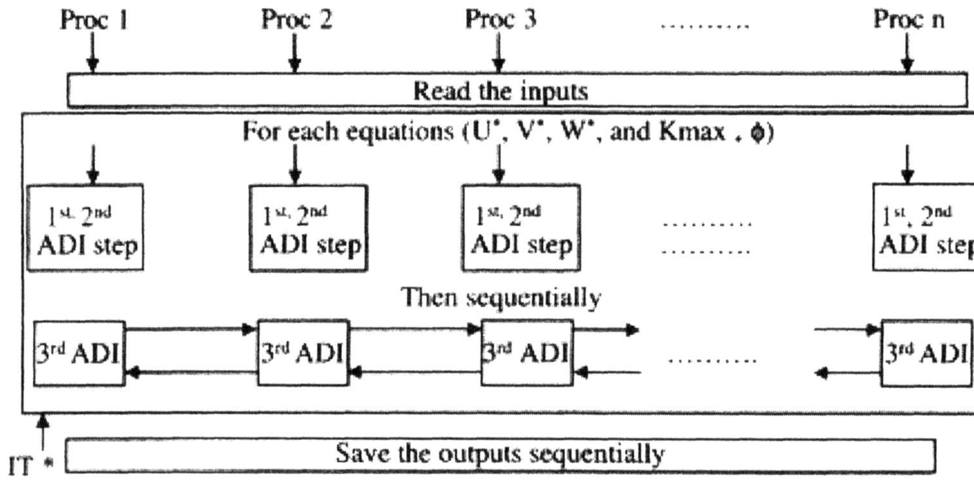

Figure 1. The parallelisation processus

iterative Choleski algorithm. The periodicity in the spanwise direction (see section 3) leads to iterate from K=1 to K=NZ and then from K=NZ to K=1 and this can't be done in a parallel way. So for each equations (U^*, V^*, W^* and ϕ) each processor starts to compute the two first ADI steps. Then we wait that all the processors finish and we solve the third step sequentially : the first processor computes its data and when finished the second can start and so on till the last processor and we do the same in the way down. When all the time iterations has been computed the data are saved sequentially.

3. Grid and boundary conditions

We aim to study the flow around a NACA12 airfoil at 20 degrees of incidence at the Reynolds number of 800 and 1200. The mesh is a structured C-type mesh of 413*70*30 points. The 2D grid (Fig 2) is extruded in the third direction.

The inlet is at 12 chords of the body, the outlet at 17 chords and the spanwise length is 4 chords.

The boundary conditions are :

- Inlet (J=NY) : Dirichlet for U,V and W; Neuman for ϕ and for U near the outlet
- Outlet : Jin & Braza [9] boundary absorption condition
  - For U, V and W : $\frac{\partial v}{\partial t} + U\frac{\partial v}{\partial x} = \nu\left(\frac{\partial^2 v}{\partial y^2} + \frac{\partial^2 v}{\partial z^2}\right)$
  - For ϕ : $\frac{1}{2}\frac{\partial^2 \Phi}{\partial x^2} + \frac{1}{U\Delta t}\frac{\partial \Phi}{\partial x} = 0$
- Spanwise Direction : periodicity conditions.

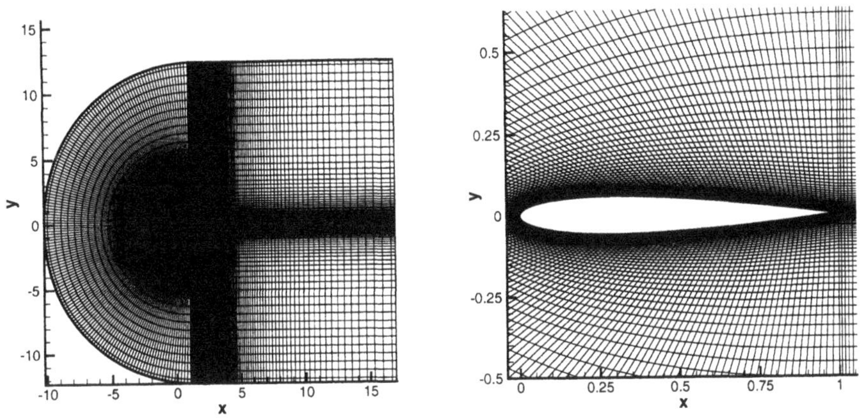

Figure 2. Mesh around the airfoil

4. The results

4.1. The First Bifurcation : the von Karman instability

The first transition step of a 2D flow around a NACA12 airfoil at 20 degrees of incidence and a Reynolds number of 800 is the von Karman instability, due to the high angle of attack and the shear stress in the wake flow. This instability leads to a quasi periodic vortex pattern : a vortex birthing at the leading edge is convected and destroyed by the mean flow as can be seen on figure 3 representing the instantaneous streamlines around the profile.

The time evolution of the drag and the lift (Fig 4) shows the periodicity of the phenomenon and by making a spectrum analysis of the drag (Fig 5), a dimensionless frequency of 0.55 can be extracted.

This initial condition is chosen to examine physically the way the 3D transition occurs form a nominally 2D configuration already submitted to the first bifurcation. So a random perturbation an order of magnitude of 1e-4 is introduced in the spanwise direction and we look how 3D occurs.

4.2. Benefits of the parallelisation

The calculation had been made on the CRAY Origin 2000 (256 processors at 300MHz) of the CINES in Montpellier.On this computer it took more than 23 hours to perform 1000 time iteration with the sequential code and around 7 hours with the parallel code using 8 processors. So we have a time improvement of more than 66%. This time earning which was the major parallelisation benefit that we wanted enables to perform the whole

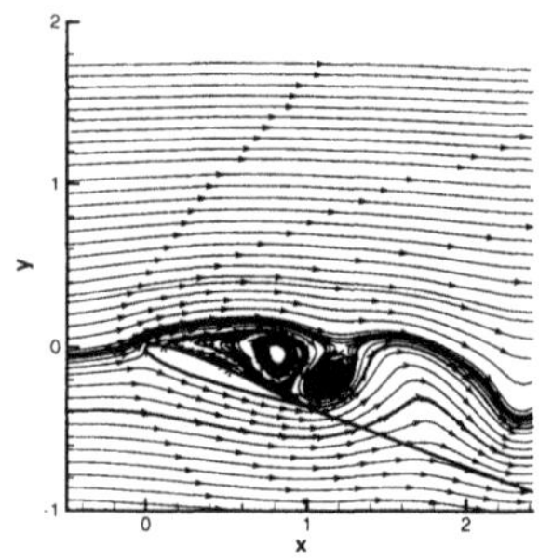

Figure 3. Instantaneous streamlines

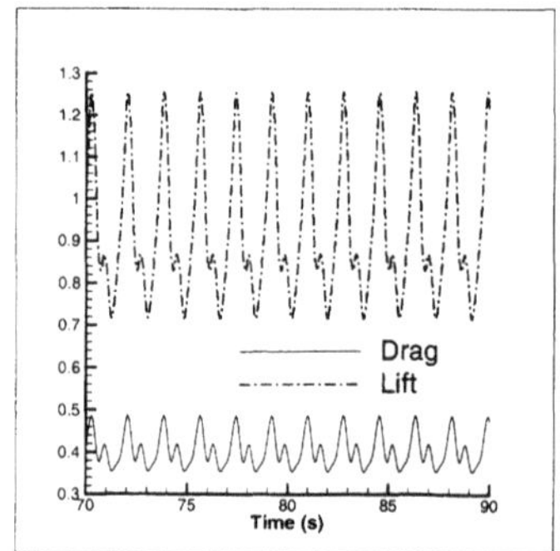

Figure 4. 2D Drag and Lift time evolution

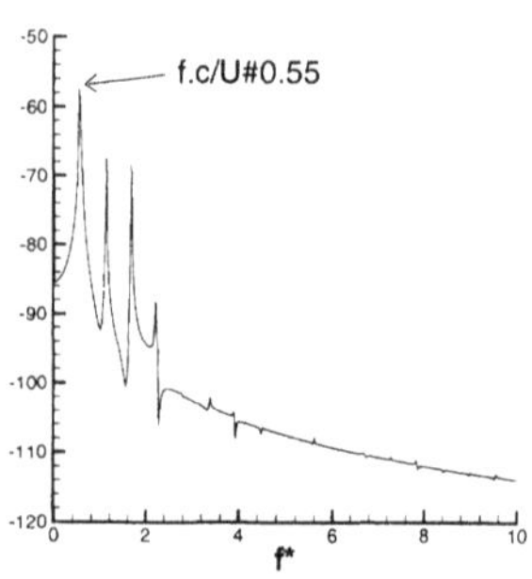

Figure 5. Drag spectrum

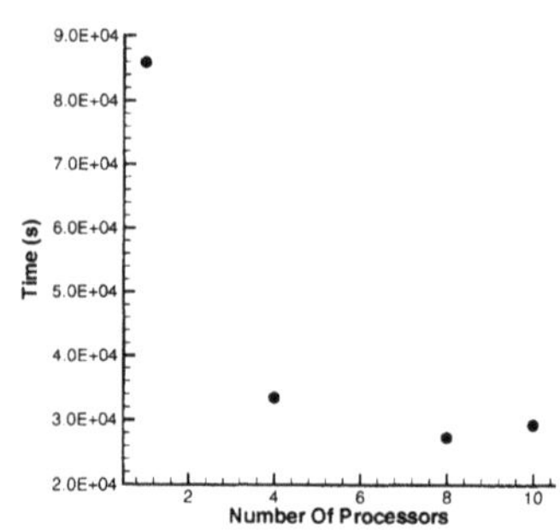

Figure 6. Speed-up

calculation within less than three weeks.

4.3. The second bifurcation

The first effect of 3D nature is the temporal amplification of the spanwise velocity (Fig 7). This amplification induces a destabilisation of the nominal 2D vortices leading to a spanwise undulation of large scale wavelength. From the spanwise evolution of the W component in the recirculation zone and in the wake flow (Fig 8) a wavelength of 0.8-0.9 can be extract and this result is in good agreement with those of Persillon & Braza [10].

The amplification of the W velocity leads to the birth and growth of vertical vorticity. This vorticity is organized in counter-rotating vortex filaments (Fig 9).Due to the vorticity conservation equation, the longitudinal (Fig 11) and the spanwise (Fig 10) vorticity appear. The spanwise undulation is well seen on figure 10.

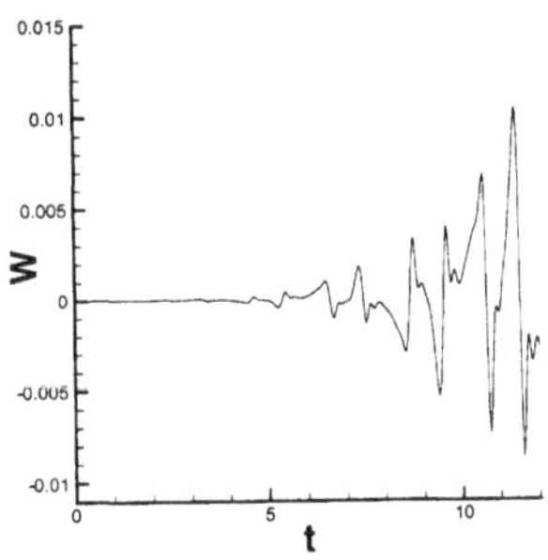

Figure 7. Time evolution of the W velocity

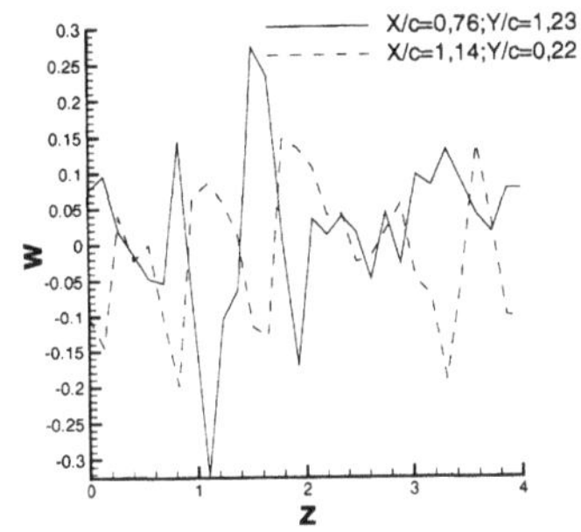

Figure 8. Spanwise evolution of W velocity

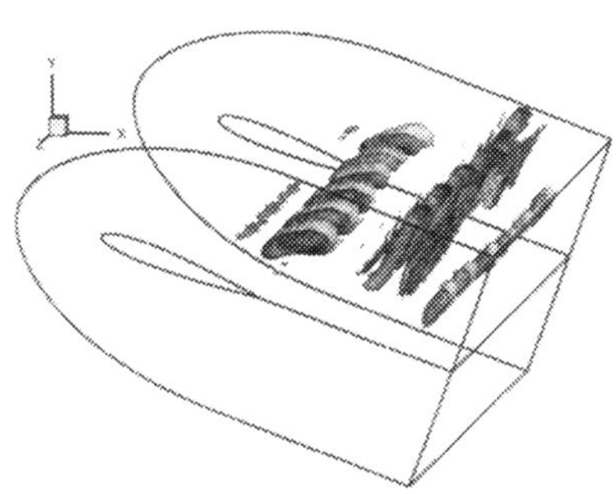

Figure 9. Vertival vorticity : $\omega_y = -2$ (blue),$\omega_y = +2$ (green)

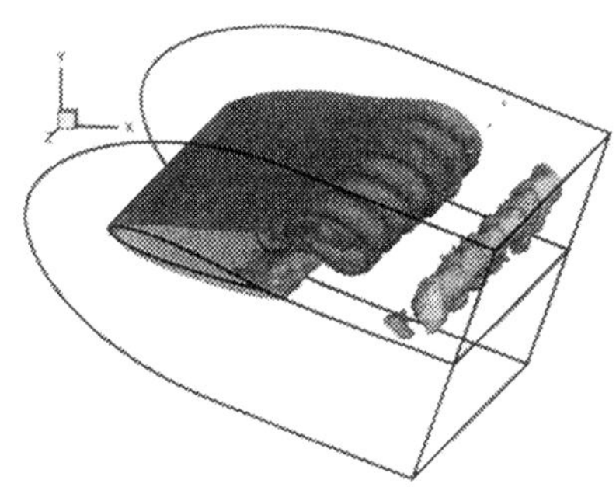

Figure 10. Spanwise vorticity : $\omega_z = -8.44$ (green), $\omega_z = 4.1$ (red)

5. Conclusion

The way to three-dimensionality of an initial 2D flow have been studied through an efficient Navier-Stokes implicit parallel solver. The benefit of the parallelisation in time and memory allows to increase by an order of magnitude the number of degrees of freedom. The data base enables to study the successive stages for birth of turbulence in wakes : the spanwise undulation of preferential large scale wavelength results from the amplification of the secondary instability, as previously mentioned through the ω_y and ω_x components. The present DNS allows the quantification of this flow feature and of the order of magnitude of the main wavelength constituting the second bifurcation of 3D nature in the transition to turbulence in wake flows.

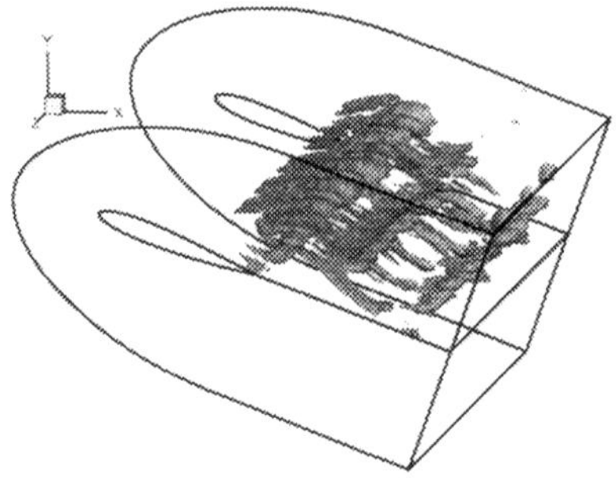

Figure 11. Longitudinal vorticity : $\omega_x = -2$ (blue), $\omega_x = 2$ (red)

REFERENCES

1. M. A. Amsden and F. H. Harlow, The SMAC Method : A Numerical Technique For Calculating Incompressible Fluid Flows, Los Alamos Scientific Laboratory Report L.A. 4370 (1970).
2. M. Braza, Etude Numérique du décollement instationnaire externe par une formulation vitesse-pression : Application à l'écoulement autour d'un cylindre, Thèse de docteur ingénieur de l'Institut National Polytechnique de Toulouse (1981).
3. M. Braza, Méthode de résolution des équations de Navier-Stokes pour des ècoulements 2D ou 3D instationnaires incompressibles : code ICARE, Rapport interne TELET-IMFT (1991)
4. H. Persillon, Analyse physique et simulation numérique de la transition laminaire-turbulente bi- et tri-dimensionnelle de l'écoulement autour d' un cylindre, Thèse de doctorat de l'Institut National Polytechnique de Toulouse (1995).
5. P. Nogues, Prédétermination d'écoulements turbulents de type sillage instationnaire 2-D et autour de configurations 3-D de sous-marin géométrie quelconque, Thèse de docteur ingénieur de l'Institut National Polytechnique de Toulouse (1995).
6. H. Persillon, Contribution à l'étude d'écoulements instationnaires transitionnels et turbulents autour d'une aile d'avion par simulation numérique et modélisation, Thèse de docteur ingénieur de l'Institut National Polytechnique de Toulouse (1998).
7. J. Allain, Analyse Physique de Mécanisme de Transition Tridimensionnelle Dans Le Sillage D'un Cylindre Circulaire Par Simulation Directe, Thèse de doctorat de l'Institut National Polytechnique de Toulouse (1999).
8. J. Douglas and J. E. Gunn, A General Formulation of Alternating Direction Method, Numerische Mathematik 6 (1964).
9. G. Jin and M. Braza, A non-reflecting outlet boundary condition for incompressible unsteady Navier-Stokes calculation, Journal of Computational Physics (1993).
10. H. Persillon and M. Braza,Physical analysis of the transition to turbulence in the wake of a circular cylinder by three-dimensional Navier-Stokes simulation, Journal of Fluid Mechanics (1998).

Parallel Computational Fluid Dynamics – Trends and Applications
C.B. Jenssen et al. (Editors)

Preliminary Studies of Parallel Large Eddy Simulation using OpenMP

W. Lo, P. S. Ong and C. A. Lin[a] *

[a] Department of Power Mechanical Engineering, National Tsing Hua University, Hsinchu 30013, TAIWAN

The present study is concerned with the parallel large eddy simulations using OpenMP. The numerical procedure is based on the finite volume approach with staggered grid arrangement and possesses second order of accuracy for both space and time. Applications are applied to the fully developed channel and pipe flows with Reynolds number of Re_τ=180 and 150, respectively. Capability of the adopted scheme is examined by comparing the predicted flow quantities with direct numerical simulation data. The preliminary results of the parallel efficiency using OpenMP are addressed.

1. Introduction

Turbulence is a phenomenon that coccus frequently in nature, which is in general a three-dimensional phenomenon. Due to the frequent occurrence of turbulence in many tech-logical applications, the predictions and control of turbulent flows have become increasingly important.

Traditionally the solutions of the turbulent flow problems involve the the applications of Reynolds averaging to the equations of motions to obtain Reynolds averaged equations, which describes the evolution of the mean quantities. The effect of the turbulent fluctuations appears in a Reynolds stress $\overline{u_i u_j}$ that must be modeled to close the system. There are many routes to model the unknown stresses, such as eddy-viscosity type model or the Reynolds stress transport models. When the models are applied to flows that are very different from the ones used for calibration, the model may not deliver accurate predictions.

The most straightforward approach to solve the equation without modeling is via direct numerical simulation of turbulence, in which the governing equations are discretised and solved numerically. However, the need to acquire accurate results require the adequate resolutions of different scales presence within the flows. To resolve all scales within the flows, the number of grid points one requires is about $Re^{9/4}$. Further more, the number of time-steps required advance the computation for a given period scales like $Re^{3/4}$. Therefore, the total cost of a direct numerical simulation is proportional to Re^3.

Large eddy simulation is a technique which lays between the DNS and Reynolds averaged approaches. In LES, the large, energy-containing structures to momentum, heat

*This research work was supported by the National Science Council of Taiwan under grant NSC-89-2212-E-007-030. and the computational facilities were provided by the National Centre for High-Performance Computing of Taiwan which the authors gratefully acknowledge.

is computed exactly, and only the effects of the smallest scales of turbulence is modeled. This has advantages potentially over the aforementioned two approaches, and is the methodology adopted here. It, however, still requires the solutions of three-dimensional time dependent equations, and this requires the usage on advance computers, such as parallel computers.

There are many message passing libraries, such as MPI and PVM, which are developed as standard platform to perform parallel computing on different machines, which primarily are distributed ones. However, the implementation of these message passing models are not straightforward, and there are many shared memory machines which scales more readily than the distributed machines. Therefore, the present study will explore the usage of the newly proposed OpenMP model[1] for shared memory programming within the large eddy simulation framework.

2. Governing Equations and Modeling

LES is based on the definition of a filtering operation: a filtered (or resolved, or large-scale) variable, is defined as,

$$f(\tilde{x}) = \int_D f(\tilde{x'})G(\tilde{x}, \tilde{x'})d\tilde{x'} \tag{1}$$

where D is the entire domain and G is the filter function. The filter function determines the size and structure of the small scales.

The governing equations are grid-filtered, incompressible continuity and Navier-Stokes equations,

$$\frac{\partial \rho \overline{u}_i}{\partial x_i} = 0 \tag{2}$$

$$\frac{\partial \rho \overline{u}_i}{\partial t} + \frac{\partial}{\partial x_j}(\rho \overline{u}_i \overline{u}_j) = -\frac{\partial \overline{p}}{\partial x_i} + \frac{\partial}{\partial x_j}[\mu(\frac{\partial \overline{u}_i}{\partial x_j} + \frac{\partial \overline{u}_j}{\partial x_i}) - \tau_{ij}] \tag{3}$$

where

$$\tau_{ij} = \rho \overline{u_i u_j} - \rho \overline{u}_i \overline{u}_j$$

is the sub-grid stress due to the effects of velocities being not resolved by the computational grids.

In the present study, the Smagorinsky model[2] has been used for the sub-grid stress, such that,

$$\tau_{ij} = -(C_s \Delta)^2 \frac{1}{\sqrt{2}} \sqrt{S_{kl} S_{kl}} S_{ij} + \frac{2}{3} \rho k_{sgs} \delta_{ij} \tag{4}$$

where $C_s = 0.1$, $S_{ij} = \frac{\partial \overline{u}_i}{\partial x_j} + \frac{\partial \overline{u}_j}{\partial x_i}$ and $\Delta = (\Delta x \Delta y \Delta z)^{1/3}$, respectively. It can be seen that in the present study the mesh size is used as the filtering operator.

A Van Driest damping function accounts for the effect of the wall on the subgrid-scale is adopted here and takes the form as,

$$l_m = \kappa y[1 - exp(-\frac{y^+}{25})] \tag{5}$$

where y is the distance to the wall and the cell size is redefined as,

$$\Delta = MIN[l_m, (\Delta x \Delta y \Delta z)^{1/3}] \tag{6}$$

3. Numerical Algorithms

The grid-filtered transport equations are discretizsed using staggered finite volume arrangement. The principle of mass-flux continuity is imposed indirectly via the solution of pressure-correction equations according to the SIMPLE algorithm[3].

For the transport equations, the spatial derivatives are approximated using second order central differencing scheme. For the purpose of temporal approximation, the momentum equations may be written as;

$$\frac{\partial \phi}{\partial t} + C_x + C_y + C_z = D_x + D_y + D_z + S \tag{7}$$

where ϕ is the dependent variable, S is the source terms and C and D are the convective and diffusive transport terms, respectively.

The temporal approximation seeks to advance ϕ from any discrete time level (n) to the following level $(n+1)$ at some specified order of accuracy. In the present work, a second-order accurate ADI scheme, originally proposed by Brian (1961)[4] for 3D heat-conduction, is applied.

The scheme involves consecutive fractional steps, with the first three steps advancing the solution by one half of the forward time interval. In either step, one of the C, D operators is treated implicitly while the other two operators only involve either known values at the old time level or such arising from a previous fractional time step. Consequently, each intermediate solution then involves an implicit inter-nodal linkage along one co-ordinate line only, and gives rise to a coupled equation system whose coefficient matrix is tri-diagonal and which can be easily and efficiently inverted. The final step advances ϕ field from the old time level to the final forward time level. These can be demonstrated below.

$$\frac{\Phi^{*} - \Phi^{n}}{\delta t/2} + C_x^{*} + C_y^{n} + C_z^{n} = D_x^{*} + D_y^{n} + D_z^{n} + S^{n+1/2} \tag{8}$$

$$\frac{\Phi^{**} - \Phi^{n}}{\delta t/2} + C_x^{*} + C_y^{**} + C_z^{n} = D_x^{*} + D_y^{**} + D_z^{n} + S^{n+1/2} \tag{9}$$

$$\frac{\Phi^{***} - \Phi^{n}}{\delta t/2} + C_x^{*} + C_y^{**} + C_z^{***} = D_x^{*} + D_y^{**} + D_z^{***} + S^{n+1/2} \tag{10}$$

$$\frac{\Phi^{n+1} - \Phi^{n}}{\delta t} + C_x^{*} + C_y^{**} + C_z^{***} = D_x^{*} + D_y^{**} + D_z^{***} + S^{n+1/2} \tag{11}$$

A simpler and computation-ease form with low storage requirements can be derived from the above equations.

4. OpenMP

The format of an OpenMP directive is as follows[1]:

```
sentinel directive_name [clause[[,] clause]...]
```

All OpenMP compiler directives must begin with a directive sentinel. Directives are case insensitive. Clauses can appear in any order after the directive name.

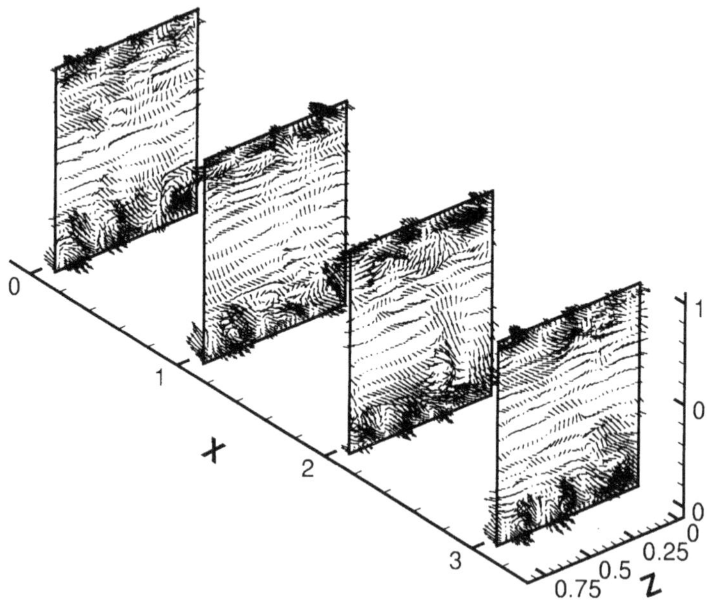

Figure 1. Instantaneous velocity field-fully developed channel flow $Re_\tau = 180$

4.1. Parallel region construct

The PARALLEL and END PARALLEL directives define a parallel region. A parallel region is a block of code that is to be executed by multiple threads in parallel. This is the fundamental parallel construct in OpenMP that starts parallel execution. These directives have the following format:

```
!$OMP PARALLEL [clause[[,] clause]...]
block
!$OMP END PARALLEL
```

When a thread encounters a parallel region, it creates a team of threads, and it becomes the master of the team. The master thread is a member of the team and it has a thread number of 0 within the team. The number of threads in the team is controlled by environment variables and/or library calls.

5. Results

Numerical simulation is first applied to a fully developed channel flow of $Re_\tau = 180$. Periodic boundary conditions are imposed in the streamwise and spanwise directions,

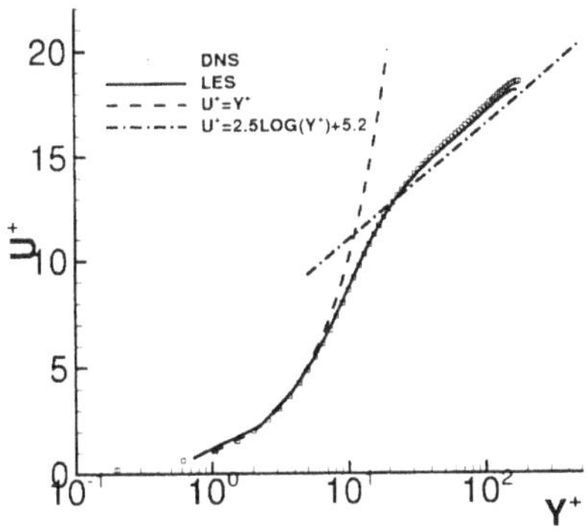

Figure 2. Log-law plot of channel flow- $Re_\tau = 180$

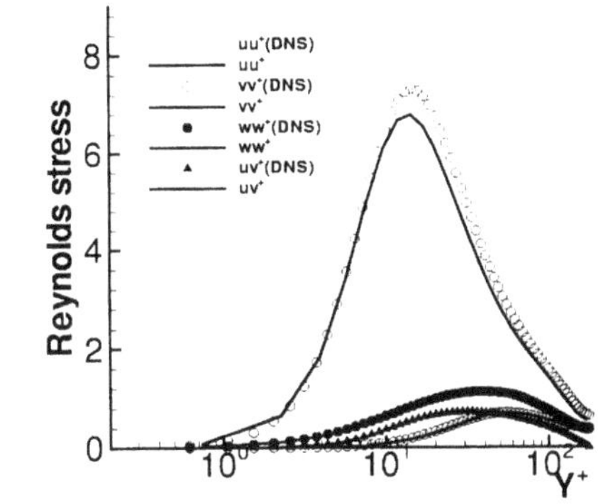

Figure 3. Predicted turbulence intensities- $Re_\tau = 180$

while the no-slip boundary is imposed in the transverse direction. The size of the computational grid adopted in the simulations is 66x64x66 in the streamwise, transverse and spanwise directions, respectively. This corresponds to the grid spacing of $\Delta x^+ \approx 18$, $\Delta y^+ \approx 1.5 \sim 20$ and $\Delta z^+ \approx 10$, respectively. The time step adopted is Δt^+ =0.001.

The instantaneous velocity field at four selected locations can be seen from Figure 1. This shows the unsteady motion of the large energy-containing eddies, which are three-dimensional in nature. The time averaged velocity distribution can be examined by looking at the log-law plot, as shown in Figure 2. It can be clearly seen that the viscous sub-layer is adequately resolved by the adopted scheme. Away from the wall, the predicted results depart slightly from the semi-empirical log-law distribution. But the predicted results agrees with the DNS data. The slight departure from the log-law is due to the lower Reynolds number adopted, and the log-law behavior is expected to prevail at elevated Reynolds numbers.

The capability of the adopted scheme can be further examined by looking at the predicted turbulence quantities, as shown in Figure 3, where the anisotropic field of the normal stresses is well represented by the adopted approach. The agreement with the DNS data is good, except the streamwise turbulence intensity is slightly under-predicted.

The numerical simulation is then applied to a fully developed pipe flows of $Re_\tau = 150$. Periodic boundary conditions are imposed in the axial and tangential directions, while the no-slip boundary is imposed in the radial direction. The size of the computational grid adopted in the simulations is 64x32x64 in the axial, radial and tangential directions, respectively. The predicted velocity distribution is shown in Figure 4. Again, the viscous sub-layer is well resolved by the present scheme. The departure from the log-law distribution at region away from the wall has also been indicated earlier.

In the present applications, the parallelism is achieved through the OpenMP[1] fortran

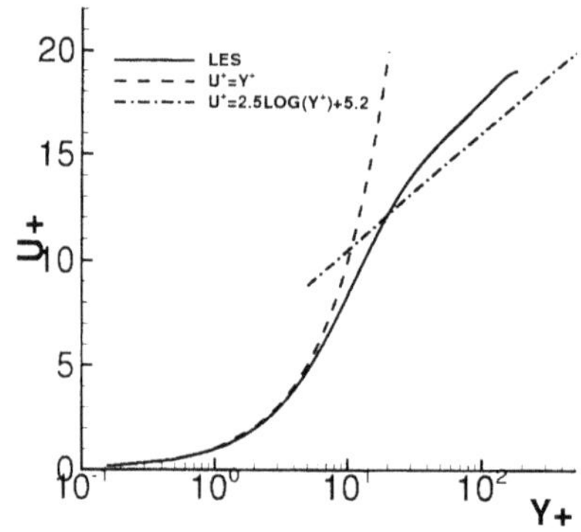

Figure 4. Log-law plot of pipe flow-$Re_\tau = 150$

implementations within the shared memory machines. Preliminary results of the performance of the TDMA algorithms using OpenMP is shown in Figure 5. Two computing platforms are adopted here. One is the SGI O2K, and the other one is the dual CPU PIII personal computers. The results indicate that the SGI O2K scales relatively well using the OpenMP, while the performance of the dual CPU personal computer does not scale as well. However, the PC represents a more cost-effective approach to achieve parallel computing. The performance of the Navier-Stokes solver is shown in Figure 6. The SGI O2K performs much better than the Dual Pentium III. However, the efficiency achieved is about 75% at 4 CPU, which is much lower than that achieved of the linear ADI solver.

6. Conclusion

The time averaged velocity distribution compared favorably with the DNS data. The viscous sub-layer is adequately resolved by the adopted scheme. The turbulence level also compares favorably with the DNS data, though the streamwise turbulence intensity is slightly under-predicted. The preliminary results of the parallel performance of ADI linear solver using OpenMP indicate the scalability of the model is good. However, the performance of the Navier-Stokes solver requires further improvement.

REFERENCES

1. OpenMP Fortran Application Program Interface (API), OpenMP Architecture Review Board, version 1.1-November-1999.
2. Smagorinsky, J., 1963, "General Circulation Experiments with the Primitive Equations. I. The Basic Experiments," Monthly Weather Review, Vol. 91, pp. 99-164.
3. Patankar, S.V. and Spalding, D.B., "A Calculation Procedure for Heat Mass Momen-

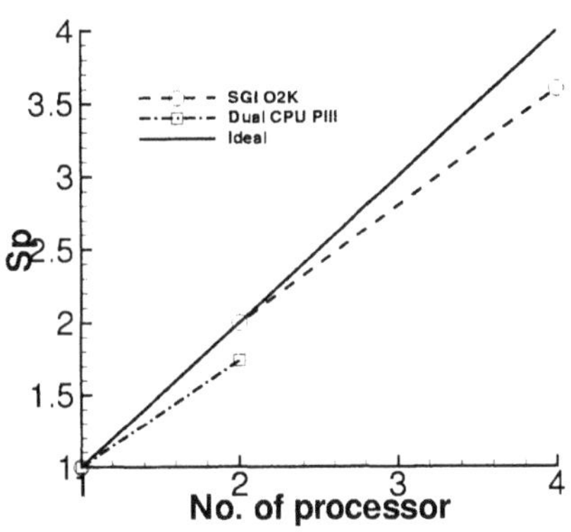

Figure 5. Performance of linear ADI solver

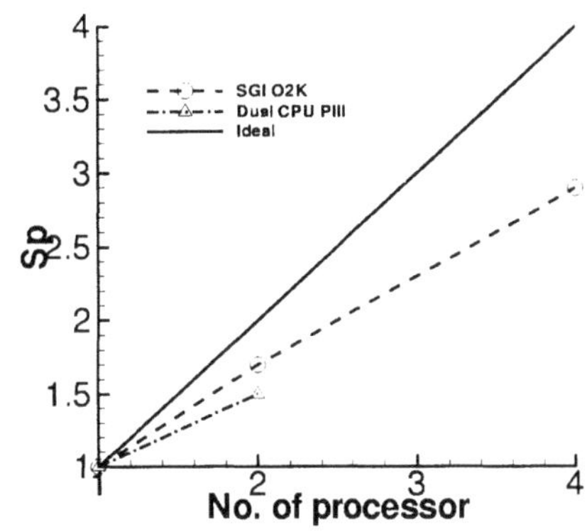

Figure 6. Performance of Navier-Stokes solver

tum Transfer in Three-Dimensional Parabolic Flow", International Journal of Heat and Mass Transfer, Vol 15, October 1972, pp. 1787-1806.

4. Brian, P. L. T., 1961, "A Finite Difference Method of High-Order Accuracy for the Solution of Three-Dimensional Transient Heat Conduction Problem," A. I. Ch. E. Journal, Vol 7, No. 3, pp. 367-370.

Parallel Computational Fluid Dynamics – Trends and Applications
C.B. Jenssen et al. (Editors)

MGLET: a parallel code for efficient DNS and LES of complex geometries

M. Manhart[a], F. Tremblay[a], R. Friedrich[a]

[a]Technische Universität, München, Germany

A code is presented for large eddy simulation (LES) and direct numerical simulation (DNS) of turbulent flows in complex geometries. Curved boundaries are approximated by 4th order (tricubic) interpolation within a Cartesian grid. The specific design of the code allows for an efficient use of vector and parallel computers. We will present a comparison of efficiency between a massively parallel (CRAY T3E) and a vector parallel (Fujitsu VPP/700) machine. As an example of application, the flow around a cylinder at $Re = 3900$ is considered. The accuracy of the results demonstrate the ability of the code to deal efficiently with large scale computational problems.

1. INTRODUCTION

The correct and reliable prediction of complicated turbulent flows with engineering methods (e.g. Reynolds averaged Navier Stokes, RANS) is still an unresolved problem in fluid mechanics. At the moment, it seems that only Direct Numerical Simulation (DNS) or Large-Eddy Simulation (LES) can provide reliable results. In a DNS, all relevant turbulent length and time scales have to be resolved. Because of limited computational power, up to now only low or moderate Reynolds numbers and simple geometries could be investigated by DNS. In a LES, higher Reynolds numbers can be simulated by resolving only the large scales of the turbulent flow and modelling the small scales by a so-called subgrid scale (SGS) model. Unfortunately, LES still requires large computational resources compared to RANS. Therefore, the efficient use of the available hardware is necessary in order to make LES affordable for industrial applications. The code presented here, MGLET, uses a number of different techniques to save memory and CPU time consumption. It runs efficiently on a number of different platforms, from workstations, to massively parallel machines (as the CRAY T3E) and vector supercomputers (as the Fujitsu VPP/700).

Being able to predict the flow over an arbitrarily shaped body with a Cartesian grid is very attractive, since typically a Cartesian code is between 10 and 30 times more economical in terms of both CPU time and memory requirements when compared to a code for general curvilinear coordinates [7]. One can thus afford to do a computation with better grid resolution and still achieve appreciable savings in computational resources. Another important aspect is the complete elimination of the need to produce a body-fitted grid, a task that is not trivial and can consume an important amount of time.

2. NUMERICAL METHOD

2.1. Basic code

The code presented here, MGLET, is based on a finite volume formulation of the Navier-Stokes equations for an incompressible fluid on a staggered Cartesian non-equidistant grid. The spatial discretization is of second order (central) for the convective and diffusive terms. For the time advancement of the momentum equations, an explicit second-order time step (leapfrog with time-lagged diffusion term) is used, i.e.:

$$u^{n+1} = u^{n-1} + 2\Delta t \left[C\left(u^{n}\right) + D\left(u^{n-1}\right) - G\left(p^{n+1}\right)\right] \tag{1}$$

where C, D and G represent the discrete convection, diffusion and gradient operators, respectively.

The pressure at the new time level p^{n+1} is evaluated by solving the Poisson equation

$$Div\left[G\left(p^{n+1}\right)\right] = \frac{1}{2\Delta t} Div\left(u^{*}\right) \tag{2}$$

where u^{*} is an intermediate velocity field, calculated by omitting the pressure term in equation 1. By applying the velocity correction

$$u^{n+1} = u^{*} - 2\Delta t G\left(p^{n+1}\right) \tag{3}$$

we arrive at the divergence-free velocity field u^{n+1} at the new time level.

The Poisson equation is solved by a multigrid method based on an iterative point-wise velocity-pressure iteration.

2.2. Arbitrary geometries in the Cartesian grid

The description of curved boundaries and complex geometries can be done by a number of different options. After testing some different possibilities (Manhart et al. [7]), we decided to extend a Cartesian code for arbitrarily curved boundaries, that has been developed and tuned for DNS and LES for more than 10 years. In our Cartesian grid approach, the noslip and impermeability conditions at curved boundaries is provided by (i) blocking the cells cut by the surface and (ii) prescibing the variables at the blocked cells as Dirichlet conditions. The values are computed by interpolation or extrapolation of the points that belong to a cell cut by the boundary. Using Lagrangian polynomials of 3rd order in three directions (tricubic), a 4th-order accurate description of the boundary is provided. Within certain restrictions concerning the geometry computed, the additional work introduced by this technique can be neglected.

2.3. Parallelisation

For running the code on parallel computers, we employed the following strategy. The original single-grid code has been extended to a block-structured code in order to manage multiple grids that arise from the multigrid algorithm and parallelisation. In this framework, parallelisation is done over the grid blocks using the original subroutines of the single-grid code. A domain decomposition technique has been employed in two directions to divide each of the grids into an arbitrary number of subgrids that are treated as independent grid blocks. The communication of neighbouring grids is done using the MPI library. In order to keep the data consistent, we employ a red-black algorithm in the

Table 1
Number of grid cells used for performance tests.

Case	#1	#2	#3	#4
N_X	256	576	1156	1156
N_Y	144	320	320	320
N_Z	96	96	96	192
N_{TOT}	$3.5 \cdot 10^6$	$17.7 \cdot 10^6$	$35.4 \cdot 10^6$	$70.8 \cdot 10^6$

velocity-pressure iteration. Therefore the convergence of the iterations is not dependent on the number of PE's used.

3. COMPUTATIONAL EFFICIENCY

3.1. Benchmark

The efficiency of the parallelisation was evaluated on two high performance computers with different architectures. The first, a CRAY T3E-900, is a massively parallel machine, whereas the second, a Fujitsu VPP700, is a vector-parallel computer. Four different numbers of grid cells corresponding to realistic actual problems (see table 1) were chosen for benchmarking. Considerable efforts have been done to optimize the single-processor performance of MGLET on scalar as well as on vector computers in order to achieve a fair comparison of the two platforms.

The performance of the vector computer VPP700 is extremely sensible on the vector length. We therefore changed the internal organization of the arrays in our Fortran77 code from (z, y, x) to (x, y, z) on the VPP in order to get the largest dimensions on consecutive memory addresses and to achieve a long vector length on the innermost loops. The domain decomposition has then been done over the y- and the z-directions, respectively. On a massively parallel computer, however, it is best to parallelize over the directions with the largest number of grid points, so we left the original organization (z, y, x) and we parallelized over y and x.

3.2. Performance

Each of the different cases has been run for 10 time steps on the two platforms with varying number of PE's used. Some observations could be made: (a) the maximal single processor performance on the VPP700 rises from 540 Mflop/s to 1021 Mflop/s with a vector length of 256 to 1156, (b) the maximal single processor performance achieved on the T3E-900 was about 70 Mflop/s, (c) a strong degradation of the single processor performance with increasing parallelisation can be found for the smallest problem on both machines. The single processor performance ratio between the VPP700 and the T3E-900 varies between 10 for small problems and 15 for large problems (i.e. long vector lengths). The resulting CPU-times spent in one time step for problem #1 are plotted in Figure 1 as a function of the number of PE's used. The small problem ($3.5 \cdot 10^6$ cells) can be computed within one CPU second/timestep. The large problem ($70.8 \cdot 10^6$ cells) takes about 10 CPU second/timestep if enough PE's are provided. In figure 2 the achieved

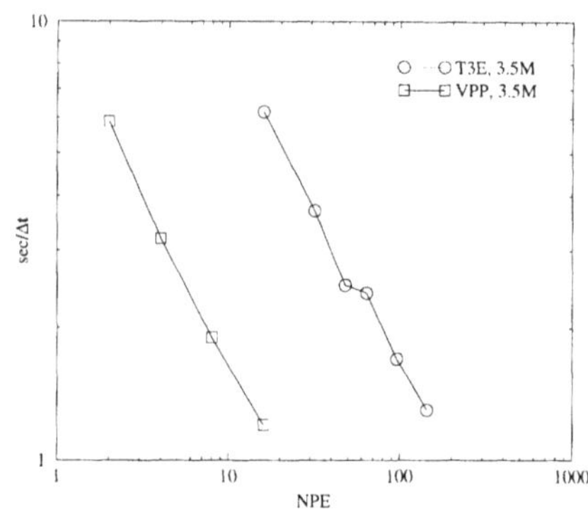

Figure 1. CPU-times for one timestep for the problem #1.

performance is plotted versus the number of PE's for the different benchmark problems. It seems that on both machines the performance scales with the problem size and number of PE's. The maximum performance lies at about 14 Gflop/s on both machines using 16 PE's on the VPP700 and 240 PE's on the T3E-900, respectively.

4. EXAMPLE

The code has been developed for more than 10 years for a number of applications. In the early versions, turbulent flows over rectangular obstacles have been treated by LES (Werner and Wengle, [11,12]). The extension to arbitrarily shaped bodies has been started by Manhart and Wengle [8] for the case of a hemisphere in a boundary layer. In this case, the body was simply blocked out of a Cartesian grid, which resulted in a first-order accuracy of the description of the body surface. A fourth-order description (tricubic) of the surface has been implemented recently. The method has been validated for a number of laminar cases. Second-order accuracy of the overall scheme has been demonstrated for the cylindrical Couette flow, and excellent agreement with other numerical experiments was obtained for steady and unsteady flows over a cylinder as studied by Schäfer et al. [10].

As an example of a current application of MGLET, the flow over a cylinder at Re=3900 is presented. The flow around a cylinder at Re=3900 has been investigated experimentally by Ong and Wallace [9], and Lourenco and Shih [5]. Recently, a DNS was performed by Ma et al. [6]. LES computations were presented by Breuer [2], Fröhlich et al. [3], Beaudan and Moin [1] and by Kravchenko and Moin [4].

Our computational domain is 20 diameters long in the streamwise direction, with the center of the cylinder being 5 diameters downstream of the inflow plane. In the normal direction, the domain size is also 20 diameters. The spanwise dimension of the domain was chosen to be πD. A uniform inflow is prescribed, and periodicity is used in the

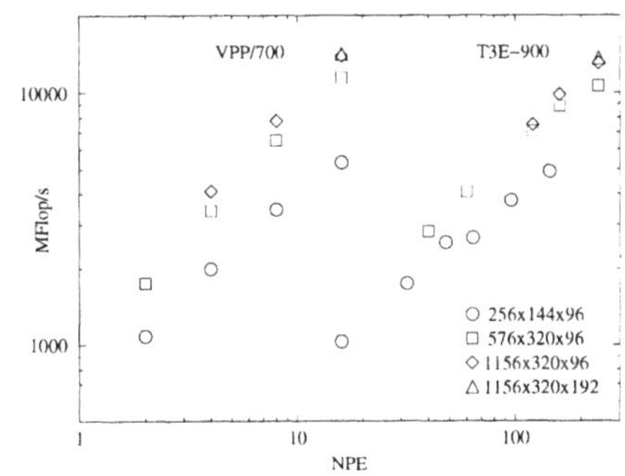

Figure 2. Mflop-rate achieved by the different test problems on the T3E-900 and VPP700.

normal and spanwise directions. A no-stress outflow condition is prescribed. The mesh was generated such that its size in the plane normal to the cylinder axis is of the same order of magnitude as the Kolmogorov length scale, which led to a total number of 48 million grid cells. The calculation was performed on 8 processors of the Fujitsu VPP700. With a mean performance of 7 GFlops, each time step requires 10 seconds.

Starting from a uniform flow field, the solution was advanced for 100 problem times, based on the diameter and the inflow velocity. Statistics were then gathered for about 300 additional problem times. The results are presented for first and second order statistics. The upper left diagram of Figure 3 contains the mean streamwise velocity a long the centerline. There is excellent agreement of our DNS data with the experiment in the near and far wake. The vertical profile of the variance of the streamwise velocity fluctuation at $X \approx D$ (upper right) reflects the proper peak values in the free shear layers and agrees well with the profile obtained by Ma et al. [6]. The Reynolds shear stress profiles at two downstream positions (lower left and right diagrams) reveal the right structural changes of the mean flow in the near wake region. At $X \approx D$, the overall shape of the shear stress is in agreement with the experiment and the results of Ma et al. [6]. The LES data of [4] on the other hand underpredict the peak Reynolds stresses.

Instantaneous spanwise velocity surfaces (Figure 4) demonstrate the complexity of the flow field consisting of large scale two-dimensional 'rolers' as well as fine grained turbulence. In the DNS results (Figure 4a), one can see that the fine structures persist over a long distance downstream. This means that a fine computational grid should be used even far downstream of the cylinder. The effect of grid coarsening can be seen in the two LES simulations of the same flow case (Figures 4b and 4c). If the grid is too coarse, the fine scales are distorted by numerical noise that affects even the large scales more downstream.

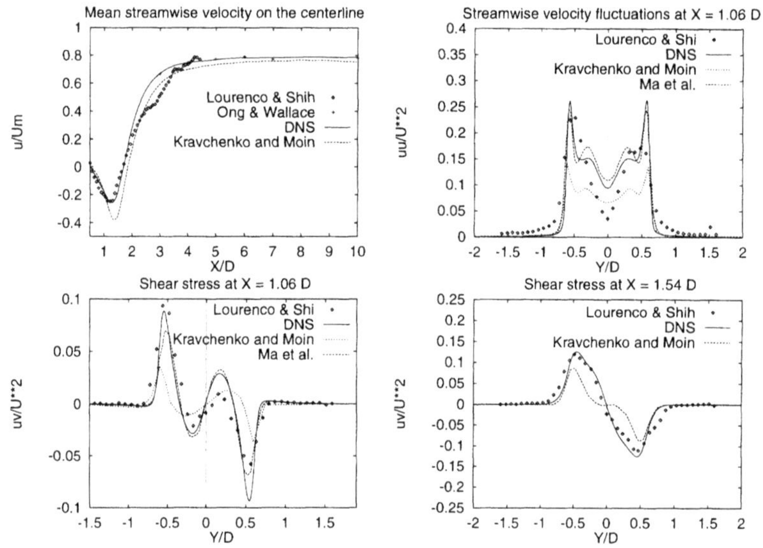

Figure 3. First and second order statistics in the near wake region

5. CONCLUSIONS

We have presented a code for DNS and LES of turbulent flows over arbitrarily shaped bodies. The code uses a Cartesian grid which results in an efficient use of computational resources. It is well suited for large scale computational problems typically done on high-performance computers.

The example showed here, the turbulent flow over a cylinder shows, that the 4th order accurate description of the surface within a Cartesian grid is an efficient way to compute such kind of flows. The results of the DNS compare well with available experiments. If on intends to use LES in order to saves computational resources, one has to be careful not to use a too coarse grid.

6. Notes and Comments

We gratefully acknowledge the support of the HLRS in Stuttgart and the LRZ in Munich. The work has been supported by the DFG under grant no. FR 478/15, Collaborative Research Centre 438 (Technical University of Munich and University Augsburg) and the European Community in context of the Alessia project.

Figure 4. Isosurfaces of instantaneous spanwise component. a) DNS b) LES ($7.7 \cdot 10^6$ cells) c) LES ($1.1 \cdot 10^6$ cells)

REFERENCES

1. P. Beaudan and P. Moin. Numerical experiments on the flow past a circular cylinder at sub-criti cal reynolds number. Report No. TF-62, Thermosciences Division, Department of mechani cal engineering, Stanford University, 1994.
2. M. Breuer. Large eddy simulation of the subcritical flow past a circular cylinder: numerical and modeling aspects. *Int. J. Numer. Meth. Fluids*, 28:1281–1302, 1998.
3. J. Fröhlich, W. Rodi, Ph Kessler, S. Parpais, J.P. Bertog lio, and D. Laurence. Large eddy simulation of flow around circular cylinders on structured a nd unstructured grids. In E.H. Hirschel, editor, *Notes on Numerical Fluid Mechanics*, pages 319–338. Vieweg-Verlag, 1998.
4. A.G. Kravchenko and P. Moin. B-spline methods and zonal grids for numerical simulations of turbulent flows. Report No. TF-73, Flow Physics and Computation Division, Department of mechanical engineering, Stanford University, 1998.
5. L.M. Lourenco and C. Shih. Characteristics of the plane turbulent near wake of a circular cylinder , a particle image velocimetry study. private communication, data taken from [4], 1993.
6. S. Ma, G.-S. Karamos, and G. Karniadakis. Dynamics and low-dimensionality of the turbulent near-wake. *J. Fluid Mech.*, to appear.
7. M. Manhart, G.B. Deng, T.J. Hüttl, F. Tremblay, A. Segal, R. Friedrich, J. Piquet, and P. Wesseling. The minimal turbulent flow unit as a test case for three different computer codes. In E.H. Hirschel, editor, *Vol. 66, Notes on numerical fluid mechanics.* Vieweg-Verlag, Braunschweig, 1998.
8. M. Manhart and H. Wengle. Large-eddy simulation of turbulent boundary layer flow over a hemisphere. In Voke P.R., L. Kleiser, and J-P. Chollet, editors, *Direct and Large-Eddy Simulation I*, pages 299–310, Dordrecht, March 27-30 1994. ERCOFTAC, Kluwer Academic Publishers.
9. J. Ong, L. & Wallace. The velocity field of the turbulent very near wake of a circular cylind er. *Experiments in Fluids*, 20:441–453, 1996.
10. M. Schäfer and S. Turek. Benchmark computations of laminar flow around a cylinder. In E.H. Hirschel, editor, *Notes on Numerical Fluid Mechanics*, pages 547–566. Vieweg-Verlag, 1996.
11. H. Werner and H. Wengle. Large-eddy simulation of turbulent flow over a square rib in a channel. In H.H. Fernholz and H.E. Fiedler, editors, *Advances in Turbulence*, volume 2, pages 418–423. Springer-Verlag, Berlin, 1989.
12. H. Werner and H. Wengle. Large-eddy simulation of turbulent flow over and around a cube in a plate channel. In F. et al. Durst, editor, *Turbulent Shear Flows 8*, Berlin, 1993. Springer.

Parallel Computational Fluid Dynamics - Trends and Applications
C.B. Jenssen et al. (Editors)

Large eddy simulation (LES) on distributed memory parallel computers using an unstructured finite volume solver

B. Ničeno and K. Hanjalić * [a]

[a]Thermofluids Section, Faculty of Applied Sciences, Delft University of Technology, Lorentzweg 1., P.O. Box 5046, 2600 GA, Delft, The Netherlands

LES is a well established technique for calculation of turbulent flows. Since it's early days, it has been basically used for calculation of turbulent flows in simple canonical geometries. Single-block finite difference or spectral methods were usual methods of choice for LES. Following the recent growth of interest for industrial application of LES, both the numerical methods and computer programs have to be changed in order to accommodate to the new, industrial, environment and to take advantage of massively parallel computers which have emerged in the meantime. In this work, we describe our approach to development of a computer program efficient enough to solve large scale fluid flow equations and yet flexible enough to deal with complex geometries. Some performance statistics and LES results obtained on the Cray T3E parallel computer are shown.

1. INTRODUCTION AND MOTIVATION

Since the early days of LES, which date to the beginning of 70's, LES has basically been used to investigate the fundamental characteristics of turbulence and guide the development of Reynolds Averaged Navier-Stokes (RANS) turbulence models. These fundamental investigations were conducted in simple canonical geometries.

In this early days of LES, the most powerful computers were the high-end vector machines. To take advantage of this high-end vector processors, computer codes had to be vectorised. Efficient vectorisation can be achieved if the data structure is regular, which led the LES practitioners to development single-block structured codes based on finite difference (of 2nd or higher order of accuracy) or on the spectral methods.

In the middle of 90's, two important things have happened, which put different demands on design of computer programs for LES of turbulence. First, the mixed success of RANS turbulence models has stimulated industry to turn its attention to LES technique. However, highly vectorised, single block codes are, in general, not flexible to deal with complex geometries (which industry demands) making such codes, although very efficient, unsuitable for industrial application. Second, distributed memory massively parallel computers have appeared which represent the future high performance computer platforms [3]. Our main goal is to develop a computer program suitable for LES of complex, industrial, applications and efficient enough on modern, distributed memory parallel computers.

*This research was sponsored by AVL AST GmbH which is gratefully acknowledged.

2. GOVERNING EQUATIONS FOR LES

We consider the incompressible Navier-Stokes equations with Smagorinsky model for sub-grid scale (SGS) terms. The Navier-Stokes equations in their integral from are:

$$\frac{d}{dt}\int_V \rho \mathbf{u} dV + \int_S \rho \mathbf{u}\mathbf{u} d\mathbf{S} = \int_S \mathbf{T}_{eff} d\mathbf{S} \tag{1}$$

These equations are valid for an arbitrary part of the continuum of the volume V bounded by surface S, with the surface vector $\mathbf{S}$ pointing outwards. Here ρ is the fluid density $\mathbf{u}$ is the velocity vector, $\mathbf{T}_{eff}$ is the effective stress tensor. The effective stress tensor is decomposed into two parts:

$$\mathbf{T}_{eff} = 2\mu_{eff}\mathbf{D} - p\mathbf{I} \tag{2}$$

where

$$\mathbf{D} = \frac{1}{2}[grad\mathbf{u} + (grad\mathbf{u})^T] \tag{3}$$

is the rate of strain tensor, μ_{eff} is the effective viscosity, p is the pressure. Effective viscosity is obtained from

$$\mu_{eff} = \mu + \mu_{sgs} \tag{4}$$

where μ is the dynamic fluid viscosity and μ_{sgs} is the turbulent viscosity, in the present approach obtained from the Smagorinsky model:

$$\mu_{sgs} = (C_s\Delta)^2(\mathbf{DD})^{1/2} \tag{5}$$

where Δ is the filter width set to $\Delta V^{1/3}$, where ΔV is the volume of computational cell and C_s is the Smagorinsky constant, usually set to 0.06–0.1 depending on the flow case.

Smagorinsky constant is reduced in the near-wall regions by taking the filter width to be the minimum between the cubic root of the cell volume and distance to the nearest wall [4].

3. NUMERICAL METHOD

When choosing a numerical method for industrial LES we seeked a compromise in terms of geometrical flexibility, accuracy and suitability for parallelisation.

The traditionally employed methods in LES, such as high order (higher then two) finite differences or spectral methods, are not suitable for complex geometries occurring in industrial applications. Spectral element methods offer higher order of accuracy and geometrical flexibility, but up to now, are only capable of dealing with geometries with one homogeneous direction, which significantly reduces their applicability for industrial LES. Furthermore, all the commercial CFD packages are based on low-order finite elements or control volume methods and is very likely that first practical LES computations will be done using such methods.

Therefore, the numerical method chosen is based on the 2nd order control volume method on unstructured grids [1]. Possible computational cells are shown in figure (1).

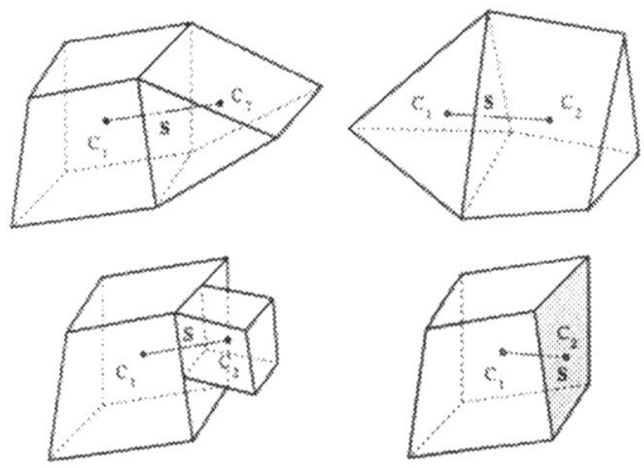

Figure 1. Supported cell shapes and associated data structure. Data structure is the same for all the cell faces (shaded).

Both the numerical method and the computer program are capable of calculating the governing equations on any type of grid. Any cell type (hexahedronal, trapezoidal, tetrahedral, hybrid) is allowed and local grid refinement can be achieved by locally splitting the computational cells. This makes our program easily applicable in industrial environment.

A very important feature of the method is that the grid is colocated, i.e. both velocities and pressure are calculated at the centres of finite volumes, which saves a great deal of memory required for storing the geometrical data. There is a price we had to pay for such saving, i.e. ccolocated variable arrangement is more prone to oscillations in the flow field, and blended differencing scheme with a small percentage of upwind (1 – 2 %) is usually needed to increase the stability of the method. Nonetheless, the savings gained by storing the cell centre values only can be used to increase the total number of cells in the domain, thus reducing the cell Peclet number and reducing the amount of upwind needed.

3.1. Cell-Face Data Structure

The applied data structure has a large impact on both the simplicity of the numerical code and on it's suitability for subsequent parallelisation. We have found it very useful to organise data around the computational cell faces [2]. In other words, for each cell face in the computational domain, we store the cell indexes adjacent to it. The typical situation is depicted in figure (1). The adjacent cell indexes for each cell face are stored which allows our code to deal with any cell type.

4. DOMAIN DECOMPOSITION

Domain decomposition is performed with the simple geometrical multi section approach. The domain can be decomposed into any number of sub-domains, which are always equally balanced in terms of number of cells per processors. The basic idea of the domain decomposition we adopted is to cut the domain along the coordinate axis with greatest dimension. This cutting can be (and usually is) recursively applied to newly formed sub-domains, until we reach the desired number of sub-domains. This procedure is very fast, since the cells are sorted with the quick-sort algorithm, and each sub-domain is stored in a separate file allowing the subsequent parallel I/O. An example domain, decomposed into ten sub-domains is shown in figure (2). It is visible that the domain was cut in coordinated directions. However, if more complex shapes of the domain are considered, this procedure might give very poor partitions and more sophisticated methods should be used.

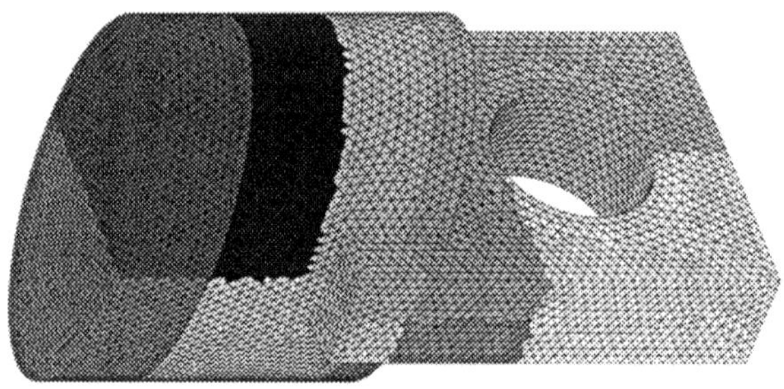

Figure 2. Grid employed for the calculation of the matrix of cubes decomposed into 32 sub-domains. Different shades of gray correspond to different partitions.

5. PARALLELISATION OF LINEAR SOLVERS

The discretised system of equations arising from the discretisation of momentum and pressure correction equations are solved using the solvers from Krylow subspace family. Conjugate gradient (CG) and conjugate gradient squared (CGS) are used for solving the pressure corrections, whereas bi-conjugate gradient (BiCG) method is used to solve the discretised momentum equations. All solvers from Krylow subspace family consist of matrix vector and vector dot products which are easy to parallelise [5]. In all the computations and results reported here, we used diagonal pre-conditioning.

6. PARALLEL COMPUTATIONAL RESULTS

When measuring the performance of the parallel code, one usually speaks about the *absolute speed-up* of the code. An absolute speed-up; S_A, is defined with:

$$S_A = \frac{T_N}{T_1} \qquad (6)$$

where T_N is the CPU time for calculation on N processors, and T_1 is the CPU time for calculation on one processor. But, due to the fact that local memory of each processor on our Cray-T3E is limited to 128 MB, and operating system consumes about 50 MB, evaluation of T_1 and T_n is limited to a very small number of cells (50000 in our case). If we decomposed the domain with 50000 cells to 64 processors, we would get less then 800 cells per sub-domain and performance of the code would be severely reduced by the

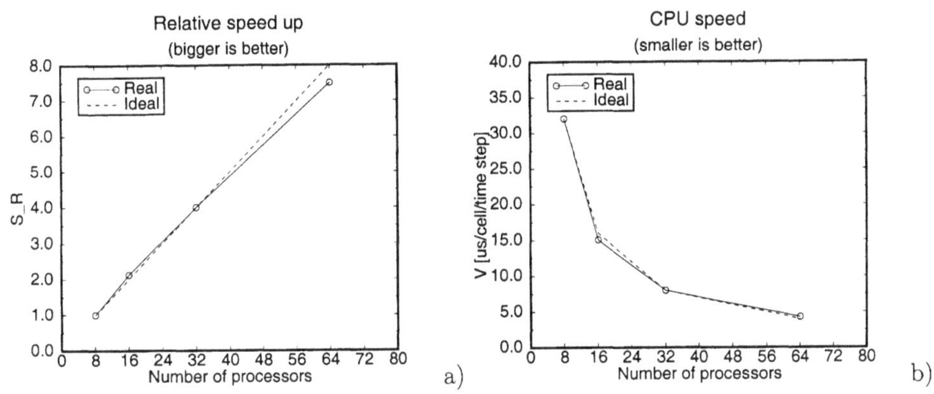

Figure 3. Performance of the code: a) relative speed-up obtained and b) absolute speed of the code measured in $[\mu s/(time\ step\ cell)]$. Both were measured on the Cray T3E.

communication overhead. As a consequence, the figure we would get for the absolute speed-up would not be very illustrative. To avoid that problem, we had to define a *relative speed-up*; S_R:

$$S_R = \frac{T_N}{T_n} \tag{7}$$

where T_N is the CPU time for calculation on N processors and T_n is the CPU time for calculation on n processors, (n is smaller then N, of course). In our case, n was 8, and N was 16, 32 and 64. The results for the relative speed-up of the code are given in figure 6. Figure (6) shows that we achieved very good, almost linear relative speed-up in the range from 8 to 64 processors. It might look surprising that we have even super-linear speed-up for 16 processor, but this is the consequence of the double buffering of the DEC-Alpha processors. Since less cells are assigned to each processor when executing on 16 then on 8 processors, more data remains in buffers, and access to data in buffers is much faster then access to data in core memory. For 32 and 64 processors even more data resides in buffers, but the communication cost is also larger, so the speed-up diminishes.

The maximum speed that we have obtained was 4.1 $[\mu s/(time\ step\ cell)]$ (micro seconds per time step and per cell) when 64 processors were used.

7. LES RESULTS

In this section we show the results obtained for the flow around the matrix of cubes. The matrix of cube of height H is mounted at the bottom of the channel wall of height h. The cubes form a rectangular array and the pitch between the cubes is $4h$. Reynolds number, based on channel height is 13000. It was experimentally investigated in [6].

The grid we was used consisted of 486782 cells and was decomposed in 32 sub-domains (figure (4). Since the problem domain is very simple (cube placed in the channel) grid was generated using hexahedronal cells only. It was stretched towards the cube faces and

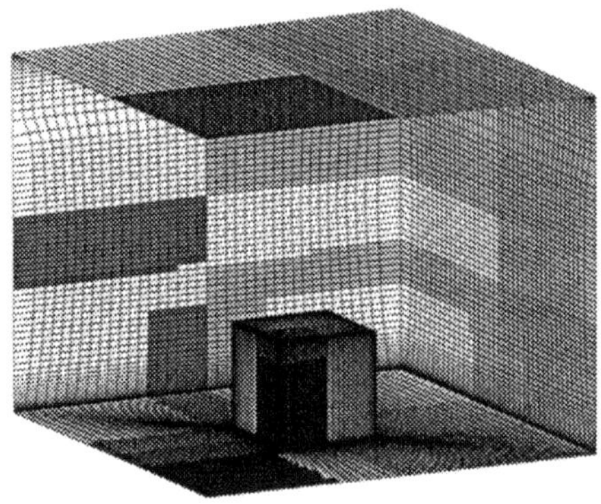

Figure 4. Hexahedronal grid used for the calculation of the matrix of cubes decomposed into 32 sub-domains. Different shades of gray correspond to different processors.

towards the channel walls to reach the y^+ of 0.5. Such a small value of y^+ was needed because the near wall regions were resolved rather then modelled by a wall function. The computation has been performed over 50000 time steps, where the last 35000 have been used to gather statistics. The time for gathering statistics corresponds to 35 shedding cycles. The entire simulation took approximately 70 hours on the Cray T3E using 32 processors.

The comparison of computed mean velocity profiles and Reynolds stresses with experiments in one characteristic plane is shown in figure (5). The agreement with experimental results is satisfactory. The comparison of our results with those obtained by other authors can be found in [7] and [8]. Structure of the flow is shown in figure (6) which shows the streamlines in the vertical plane and in figure (7) which shows the streamlines in the horizontal plane.

8. CONCLUDING REMARKS

LES with unstructured solvers is relatively new topic and there are many open questions associated with it. In this work, however, we report our approach towards that goal which so far consist in development of the basic tool which was an unstructured finite volume solver parallelised for the distributed memory machines. There are many issues which require more attention in the future work. Efficient parallel pre-conditioning, more flexible domain decomposition techniques, thorough examination of filers and implementation of more sophisticated models to name just the few. Nonetheless, in this work we have shown that LES with an unstructured solver is feasible on modern computational platforms and might find it's place in the arsenal of tools applied in industrial research of turbulent flows.

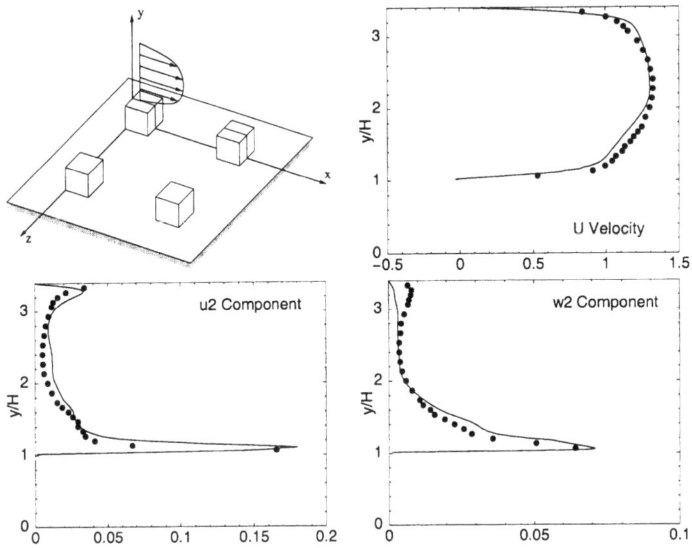

Figure 5. Mean velocity profile normalised with bulk velocity in $x - y$ plane at $x = 0.3H$. Dotted line are the experimental results from Meinders et.al. [6], continuous line are present results.

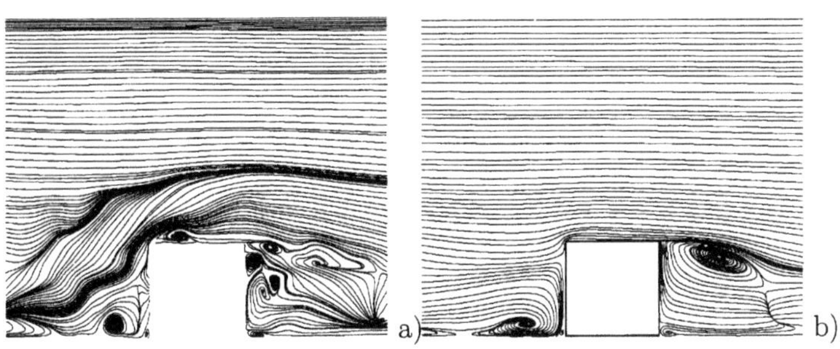

Figure 6. Streamlines in the $x - y$ plane at $z = 0$: a) Instantaneous b) Time averaged.

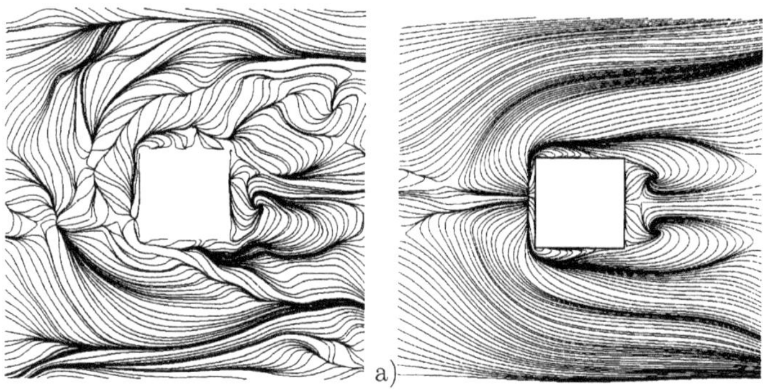

Figure 7. Streamlines in the $x-z$ plane at $y=0.5H$: a) Instantaneous b) Time averaged.

REFERENCES

1. I. Demirdžić, S. Muzaferija and M. Perić, "Advances in computation of heat transfer, fluid flow and solid body deformation using finite volume approaches", *Advances in numerical heat transfer* Vol. 1, pp. 59-96, 1997.
2. T. J. Barth, "Apsects of unstructured grids and finite volume solvers for the Euler and Navier-Stokes equations", *von Karman Institute lecture Series 1994-05.* (1994).
3. P. H. Michielse, "High Performance Computing: Trends and Expectations", *High Performance Computing in Fluid Dynamics*, P. Wesseling (Ed.), Delft, 1996.
4. P. R. Spalart, W. H. Jou, M. Strelets and S. R. Allmaras, "Comments on the feasibility of LES for wings, and on a hybrid RANS/LES approach", *Numer. Heat Transfer, Part B*, Vol. 27, pp. 323-336, 1995.
5. Henk A. Van der Vorst, "Parallel aspects of iterative methods", *Proc. IMA Conf. on Parallel Computing*, A.E. Fincham, and B. Ford (Ed.), Oxford University Press, Oxford, UK, 1993, pp. 175-186.
6. E.R. Meinders and K. Hanjalić, "Fully developed flow and heat transfer in a matrix of surface–mounted cubes", *Proceedings of the* 6th *ERCOFTAC/IAHR/COST Workshop on refined flow modelling.*, K. Hanjalić and S. Obi (Ed.), Delft, June, 1997.
7. Van der Velde, R.M., Verstappen, R.W.C.P. and Veldman, A.E.P., "Description of Numerical Methodology for Test Case 6.2", *Proceedings 8th ERCOFTAC/IAHR/COST Workshop on Refined Turbulence Modelling*, Report 127, pp 39-45, Helsinki University of Technology, 17-18 June 1999.
8. Mathey, F., Fröhlich, J., Rodi. W., Description of Numerical Methodology for Test Case 6.2 *Proceedings,* 8th *ERCOFTAC/IAHR/COST Workshop on Refined Turbulence Modelling*, Report 127, pp 46-49, Helsinki University of Technology, 17-18 June, 1999.

Parallel Computational Fluid Dynamics – Trends and Applications
C.B. Jenssen et al. (Editors)

LES Applications on Parallel Systems

L. Temmerman[a], M. A. Leschziner[a], M. Ashworth[b] and D. R. Emerson[b]

[a]Engineering Department, Queen Mary College, University of London, Mile End Road, London E1 4NS, United Kingdom

[b]CLRC Daresbury Laboratory, Warrington WA4 4AD, United Kingdom

A Large Eddy Simulation code based on a non-orthogonal, multiblock, finite volume approach with co-located variable storage, was ported to three different parallel architectures: a Cray T3E/1200E, an Alpha cluster and a PC Beowulf cluster. Scalability and parallelisation issues have been investigated, and merits as well as limitations of the three implementations are reported. Representative LES results for three flows are also presented.

1 INTRODUCTION

With computing power continuing to increase at a rate exceeding most conservative estimates, the high computational costs of Large Eddy Simulation (LES), relative to those required by statistical turbulence models, no longer represent one of the principal obstacles to LES becoming a viable approach to predicting industrial flows.

In LES, filtered forms of the Navier-Stokes equations are used to simulate the large-scale turbulent motions of a flow in a time-accurate manner. The small-scale motion, which cannot be resolved due to the coarseness of the mesh, is represented by a suitable "subgrid-scale model". Fundamentally, this method is superior to that based on the Reynolds-Averaged Navier-Stokes (RANS) approach because it is insensitive to disparities between periodic, coherent motion and stochastic turbulence. Moreover, it captures the turbulence dynamics and gives a better representation of the effects of the large-scale, energetic eddies characteristic of complex separated flows.

LES, however, also poses a number of specific challenges. The nature of the subgrid-scale model can have a strong influence on the accuracy; the treatment of wall effects can be very problematic in flows separating from a continuous surface; and grid quality, in terms of aspect ratio, expansion ratio and skewness, is much more important than in RANS computations. Last, but not least, LES requires very high cpu resources due to the (invariably) 3-d nature of the computation, the fineness of the mesh and the many thousands of time steps required for converged turbulence statistics to be obtained.

As with many numerical approaches, parallelisation offers an effective means of reducing run-times substantially. This paper describes one such approach in LES computations using a non-orthogonal, multiblock, finite-volume, pressure-based scheme developed for simulating incompressible flows separating from curved walls. The code was ported to three quite different architectures: a Cray T3E, an Alpha cluster and a PC Beowulf cluster. The paper focuses primarily on parallel performance and scalability issues, but also reports some representative simulation results for fully-developed channel flow, a separated flow in a duct with a wavy lower wall, and a high-lift single-element aerofoil.

2 SOLUTION STRATEGY

The code [1] is based on a finite-volume approach and fully co-located arrangement for the variables. A second-order fractional step method is employed for the velocity in conjunction with domain decomposition, multigrid acceleration, Rhie-and-Chow [2] interpolation and partial diagonalisation [3] in one direction, if that direction is statistically homogeneous and orthogonal to the other two. In the first step, the convection and diffusion terms are advanced using the Adams-Bashforth method, while the cell-centred velocities are advanced using a second-order backward scheme *viz.*:

$$\frac{3u^* - 4u^n + u^{n-1}}{2\Delta t} = 2CD^n - CD^{n-1} \tag{1}$$

In (1), u^* represents an intermediate velocity and C and D are the convective and diffusive operators, respectively. Spatial derivatives are evaluated using a second-order central differencing scheme. This first step, given by (1), is fully explicit. The second step consists of solving a pressure-Poisson problem, obtained by projecting the intermediate contravariant velocities onto the space of the divergence-free vector field, and applying the mass-conservation equation. The pressure equation arises as:

$$\begin{aligned} &\Delta p^{n+1} = \frac{3}{2\Delta t}\vec{\nabla} \bullet \vec{C}^* \\ &\vec{\nabla} p^{n+1} \bullet \vec{n} = 0 \text{ on the boundary} \end{aligned} \tag{2}$$

where $\vec{C}^*$ is the intermediate contravariant velocity. Finally, cell-centred velocities and face fluxes are updated with two different formulations of the discrete pressure gradients using the Rhie-and-Chow [2] interpolation.

The method has been successfully applied to periodic channel flow and separated flow in a duct with periodic hills. Work in progress focuses on separated aerofoil flow at a Reynolds number, based on the chord length, of $2.2\text{x}10^6$.

3 DOMAIN DECOMPOSITION

The present approach uses block decomposition with halo data. Due to the elliptic nature of the Poisson equation, each block has an influence on all others. To reduce the amount of communication between blocks, partial diagonalisation [3] is employed to accelerate the convergence of the Poisson equation. This decomposes the 3-d problem into a series of 2-d problems, each consisting of one spanwise plane. The interdependence between blocks is reduced, and a 2-d multigrid solver is used to solve the pressure-Poisson equation across spanwise planes. The current algorithm combines a Successive-Line-Over-Relaxation (SLOR) on alternate directions and a V-cycle multigrid scheme. This approach is very efficient, but partial diagonalisation limits the applicability of the code to problems for which one of the directions is orthogonal to the two others. For fundamental LES studies, this is not a serious restriction, because of the statistically 2-d nature of many key laboratory flows for which extensive measurements or DNS data have been obtained and which are used to assess the capabilities of LES. Examples include high-aspect-ratio channel and aerofoil flows in which the spanwise direction may be regarded as statistically homogeneous.

4 DESCRIPTION OF COMPUTING ARCHITECTURE

It is widely agreed that Beowulf systems offer very cost-effective computing platforms. However, the weak point of these systems is their communications, which is usually effected through Ethernet. Other options are available that give very good performance, but their cost has generally been too high for modest systems. The Beowulf facilities at Daresbury that have been used in the present investigation are as follows:

- a 32-processor Intel Pentium III system (Beowulf II), with each processor having a cycle rate of 450 MHz with 256 MB of memory, communications is via Fast Ethernet;
- a 16-node Compaq system (Loki) with each node having a dual-processor Alpha EV67 (21264A) with a clock cycle of 667 MHz, 512 MB of memory, and communication is via the QSW high performance interconnect.

The total cost per processor of these systems is far lower than the UK's current flagship facility, the Cray T3E/1200E. This machine has 788 application processors running at 600 MHz and the peak performance is just under 1 Tflop/s. Each PE has 256 MB of memory. Tests performed using MPI for communication indicate that the latency of the Cray T3E system is approximately 10 μs. The latency on the Loki system, using MPICH, is approximately 20 μs and on the Beowulf using LAM it is 100 μs. The maximum bandwidth achieved on the T3E was around 220 MB/s, as against 160 MB/s on Loki and 10 MB/s on the Beowulf II system. Iterative algorithms on parallel systems, particularly those using multigrid schemes, require fast low-latency communications to work efficiently.

5 PARALLEL PERFORMANCE FOR CHANNEL FLOW

The test case selected for examining parallel performance was a periodic channel flow, which is a typical initial LES validation case. The size of the computational domain was $2\pi h \times 2h \times \pi h$. The number of time steps was fixed at 1000. To minimise any algorithmic effects, the number of sweeps in the multigrid routine was kept constant. This is referred to later as a 'fixed problem size'. The total number of iterations required for the multigrid algorithm to converge to a given tolerance depends, of course, on the grid size and tends to increase as the grid is refined. A restriction of the current code is that the minimum number of processors required must be greater than or at least equal to the number of cells in the spanwise direction.

The first set of results gives the time to solution for a fixed problem size, as described above. The grid contained 96×48×4 points. Figure 1 shows the solution time for 1000 time steps, in seconds, on 4 to 32 processors on the Beowulf II, Loki and the Cray T3E.

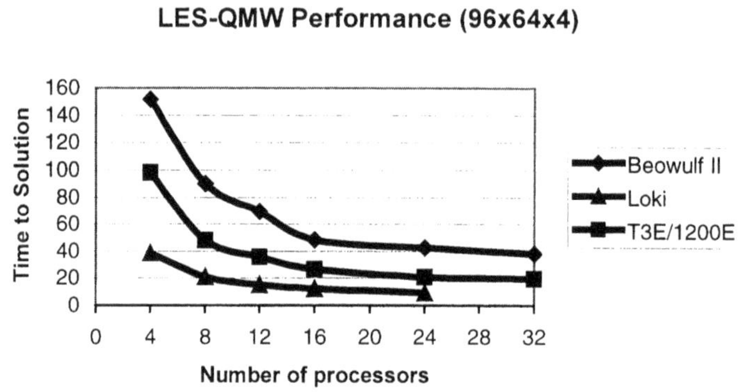

Figure 1. Time to solution for periodic channel flow (96×48×4 mesh, 1000 steps)

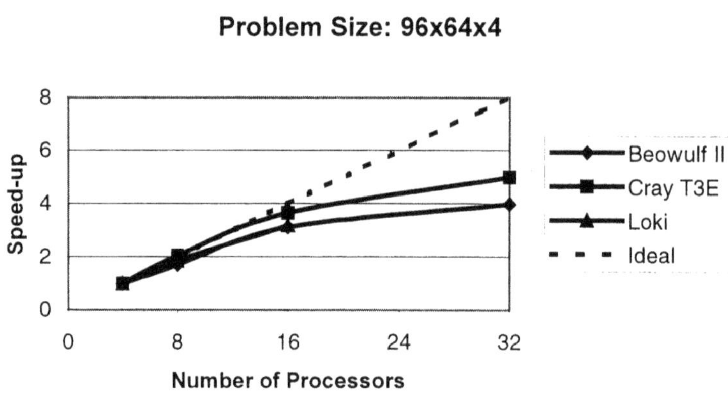

Figure 2. Speed-up comparison between all three systems for periodic channel flow

The Pentium system, whilst the slowest, performs very well and shows that the code is scaling satisfactorily. The performance of the new Compaq system is clearly superior to that of both the Pentium cluster and, significantly, also to the Cray T3E. For technical reasons, it was not possible to run the 32-processor case on Loki. Figure 2 compares the speed-up of the Pentium system and the Cray T3E for the same modest fixed-size problem. For this case, it is clear that the better communication network of the Cray T3E allows better scalability. Figure 3 shows that for larger problems the Beowulf system scales as well as the Cray T3E. This figure also indicates a super-linear speed-up of the Cray. This feature is quite common on such machines and is the result of effective cache utilisation.

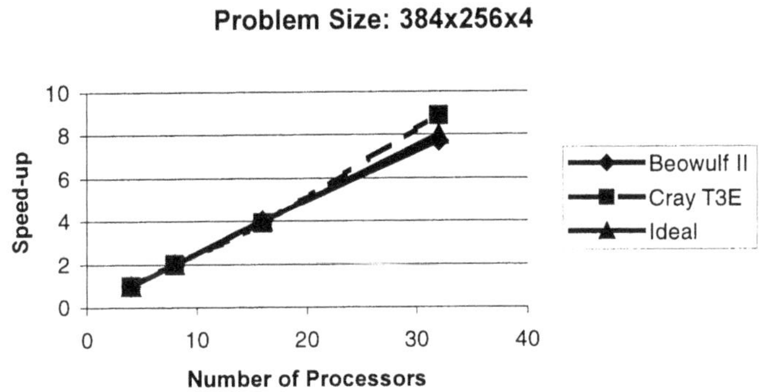

Figure 3. Scaling of fine-grid channel-flow solution on Beowulf II and the Cray T3E

6 SIMULATION RESULTS

Results given below demonstrate the ability of the code to perform realistic Large Eddy Simulation for the benchmark geometry (the periodic channel) as well as for more complex configurations.

6.1 Periodic channel case

Simulations for the channel flow were performed with a 96×64×64 mesh covering the box $2\pi h \times 2h \times \pi h$. Only one of many computations performed with different subgrid-scale models and near-wall treatments is reported here. The simulation was carried out with 64 processors on the Cray T3E, the domain being decomposed into 4×4×4 blocks. The Reynolds number is 10,935, based on bulk velocity and channel half-width, h, for which statistics obtained by DNS are available for comparison [4]. Subgrid-scale processes were represented by the Smagorinsky model [8] coupled with the van-Driest damping function. Figures 4-6 show, respectively, statistics of streamwise velocity, turbulence intensities and shear stress, averaged over a period of 12 flow-through time, in comparison with DNS data [4].

6.2 Periodic hill flow

This geometry (see Figure 7) is a periodic segment of a channel with an infinite number of 'hills' on the lower wall. Periodicity allows the simulation domain to be restricted to a section extending from one hillcrest to the next. The Reynolds number, based on hill height, h, and bulk velocity above the hillcrest, is 10,595. The flow was computed using the WALE subgrid-scale model [5] and the Werner-Wengle near-wall treatment [9], implying the existence of a $1/7^{th}$ power law for the instantaneous velocity in the near-wall region. The solution domain is 9h×3.036h×4.5h and is covered with 112×64×92 cells. The simulation was performed using 92 processors of the Cray T3E. Statistics were collected over a period of 27.5 flow-through times, with a time step equal to 0.006 s, and this required approximately 289 cpu hours.

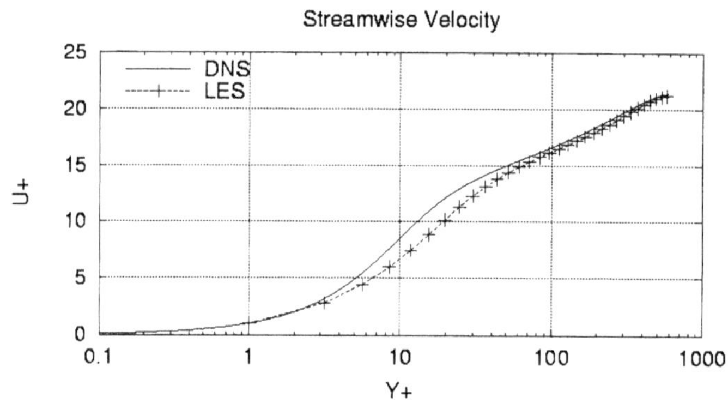

Figure 4. Streamwise velocity profile in wall co-ordinates in channel flow

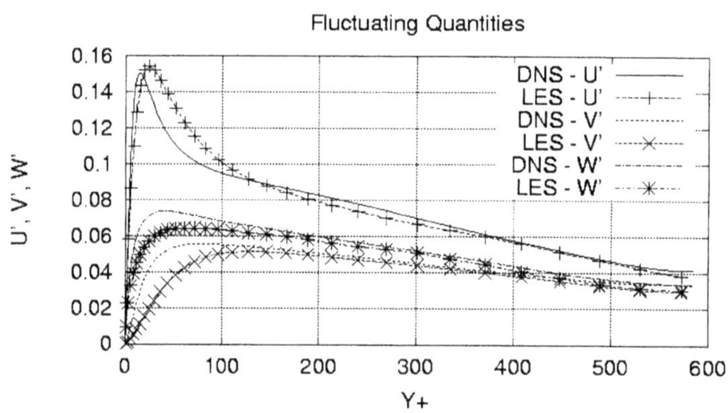

Figure 5. Turbulence intensities in channel flow

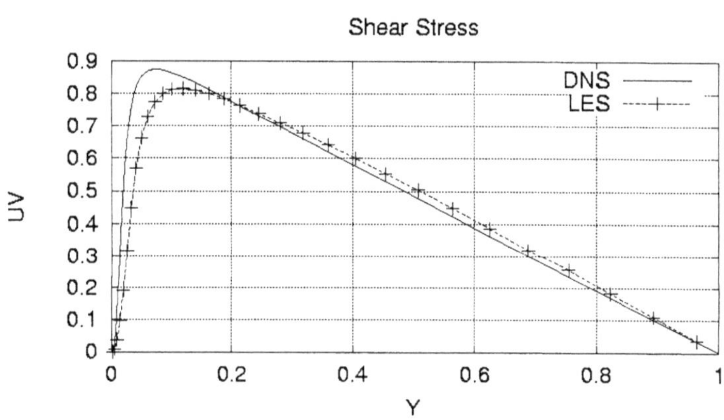

Figure 6. Shear stress in channel flow

Figure 7(a) shows the grid normal to the spanwise direction, and Figure 7(b) gives a view of the mean flow, represented by streamlines. This result is part of a much larger study in which the performance of several subgrid-scale models and near-wall treatments have been compared to a nearly fully-resolved computation on a grid of 6 million nodes, this reference computation requiring about 30,000 cpu hours, corresponding to about 150 wall-clock hours.

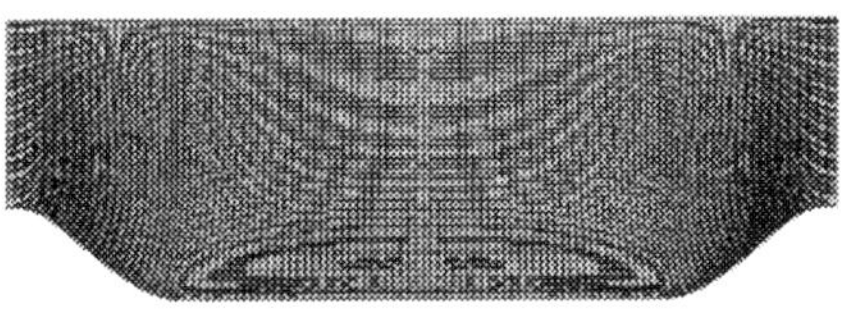

(a)

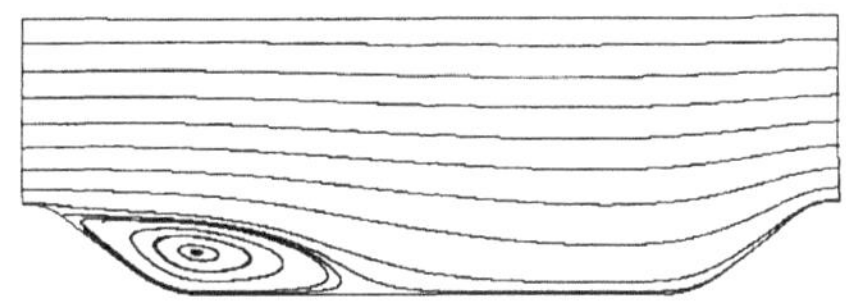

(b)

Figure 7. (a) Grid and (b) time-averaged streamfunction contours for the periodic hill flow

6.3 Aerofoil flow

This last case illustrates work in progress. The geometry, shown in Figure 8(a), is a single-element high-lift aerofoil ("Aerospatiale-A") at a 13.3° angle of attack and Reynolds number of 2.0×10^5, based on the chord and free-stream velocity. The flow is marginally separated on the rear part of the suction side.

Figure 8(a) gives a greyscale plot of instantaneous streamwise velocity obtained on a 320×64×32 grid. Of greater interest than the physical interpretation, in the context of this paper, is the parallel performance achieved on partitions of 32 to 256 processors. Figure 8(b) shows the speed-up obtained on a Cray T3D with a 320×64×32 grid used at a preliminary stage of the investigation. The results demonstrate good scalability characteristics of the code for this complex configuration and challenging flow conditions.

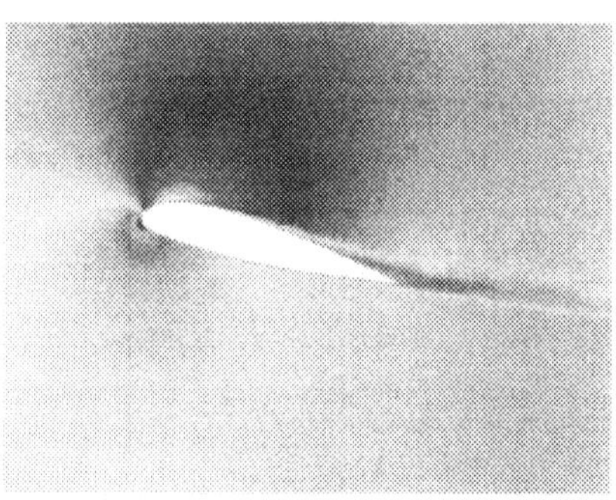

(a)

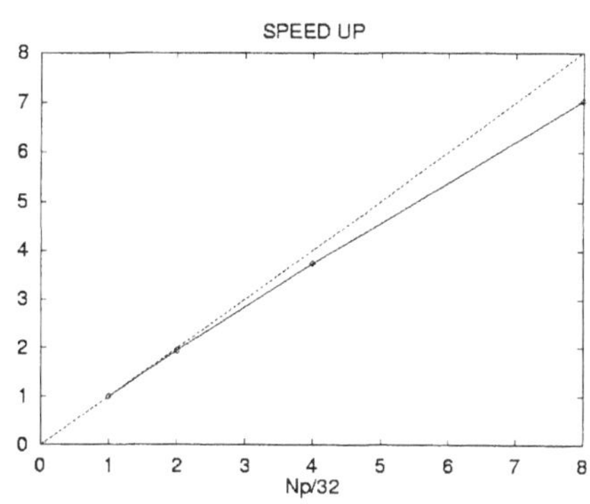

(b)

Figure 8. (a) Instantaneous streamwise velocity and (b) speed-up relative to 32 T3E processors for the flow around the high-lift "Aerospatiale-A" aerofoil.

7 CONCLUDING REMARKS

A parallel LES code has been successfully ported to three different parallel architectures. The code was shown to scale well on all three machines when the problem size is appropriate to the particular architecture being used. The relationship between problem size per cpu, cpu performances and network speed is shown to be complex and of considerable influence on performance and scaling.

Overall, the Compaq-based Loki configuration gave the best performance, the Cray T3E having better scalability for smaller problems. The Pentium-based Beowulf system was shown to be very competitive, giving a similar speed-up to the Loki system. The LES results included for geometrically and physically more challenging flows demonstrate that parallel systems can be used for such simulations at relatively low cost flow and very modest wall-clock times.

8 ACKNOWLEDGEMENTS

Some of the results reported herein have emerged from research done within the CEC-funded project LESFOIL (BRPR-CT-0565) in which the first two authors participate. The authors are grateful for the financial support provided by the CEC and also to EPSRC for support allowing the use of the CSAR Cray T3E facility at the University of Manchester and the Beowulf facilities at Daresbury Laboratory.

REFERENCES

1. R. Lardat and M. A. Leschziner, *A Navier-Stokes Solver for LES on Parallel Computers*, UMIST Internal Report (1998).
2. C. M. Rhie and W. L. Chow, *Numerical Study of the Turbulent Flow Past an Airfoil with Trailing Edge Separation*, AIAAJ, 21, No. 11, 1983, pp. 1525-1532.
3. U. Schumann and R. A. Sweet, *Fast Fourier Transforms for Direct Solution of Poisson's Equation with Staggered Boundary Conditions*, JCP, 75, 1988, pp.123-137.
4. R. D. Moser, J. Kim and N. N. Mansour, *A Selection of Test Cases for the Validation of Large Eddy Simulations of Turbulent Flows*, AGARD-AR-345, 1998, pp. 119-120.
5. F. Ducros, F. Nicoud and T. Poinsot, *Wall-Adapting Local Eddy-Viscosity Models for Simulations in Complex Geometries*, in Proceedings of 6th ICFD Conference on Numerical Methods for Fluid Dynamics, 1998, pp. 293-299.
6. M. Germano, U. Piomelli, P. Moin and W. H. Cabot, *A Dynamic Subgrid-Scale Eddy Viscosity Model*, Physics of Fluids A3 (7), 1991, pp. 1760-1765.
7. D. K. Lilly, *A Proposed Modification of the Germano Subgrid-Scale Closure Method*, Physics of Fluids A4 (3), 1992, pp. 633-635.
8. J. Smagorinsky, *General Circulation Experiments with the Primitive Equations: I The Basic Experiment*, Mon. Weather Review, 91, 1991, pp. 99-163.
9. H. Werner and H. Wengle, *Large-Eddy Simulation of Turbulent Flow over and around a Square Cube in a Plate Channel*, 8th Symposium on Turbulent Shear Flows, 1991, pp. 155-168.

10. Fluid-Structure Interaction

Parallel Computational Fluid Dynamics – Trends and Applications
C.B. Jenssen et al. (Editors)

Parallel Application in Ocean Engineering. Computation of Vortex Shedding Response of Marine Risers

Kjell Herfjord[a] Trond Kvamsdal[b] and Kjell Randa[c]

[a]Norsk Hydro E&P Research Centre,
P.O.Box 7190, N-5020 Bergen, Norway

[b]Sintef Applied Mathematics,
N-7491 Trondheim, Norway

[c]Norsk Hydro Data,
P.O.Box 7190, N-5020 Bergen, Norway

In ocean engineering, inviscid solutions based on potential theory have been dominating for computing wave effects. Forces dominated by viscous effects, as for the loading on slender bodies as risers, have been computed by the use of empirical coefficients. This paper is describing a strategy and procedure for consistent computation of the fluid-structure interaction (FSI) response of long risers. The fluid flow (CFD) is solved in 2D on sections along the riser. The riser response (CSD) is computed in 3D by a nonlinear finite element program. The two parts (CFD/CSD) are self-contained programs that are connected through a coupler. The computations are administrated by the coupler which is communicating with the modules using PVM. The package of program modules as a unit is referred to as the FSI tool. The CFD/CSD modules are described briefly. The coupler is reported more thoroughly. Examples from the use of the FSI tool are presented.

1. INTRODUCTION

The engineering tools for design of risers in ocean engineering have been based on the finite element method for modeling of the structure, and empirical coefficients for the hydrodynamic forces. The riser is modeled by beam elements of a certain number. Each beam element is loaded with a force according to the water particle motion at the mean coordinate of the element. The forces are assembled to give the load vector which forms the right hand side of the system of equations each time step. The force coefficients are empirical quantities from two-dimensional idealised experiments. The assembling of the force vector is performed according to a so-called strip theory, i.e. there is no interaction from one element to the other hydrodynamically. The loading is typically due to ocean current, wave particle motion as well as top end motion from the platform. The ocean current is producing a mean force in the flow direction, and a corresponding mean offset. The wave motion is producing forces which are approximated by Morison's equation, which also involves a mass coefficient, giving a force proportional to the acceleration. The dynamic force is thus produced by the dynamic wave particle motion velocity as well as the dynamic platform motion. Only forces in-line with the flow is produced by the traditional methods described above. It is well known that the vortex shedding from blunt bodies produces alternating forces even in constant current. The forces act in-line with the current as well as transverse to it. These forces produce the vortex induced vibrations (VIV) experienced on e.g. risers. The pressure change due to the viscosity produces in addition a mean force in-line with the current. This is the force which is modeled by the drag coefficient used in the load model in the methodology described above.

The only stringent way of computing the forces due to vortex shedding, is by solving the Navier-Stokes equations. However, the solution of a full length riser with a length to diameter ratio of about one thousand, is not feasible to solve in complete 3D. By the use of two dimensional loading and strip theory, as for the classical riser programs, it is possible to do feasible computaitons, especially as parallelisation is employed. The present paper is reporting a method that is doing this.

While the CFD computations are done in 2D sections along the riser, the computation of riser response (CSD) is done in 3D by a non-linear finite element code. The motion of the riser at each section is influencing the flow at that position, which is considered by the CFD program. Thus the flow and force will develop individually at each section, however coupled through the motions of the riser. The parallelization is performed by organizing the computation of each section as a dedicated process, either on a dedicated CPU, or as different processes on powerful CPUs. The CSD computation is one single process. The communication between the different processes is being done by the use of the programming library PVM [1] (Parallel Virtual Machine). The setup of the processes and organizing of the communication is performed by a special coupling program. The strategy of parallelization described here is based on the philosophy that the rather demanding computations can be performed on existing workstations rather than a supercomputer. A cluster of workstations is the hardware environment needed.

In this paper the programs handling the physics (CFD and CSD programs) are presented briefly. The main part will be dedicated to the presentation og the coupling module and how the communication is treated. Examples of the use of the program system are given.

2. THE NUMERICAL METHODS

2.1. The fluid dynamics program

The CFD program is solving Navier-Stokes equations by a finite element method. The program is presented and validated in Herfjord (1996) [2]. Here a short summary of the implementation is given. The equations of motions are solved in 3 steps every timestep. The method is referred to as fractional step method, and dates back to Chorin (1968) [3], who called it a split operator technique. The setup, including the variational formulation for the finite element method is given in Herfjord *et al.* (1999) [4]. The first step solves the advection and diffusion part of the equation. In this step the pressure term is ignored. The second step is a Poisson equation for the pressure. The third step is a correction step on the velocity. In this step, the incompressibility constraint is satisfied implicitly. There are no further iterations for obtaining this constraint. In the first and the third steps, the equations are solved by a so-called lumped mass. The pressure equation is solved unmodified.

The equation of motion (Navier-Stokes) is discretised through an element method. The motions of the riser is solved in an accelerated frame of reference. The riser moves several diameters transverse to its axis. This means that the deformation cannot be absorbed by a deforming grid. The phrase accelerated coordinate system means that the grid is kept undeformed throughout the simulation. The velocity of the grid (the riser) is taken care of by an appropriate term in the equation of motion. This methodology can strictly not be used when there are two risers. In that case, the relative motion between the risers will have to be absorbed by deformation of the grid.

2.2. The mechanical response module, USFOS

The mechanical response of the fluid-structure integrated analysis is handled by the computer code USFOS, (Ultimate Strength of Framed Offshore Structures). USFOS is a non-linear 3D finite element code capturing geometrical non-linearity as well as non-linear material behaviour. USFOS was originally developed as a specialised computer program for progressive collapse analysis of steel offshore platforms under accidental loads (extreme waves, earthquake, accidental fire, collision, etc.) and damage conditions. USFOS is used by the oil industry world wide in design and during operation[5–7]. USFOS is based on an advanced beam-column element, capturing local buckling as well as column buckling, temperature effects and material non-linear behaviour. The formulation is based on Navier beam teory, and an updated Lagrangian formulation (co-rotational) is used to describe motion of the material.

In connection with the fluid-structure interaction, the Navsim (CFD) simulations are treated as a special load routine as seen from USFOS. In each "Navsim-node", a special plane (or disc) is inserted representing the fluid behaviour at this section of the pipeline. These "Navsim discs" are oriented perpendicular to the pipeline configuration, and the discs are updated during the simulations, always following the rotations at the actual nodal points.

3. THE COUPLING MODULE

The coupling between the CFD and the CSD programs are performed according to a so-called staggered time stepping procedure. This means that the forces at a certain time step is transferred to the CSD program after the time integration step is finished by the CFD program. The CSD program then computes the deformation related to that particular time step. The deformation is fed back to the CFD program, who uses the information for computing the force one step forward. A procedure where both CFD and CSD are stepped forward in time simultaneously as an integrated process is called concurrent time discretization [8]. Since the two tasks normally are performed by two different program executables, possibly even on different computer architecture, the staggered procedure is the one that is practical to implement. This approach also means that the CFD and CSD codes may be considered as modules to be connected to the coupler without doing major modifications. This modular architecture makes it feasible to replace them. As it has turned out, the computation of the fluid flow is controlling the time step, due to the variations of the flow that need to be captured. Any non-linearitiy in the structural response will be captured by the time step decided by the CFD program. Do to this, the CFD and CSD problems does not need to be solved concurrently.

The coupler program uses the PVM programming library to implement the communication between the CFD and CSD program. PVM consists of an integrated set of software tools and libraries that emulates a general purpose heterogeneous concurrent computing framework on interconnected computers of varied architecture. The PVM system contains two main parts. The first is a daemon that resides on all computers making up the virtual machine. One of the jobs for the daemon is to preserve the consistency of the parallel virtual machine. The second part of the system is a library of PVM interface routines. This library contains user callable routines for message passing, spawning processes, coordinating and modifying the virtual machine. The PVM system can be used with C, C++ and Fortran. It supports both functional parallelism and data parallelism (SPMD).

The coupler, CFD and CSD programs are designed to run in a heterogenous computer environment and all programs can run on any computer architecture supporting PVM, Fortran and C. During testing, the coupler and CFD was developed and tested on DEC/Alpha running OSF/1 operating system and USFOS on a SGI computer. Later the coupler has been ported to RS/6000 running AIX and SGI running IRIX. The CFD program is currently running on DEC/Alpha (OSF/1), RS/6000 (AIX), SGI (IRIX) and SUN (Solaris). The CSD program is still only running on SGI.

As the computation of each CFD plane is independent of the other planes, these can be computed in parallel by running each plane on separate CPUs. By using PVM, the program can be run on either a network of workstations or on a dedicated parallel computer. The performance and the scalability will of cause be better on a dedicated computer than on a network of workstations. The computation time is totally dominated by the CFD computation, but of cause as the number of CFD planes increases, the communication overhead increases too. This fact also favour a dedicated parallel computer which also have a dedicated highspeed interconnect between the CPUs as opposed to workstations that are connected by a 10 or 100 Mbit ethernet, alternatively a 100 Mbit FDDI network. Another complicating factor when the scalability and performance is to be measured is that a farm of workstations usually consists of hosts on different speed and the computation speed and scalability will be limited by the slowest workstation. These workstations are also used to perform other computation at least during daytime and this may interfer heavily with the CFD computations and the loadbalansing of the system. On a dedicated parallel computer all CPUs are generally of the same type and dedicated to a single job and is a much more controlled environment for running parallel programs. However, this may not be a dominating issue when a production run is being made. If the computations are arranged in such a manner that the slowest CPUs and those with smallest memory are given only a limited part of the work, a simulation through the night will be ready for postprocessing the next morning anyway.

On start-up, the coupler reads two input files. One file describing the riser model as well as the number of CFD planes to be used and their positions along the riser. In addition, some parameters for the simulation are being read. The other file contains the names of the hosts were the CFD program is to be executed and the number of CFD planes to run on each host. The host that shall run the CSD program is given as an input parameter to the coupler. The coupler exchanges information between the CFD and CSD modules according to Fig. 1.

When the CFD slaves has finished their last timestep, they send a message to the

Table 1
Results from a standard benchmark test of 200 timesteps run on a network of workstations/servers. Wallclock time is in seconds.

CFD planes	1	2	4	8	16	32
Wall Clock	174	175	190	193	210	241
Efficiency	1.00	0.99	0.92	0.90	0.83	0.72
Speed-up	1.00	1.98	3.68	7.20	13.28	23.04

coupler and terminate. When this message has been received from all slaves, the coupler sends a message to the CSD program that the simulation has finished. The CSD program then terminate in a standard way and closes all its output files.

The benchmarks were run on a heterogenous network of workstations/servers connected by 10 or 100 Mbit ethernet and some servers on a 100 Mbit FDDI network. The benchmarks were run with a single CFD slave on each CPU. On multi CPU hosts, several CFD slaves could be run. The job with one single CFD plane was run on one of the slowest workstations. By adding more hosts of equal and faster CPU speed, the increase in wall clock time is mostly due to communication overhead. As these workstations/servers were not dedicated to run this application, and the network traffic was not measured, this may influence on how the application scales. Still the use of heterogenous networked workstation/servers show a good speedup as the number of CFD slaves increase. The results of the benchmark tests are summarized in Table 1.

4. VALIDATION OF THE FSI TOOL

The FSI tool has been validated versus measured results in ealier publications, see [9–11]. In this paper we will demonstrate the capability of the tool by showing an example of computation of a flexible riser in a current. The riser is a standard flexible riser used in the North Sea for oil production. The riser has a diameter of 0.5 m, and the water depth is 300 m. The shape of the riser is shown in Fig. 2. The top end is fixed to the floating platform. The lower end is resting on the sea floor. One part of the riser is equipped with buoyancy elements, making a hog bend, in order to reduce the loads at the contact with the sea bottom. Again refering to Fig. 2, the current is flowing from left to right. The equilibrium position is depicted in blue, while the updated mean position in a current of 1 m/s is depicted in red. On the right hand side of the figure, the deflection od a point between the two bends of the riser is shown. The in-line deflection is as much as 20 meters, the amplitude of the transverse motion is in the order of 1 diameter (i.e. 0.5 m). In Fig. 3, the transverse oscillating motion is given together with the non-dimensional forces for two points. At the left hand side, transverse motion and forces near the highest point of the hog bend are given. At this point the flow velocity perpendicular to the riser is small, and the diameter of the riser is larger due to the buoyancy elements. This is the reasons for the small motions. At the right hand side, the same quantities at a point near the sea surface are depicted. A different pattern of motions is shown. The results presented here are not a true validation, since there are no measurements to compare with. However,

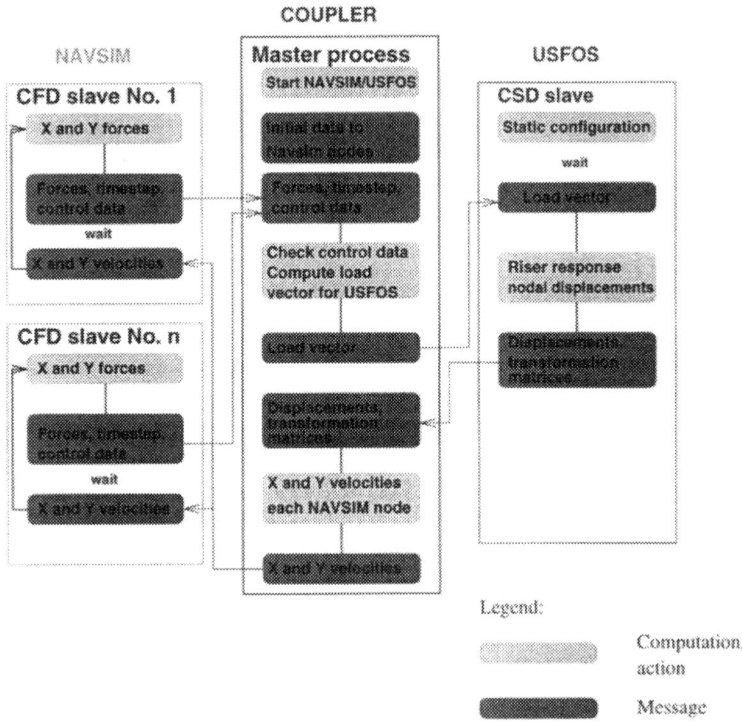

Figure 1. Schematic presentation of the coupler.

the capability of handling a general shape of a riser is demonstrated. It is to be hoped that good measurements of the behavior of such risers can be provided.

5. SUMMARY AND CONCLUSIONS

The FSI tool has been made in order to enable computations of vortex induced vibrations on risers and other slender and flexing bodies. The objectives behind the construction of the tool, with the coupler centrally positioned, can be summarized as follows:

- Acceptable accuracy and simulations of realistic cases within acceptable computing times. In addition, the program should be easy to use.
- Modular make with versatility in accepting different computer architecture.
- Parallelization with efficient communication and good scalability.

The simulations presented in this paper has been carried through with computing times in the order of hours (5 to 10 h). The analysis programs used are self-containing programs

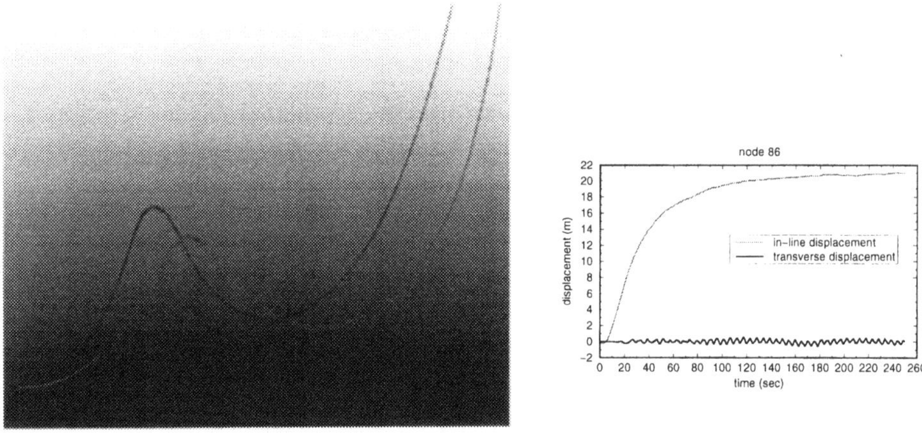

Figure 2. Flexible riser in current. On the left hand side, the shape of the riser in equlibrium without current, as well as the mean shape in a current from the left is depicted. On the right hand side, the displacements of a point between the upper and lower bends are shown.

that are linked to the coupler with only minor modifications, and the FSI tool may be executed on a wide variety of computer architectures. The use of other programs as analysis modules are in this way facilitated in a good manner. In addition, other facilities as error estimation and grid updating may be connected as new modules at very reasonable costs. The parallelization is done by doing the CFD computations in 2D planes along the riser and performing the work on each planes as independent processes on many CPUs. The computations are influenced by the motions of the riser, and are in this way coupled. The communication between the different processes and the coupler is made by PVM and with very restricted lengths of the messages. In this way the efficiency is high and it is demonstrated that the problem scales well with increasing number of CFD planes.

ACKNOWLEDGEMENTS

The development of the coupler presented here has been supported by the European Commission under the contract Esprit IV 20111.

REFERENCES

1. Geist, A., Beguelin, A., Dongarra, J., Jiang, W., Manchek, R. and Sundaram, V., PVM: Parallel Virtual Machine. A User's Guide and Tutorial for Networked Parallel Computing, MIT Press, Cambridge, (1994).
2. Herfjord, K., A Study of Two-dimensional Separated Flow by a Combination of the Finite Element Method and Navier-Stokes Equations, PhD thesis, Norwegian Institute of Technology, (1996).

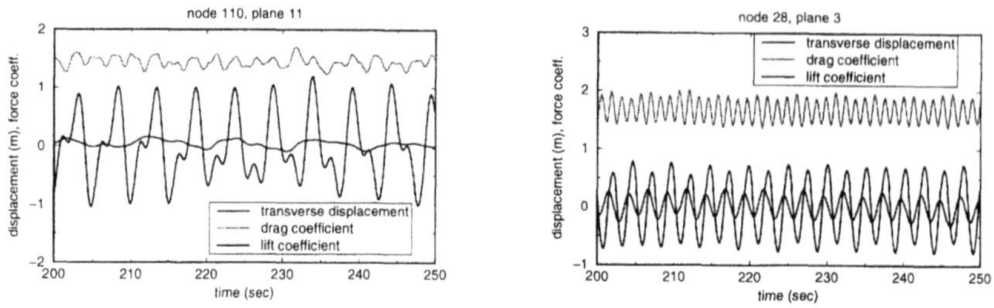

Figure 3. Displacement transverse to the flow and non-dimensional forces for two points along the riser. On the left hand side, motion and forces at a point near the top of the hog bend is given. To the right, the same quantities at a point near the sea surface are given.

3. Chorin, J.C., Numerical Solution of the Navier-Stokes Equations, Math. Comp. American Mathematical Society, Vol. 22, pp 449-464, (1968).
4. Herfjord, K., Drange, S.O. and Kvamsdal, T., Assessment of Vortex-Induced Vibrations on Deepwater Risers by Considering Fluid-Structure Interaction, Journal of Offshore Mechanics and Arctic Engineering, Vol. 121, pp 207-212, (1999).
5. Søreide, T., Amdahl, J., Eberg, E., Holmås. T. and Hellan, Ø, USFOS - Ultimate Stength of Offshore Structures, Theory Manual SINTEF Report F88038.
6. Hellan, Ø., Moan, T. and Drange, S.O., Use of Nonlinear Pushover Analysis in Ultimate Limit State Design and Integrity Assessment of Jacket Structures, 7th International Conference on the Behaviour of Offshore Structures, BOSS'94, (1994).
7. Eberg, E., Hellan, Ø. and Amdahl, J., Nonlinear Re-assessment of Jacket Structures under Extreme Storm Cyclic Loading, 12th International Conference on Offshore Mechanocs and Arctic Engineering, OMAE'93, (1993).
8. Pegon, P. and Mehr, K., Report and Algorithm for the Coupling procedure, R4.3.1, ESPRIT 20111 FSI-SD, (1997).
9. Herfjord, K., Holmås, T. and Randa, K., A Parallel Approach for Numerical Solution of Vortex-Induced Vibrations of Very Long Risers, Fourth World Congress on Computational Mechanics, WCCM'98, Boenos Aires, Argentina, (1998).
10. Herfjord, K., Larsen, C.M., Furnes, G., Holmås, T. and Randa, K., FSI-Simulation of Vortex-Induced Vibrations of Offshore Structures, In: Computational Methods for Fluid-Structure Interaction, Kvamsdal *et al.* (eds.), Tapir Publisher, Trondheim Norway, (1999).
11. Kvamsdal, T., Herfjord, H. and Okstad, K.M., Coupled Simulation of Vortex-Induced Vibration of Slender Structures as Suspention Bridges and Offshore Risers, Third International Symposium on Cable Dynamics, Trondheim, Norway, (1999).

Parallel Computational Fluid Dynamics – Trends and Applications
C.B. Jenssen et al. (Editors)

Experimental and numerical investigation into the effect of Vortex induced vibrations on the motions and loads on circular cylinders in tandem.

By: R.H.M. Huijsmans[a], J.J. de Wilde[a] and J. Buist[b]

[a] Maritime Research Institute Netherlands,
P.O. Box 28, 6700 AA Wageningen, The Netherlands:

[b] BuNovaDevelopment,
Postbus 40023, 8004 DA Zwolle, The Netherlands

ABSTRACT

In this paper a study of the flow around fixed mounted cylinders will be presented. The aim of the study is to set up a method for the computation of the flow around a bundle of flexible cylinders. The flow is assumed to be two-dimensional. The Reynolds number of the flow ranges from 20,000 to 550,000. The calculations for the flow around the fixed circular cylinder were based on commercial available CFD codes such as STAR-CD and CFX 4.2. For the validation of the CFD codes for this application model test experiments were performed on fixed and flexible mounted cylinders. The cylinders were mounted as single cylinder or in pairs. The model test experiments consisted of force measurements in stationary flow as well as detailed Particle Image Velocimetry measurements.

1. INTRODUCTION

One of the grand challenges in the offshore industry is still the assessment of the motions of a circular cylinder in waves and current for application to riser bundles up to 10,000 feet water depth. Here the fatigue life of riser systems is dominated by the VIV phenomena. Also the possibility of riser collision is governed by VIV effects. The nature of vortex induced vibration (VIV) problem relate to a hydro-elastic problem, i.e. the vibration of the cylindrical riser system is triggered by force fluctuations due to the generation of vortices. Force fluctuations on the cylinder are strongly influenced by the subsequent motions of the riser system.

As is already known, the vortex shedding is a three dimensional phenomena. However the three dimensionality of the flow around the cylinder also stems from the fact that the cylinder is excited in a few normal modes. The actual fluid loading, as a first approximation, is often regarded as two-dimensional.

The proximity of another circular cylinder will influence the flow drastically. By varying the spacing between the two cylinders several regimes of flow characteristics can be distinguished [1,2]. An experimental study into VIV has been performed, where both flexible mounted rigid cylinders as well as fixed rigid cylinders have been investigated. The flexible mounted cylinder was segmented into three parts in order to identify the influence of the 3-D wake effects behind the cylinder. In order to quantify flow characteristics around the cylinder special Particle Image Velocimetry measurements have been performed [3]. The drag and lift forces on the cylinders in tandem operation were measured. The measured forces, the resulting motions of the cylinder and the flow field around the cylinder are correlated with results of Navier-Stokes calculations. The Navier-Stokes computations are based on the RANSE model in CFX 4.2, were the turbulence was modeled using a k-ε model and alternatively a k-ω model. Navier stokes solvers which are build specifically for flows around circular cylinder shaped bodies are amongst others also based on spectral - or FEM type of methods [4,5].

2. DESCRIPTION OF EXPERIMENTS

The experiments were conducted at MARIN's Basin for Unconventional Maritime Constructions, consisting of a 4 by 4 m rectangular channel of 200 m length and an overhead towing carriage. The circular test cylinder of 206 mm in diameter and 3.87 m in length was suspended from the towing carriage on two streamlined vertical struts at a submergence of 1.7 m, as depicted in figure 1.

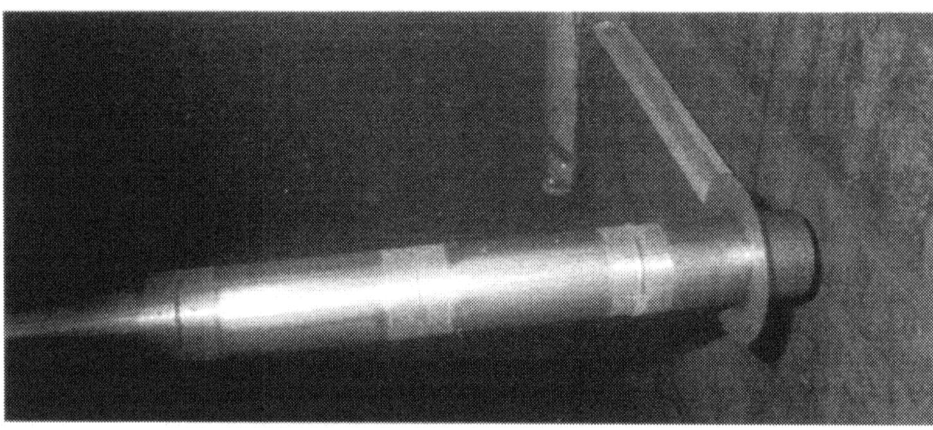

Figure 1: Test cylinder

Stiff horizontal beams were used to push the cylinder forward at a distance of 0.7 m in front of the struts, in order to minimize the possible blockage effects of the struts. The clearance between the basin walls and the cylinder ends was 0.08 m. Circular end plates of 400 mm in diameter were mounted at a distance of 178 mm from the cylinder ends. The surface roughness of the stainless steel cylinder was estimated to be less than 0.1 mm.

The test cylinder was constructed with a rigid circular backbone on which three instrumented cylinder segments of 1.0 m in length were mounted. With the two end segments of 0.44 m, the total cylinder lengths was equal to the above mentioned 3.87 m. Also the side-by-side configuration with a second rigid circular cylinder parallel mounted above the original cylinder was tested. A 400 mm pitch between the two cylinders was tested. The tests were conducted by towing the cylinder at a constant speed over the full length of the tank, meaning at least 50 vortex shedding cycles in one run (up to 2 m/s towing speed).

3. NUMERICAL MODEL

3.1. Mesh

The grid is a simple grid of hexahedral elements. An impression of the grid is given in figure 2. This grid is used for both the simulations with the LRN k-ε as well as the Wilcox LRN k-ω turbulence model. The grid can be viewed as build in two steps: firstly, a radial grid was designed around the cylinder. Secondly, the grid is extruded downstream in order to be able to follow the behavior of the vortices over a number of cylinder diameters. The distance between the cylinder and one of the symmetry planes is about 5 diameters. The number of cells in the grid is around 17.000.

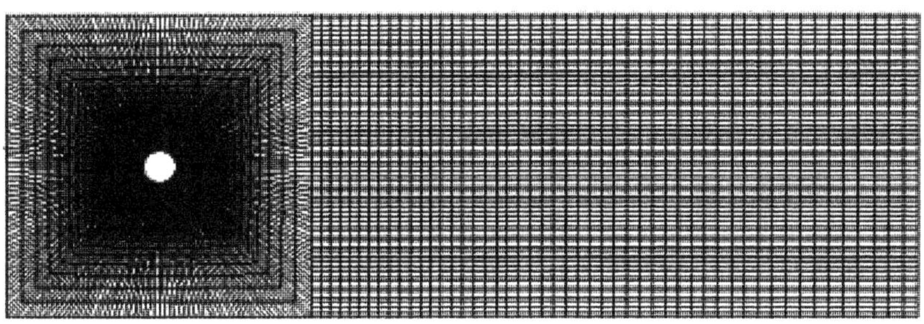

Figure 2: Impression of the grid

The strong refinement towards the wall is needed for the k-ω turbulence model, because the equations are integrated into the viscous sub-layer near the wall. The near wall region is usually described in terms of the dimensionless wall co-ordinate y^+, defined by:

$$y^+ = \frac{y \cdot U_\tau}{\nu} \quad \text{, with } U_\tau = \sqrt{\frac{\tau_w}{\rho}} \quad \text{being the friction velocity.}$$

y is the real, physical distance to the wall. Oskam [6] has shown by an analysis of the near wall grid dependency that as long as one or two cell centers are located within the viscous sublayer, i.e. $y^+ < 5$, the solution will be independent of the near wall grid spacing. An impression of the grid refinement close to the cylinder wall is given in figure 3.

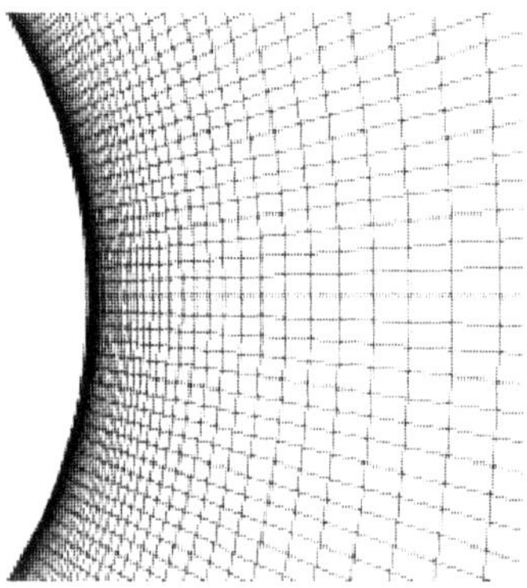

Figure 3: Grid refinement towards the wall

In the simulations treated in this article , y^+ values in the range between 0.1 and 3 have been found. This satisfies the criterion.

3.2. Analysis of the model cylinder used for VIV measurements

Strouhal number and C_D prediction

The use of the Wilcox LRN k-ω model should also give a better flow prediction for flows in which the near wall flow behavior has a large influence on the flow field as a whole. This is the case in the analysis of vortex-shedding behind a cylinder. The attachment of the flow has a dominant influence on the flow field behind the cylinder, even if the Reynolds number of the flow is high.

Strouhal number

The experiments and the simulations discussed in this subsection concern the flow around a stiff, submerged cylinder having a diameter of 0.206 m. Both experiments and simulations have been done with the same geometry. In the experiments the flow field is analyzed for flow velocities of 0.2, 1.0 and 2.5 m/s. The same is done by simulations with CFD code CFX-4. The findings have been summarized in the following tables.

Table 1: Results of the experiments

U (m/s)	D (m)	Re (-)	f (Hz)	Str (-)
0.2	0.206	$3.75 \cdot 10^4$	0.19	0.195
1.0	0.206	$1.87 \cdot 10^5$	0.87	0.178
2.5	0.206	$4.68 \cdot 10^5$	-	-

Table 2: Results of the simulations

Simulation	U (m/s)	D (m)	Re (-)	f (Hz)	θ_s (°)	Str (-)
1	0.2	0.206	$3.75 \cdot 10^4$	0.235	80 ±4	0.205
2	1.0	0.206	$1.87 \cdot 10^5$	1.262	74 ±6	0.211
3	2.5	0.206	$4.68 \cdot 10^5$	3.143	73 ±7	0.208

Here θ_s displays the mean shedding angle of the flow from the cylinder.

Resistance coefficients

$$C_D = \frac{F_{x,mean}}{0.5 \cdot \rho \cdot U^2 \cdot D}$$

Table 3: Results of the experiments

U (m/s)	ν (m^2/s)	Re (-)	$F_{x,\,mean}$ (N)	C_D (-)
0.2	$1.1 \cdot 10^{-6}$	$3.75 \cdot 10^4$	4.3	1.04
1.0	$1.1 \cdot 10^{-6}$	$1.87 \cdot 10^5$	84.2	0.83
2.5	$1.1 \cdot 10^{-6}$	$4.68 \cdot 10^5$	323.4	0.51

Table 4: Results of the simulations

Simulation	U (m/s)	ν (m^2/s)	Re (-)	$F_{x,\,mean}$ (N)	C_L (-)	C_D (-)
1	0.2	$1.43 \cdot 10^{-6}$	$3.75 \cdot 10^4$	4	1.02	0.774
2	1.0	$1.43 \cdot 10^{-6}$	$1.87 \cdot 10^5$	89	0.97	0.640
3	2.5	$1.43 \cdot 10^{-6}$	$4.68 \cdot 10^5$	570	1.01	0.637

4. COMPUTATIONAL ASPECTS

4.1. Remarks on parallel computing

All simulations treated in this paper are performed as single processor jobs. Also a performance test for parallel computing was carried out. The simulated time was such that at least 30 full cycles of vortex-shedding are simulated after the start-up phenomenon. For the parallel run, the flow domain was divided in two sub-domains having an equal number of cells. The simulations were performed on a single SGI R10000 processor and on two of these in parallel respectively. The CPU time for both equal jobs was as follows:

Table 5: CPU times for single and parallel run

CPU times	
Single processor	20.5 hrs
Two processors	9.0 hrs

The speed-up is larger than a factor of two. a probable explanation for this phenomenon is that the increase in the amount of cache that is available for the floating point operations on a dual processor run outweighs the slow-down of the calculation because of the communication between the two processors. The start-up behaviour of the two simulations differs. The dual processor simulation shows a faster increase of the amplitude than the single processor simulation. However, both simulations reach a state of steady cycling at the same time. From this time forward, the results of both simulations are equal, apart from a phase shift. The amplitude and the frequency of the velocity components, pressure, turbulent viscosity and turbulent kinetic energy are equal. As a consequence, the predicted Strouhal number of the parallel run equals the Strouhal number in the single run.

5. DISCUSSION

5.1. Measured drag loads and vortex shedding frequencies for a single cylinder

The measured drag coefficient C_d and Strouhal number St of the single cylinder are presented in the figure 4, for Reynolds numbers between 2.0×10^4 and 5.5×10^5. Also presented are the measured drag coefficients by Güven et al. [7], for a smooth cylinder and a cylinder with a surface roughness of $k/D = 1.59 \times 10^{-3}$.

The present measurements confirm the earlier measurement by Güven. The well-known drop in drag coefficient in the critical Reynolds regime ($2 \times 10^5 < Re < 5 \times 10^5$) is clearly observed. The results suggest an effective surface roughness of the cylinder between smooth and $k/D = 1.59 \times 10^{-3}$. The measurements also confirm the vortex shedding frequencies of a smooth cylinder, as found by other investigators. The commonly accepted upper and lower boundary values of the Strouhal number are schematically depicted in figure 4 for reference. The Strouhal number in the present experiments for the sub-critical Reynolds regime was approximately St = 0.195. For Reynolds numbers between $1.5 \times 10^5 < Re < 2.5 \times 10^5$ a small decrease in Strouhal number as function of the Reynolds number was observed. For Reynolds numbers above 2.5×10^5 it was found that a single vortex shedding could not well be determined.

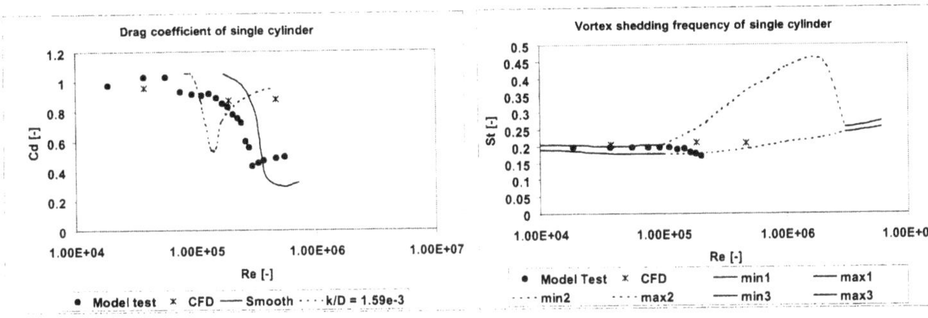

Figure 4: Drag coefficients and Vortex shedding frequency of single cylinder

5.2. Measured drag loads and vortex shedding frequencies for two cylinders side-by-side

The measured drag coefficients and Strouhal numbers for the side-by-side situation are presented in figure 5:

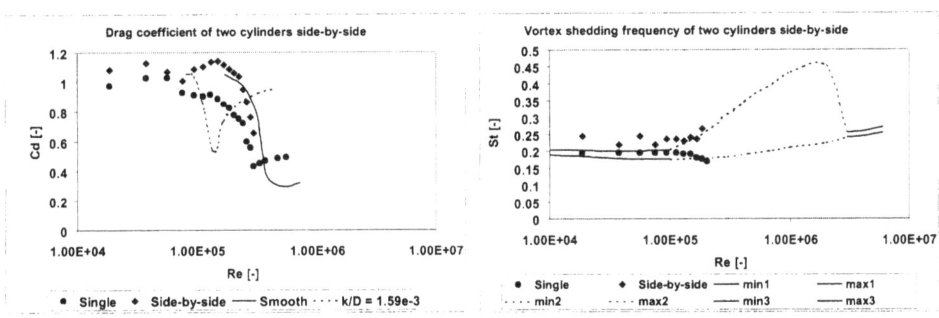

Figure 5: Drag coefficients and Vortex shedding frequency of two cylinders side-by-side

Clear differences are observed between the side-by-side situation and the situation of the single cylinder. For the side-by-side situation in the sub-critical Reynolds regime, a slightly higher mean drag and vortex shedding frequency was found. Also the behaviour of the C_d-values in the critical regime is clearly different. The drag coefficient for the side-by-side situation is initially larger and than drops much more rapidly as a function of the Reynolds number. Regarding the vortex shedding frequency it can be observed that the Strouhal number has a tendency to increase as a function of the Reynolds number in the side-by side situation, whereas the opposite is observed for the single cylinder.

6. FUTURE CFD VALIDATION

Future CFD validation will concern the freely vibrating cylinder in steady flow as well as the flow around a pair of cylinders. Here the CFD codes have to be able to handle the grid near the cylinder walls in a dynamic way.

7. CONCLUDING REMARKS

This analysis of vortex-shedding behind a cylinder has shown that commercial CFD codes can assist in the simulation the flow behaviour.

From our study we found:

- The LRN k-ε turbulence model is less robust than the Wilcox LRN k-ω model. When using the LRN k-ε model, more time steps per cycle and also more iterations per time step are needed. Convergence appeared to be troublesome with the LRN k-ε model.
- The results of the simulations of a model cylinder (D = 48 mm, U = 0.4 m/s) with the Wilcox LRN k-ω model compare reasonably well with experimental data. U and V components of the velocity vector and vorticity have been compared. Field data on a sampling line downstream the cylinder show that there is agreement between the simulation and the experiment on the amplitude of the oscillation just downstream the cylinder. A difference was found between the predicted and measured frequency.
- The analysis of the Strouhal number at different Reynolds number shows that the simulations with the Wilcox LRN k-ω model are well capable of predicting the trend of the Strouhal number as given in literature.

REFERENCES

1 P.Bearman, A. Wadcock: The interaction between a pair of circular cylinders normal to a stream. J. of Fluid Mech. Vol.61 1973.
2 C.Siqueira,J.Meneghini,F.Saltara, J.Ferrari: Numerical simulation of flow interference between two circular cylinders in tandem and side by side arrangement. Proceedings of the 18th int Conf. On Offshore Mech. And Arctic Eng. 1999 St John's New Foundland.
3 J.Tukker, J.J.Blok,R.H.M.Huijsmans,G.Kuiper: Wake flow measurements in towing tanks with PIV. 9th Int. Symp. On Flow Visualisation. Edinborough 2000.
4 J.J. van der Vegt. A variationally optimized vortex tracing algorithm for three dimensional flows around solid bodies. PhD thesis Delft 1988.
5 K.W.Schultz and J.Kallenderis: Unsteady flow structure interaction for incompressible flows using deformable hybrid grids. J.Compt. Physics, vol 143, 569 (1998).
6 Oskam, A. : Flow and heat transfer in residential heating systems, MSc thesis University of Twente, Enschede 1999.
7 Güven, O., et al., "Surface Roughness Effects on the Mean Flow Past Circular Cylinders", Iowa Inst. of Hydraulics Research Rept. No. 175, Iowa City, 1975.

Parallel Computational Fluid Dynamics – Trends and Applications
C.B. Jenssen et al. (Editors)

Meta-computing for Fluid-Structure Coupled Simulation

Hiroshi Takemiya[a,b], Toshiya Kimura[c]

[a]Center for Promotion of Computational Science and Engineering, Japan Atomic Energy Research Institute
2-2-54, Nakameguro, Meguro-ku, Tokyo, 153-0061, JAPAN

[b]Hitachi Tohoku Software, Ltd.
2-16-10, Honcho, Aobaku, Sendai, 980-0014, JAPAN

[c]Kakuda Research Center, National Aerospace Laboratory
Kimigaya, Kakuda, Miyagi, 981-1525, JAPAN

Metacomputing for a fluid-structure coupled simulation has been performed on a heterogeneous parallel computer cluster. The fluid and the structure simulation codes are executed on parallel computers of different architectures connected by a high-speed network. These codes are linked by a loose coupling method exchanging the boundary data between fluid and structure domains. The performance evaluation has shown that metacomputing for fluid-structure coupled simulations attains better performance compared with the parallel computing on a single parallel computer.

1. Introduction

The progress of high-speed networks and computers is expected to realize a new computing style, called metacomputing[1]. Metacomputing enables to use computers, networks, and information resources as a single virtual computer. It is said that there are five kinds of representatives in metacomputing[2]. Among them, distributed supercomputing, which tries to solve a single problem by using networked supercomputers, has a possibility of performing very large and complex simulations in the scientific computing field.

There are two kinds of merits in distributed supercomputing. The first is called scale merit.

When we execute a simulation on a single supercomputer, the number of processors and the size of memory are restricted by the hardware architecture of the computer. Distributed supercomputing can alleviate these restrictions to execute larger or more detailed simulations.

The second is called architecture merit. As the numerical simulation technique advances, it becomes possible to simulate more complex phenomena. Codes of these simulations are often constructed based on multiple disciplines. In executing these simulations, some parts of the code can be executed efficiently on a computer with a particular architecture, but others can not. Distributed supercomputing enables to allocate portions of the code on computers with architecture appropriate for processing them.

Although we can take advantages of these merits in distributed supercomputing, it is not obvious whether real programs can be executed efficiently. The reason is that architecture of the virtual supercomputer is quite heterogeneous. For example, data transfer speeds will be typically different by orders of magnitude and processing speeds will also be different by some factors. Therefore, it is very difficult to simulate efficiently on such a computer.

In order to verify the effectiveness of metacomputing, we have developed a fluid-structure coupled simulation code for metacomputing and evaluated the performance. In this paper, we describe the result of performance evaluation.

2. Fluid-Structure Coupled Simulation Code

In the present work, the aeroelastic response of a 3-D wing in a transonic flow is calculated as one of typical fluid-structure interaction problems. Hence, our code is constructed by integrating a computational fluid dynamics (CFD) solver, a computational structure dynamics (CSD) solver, and a grid generator.

To simulate the flow field around the wing in a transonic flow, the dynamics of the compressible gas flow are numerically examined by solving the 3-D Euler equations. Chakravarthy and Osher's TVD method [3] is used as a finite difference scheme for solving the Euler equations. Time integration is explicitly done by the second-order Runge-Kutta method[4]. The CFD code is parallelized by a domain decomposition method.

The elastic motion of the wing structure is numerically simulated by solving the structure equation of motion. The equation is solved by ITAS-Dynamic code[5], which is based on the finite element method. The time integration is explicitly performed by the central difference method. Task decomposition method is adopted to parallelize the CSD code. The index of main DO loops in the hot spots of the CSD solver is decomposed, and each decomposed DO loop is calculated in parallel with corresponding index ranges on each processor, each of which has the whole grid data of all node points.

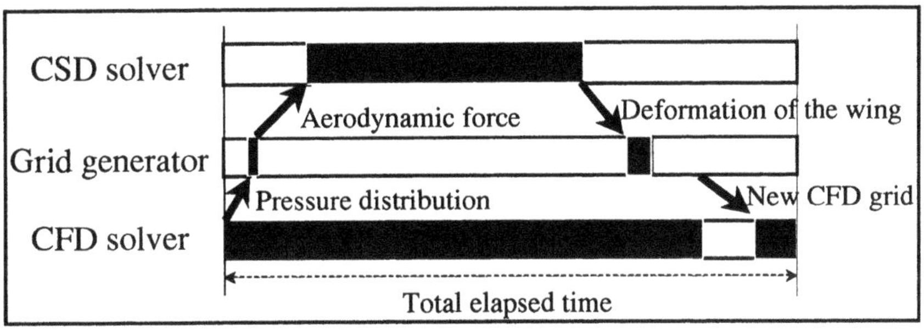

Figure 1. The execution timing and the data flow of the code

The grid generator is also parallelized and produces grid for the CFD simulation algebraically. The fluid domain is made of the C-H type numerical grid; C-type in the chord direction and H-type in the span direction.

We adopted loose coupling method to link CFD and CSD computations. In loose coupling, the fluid equations and the structure equations are solved independently in different domains using CFD and CSD numerical methods. These dynamics are coupled by exchanging the boundary data at the interface between the fluid and the structure domains.

In this simulation, the aeroelastic response of a wing is calculated by three components in the following manner (see Figure1). The CFD code calculates flow field around the wing by using a grid data sent from the grid generator. Then, it sends pressure distribution around the wing to grid generator. The grid generator transforms them into force distribution. The CSD code receives the data to calculate wing deformation and returns surface displacement to the grid generator. (It should be noted that both fluid field and wing deformation are calculated simultaneously in our implementation[6].) Finally, grid generator produces coordinates based on the displacement data. The simulation proceeds by repeating this calculation cycle.

3. Communication Library

In order to execute our code on a heterogeneous parallel computer cluster, we have developed a new communication library called Stampi[7]. Stampi is an implementation of MPI and MPI2 specification and is designed to perform efficient communication in a heterogeneous environment.

Main features of Stampi are the following:

- Stampi uses different mechanisms for intra- and inter-computer communication. In general, a parallel computer has a vender specific communication mechanism for better communication performance. Stampi uses the vender specific communication mechanism for intra-computer communication through the vender specific MPI library. On the other hand, inter-computer communication is realized by using TCP/IP, because it requires a common communication mechanism for both computers.

- In case of inter-computer communication, Stampi sends messages through message routers. If all processes are connected directly, so many connections have to be established between parallel computers. For example, if there are hundreds of processes on both sides, thousands of connections are required. Many of parallel computers can not establish so many connections. Indirect communication through message routers can reduce the number of connections.
- The number of message routers, through which the inter-computer communication is performed, can be varied. This function is effective for efficient communication because the number of routers realizing the best performance depends on the computer architecture, network speed, the number of processes, and algorithms used in a program.
- The byte orders and the format of the data can be automatically transformed.

4. Performance Evaluation

4.1 Parallel computing Experiments

We have executed our code on a single parallel computer and evaluated performance as a benchmark. Two kinds of computers, Fujitsu VPP300 vector parallel computer and Hitachi SR2201 scalar parallel computer, have been used for the experiment. They have 15 and 64 processors respectively.

The number of mesh around the wing is 101, 100 and 100 along each axis and 4,500 nodes are used for CSD simulation. Performance results of the experiment are shown in the first and the second columns of table 1. Elapsed time for 1 time step for each solver and the total are presented. The numbers of processors used for each solver are determined to bring the best perfor-

Table 1
Performance results of parallel computing and local area metacomputing

	parallel computing		local area metacomputing
	SR2201 (48PE)	VPP300 (15PE)	VPP300 (15PE)+SR2201 (4PE)
CFD	2.818 (44PE)	1.651 (8PE)	1.326 (14PE)
Grid	0.884 (1PE)	0.057 (1PE)	0.058 (1PE)
CSD	1.250 (3PE)	1.773 (6PE)	0.896 (4PE)
Total	4.345	1.995	1.408

mance. When using VPP300, total time is amounted to 1.995 sec. SR2201 requires 4.345 sec to simulate the same problem.

Both the CFD and the grid code can be executed efficiently on a vector parallel computer, because these codes can be highly vectorized. On the other hand, the CSD simulation results in a better performance on a scalar parallel computer. The reason is that this code uses list vectors and, in addition, vector length is very short.

4.1 Local-area Metacomputing Experiments

Based on the result of parallel computing experiments, we have selected computers on which each code should be allocated for metacomputing. In deciding computers, we have considered two factors. The first is how well the code is suited for the computer architecture. The second is how much data is transferred between codes.

According to the result of the parallel computing experiment, both the CFD and the grid code are well suited for a vector parallel computer, while the CSD code should be allocated on a scalar parallel computer due to its low vectorization.

From the aspect of communication cost, the CFD and the grid code will be better to be allocated on the same parallel computer. When considering the data transferred between the CFD and the grid codes, it is amounted to 24M bytes, because the CFD code needs whole 3D grid data around the wing. Therefore, if we allocate the CFD and the grid codes on different computers, we have to transfer the data within a few hnudred milli second. On the other hand, the CSD code needs only 2D data on the wing surface, which is amounted to only 100K bytes. Therefore, communication cost between the CSD and the grid code will be expected not to degrade total performance so much even if these cods are allocated on different computers.

We have, therefore, decided to allocate the CFD and the grid code on Fujitsu VPP300, and the CSD code on SR2201. These computers are connected by an ATM network with a data transfer rate of 18Mbit/sec.

The third column of table 1 shows the best total performance among the experiments. Total performance of metacomputing case is improved about 30 % compared with the second case, and 70% compared with the third case.

Comparison between the result of the metacomputing case and the parallel computing case (using VPP300) shows that CFD performance of the former case is about 20% better than that of the latter case. This can be interpreted as the scale merit. The metacomputing case can use 14 processors for CFD simulation, while the parallel computing case can use only 8 processors due to the hardware resource limitation.

Moreover, CSD performance of the metacomputing case is 60% better than that of the parallel computing case. This can be interpreted as the architecture merit. The metacomputing case can execute the code on SR2201, while the second case has to execute it on VPP300.

Although communication cost between the CSD and the grid code in the metacomputing case is about two orders larger than that in the parallel computing case, both merits can surpass this drawback.

4.2 Wide-area Metacomputing Experiments

We have conducted another metacomputing experiment, which uses widely distributed parallel computers. Wide area metacomputing is harder than the local one, because it is suffered from large communication cost. We have used AP3000 scalar parallel computer and VPP300 vector parallel computer, which are about 100 km apart from each other and connected by ATM with 15Mbit/sec data transfer speed.

In order to check the effect of communication cost on total elapsed time, we have used the same number of processors as the local area metacomputing experiment. Table 2 shows the performance results of the experiments. Columns show the result of the local and wide area metacomputing, respectively.

Although the results show excellent performance compared with the parallel computing case, the wide area metacomputing case (WAN case) needs somewhat longer total time compared with the local area metacomputing case (LAN case). The reason of the increased total time is as follows.

Figure three shows time charts of the experiment. The upper diagram shows the result of the WAN case and the lower shows that of the LAN case. Computation time of each code is about the same in both cases. The increased total time is caused by high communication cost between the CSD and the grid codes. It is amounted to 0.308 second and is about three times larger than that in the LAN case.

Large communication cost puts off the start of the grid computation. As a result, CFD code has

Table 2
Performance results of both local and wide area metacomputing

	local area metacomputing VPP300 (15PE)+SR2201 (4PE)	wide area metacomputing VPP300 (15PE)+AP3000 (4PE)
CFD	1.326 (14PE)	1.299 (14PE)
Grid	0.058 (1PE)	0.059 (1PE)
CSD	0.896 (4PE)	0.739 (4PE)
Total	1.408	1.573

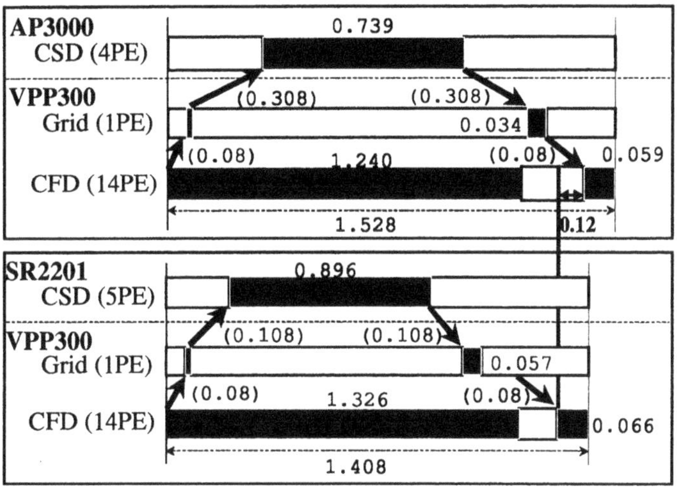

Figure 1. Timing charts of the wide-area metacomputing (upper) and the local-area metacomputing (lower)

to wait to start metric computation by about 0.12 second.

Although this communication cost can not be decreased directly, it can be compensated by decreasing CSD computation time. In this experiment, we have used only four processors of AP3000 for CSD computation. If we use more processors to shorten its computation time by more than about 0.12 second, we can expect to get the same total performance in the WAN and the LAN cases.

Based on this consideration, we have increased the number of processors for the CSD code up to twenty. As a result, the CSD computation time has been decreased to 0.381 second and the total performance has become comparable to that in the LAN case (see table 3).

Table 3
Performance results of wide area metacomputing

	VPP300 (15PE)+AP3000 (4PE)	VPP300 (15PE)+AP3000(20PE)
CFD	1.299 (14PE)	1.265(14PE)
Grid	0.059 (1PE)	0.035 (1PE)
CSD	0.739 (4PE)	0.381 (20PE)
Total	1.573	1.345

5. Conclusion

In the present work, we have conducted the experiments of both the local and the wide area metacomputing for fluid-structure coupled simulation. Loose coupling method has been used to link the CFD and the CSD codes. Newly developed communication library Stampi has been used to enable communication among processors on a heterogeneous parallel computer cluster.

Our metacomputing experiments have shown higher total performance than calculations on a single parallel computer. In particular, although experiments on a wide area network are suffered from large communication cost, it can be hidden behind the CFD computation.

References

[1] L. Smarr and C. Catlett: Metacomputing, Communications of ACM, Vol. 35, No.6, pp.45-52 (1992)

[2] I. Foster and C. Kesselman: The Grid: Blueprint for a New Computing Infrastructure, Morgan Kaufmann Pub. (1998).

[3] S. R. Chakravarthy and S. Osher: A new class of high accuracy TVD schemes for hyperbolic conservation laws, AIAA Paper No. 86-0363 (1985)

[4] C. Hirsh: Numerical computation of internal and external flows: Vol. 1. Fundamentals of numerical discretization, New York: John Wiley. (1992)

[5] T. Huo and E. Nakamachi: 3-D dynamic explicit finite element simulation of sheet forming, In Advanced technology of plasticity, pp. 1828-33 (1993)

[6] T. Kimura, R. Onishi, T. Ohta, and Z. Guo: Parallel Computing for Fluid/Structure Coupled Simulation, Parallel Computational Fluid Dynamics -Development and Applications of Parallel Technology, North-Holland Pub., pp. 267-274

[7] T. Imamura, Y. Tsujita, H. Koide, and H. Takemiya: An architecture of Stampi: MPI Library on a Cluster of Parallel Computers, in Proc. of 7th European PVM/MPI User's Group Meeting (2000).

11. Industrial Applications

Parallel Computational Fluid Dynamics – Trends and Applications
C.B. Jenssen et al. (Editors)

A Parallel Fully Implicit Sliding Mesh Method for Industrial CFD Applications

G. Bachler, H. Schiffermüller, A. Bregant

AVL List GmbH, Advanced Simulation Technologies
Hans-List-Platz 1, A-8020 Graz, Austria

1. INTRODUCTION

In the past decade the computational fluid dynamics package FIRE has been developed for the simulation of unsteady engine flows with arbitrary moving parts in the computational domain. At certain stages of the grid movement, the solution of the discretized transport equations has to be mapped from one mesh to another. The corresponding mapping technique is called rezoning or re-meshing. The rezoning technique is very general and, therefore, has also been applied to rotational grid movement in fans and water pumps with strong rotor-stator interactions.

Since unsteady applications with moving grids are CPU demanding tasks, a parallel local memory version of rezoning has already been implemented in the early nineties /1/. For this purpose, an nCUBE2 system with up to 128 processors, an IBM workstation cluster and a SP system with up to 64 PowerX processors have been used /2/. The communication was performed with the nCUBE vertex and IBM PVMe message passing libraries, respectively.

Unfortunately, rezoning techniques are always accompanied by mesh distortion between the rezoning events. In engine flows, the influence of mesh distortion on numerical accuracy is less critical than in rotating fan flows. The reason is, that the dominating pressure changes, caused by compression and expansion, are uniformly distributed over the combustion chamber and the local pressure gradients become negligible - at least, as long as the intake and exhaust valves stay closed. In contrast to internal engine flows, rotating fan flows behave like external flows. The major driving forces arise from local pressure gradients and gradients of shear stresses. Their accurate computation is strongly dependent on the grid quality in the vicinity of the fan blades.

Another drawback of the rezoning technique for rotating fan flows is the lack of numerical stability, which is not observed in engine calculations. Although the reason is not yet totally clear, it seems to be related the to re-construction of cell-face gradients from the cell-centred solution.

In order to meet all accuracy and stability requirements, the rezoning technique has been replaced by a sliding mesh technique which does *not* show any distortion and numerical instability during grid movement.

In what follows, a survey of FIRE and the basic principles of the implicit sliding mesh technique will be presented. Subsequently, the parallel strategy and the domain decomposition methods will be discussed. The results of rotating fan flows, obtained with rezoning and sliding meshes will be compared with respect to predictive capability and parallel performance. As will be demonstrated, the sliding mesh technique is superior to the rezoning technique in both respects.

2. SURVEY OF FIRE

FIRE solves the governing partial differential equations of fluid flow guided by the physical principles: (1) conservation of mass; (2) F = ma (Newton's second law); and (3) conservation of energy /3/.

A *finite-volume* method is used for the numerical solution of the unsteady, Reynolds-averaged transport equations of momentum (Navier-Stokes, NS), mass conservation (continuity) and conservation of thermal energy. Turbulence phenomena are taken into account via the two-equation k-ε or higher order Reynolds stress turbulence model, whereby the k-equation is replaced by 6 equations for the mean turbulent stresses $<u_i'u_j'>$ /4/.

The governing fluid flow equations (NS, turbulence, enthalpy) can be represented by a single generic transport equation for a general, scalar variable ϕ. The integral form of the generic equation is given by

$$\frac{\partial}{\partial t}\int_V \rho\phi\, dV + \int_S \rho\phi\vec{u}\, d\vec{S} + \int_S \Gamma_\phi \nabla\phi\, d\vec{S} = \int_V S_\phi\, dV \tag{1}$$

Applying Gauss' divergence theorem to the surface integrals the coordinate-free vector form of (1) can be obtained

$$\frac{\partial\rho\phi}{\partial t} + \nabla(\rho\vec{u}\phi) + \nabla(\Gamma_\phi\nabla\phi) = S_\phi \tag{2}$$

The variable $\phi = \{1, u_i, k, \varepsilon, h, ..\}$ stands for the actual transport variable considered, e.g. $\phi = 1$ results in the continuity equation. ρ and Γ_ϕ represent the mean fluid density and the effective diffusivity, respectively. The source term S_ϕ on the right-hand side of (1) and (2) describes all explicit dependencies from the main solution variable and all effects of external, volumetric forces e.g. gravity, electromagnetic forces etc.. The left-hand side describes the time-rate-of-change, the convection and the diffusion transport of ϕ.

The numerical solution of equation (1) is conducted with the *finite-volume* method. As a starting point, the solution domain will be sub-divided into a finite number of computational cells, the control volumes (CV). The primary flow variables are stored in the centres of the CV's. The surface and volume integrals are approximated from the centre values by interpolation between the nodal values of neighbouring CV's. From the transformation and discretization in 3-dimensional non-orthogonal co-ordinate space, a system of non-linear algebraic equations $A\phi = b$ can be derived. The system matrix A contains eight coefficients in off-diagonal positions together with the strictly positive, non-zero pole coefficient in the main diagonal. The vector b stands for the discretized source vector S_ϕ.

For an implicit solution of equation (1) all values of ϕ must be known at time step n+1, which requires the solution of large simultaneous algebraic equation systems for all control volumes of the grid. The biggest advantage of the implicit approach is, that stability can be maintained over much larger values of Δt than for an explicit approach. This results in less computer time /5/.

The numerical solution of the simultaneous non-linear equation systems is performed with iterative techniques. During each iteration, a linearization step and a correction step of the linearized solution is conducted. The process will be repeated until the equation residuals, defined by the normalized sum of the local solution errors, fall below a small value, typically of 10^{-5}. One such iteration step is called an outer or non-linear iteration.

The inner iteration process consists of the solution of the linearized equation systems for each transport variable. It will be performed by state-of-the-art numerical solution methods for sparse linear systems, e.g. the truncated Krylov sub-space methods ORTHOMIN or Bi-CG with parallel pre-conditioning /6/.

3. IMPLICIT SLIDING MESH METHOD

The sliding mesh method described in this section satisfies the requirements of the implicit approach in the whole computational domain. The basic solution process starts from a single computational mesh, which is sub-divided into a moving and a static part, with respect to the basic frame of reference.

The moving and the static parts are separated by the sliding interface, which consists of a set of identical surface elements (patches), accessible from both sides of the interface. In a single movement step, the mesh in the moving part slides with a predefined velocity across the mesh in the static part. After each step, the interface vertices (= corners of the surface patches) in the moving and static parts will be re-attached according to the initially computed *vertex map* list.

Due to the implicit approach the grid nodes will be rotated into their final position already at the beginning of each calculation time step. For the integration of the fluid flow equations, the grid nodes will remain attached in order to ensure strong implicit coupling across the interface. At the beginning of a new calculation time step, the grid movement mechanism will be repeated and the vertices at the interface will be again mapped into their final position.

From a cell-centre point of view, the sliding interface consists of three different types of cells, the *parent*, the *child* and the *ghost* cells. As shown in Figure 1, the parent and child cells belong to the cell layer adjacent to the sliding interface. The parent cells are associated with the static part and the child cells are associated with the sliding part. They are linked to each other via the cell *connectivity* list. The ghost cells are virtual boundary cells, which are located between the parent and child cells.

Both, the *vertex map* and the cell-to-cell *connectivity* list are set up once at the beginning of the calculation. The lists are required for the management of grid movement and interface data exchange.

As an advantage of implicit coupling, the algorithms for data exchange and interface re-construction are purely based on integer arithmetic and therefore do not suffer from expensive floating point computations, as will be the case for explicit approaches.

4. PARALLEL STRATEGY

The parallel strategy of FIRE is based on a data parallel approach, whereby different domain decomposition methods can be selected to partition the computational meshes into a prescribed number of non-overlapping sub-domains. The number of sub-domains is usually equal to the number of processors attached. In order to ensure the quality of the mesh partitioning the following criteria have to be considered:

- ⇨ Optimum load balance, i.e. the number of computational cells has to be uniformly distributed over the processor array.
- ⇨ Minimum surface-to-volume ratio, i.e. the number of communication cells (size of communication surface) should be small compared with the sub-domain size.
- ⇨ Homogeneous distribution of the communication load.

The mesh partitioning process is applied prior to the calculation process. Various partitioning techniques ranging from simple data decomposition (DD) or coordinate bisection (CB) methods up to sophisticated spectral bisection methods (RSB) /7/ are available. Due to the complexity of the application the spectral bisection method has been selected for the optimum partitioning of the computational mesh.

The standard version of the spectral bisection method results in a minimum surface-to-volume ratio, but it can not be avoided, that cells belonging to the sliding interface are assigned to two or more processors. In such a case, a time-consuming, repetitive computation of the send and receive lists is required during runtime. In order to overcome this deficiency, and still minimizing the computational effort, the spectral bisection method has been modified such, that the cells belonging to the sliding interface are strictly assigned to a single sub-domain.

An additional benefit obtained by this decomposition strategy is, that the data transfer across the sliding interface is completely performed by one processor, so that no further effort is required to parallelize the vertex map and connectivity lists.

The rezoning facility on the other hand, requires a high parallelization effort, because of the computation of the cross reference list, which contains the connectivity between the old and the new mesh. The inherent problem is, that the cross reference list may point to cells that are located on different processors. In the worst case a totally irregular sub-domain distribution will result in a tremendous amount of communication load during the rezoning process, which then becomes the major performance bottleneck.

Two basic communication concepts are found in the FIRE kernel: the *local* and the *global* data exchange.

Local data exchange refers to all kind of communication, that has to be done between two different processors (point-to-point communication). The amount of exchanged data depends on the number of send and receive cells and the number of neighbouring sub-domains. Therefore, the communication effort depends strongly on the quality of the mesh partitioning, especially when the number of sub-domains becomes large. Local data exchange is implemented as non-blocking all-send/all-receive strategy.

The second type of data exchange is the global communication. This kind of data exchange is necessary when global values over all computational cells and all processors have to be computed, e.g. the computation of an inner product of two vectors. The data packages submitted into the network are extremely small (in most cases just one number) and the data exchange takes place between all processors. The speed of the global data exchange depends strongly on the network latency and on the number of processors used; but it is widely independent of the mesh partitioning method. As a rule of dumb, the time for global sum operations increases linearly with the number of processors.

5. RESULTS

The analysis of the rezoning and sliding mesh methods has been performed by simulating the air flow in the under-body of a laundry drying machine /8/. The drying process consists of two separated air circuits, one for the cooling air and one for the process air. The air circuits are thermally coupled via the condenser. Figure 2 displays the layout of the complete under-body system.

The present analysis will be focused on the cooling air component (light grey part), which consists of a conical inflow section, the rotating fan and the condenser. The rotating fan and the condenser are connected by a diffuser element in order to achieve a homogeneous load at the entrance section of the condenser.

The rezoning and the implicit sliding mesh techniques are used to resolve the air flow in the rotating fan part. Figure 3 presents the computational grid of the complete cooling air component with 585.448 active cells. In the zoomed cross section through the fan housing the computational mesh, partitioned by RSB, is displayed.

The sliding interface is represented by the cylindrical surface, located between the fan blades and the outside wall of the fan housing. The computational cells of the static and the moving part of the interface are contained in a single processors sub-domain.

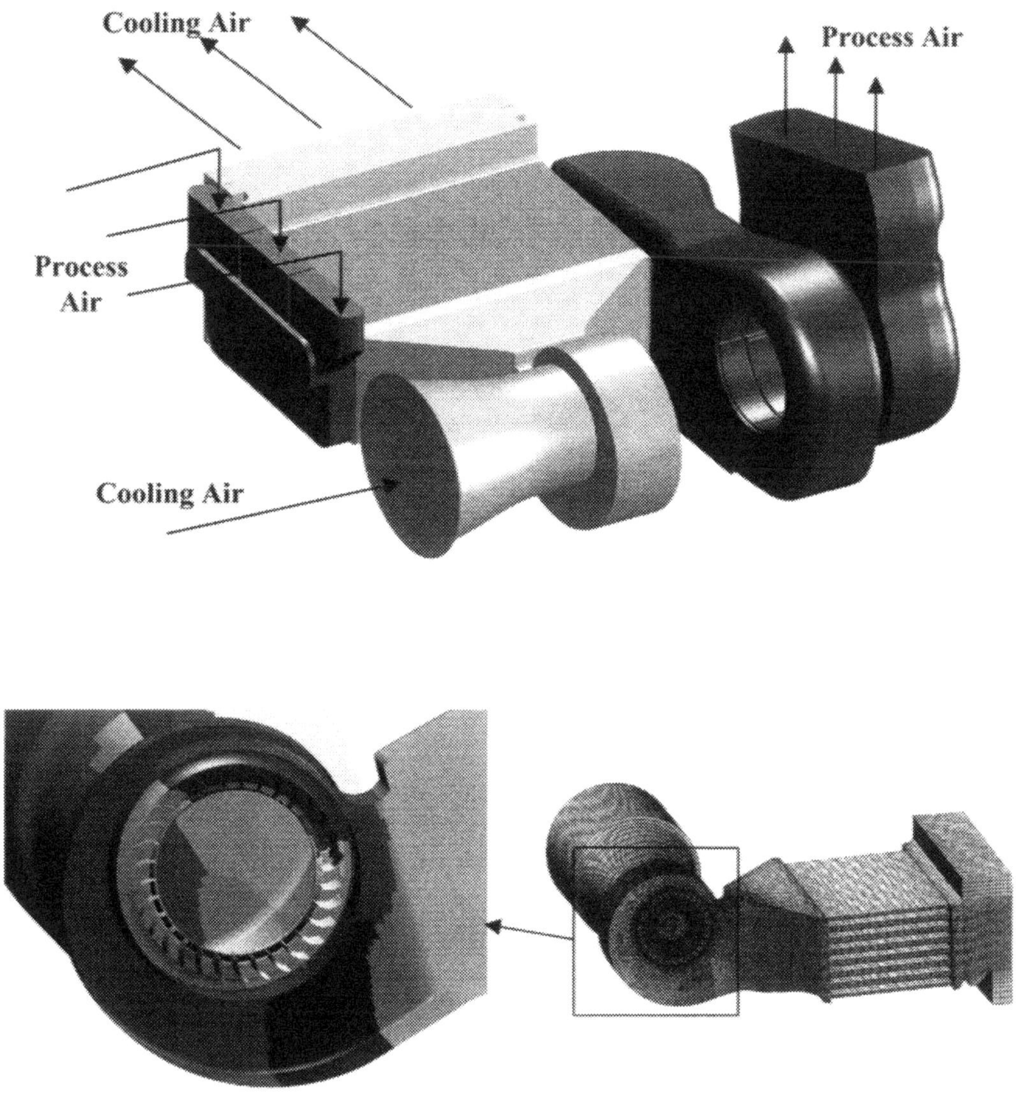

Figure 3. Computational mesh of the cooling circuit ; RSB domain decomposition

In order to justify the quality of the RSB mesh partitioning method, the load balance, the surface-to-volume ratio and the inter-processor connectivity are presented in Table 1 for the 8 processor case.

Sub-domain	Active Cells	Comm. Cells	Surf/Vol Ratio	#Neigh Subdom	Connect.
1	73.181	11.220	0.153	5	2-3-4-5-8
2	73.181	3.536	0.048	2	1-3
2	73.181	5.607	0.077	5	1-2-6-7-8
4	73.181	6.609	0.090	5	1-5-6-7-8
5	73.181	6.320	0.086	4	4-6-7-8
6	73.181	6.707	0.092	5	1-3-4-5-7
7	73.181	6.467	0.088	5	3-4-5-6-8
8	73.181	6.591	0.090	5	1-3-4-5-7

Table 1. Domain decomposition profile

The number of active cells is exactly the same for all sub-domains, therefore an optimum load balance has been achieved. The number of communication cells is well balanced for sub-domains 2 to 8, but sub-domain 1 consists of a higher number of communication cells. This is due to the enclosed sliding interface cells. The surface-to-volume ratios (Surf/Vol Ratio) of the sub-domains 2-8 are always less than 10 percent. Only sub-domain 1 gets 15 percent. Therefore, sub-domain 1 plays the role of the limiting factor for the total communication effort. In the remaining two columns the number of neighbouring sub-domains and the sub-domain connectivity are presented. A uniform distribution of both quantities over the processor array is desirable for a homogeneous load of the network.

It is important to note, that the numbers displayed above have to be related with the system architecture. Provided that the communication network is fast enough, as is the case on IBM SP, the communication load imbalance will be easily compensated for. Typically, the amount of communication time for 8 processors is about 10-15 percent of the total calculation time. In contrast, a load imbalance will directly increase the execution time in proportion to the difference of the maximum and minimum number of cells.

The performance evaluations of the rezoning and sliding mesh techniques have been conducted on a 28 Processor IBM RS6000 Power3 SP system with 200 MHz clock rate. All calculations have been performed over a period of 10 fan revolutions, whereby the fan was rotating with 2750 rpm. In total, 900 time steps, with size of 3.6 degrees each, have been performed. The required number of rezoning events was 90. In case of rezoning, a single processor execution time of 95 hours and, in case of sliding mesh 85 hours have been measured to achieve a periodic stable solution. As demonstrated in Figure 4, the speed-up of the rezoning method drops significantly for more than 2 processors. This is due to the dominant serial portion of the rezoning algorithm which remains constant and, therefore, is independent of the number of processors. In contrast, the sliding mesh method maintains scalability up to 16 processors, where the performance of 16 processors is already three times higher than that for rezoning.

The described methods have also been used to investigate the mass flow rates and pressure increase obtained with different shapes of fan blades. After 10 revolutions, the rezoning method ends up with an oscillating pressure field of constant amplitude. The frequency of the oscillation is coupled with the rezoning frequency and could not be related to any characteristic acoustics frequency in the system. The amplitude of the pressure oscillation is about 30 percent of the overall pressure drop. The 'numerical' oscillations, together with the performance issues mentioned above, were the main reasons for the replacement of rezoning by sliding meshes for the simulation of rotating fan flows. Nevertheless, similar effects could never be observed in engine applications.

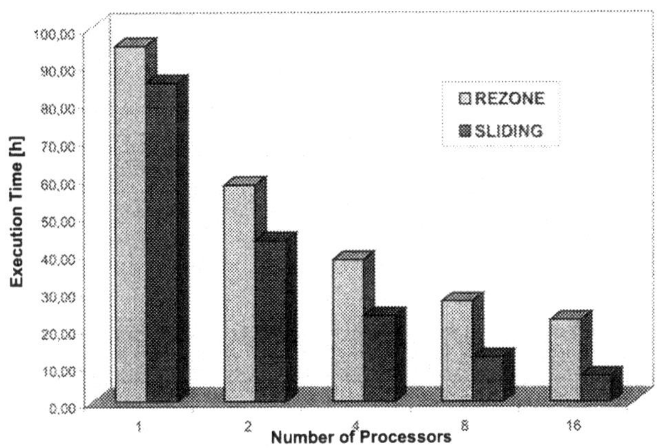

Figure 4. Performance of rezoning vs. sliding mesh on IBM SP

Figure 5 shows a comparison of the mass flow computed with the sliding mesh technique for three types of fan blades: straight, curved and tangential. The measured mass flow of the straight fan is given by 0.067 kg/sec, which agrees well with the computed value of 0.071 kg/sec. The deviation of 5 percents is a result of the coarse mesh in the vicinity of the leading and the trailing edges of the moving blades.

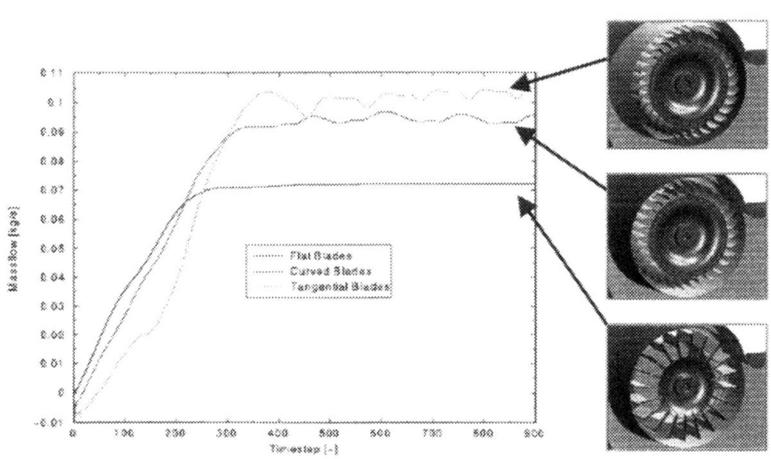

Figure 5. Mass flow for different fan configurations

The mass flow measured for the curved fan resulted in a 40 percent higher value than for the straight fan. The calculation showed a similar increase. Another increase of 10 percent could be achieved by extending the curved blades in the exit section until they become tangential to the circumference circle.

6. CONCLUSIONS

⇨ The parallel implicit sliding mesh method is superior to the partly serial rezoning techniques for the accurate computation of unsteady fluid flow with rotor-stator interaction.

⇨ By using the sliding mesh technique together with the MPI version the execution times for the fluid flow analysis can be significantly reduced.

⇨ The implicit sliding mesh method is based on a single start mesh. All transformations required for grid movement are performed inside the flow solver.

REFERENCES

/1/ Bachler G., Greimel, R., *Parallel CFD in the Industrial Environment*, UNICOM Seminars, London, 1994.

/2/ Bernaschi M., Greimel R., Papetti F., Schiffermüller H., Succi S.: *Numerical Combustion on a Scalable Platform*, SIAM News, Vol. 29, No. 5, June 1996.

/3/ Anderson D.A, Tannehill J.C. and Pletcher R.H., *Computational Fluid Dynamics and Heat Transfer*, Second Edition, Taylor & Francis, 1997.

/4/ Schiffermüller H., Basara B., Bachler G., *Predictions of External Car Aerodynamics on Distributed Memory Machines*, Proc. of the Par. CFD'97 Conf., Manchester, UK, Elsevier, 1998.

/5/ Anderson, J. D. Jr, *Computational Fluid Dynamics*, Editor: Wendt J. F., Second Edition, A v. Karman Institute Book, Springer, 1995.

/6/ Vinsome P.K.W., *ORTHOMIN, an iterative method for solving sparse sets of simultaneous linear equations*. Proc. Fourth Symp. On Reservoir Simulations, Society of Petroleum Engineers of AIME, pp 149-159, 1976.

/7/ Barnard S.T., Pothen A., Simon H.D., *A Spectral Algorithm for Envelope Reduction of Sparse Matrices*, NASA Rep. ARC 275, 1993.

/8/ Bregant A., *CFD Simulation for Laundry Drying Machines*, Proc. of the Simulwhite Conf., CINECA, Bologna, Italy, 1999.

Parallel Computational Fluid Dynamics – Trends and Applications
C.B. Jenssen et al. (Editors)

Using massively parallel computer systems for numerical simulation of 3D viscous gas flows

Boris N. Chetverushkin[a*], Eugene V. Shilnikov[a*] and Mikhail A. Shoomkov[a*]

[a]Institute for Mathematical Modeling, Russian Academy of Sciences, Miusskaya Sq.4, Moscow 125047, Russia

Numerical simulation of oscillating regimes of supersonic viscous compressible gas flow over the 3D cavity with the aid of the explicit kinetically consistent finite difference schemes was fulfilled using different multiprocessor computer systems. The essential features of flow structure and properties of pressure oscillations in critical body points were studied.

1. INTRODUCTION

The problem is extremely actual for modern aerospace applications of detailed investigation of the oscillating regimes in transsonic and supersonic viscous gas flows over various bodies. This is connected in the first turn with the possible destructive influence of the acoustic pressure oscillations upon mechanical properties of the different aircraft parts especially in the resonant case. From mathematical point of view such 3D problems are quite difficult for numerical simulation. This work is dedicated to studying such a flow around rectangular cavity. Under certain freestream conditions such flows may be characterized by regular self-induced pressure oscillations. Their frequency, amplitude and harmonic properties depend upon the body geometry and external flow conditions. Such a task was studied by many scientific laboratories which used modern high performance parallel computers.

In this work the original algorithms were used named kinetically consistent finite difference (KCFD) schemes [1]. There is close connection between them and quasigasdynamic (QGD) equation system [2]. QGD equation system may be considered as some kind of differential approximation for KCFD schemes [3]. The basic assumptions used for the construction of as KCFD schemes as QGD system are that one particle distribution function (and the macroscopic gas dynamic parameters too) have small variations on the distances compatible with the average free path length l and the distribution function has Maxwellian form after molecular collisions. So the QGD system has the inherent correctness from the practical point of view. This correctness of QGD system gives the real opportunity for the simulation of unsteady viscous gas flows in transsonic and supersonic regimes. QGD system and KCFD schemes give the same results as Navier Stokes equations, where the latter are applicable, but have another mathematical form. It's also

*This work was supported by RFBR (grant No. 99-07-90388).

must be mentioned that the numerical algorithms for QGD system and KCFD schemes are very convenient for the adaptation on the massively parallel computer systems with distributed memory architecture. This fact gives the opportunity to use very fine meshes, which permit to study the fine structure of flow. Some results of such calculations are demonstrated in this paper.

2. THE TEST PROBLEM DESCRIPTION

Supersonic flow near an open rectangular cavity is numerically investigated in this work. Such a flow is characterized by a complex unsteady flowfields. The computational region is presented on Figure 1. The geometrical parameters of cavity are: the ratio of the cavity length l to cavity depth h was $l/h = 2.1$ ($l = 6.3$ mm, $h = 3$ mm). The inflow is parallel to the XY-plane and makes angle ψ with the X direction. Let us consider the following time-constant freestream parameters which were taken in accordance with the experimental data of [4]: freestream Mach number $M_\infty = 1.35$, Reynolds number based on freestream parameters and cavity depth $Re_h = 3.3 \times 10^4$, Prandtl number - $Pr = 0.72$, specific ratio $\gamma = 1.4$ and the thickness of the boundary layer was $\delta/h = 0.041$. The intensive pressure pulsations in the cavity take place for such parameters. It was supposed in the experiments that $\psi = 0$, but it seems to be quite difficult to be sure of exact zero angle. That's why the calculations were fulfilled as for $\psi = 0$ as for small incident angles $\psi = 1°$, $2°$ and $4°$. The beginning distribution corresponds to share layer over the cavity and immobile gas with the stagnation parameters inside it.

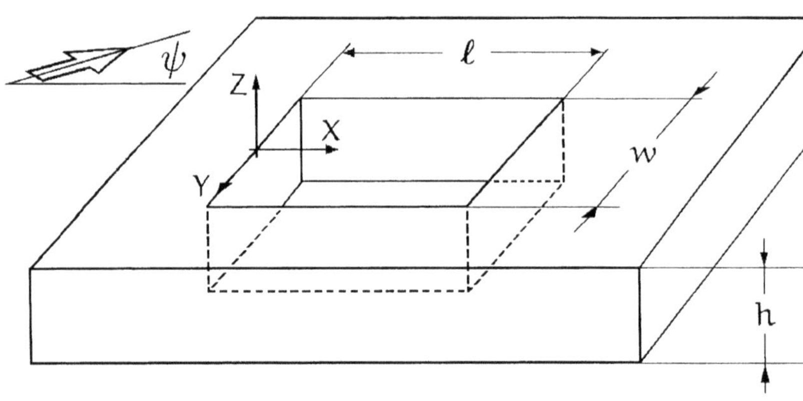

Figure 1. The scheme of computational region.

To predict a detailed structure of unsteady viscous compressible flows we need to use high performance parallel computer systems. KCFD schemes can be easily adapted to parallel computers with MIMD architecture. These schemes are homogeneous schemes i.e. one type of algorithm describe as viscous as inviscous parts of the flow. We used the explicit schemes which have soft stability condition.

The geometrical parallelism principle have been implemented for parallel realization. This means that each processor provides calculation in its own subdomain. The explicit form of schemes allows to minimize the exchange of information between processors. Having equal number of nodes in each subdomain the homogeneity of algorithm automatically provides load balance of processors. The real efficiency of parallelization for explicit schemes is close to **100%** and practically do not depend on processors number (see [5]).

We used the Parsytec CC and HP V2250 multiprocessor RISC computer systems. Distributed memory Parsytec CC is equipped with PowerPC-604 133MHz Motorola microprocessors. Fast communication links gave 40 MB/Sec data transmission rate. Shared memory HP V2250 is equipped with PA-8200 240 MHz HP microprocessors. C and Fortran programming languages were used to develop our applied distributed software. All needed parallel functions are contained in special parallel libraries (MPI standard).

As cavity geometry as splitting of whole computational area to subareas, every of which is loaded to separate processor, are described by special auxiliary language in the text files. Specific subroutine transforms content of these files to format known to computational modules. Fast system of communication links is created on the base of content of these files when distributed task started. Our software takes the possibility to split whole computational area by arbitrary subareas of parallelepiped shape.

3. THE OBTAINED RESULTS

The calculations were accomplished on rectangular grid with the total number of cells near 640000. Detailed information of 3D gas flow around the open cavity was obtained for different angles of incidence.

For $\psi = 0$ the 3D gas flow structure in the middle part of the cavity was approximately the same as for the 2D problem. Gas behaviour in other cavity regions was essentially three dimensional. The most interesting 3D motion was observed in the vicinity of output cavity corner and edge of the long cavity side. Lengthwise gas movement was combined with traverse one in these regions resulting in the gas vortices and swirls appearance. Periodical processes of gas input and output through the side cavity edges occurred.

The analysis of flow structure for low values of incidence angle was fulfilled. The intensive traverse oscillations occur in the cavity for such inflow in addition to previous ones observed in the case of zero angle. Nonzero incident angle leads to appearance of traverse vortical motion over whole cavity (oscillation of longwise swirls) and some vortices in the XY-plane inside the cavity. One can see very complicated asymmetric gas flow behaviour in the middle part of cavity and practically stationary flow in its down forward upwind corner. The fact which seems to be very interesting is the disappearance of boundary layer separation on the forward cavity edge. This effect may be explained by the weakening of feedback between cavity rear and forward bulkheads in the case of nonzero ψ. Because of flow side-drift a compression wave coming to the forward cavity edge is less intensive then for $\psi = 0$ pressure difference doesn't exceed the critical value and it can't initialize the boundary layer separation.

The Figure 2 presents the picture of flow fields in the traverse sections of the cavity. The periodical motion of vortices accompanied by the transformation of their shape may

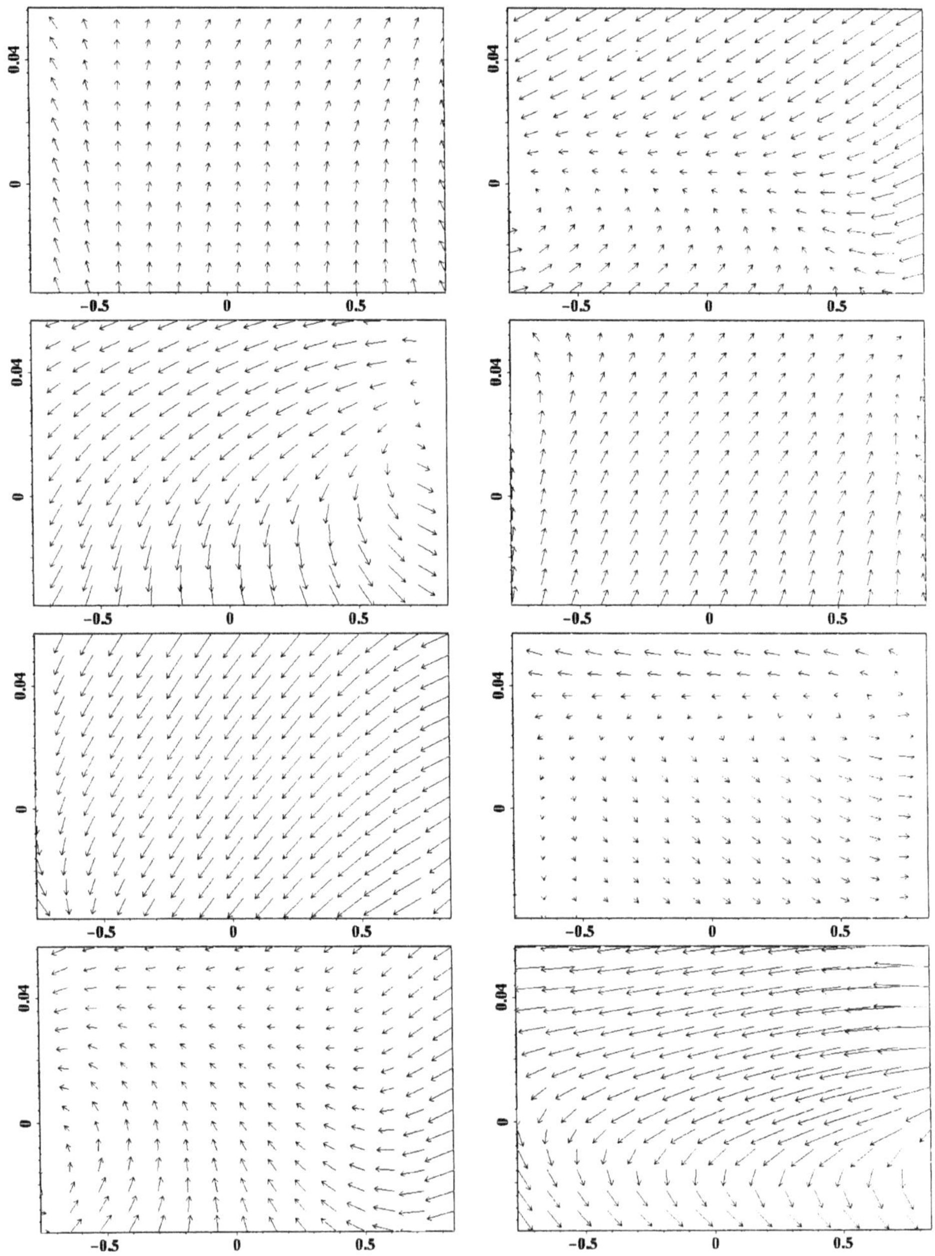

Figure 2. Velocity vector fields for the YZ-section in the middle of the cavity (left column) and near rear bulkhead (right column) for the different time moments

be observed in these regions. This effect corresponds to the longwise swirls oscillations. The swirl motion is correlated with periods of inflow and outflow over lateral cavity edges. Thus first two rows correspond to outflows and the third row — to inflow over both cavity sides. The inflow over upwind edge and outflow over downwind one take place at the time moment corresponding to last row. The duration of inflow period on the upwind side is longer then on the downwind one.

Properties of pressure oscillations in critical cavity points were studied. The spectrum analysis of these oscillations was carried out. This analysis showed the presence of the intensive high frequency discrete components. They had the most amplitudes close to cavity bulkhead and were absent in the cavity central zone. Areas of the most probable wreckage on the cavity surface were revealed. The main modes of pressure pulsation were obtained. The values of Strouhal numbers calculated are well agreed with the experimental values [4]. The first fluctuation mode results from the interaction between the over-cavity shear layer and large vortices formed inside the cavity and comparable-sized with it.

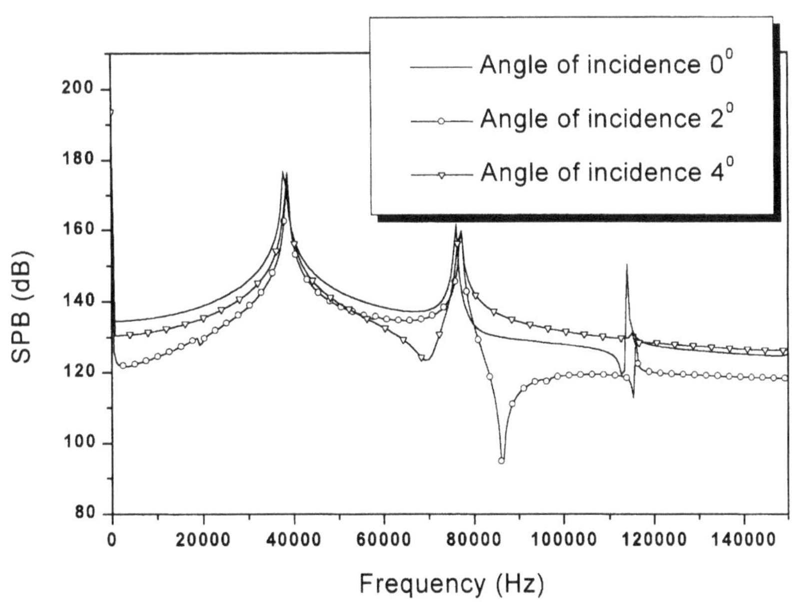

Figure 3. Pressure oscillations spectra in the middle point of the rear cavity edge.

The pressure oscillations spectra in the various cavity points are presented on Figure 3 and Figure 4. One can see the diminishing of SPL for $\psi = 2°$ in comparison with $\psi = 0°$ case. This effect is in the accordance with our hypothesis of weakening of feedback between

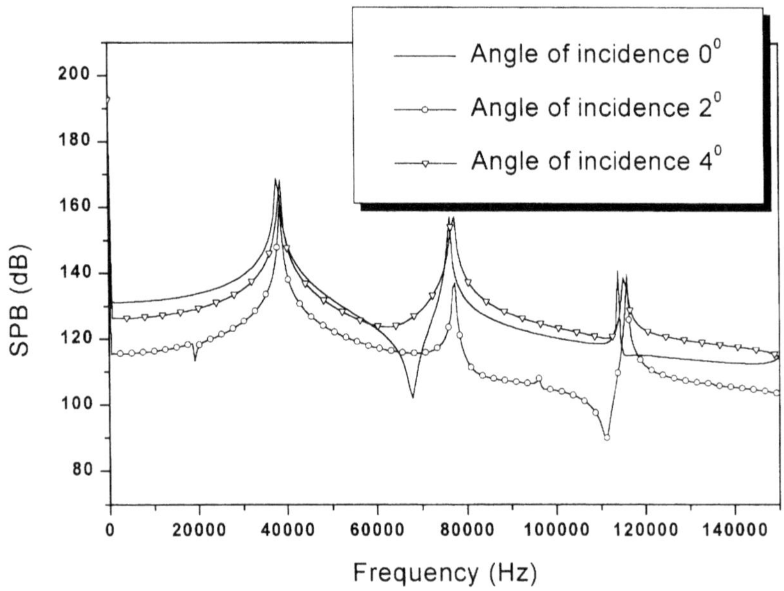

Figure 4. Pressure oscillations spectra in the corner of the rear cavity edge.

cavity bulkheads. The phenomenon of small displacement of spectra discrete components may be observed while changing inflow incident angle.

As it is customary, spectral characteristics of pressure oscillations are represented as a sound pressure level (SPL) in decibels (dB) which is defined as the following

$$SPL = 20 \log_{10} \left(\frac{\sigma}{\sigma_0} \frac{p_s}{p_\infty} \right), \tag{1}$$

where σ_0 — the acoustic sound reference level of 2×10^{-5} Pa, σ — root-mean-square value of pressure pulsation amplitude, $p_s = 101.325$ kPa — standard pressure, p_∞ — static pressure.

Thus, using of detailed spatial mesh allow to calculate flowfield in the cavity and visualize middle scale structures. One can hope that more detailed grid will make possible to receive the whole structure of flow in the cavity in transient case. There is also perspective from our point of view to use some kinetical analogue of K-$\mathcal{E}$ model of turbulence. We intend to combine in our future activity the direct modelling of large-scale and middle-scale flow structures with the description of small-scale turbulent structures basing on K-$\mathcal{E}$ model.

REFERENCES

1. B.N. Chetverushkin, Kinetically consistent finite difference schemes and simulation of unsteady flows, in: Computational Fluid Dynamics 96, Proceedings of III ECCOMAS, Wiley, Paris, 1996.
2. T.G. Elizarova, B.N. Chetverushkin, Using of kinetic models for the computation of gasdynamic flows, in: Mathematical Modelling. Processes in Nonlinear Media, Nauka, Moscow, 1986, (in Russian).
3. B. Chetverushkin, On improvement of gas flow description via kinetically-consistent difference schemes, in: Experimentation, Modelling and Computation in Flow, Turbulence and Combustion, Vol 2, ed. B.N. Chetverushkin, J.A. Desideri et al, Wiley, Chichester, 1997.
4. Antonov A., Kupzov V., Komarov V. (eds.), Pressure oscillation in jets and in separated flows, Mashinostroeniye, Moscow, 1990, (in Russian).
5. B.N.Chetverushkin, E.V.Shilnikov, Unsteady viscous flow simulation based on QGD system, in: Mathematical Models of Non-Linear Excitations, Transfer, Dynamics, and Control in Condensed Systems and Other Media, (Eds. L.A. Uvarova, A.E. Arinshtein, and A.V. Latyshev), pp. 137 - 146, Plenum press, New York, 1999.

Parallel Computational Fluid Dynamics – Trends and Applications
C.B. Jenssen et al. (Editors)

Explosion risk analysis - Development of a general method for gas dispersion analyses on offshore platforms

Asmund Huser and Oddmund Kvernvold

Det Norske Veritas,

Veritasveien 1, N-1322 Høvik, Norway

A general method to determine the probabilistic distribution of gas cloud sizes from accidental gas leaks on offshore installations has been developed. The main motivations for this development have been to improve accuracy and to reduce the analysis time for new explosion risk analyses. This has been achieved by performing the present very detailed analysis once and for all. Results from this detailed analysis comprise functional relationships (response surfaces) which gives the gas cloud size as a function of the most important dependant variables. For the next analysis that should be performed, only a few critical CFD dispersion simulations (typically 10) needs to be performed in order to obtain the probabilistic distribution of gas cloud sizes. Physical evidence, dimensional analysis, engineering judgement together with a large number of CFD simulations (64 scenarios) has been applied to obtain the "universal" response surfaces. The analysis has been performed on two different installations (the wellhead platform Huldra and the production ship Norne) and the results indicate that the present response surfaces will work also for other installations.

1 INTRODUCTION

The problem considered in this paper is a part of the explosion risk analyses, which is routinely performed for offshore installations. The overall objectives of the explosion risk analysis are to determine design accidental loads and to obtain improvements to design and operations against explosions. In the offshore industry, an increased focus was placed on the explosion loads after the Piper Alpha accident and the subsequent full-scale experiments /1/. The experiments showed that commonly used explosion simulation programmes under-predicted pressures dramatically /4/. As a result, great efforts has been placed in improvement of the prediction tools and the calculation procedures /2/3/4/. Improvements have resulted in a more detailed analysis, where also the effects of gas dispersion are included. The complex flow that occurs in a naturally ventilated process area has been investigated in a full-scale experiment by Claever *et. al.* /5/. With the present work, the main effects of gas dispersion have been reviled and included systematically in the explosion risk analysis applying CFD tools. The motivation for the present work has been to improve the accuracy of the analysis, and to reduced the analysis cost by reducing the number of new CFD simulations. A separate tool (called EXPRESS) has been developed, which uses the present results and a few new CFD simulations for each new installation in a Monte Carlo Simulation routine to obtain the probabilistic distribution of gas cloud sizes and explosion pressures.

2 PROBLEM DEFINITION AND DIMENSIONAL ANALYSIS

In order to calculate the probabilistic distribution of accidental gas clouds on an offshore installation, all combinations of accidental leak situations and typical weather conditions needs to be determined. The geometric model of the complex geometry is made applying a CAD model of the installation. Applying the CFD programme FLACS /6/, which is based on a Cartesian structured grid system, the geometry is included as porous regions when the grid can not resolve the details. For the dispersion simulations, all geometry of the platform is included in order to model the external wind as well as the wind field internally in the module.

A gas cloud is created by an accidental gas leak as schematically pictured in Figure 1. The size of the cloud is written as a general function of the following variables:

$$V_f = f(\overline{u}, \dot{m}, L, \rho_g \mid \alpha, \beta, leaklocation), \quad \textbf{(1)}$$

where, V_f is the volume of the explosive gas cloud (m³) (the volume and mass of the explosive HydroCarbon (HC) gas in the cloud is defined from the CFD results), $\dot{m}$ is the leak rate (kg/s), $\overline{u} = Q_a/L^2$ is the mean wind speed in the module (m/s), Q_a (m³/s) is the air ventilation rate in the module before the leak starts, L (m) is the mean dimension of the module ($L=V^{1/3}$), V is the module volume (m³), ρ_g is the leak gas density at release conditions (1 bara) (kg/m³), α is the wind direction, and β is the leak direction.

When the wind direction, leak direction and leak location is fixed, the normalised cloud size can be written as a function of one non-dimensional variable when performing a dimensional analysis /5/:

$$V_f / V = f(R \mid \alpha, \beta, leaklocation). \quad \textbf{(2)}$$

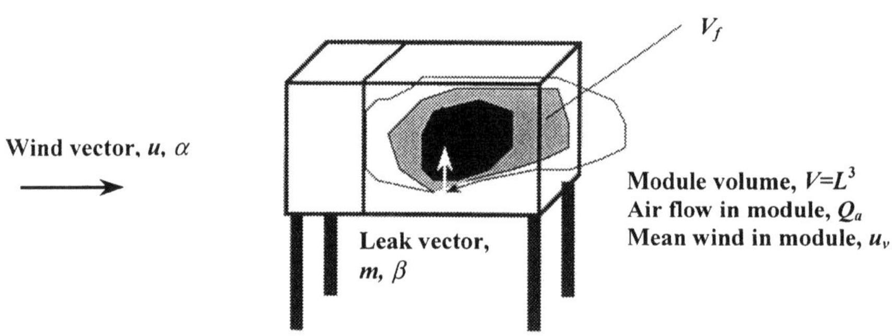

Figure 1 Definition of variables in leak scenario in a typical offshore module. The outer white cloud has concentration below Lower Explosion Limit (LEL), grey cloud is explosive and black cloud has concentration higher than Upper Explosion Limit (UEL).

Here, the new variable R is defined as follow:

$$R = \frac{Q_g}{Q_a} = \frac{\dfrac{\dot{m}}{\rho_g}}{\dfrac{Q_{aref}}{u_{ref}} u}. \tag{3}$$

Definitions of the variables are further:

- u is the wind speed (m/s),
- Q_{aref}/u_{ref} is the reference ventilation rate, normalised by the reference wind speed (determined from CFD analysis). When buoyancy forces are small, the ventilation rate is proportional to the wind speed.
- Q_a is the air volume flowrate (m³/s) through the module before the leak starts, and
- Q_g is the gas volume flowrate (m³/s)

These results from the dimensional analysis suggest that the variable R should be used instead of the leakrate and the windspeed.

3 EFFECTS OF DISPERSION PARAMETERS

The discussion in this section is aimed to give the typical physical effects that are observed. The most important effects are applied to develop the response surface formulas.

The effects on the explosive gas cloud size by changing the dependent variables one by one have been determined by analysing in total 64 leak scenarios. The full scale analysis by Cleaver *et. al.* /5/ also indicate similar effects.

3.1 Effect of wind speed and leak rate

The effects of wind speed and leak rate have been combined in the non-dimensional variable, R, (see Eq. **(3)**). A constant R will give a constant cloud size as long as the velocity field in the module is created by the wind field outside the platform and not created by buoyancy effects. The reason for the similarity between two cloud sizes with equal R is that the gas concentration at any point away from the source is approximately proportional to the leakrate and inverse proportional to the wind speed. This effect has also been applied in the "frozen cloud" assumption, which has been applied to analyse each CFD simulation in detail. This effect is demonstrated from the CFD simulations at Huldra, where for different scenarios with large variations in leakrate and wind speed, but with constant R, the same cloud size is obtained. For the 3 Scenarios in Figure 2, the values of the wind speed and leak rate are different, however, the value of R is approximately the same ($R \approx 0.1$), for these cases the cloud size is found to be approximately the same. These 3 scenarios are also plotted as the 3 spot values near the maximum point in Figure 3.

When R is small (i.e. small leakrate and/or large wind speed), a small cloud is obtained due to quick dilution of the cloud (see Figure 3 which show the degree of filling, V_f/V, as a function of R). For large R, the explosive cloud size is reduced because most of the gas will have concentrations higher than the Upper Explosion Limit (UEL). The cloud size reaches a maximum for an intermediate R when most of the gas-air mixture has concentration between

Lower and Upper Explosion Limit (LEL and UEL). Based on this general behaviour, the following formula (response surface) is derived for a general gas cloud size:

$$\frac{V_f}{V} = \frac{C_3}{1 + \frac{1}{C_1 R^{P_1}} + C_2 R^{P_2}}. \quad \textbf{(4)}$$

Here,

- $P_1 = 3/2$ has been found to fit well for the present cases. For small values of R (e.g. small leakrates), the cloud size is independent of the module volume. This is fulfilled only with P_1=3/2, ref. /5/.
- C_1 and P_2 are constants to be determined from CFD analysis. One set of constants is to be found for each *leak location.*
- C_3 is the total calibration constant, which is adjusted to obtain the best fit to the CFD data.
- The parameter C_2 is derived from Eq. (4) as a function of the maximum cloud size, $V_{f\max}/V$.

Hence, the parameters C_1, P_2 and $V_{f\max}/V$, are enough to describe the cloud size as function of R. These parameters are to be determined from a few new CFD simulations for new installations. The most important parameter of these three is $V_{f\max}/V$. This variable is strongly dependent of the wind- and leak-directions, and a functional relationship has been derived in the next section.

Buoyancy effects are created by buoyancy of the light HC gas and by hot equipment creating buoyant air. These effects only become dominant at low wind speeds hence, in the present analysis, buoyancy effects have not been investigated further.

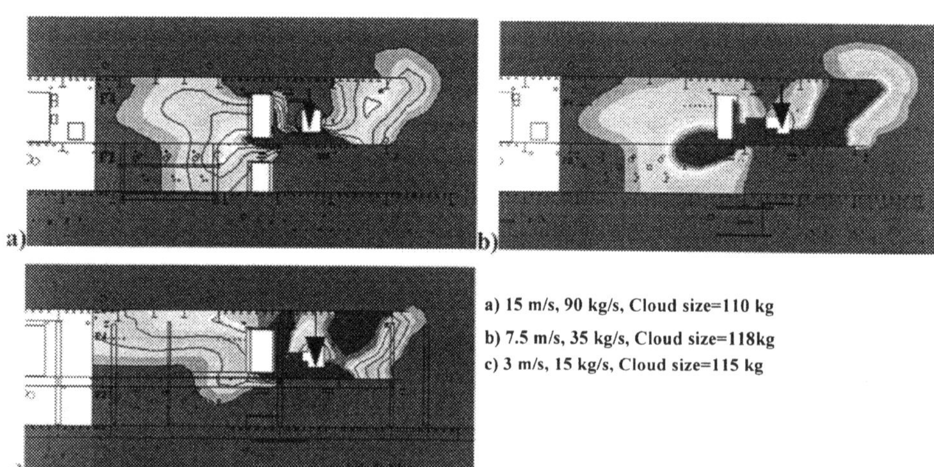

Figure 2 Contour plots of gas cloud in a vertical E-W plane for 3 different wind speeds and leak rates with $R \approx 0.1$. For all scenarios is applied; leak direction, down (arrow); wind direction, NE; leak location, const. Light grey, explosive gas; dark grey, not explosive.

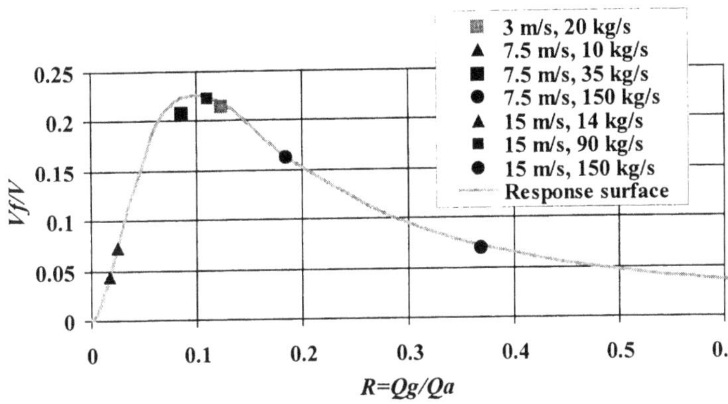

Figure 3 Example of degree of filling (V_f/V) as a function of the non-dimensional parameter R. Cases in the plot are all for releases from the same leak location, leak direction and wind direction, at steady state cloud size.

3.2 Effect of wind- and leak-direction

The wind direction has a large effect on the gas cloud size and location in a naturally ventilated module, shown in Figure 4. For winds from N, NW and W, most of the gas is diluted before the cloud is leaving the module, resulting in a large explosive cloud. For wind from E, most of the gas is blown out of the module causing a smaller cloud inside the module. Note that explosive gas outside the module is not contributing to increase in explosion pressures. Hence, only gas inside the module is accounted for when determining the cloud size.

The effects of leak- and wind- direction are in general that wind-direction opposite and equal the leak-jet direction result in large and small clouds, respectively. This is also illustrated in Figure 4, typically the largest clouds occur when the wind comes from West and North, and smaller clouds with wind from East.

A functional relationship of this behaviour is given in Eq. (5). The variable α_{max} is defined as the wind direction where the cloud has its largest size, *except* for "wake wind". Applying the "cos." function, the maximum value will occur when $\alpha=\alpha_{max}$. The wind direction that is towards the leak, is in general close to α_{max}. Hence, there will be one α_{max} for each leak direction. When the leak is directed towards a large structure, a wall or a deck, a diffusive jet occurs, and the effect of the leak direction (the value of B) is smaller.

$$\frac{V_{f\,max}}{V} = \begin{cases} A + B\cos(\alpha - \alpha_{max}), & \alpha \neq \alpha_W \\ \dfrac{V_{f\,maxW}}{V}, & \alpha = \alpha_W \end{cases} \tag{5}$$

Definition of parameters in Eq. (5):

- A and B are constants dependant of the leak direction, β, and determined from CFD simulations.

- α_{max} is the wind direction that gives the largest gas cloud, *except* for wake wind. Typically wind directed toward the leak-jet gives the largest clouds, i.e. $\alpha_{max} \approx \beta$.
- α_W is the wake wind direction (Wind from west for Huldra). Causes a larger cloud size due to large re-circulating zones. Large re-circulating zones in the module are created by the large firewall at the West wall of the Huldra module. In general, the re-circulating zones results in larger gas clouds.

3.3 Fit of response surface to CFD results

The fit of the response surface with the CFD results has been plotted on **Figure 5**. Here it is shown that most of the scenarios are within ±30%. It is found that the *rms* value of the normalised difference between the response surface and the CFD result is 29%.

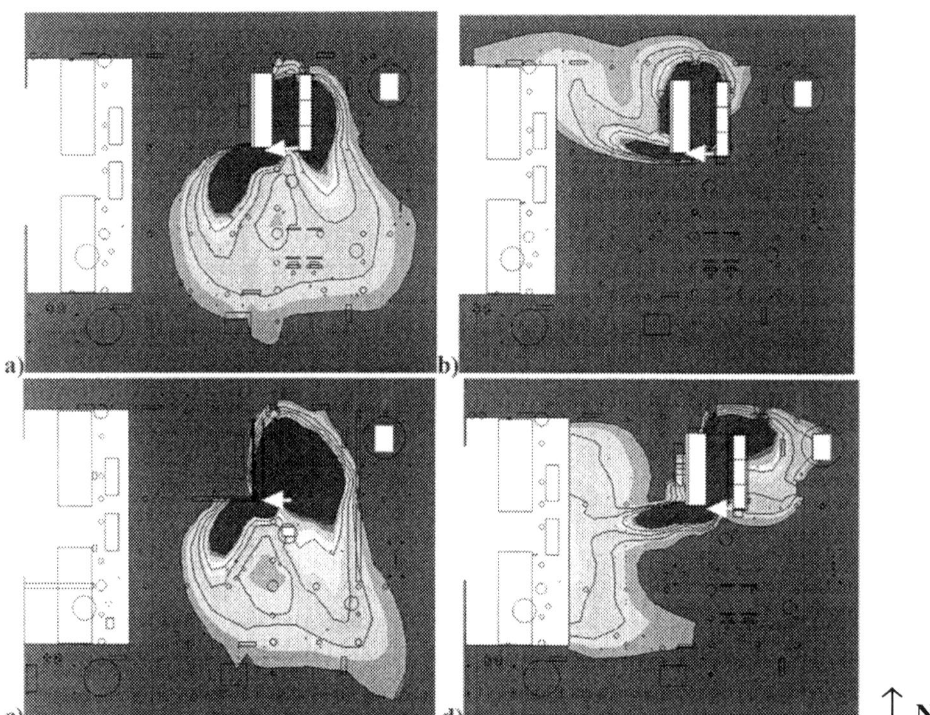

Figure 4 Effect of wind direction. Contour plots of horizontal cuts through leak for 4 wind directions; a), wind from N; b), wind from E; c), wind from NW; and d), wind from W. For all scenarios is applied: Leak direction, W (arrow); leak rate, 20 kg/s; leak location, const.; and wind speed, 7.5 m/s. Light grey, explosive gas; dark grey, not explosive.

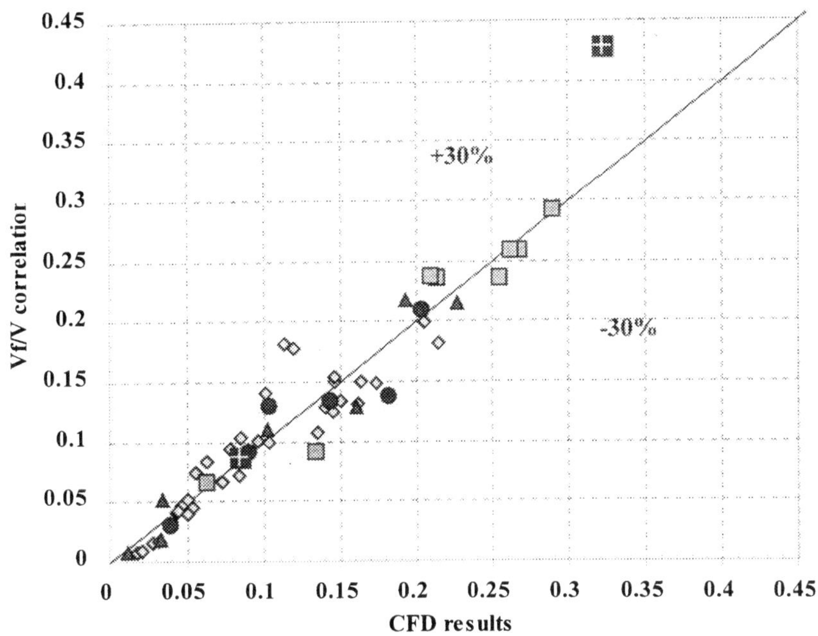

Figure 5 Comparing results from CFD simulations with response surface formula for all CFD scenarios. Different symbols represent different leak locations.

4 CONCLUSIONS

The following main conclusions have been obtained:

- Cost effective design and operation of new and existing offshore installations may be obtained due to the increased accuracy and level of detail of the analysis.
- General trends and typical behaviour of the gas clouds have been determined in the present paper, reducing the need for extensive analyses for new installations.
- The new method is a part of a general and *uniform* procedure to perform explosion risk analyses. This will ensure consistency among analysis and when one analysis is updated. The analysis also becomes person independent.
- The sizes of the explosive gas clouds have been expressed by explicit formulas (response surfaces). The dependent variables in the response surfaces are wind speed, wind direction, leak rate, leak direction, leak location and time.
- The response surfaces are developed by studying the effect of the dependent variables, and by findings from a full-scale test project /5/. The Huldra WHP geometry has been used for most of the analysis. Also a few cases applying the Norne FPSO geometry is included. Comparisons between Norne and Huldra results indicate that the method can be applied for different geometry configurations.
- By a dimensional analysis, the total number of dependent variables has been reduced.

- A power law dependency is obtained between cloud size and the relation, leakrate/windspeed. For small values of leakrate/windspeed, a constant power is obtained. For large values of leakrate/windspeed, the power must be obtained for each geometry and leak configuration.
- Combining the effects of wind direction and leak direction further reduce the number of dependant variables.
- The *rms* value of the normalised difference between the response surface and the CFD result is 29%.
- The response surface technique represents a quick method to calculate cloud sizes, and is well suited for Monte Carlo simulations. A new risk analysis tool called EXPRESS has been developed where these techniques are implemented.
- A requirement for the present development method is the need to isolate the effect of single parameters. This can quickly and easily be done with a CFD model set-up.
- High gas velocity at the leak requires small timesteps and hence long computer times. Also the high number of scenarios in this work resulted in long computer times. By the use of more nodes on each computer, the total computer time has been reduced to manageable times.

REFERENCES

/1/ C.A. Selby & B.A. Burgan "Blast and Fire Engineering for Topside Structures – Phase 2" The Steel Construction Institute Publication no. 253, (1998)

/2/ J. Pappas "Operatørselskapenes nye prosedyre for å bestemme eksplosjonsrisiko" Conference arranged by "Norske Sivilingeniørers Forening" on fire and explosion safety. March 1999.

/3/ O. Talberg, O.R. Hansen & J.R. Bakke "Explosion Risk Analysis Using Flacs" Proc. From 8th annual conference on offshore installations: Fire and explosion engineering. (1999)

/4/ J. Wiklund & I. Fossan "Model for Explosion Risk Quantification" Proc. From 8th annual conference on offshore installations: Fire and explosion engineering. (1999)

/5/ R.P. Claever, S. Burgess, G.Y. Buss, C. Savvides, S. connolly, R.E. Ritter "Analysis of gas build-up from high pressure natural gas releases in naturally ventilated offshore modules" Proc. From 8th annual conference on offshore installations: Fire and explosion engineering. (1999)

/6/ FLACS 98 User's Guide, GexCon 1999

Parallel Computational Fluid Dynamics – Trends and Applications
C.B. Jenssen et al. (Editors)

Parallel multiblock CFD computations applied to industrial cases

H. Nilsson, S. Dahlström and L. Davidson

Chalmers University of Technology, Department of Thermo and Fluid Dynamics,
SE-412 96 Göteborg, Sweden

A parallel multiblock finite volume CFD (Computational Fluid Dynamics) code CALC-PMB [3,6,7] (Parallel MultiBlock) for computations of turbulent flow in complex domains has been developed. The main features of the code are the use of conformal block structured boundary fitted coordinates, a pressure correction scheme (SIMPLEC [4] or PISO [5]), cartesian velocity components as principal unknowns, and collocated grid arrangement together with Rhie and Chow interpolation. In the parallel multiblock algorithm, two ghost cell planes are employed at the block interfaces. The message passing at the interfaces is performed using either PVM (Parallel Virtual Machine) or MPI (Message Passing Interface).

This work was performed on a 64-processor shared memory SUN Enterprise 10000 at Chalmers and on a 170-node distributed memory IBM SP at the Center for Parallel Computing at KTH (Royal Institute of Technology).

Parallel aspects of computations from two different industrial research areas, hydraulic machinery and aerospace, and an academic test case are presented. The parallel efficiency is excellent, with super scalar speed-up for load balanced applications using the best configuration of computer architecture and message passing interface [2].

1. Industrial cases

This work presents the parallel aspects of computations from two different industrial research areas, hydraulic machinery - *Numerical investigations of turbulent flow in water turbines* [6] and aerospace - *Large eddy simulation of the flow around a high-lift airfoil* [1]. The background of each case is briefly described below.

1.1. Hydraulic machinery

This work is focused on tip clearance losses in Kaplan water turbines, which reduce the efficiency of the turbines by about 0.5%. The work is part of a Swedish water turbine program financed by a collaboration between the Swedish power industry via ELFORSK (Swedish Electrical Utilities Research and Development Company), the Swedish National Energy Administration and GE (Sweden) AB. The turbine investigated (fig. 1) is a test rig with a runner diameter of $0.5m$. It has four runner blades and 24 guide vanes (fig. 2). The GAMM Francis runner [8] (fig. 3) is used for validation of the computational code, since there are no detailed measurements for the Kaplan runner. The tip clearance between the Kaplan runner blades and the shroud is $0.25mm$. In order to resolve the turbulent

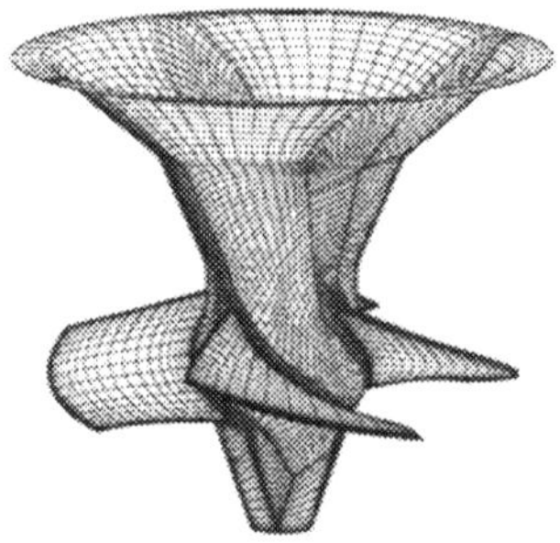

Figure 1. A Kaplan turbine runner multiblock grid.

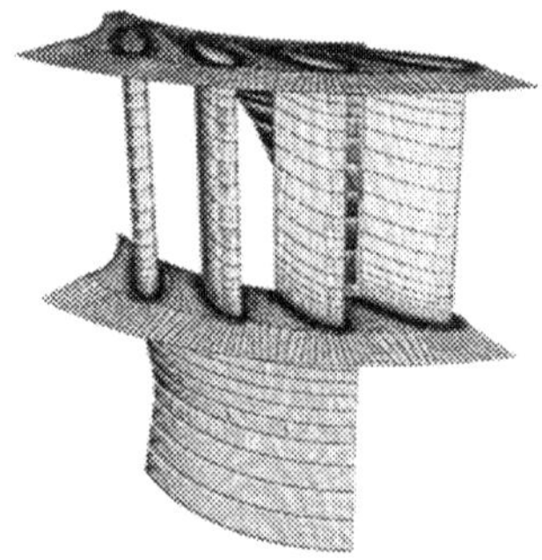

Figure 2. Four (of 24) guide vanes.

Figure 3. A Francis turbine runner multiblock grid.

flow in the tip clearance and in the boundary layers, a low Reynolds number turbulence model is used. Because of computational restrictions, complete turbine simulations usually use wall functions instead of resolving the boundary layers, which makes tip clearance investigations impossible.

Although the computations assume that the flow is periodic, which allows only one blade passage to be computed, and that the boundary conditions are assumed to be axisymmetric, these kinds of computations tend to become computationally heavy and numerically challenging. This, together with the complicated geometry requiring complex multiblock topologies, makes a parallel multiblock CFD solver a suitable tool.

1.2. Aerospace

This work is part of the ongoing Brite-Euram project called LESFOIL (Large Eddy Simulations of Flows around Airfoils). One of the main objectives of the project is a demonstration of the feasibility of LES for simple 2D airfoils (fig. 4). The test case chosen is the flow around the Aerospatiale A-airfoil at an angle of attack equal to 13.3° and where the chord Reynolds number is $2.1 \cdot 10^6$. This is a challenging case for LES because of the high Reynolds number and because of the different flow situations around the airfoil, including transition from the laminar flow near the leading egde and separation near the trailing edge (fig. 5). Even at the low Reynolds number, from an aeronautical point of view, a wall-resolved LES is too expensive. The use of approximate boundary conditions in the near-wall region is thus necessary. By using a *20-nodes per boundary-layer thickness estimate* [10] in each direction, $50-100$ million nodes are needed for this case [9]. However, with a good method of prescribing and controlling the transition we hope that a 2 million node mesh is sufficient for LES with wall-functions. Still, the requirements on the mesh are demanding and result in meshes with a large number of nodes. For this reason, an efficient numerical method with an effective parallelization is needed.

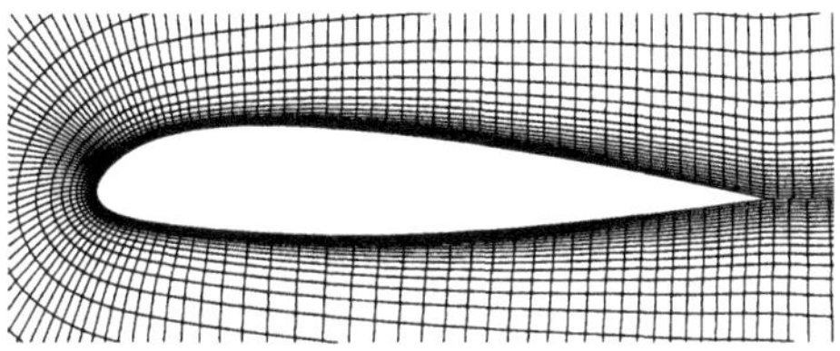

Figure 4. Zoom of mesh around the Aerospatiale A-airfoil (every 4th node in the wrap around direction and every 2nd node in the surface normal direction is plotted).

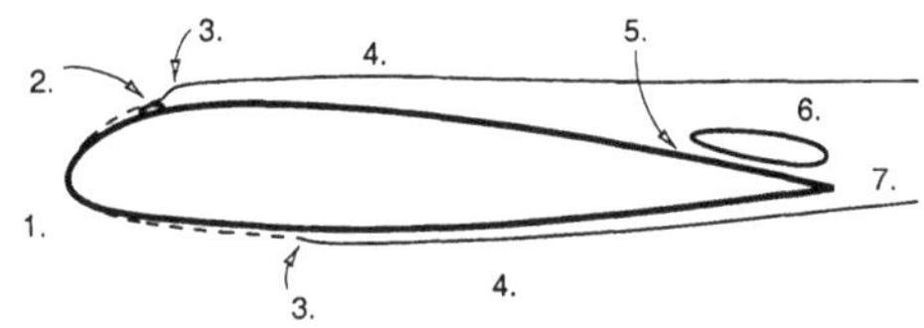

Figure 5. Schematic sketch of the flow regimes around the Aerospatiale A-profile: 1. laminar boundary layer, 2. laminar separation bubble, 3. transition region, 4. turbulent boundary layer, 5. separation point, 6. separation region, 7. wake region.

2. CALC-PMB: The Parallel MultiBlock CFD solver

A single structured block sequential finite volume CFD solver, CALC-BFC (Boundary Fitted Coordinates) [3], has been extended with message passing utilities (PVM or MPI) for parallel computations of turbulent flow in complex multiblock domains [6,7] (fig. 6). The main features of the resulting SPMD (Single-Program-Multiple-Data, all the processes run the same executable on different data) code, CALC-PMB, are the use of conformal block structured boundary fitted coordinates, a pressure correction scheme (SIMPLEC [4] or PISO [5]), cartesian velocity components as principal unknowns, and collocated grid arrangement together with Rhie and Chow interpolation. In the parallel multiblock algorithm, two ghost cell planes are employed at the block interfaces. The message passing at the interfaces is hidden behind a high level parallel multiblock library with data structures that fit CALC-PMB (fig. 7) and the underlying message passing interface (PVM or MPI) is chosen at compile time. The calls for parallel multiblock routines in the code are thus completely independent of the message passing interface that is used. Thus most of the parallelism is hidden from the user, who can easily manipulate the code for his/her purposes using the high level parallel multiblock library if necessary. The advanced user may easily add optional message passing interfaces if needed. The code may be run on everything from inhomogenous NOW (Network Of Workstations) to Beowulf Linux clusters and distributed and shared memory supercomputers. It may read a predefined multiblock topology with connectivity information from the disk or subdivide single block domains into equal sized sub-blocks for load balanced parallel computation. The gains of this operation are several: the computational speed may be increased, larger problems may be solved since the memory requirement is divided between the processors (when using distributed machines), more exact solutions may be obtained because of the extra memory available and parallel supercomputers may be employed.

2.1. Numerical procedure

The parallel SIMPLEC [4] numerical procedure can briefly be summarized as follows (see [6] for a more thorough description, or see [1] for a description of the parallel PISO [5]

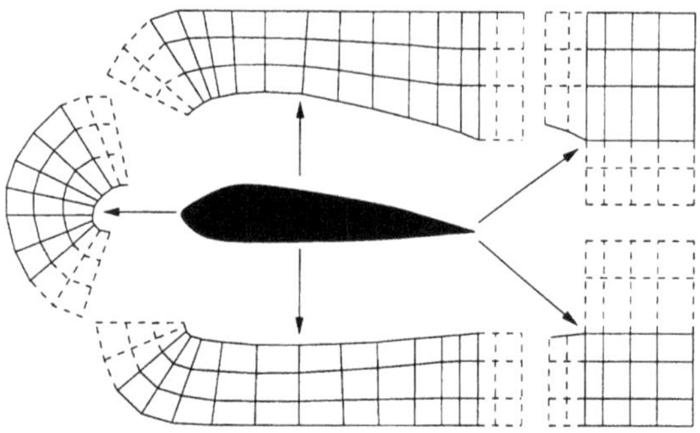

Figure 6. An example of an airfoil multiblock topology that can be computed using CALC-PMB. The blocks overlap two ghost cell planes (dashed lines).

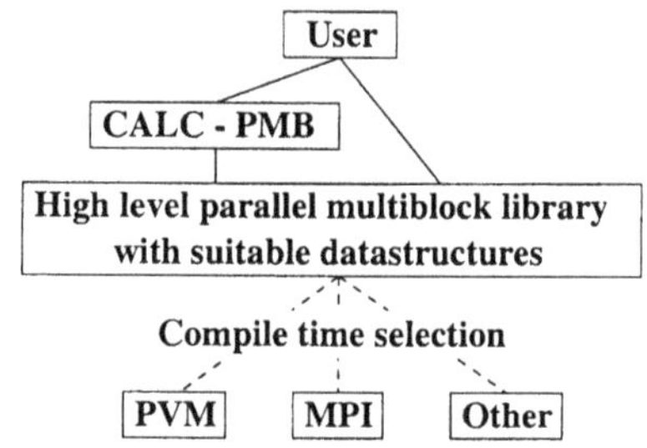

Figure 7. The message passing is hidden behind a high level parallel multiblock library with data structures that fit CALC-PMB.

algorithm).

Iterate through pts. I - X until convergence.

I The discretized momentum equations are solved.

II The inter-block boundary conditions for the diagonal coefficient from the discretized momentum equations are exchanged since they are needed for the Rhie & Chow interpolation.

III The convections are calculated using Rhie & Chow interpolation.

IV The continuity error, needed for the source term in the pressure correction equation, is calculated from these convections.

V The discretized pressure correction equation is solved.

VI The inter-block boundary conditions for the pressure correction are exchanged since they are needed for the correction of the convections.

VII The pressure, convections and velocities are corrected and the pressure field level is adjusted to a reference pressure in one point of the global computational domain. The velocity correction is actually not necessary but has proven to increase the convergence rate.

VIII Inter-block boundary conditions for all variables are exchanged.

IX Other discretized transport equations are solved.

X The residuals are calculated and compared with the convergence criteria. When converged, update the *old* (previous time step) variables and continue with the next time step for transient computations or conclude the computations for stationary computations.

3. Parallel aspects

CALC-PMB computes multiblock cases in parallel by assigning the blocks to separate processes. The computational grid may be given directly as a multiblock grid where the given block size distribution will be kept during the computations. If possible, it may also be given as a single structured block allowing for load balanced (user defined) domain decomposition in CALC-PMB. This work presents parallel aspects from two industrial research areas and one academic test case that has been used for validation and parallel efficiency tests.

Some comments on parallel efficiency should first be given. Commonly, the parallel efficiency is displayed on a 'per iteration' basis. However, for some methods such as the domain decomposition method used in this work, the number of iterations to convergence changes with the number of blocks/processors used. The time to convergence used in this work is then a better measure of the parallel efficiency.

3.1. Academic test cases - parallel aspects

The code has been validated and the parallel efficiency investigated for common academic test cases [7], such as the backward-facing step. Here, parallel aspects from the investigations of both 2D and 3D backward-facing step flow are presented. The 2D case has 22 512 nodes (134x42x4) and a Reynolds number of $Re = 24\,000$, and the 3D case has 874 120 nodes (130x82x82) and a Reynolds number of $Re = 5\,000$. Both cases use the $k - \epsilon$ turbulence model. The computational grid is decomposed in a load balanced way in CALC-PMB. The results of the investigations are presented in table 1. Since the 2D

Table 1
Speed-up and efficiency of 2D and 3D backward-facing step computations. The numbers are based on elapsed wall time to convergence, normalized by the respective single block computation. The domain decompositions are specified by the number of blocks in each direction.

Case	# processors	Domain decomposition	Speed-up	Efficiency
2D	1	1x1x1	1	100%
2D	2	2x1x1	2.0	100%
2D	4	2x2x1	3.1	78%
2D	8	4x2x1	5.7	71%
3D	1	1x1x1	1	100%
3D	4	2x1x2	4.3	108%
3D	8	4x1x2	8.3	104%

case is a rather small problem, with large block surface to volume ratios, the delay times caused by communication are already apparent for four processors. For larger problems, such as the 3D case or common industrial CFD applications, this is not a problem and the parallel efficiency is excellent for load balanced applications. For the 3D case shown in table 1 the speed-up is actually superscalar, probably because of a reduction of cache misses in the smaller subproblems. It is important to notice that, since the computational times are normalized with the computational times of a non-decomposed grid, both domain decomposition and communication effects are included. However, the convergence rates for these cases were affected only slightly by the domain decomposition. The computations were performed on a SUN Enterprise 10000 machine at Chalmers.

3.2. Hydraulic machinery - parallel aspects

The geometries in hydraulic machinery are generally very complicated and several regions of turbulent boundary layers must be resolved. For computational and numerical reasons, the grid size should be kept as small as possible and the control volumes as orthogonal as possible. This requires a complicated multiblock topology. Further challenges arise when there are large differences in geometrical scales, such as in tip clearance computations where the tip clearance block is an order of magnitude smaller than the largest block. If the blocks assigned to separate processes are not of equal size some blocks will wait for others to finish. Thus processors will be temporarily idle and the parallel efficiency will decrease rapidly. The level of parallelization is therefore determined by the block size distribution and the distribution of the processes over the available processors. This is the major problem in these kinds of computations. A load balancing procedure for these cases requires re-distribution of the multiblock topology and re-meshing, which are very time consuming. To a certain extent, this can be avoided by distributing the large blocks on separate processors and the small blocks on shared processors. On a shared memory supercomputer with many processors available, the computational speed will simply adjust to the computational speed of the largest block since the smaller blocks will be run using time sharing.

The CPU usage of each block is thus determined by the block sizes. Summing up the CPU usage from the different processes and normalizing it with the CPU usage of the largest block (which is a measure of the total load of the machine), this 12-block Kaplan runner tip clearance computation (with large differences in block sizes) runs in average on about 7.8 processors. If these computations are to be performed on a distributed system, the blocks must be distributed on eight processors in a way that guarantees load balancing.

An example of the load balancing problem is given in table 2, where the CPU usage of the processors is compared with the block sizes of the GAMM Francis runner computations. The distribution of the CPU usage is quite similar to the distribution of gridpoints, except for some overhead CPU usage that might arise from the message passing procedures.

3.3. Aerospace - parallel aspects

In the airfoil computations presented in this work, the computational grid may be represented as a single structured C-grid wrapped around the airfoil (fig. 4). This configuration allows the user to decide upon the multiblock topology using some parameters in CALC-PMB. The resulting domain decomposition is load balanced, and the multiblock topology and number of blocks may easily be changed without re-meshing.

The computations were done on a 64-processor SUN Enterprise 10000 at Chalmers and

Table 2
Load balance problem (example from a Francis runner computation)

Block	#gridpoints	Normalized #gridpoints (%)	Normalized Instant. CPU %
1	126,360	87.8	94.4
2	144,000	100.0	100.0
3	144,000	100.0	100.0
4	114,660	79.6	86.6
5	72,150	50.1	59.2
Total	601,170	417.5	440.2

on a 170-node IBM SP at the Center for Parallel Computing at KTH. Table 3 compares the elapsed time per time step for different combinations of domain decompositions, message passing interfaces and computers. A comparison of mesh size effects was also made. Two

Table 3
Elapsed time per time step on the coarse and fine meshes (722568 and 1617924 computational nodes, respectively).

Mesh	Computer & message passing system	Number of processors 8	16	32
Coarse	SUN, PVM, socket based	$48s$	$38s$	$36s$
	SUN, PVM, shared memory based	$24s$	$12s$	$6s$
	SUN, MPI	$24s$	-	-
	IBM SP, PVM (based on MPI)	$12s$	-	-
	IBM SP, MPI	-	$5.4s$	$2.8s$
Fine	IBM SP, MPI	-	-	$6.0s$

different versions of PVM are available on the SUN computer: a shared memory-based PVM and a socket-based PVM. When eight processors are used for the present case, the shared memory-based PVM is twice as fast as the socket-based PVM. Some preliminary tests have shown that, when using the shared memory-based PVM, CALC-PMB may scale linearly at least up to 32 processors but that it does not scale at all when the socket based PVM is used. The use of MPI yields the same execution time as the shared memory PVM for the eight processor case, and it is reasonable to believe that it will scale linearly as well. However, since it did not perform better than the shared memory-based PVM for eight processors, the tests were moved to another computer architecture.

Using PVM on the distributed memory IBM SP computer, the eight processor case was twice as fast as when the SUN shared memory PVM was used. Since the IBM SP PVM is based on MPI, using MPI should yield the same computational time as when using PVM. The 16 and 32 processor cases were performed using MPI, and approximately linear speed-up is obtained between the 8 and 32 processor cases.

When the mesh size was scaled up by a factor of 2.3, the computational time was increased by a factor of 2.1 for the 32 processor IBM SP MPI case, which shows that CALC-PMB scales very well for large CFD problems using the most appropriate combination of message passing interface and computer architecture.

4. Conclusions

A parallel multiblock finite volume CFD code, CALC-PMB, for computations of turbulent flow in complex domains has been developed. Most of the parallelism is hidden from the user, who can easily manipulate the code for his/her purposes using a high level parallel multiblock library if necessary. The advanced user may easily add optional message passing interfaces, besides PVM and MPI, if needed.

The parallel efficiency of the code is excellent, with super scalar speed-up at least up to 32 processors for large 3D load balanced applications using the best configuration of computer architecture and message passing interface. However, it has been shown that the parallel efficiency may decrease drastically if the problem size is small, the load balancing poor or the configuration of computer architecture and message passing interface is not good.

REFERENCES

1. S. Dahlström. Large Eddy Simulation of the Flow Around a High Lift Airfoil. Thesis for the degree of Licentiate of Engineering 00/5, Dept. of Thermo and Fluid Dynamics, Chalmers University of Technology, Gothenburg, 2000.
2. S. Dahlström, H. Nilsson, and L. Davidson. LESFOIL: 6-months progress report by Chalmers. Technical report, Dept. of Thermo and Fluid Dynamics, Chalmers University of Technology, Gothenburg, 1998.
3. L. Davidson and B. Farhanieh. CALC-BFC: A Finite-Volume Code Employing Collocated Variable Arrangement and Cartesian Velocity Components for Computation of Fluid Flow and Heat Transfer in Complex Three-Dimensional Geometries. Rept. 92/4, Thermo and Fluid Dynamics, Chalmers University of Technology, Gothenburg, 1992.
4. J.P. Van Doormaal and G.D. Raithby. Enhancements of the SIMPLE method for predicting incompressible fluid flows. *Num. Heat Transfer*, 7:147–163, 1984.
5. R. I. Issa. Solution of Implicitly Discretised Fluid Flow Equations by Operator-Splitting. *J. Comp. Physics*, 62:40–65, 1986.
6. H. Nilsson. A Numerical Investigation of the Turbulent Flow in a Kaplan Water Turbine Runner. Thesis for the degree of Licentiate of Engineering 99/5, Dept. of Thermo and Fluid Dynamics, Chalmers University of Technology, Gothenburg, 1999.
7. H. Nilsson and L. Davidson. CALC-PVM: A parallel SIMPLEC multiblock solver for turbulent flow in complex domains. Int.rep. 98/12, Dept. of Thermo and Fluid Dynamics, Chalmers University of Technology, Gothenburg, 1998.
8. G. Sottas and I. L. Ryhming, editors. *3D-Computations of Incompressible Internal Flows - Proceedings of the GAMM Workshop at EPFL, September 1989, Lausanne - Notes on Numerical Fluid Mechanics.* Vieweg, Braunschweig, 1993.
9. P. R. Spalart. Private communication. Boeing Commercial Airplanes, 2000.
10. P.R. Spalart, W-H. Jou, M. Strelets, and S.R. Allmaras. Comments on the feasibility of LES for wings, and on a hybrid RANS/LES approach. *1st AFOSR Int. Conf. on DNS/LES, Aug. 4-8, Ruston, LA. In Advances in DNS/LES, C. Liu & Z. Liu Eds., Greyden Press, Columbus, OH*, 1997.

Parallel Computational Fluid Dynamics – Trends and Applications
C.B. Jenssen et al. (Editors)

Parallel and Adaptive 3D Flow Solution Using Unstructured Grids

E. Yilmaz [a], H.U. Akay [a], M.S. Kavsaoglu [b], and I.S. Akmandor [b]

[a] Department of Mechanical Engineering
Indiana University-Purdue University Indianapolis
723 W. Michigan St., Indianapolis, IN 46202, USA

[b] Aeronautical Engineering Department
Middle East Technical University, Ankara 06531, Turkey

A parallel adaptive Euler flow solution algorithm is developed for 3D applications on distributed memory computers. The contribution of this research is the development and implementation of a parallel grid adaptation scheme together with an explicit cell-vertex based finite volume 3D flow solver on unstructured tetrahedral grids. Parallel adaptation of grids is based on grid-regeneration approach by using an existing serial grid generation program. Then, a general partitioner repartitions the grid. An adaptive sensor value, which is a measure to refine or coarsen grids at each node, is calculated considering the pressure gradients in all partitioned blocks of grids. The parallel performance of the present study was tested. Parallel computations were performed on Unix workstations and a Linux cluster using MPI communication library. It is proved that parallel adaptation is very effective for obtaining accurate and efficient flow solutions.

1. INTRODUCTION

Parallel implementation of adaptive grid algorithms has received more attention during the recent years in CFD, [1-3], after the introduction of various parallel computing environments. In those studies, grid adaptation was done mostly by grid refinement. Therefore, an imbalance of of mesh block size among processors is unavoidable. This makes some kind of grid migration process necessary to balance block sizes. Obviously, this group of applications is more complicated in implementation and application.

The present research effort has focused on bringing a more practical approach to parallel grid adaptive solutions with an affordable cost. A good way for this goal is to separate all individual tools required for parallel adaptive flow solution and then to couple them in the adaptation cycle. In such an approach, four individual tools are needed; flow solver, grid adapter, grid generator, and grid partitioner. First, the parallel flow solver calculates solution variables at each grid point using initial partitioned grid. Then, the parallel grid adapter uses solution already obtained by flow solver and generates grid spacing values needed for grid generator to define new grid density. Each partition performs its own sensor calculations. Several communication steps are followed at interfaces. To preserve the global compatibility of the adaptation, communications between processors associated with global gradient limitations are achieved as required. After that, grid generator generates a new grid based on new grid spacing. The final step in adaptation cycle is to partition new grid into several blocks for parallel flow solution. The present approach is obviously modular and gives more freedom to use different individual tools, either commercial or in-house programs, based on different algorithms.

In this study, a flow solver and grid adapter is developed and implemented. Existing tools were used for grid generation and partitioning. The flow solver is the implementation of 3D Euler flow equations by using cell-vertex based finite volume discretization on unstructured grids with explicit time stepping. Parallel adaptive program calculates adaptive sensor values, which is a measure to refine or coarsen grids, based on pressure gradient for grid points and then converts them to source value, which is a measure of grid spacing definition, for grid generator. Details of present study are given in [4].

The 3D grids are generated using the grid generator program, VGRID [5]. The General Divider (GD) program prepares grid partitions [6]. The program GD uses a partitioner program called METIS, [7], to generate interface information for flow solvers.

2. EULER FLOW EQUATIONS

For 3D compressible flow, Euler flow equations can be expressed in integral form as:

$$\frac{\partial}{\partial t}\iiint_{\Omega} \mathbf{Q}\, d\Omega + \iint_{\delta\Omega} \vec{\mathbf{F}} \cdot \hat{\mathbf{n}}\, dA = 0 \tag{1}$$

where Ω is the control volume, A is the surface area that surrounds Ω, $\hat{\mathbf{n}}$ is the unit normal vector to dA, $\mathbf{Q}$ is called solution vector and represents the conserved quantities and $\vec{\mathbf{F}}$ is the flux tensor having Cartesian components with

$$\mathbf{Q} = \begin{pmatrix} \rho \\ \rho u \\ \rho v \\ \rho w \\ \rho E \end{pmatrix} \quad \mathbf{F}_x = \begin{pmatrix} \rho u \\ \rho u^2 + p \\ \rho uv \\ \rho uw \\ \rho uH \end{pmatrix} \quad \mathbf{F}_y = \begin{pmatrix} \rho v \\ \rho vu \\ \rho v^2 + p \\ \rho vw \\ \rho vH \end{pmatrix} \quad \mathbf{F}_z = \begin{pmatrix} \rho w \\ \rho wu \\ \rho wv \\ \rho w^2 + p \\ \rho wH \end{pmatrix} \tag{2a}$$

and

$$E = \frac{p}{(\gamma - 1)\rho} + \frac{u^2 + v^2 + w^2}{2} \quad , \quad H = E + \frac{p}{\rho} \tag{2b}$$

where ρ is density, u, v and w are Cartesian velocity components, p is pressure, H is total enthalpy, and E is total energy per unit mass.

3. NUMERICAL SOLUTION OF FLOW EQUATIONS

The flow variables are stored at the vertices of the tetrahedral. The control volume, Ω, for a particular point K is taken as the union of all tetrahedrals with a vertex at K as shown in Figure 1. When the integral form of the Euler equation is applied to a finite control volume at K, one obtains the following coupled ordinary differential equations

$$\frac{\partial}{\partial t}(\mathbf{Q}_K \Omega_K) + \sum_{i}^{Nface} \left(\mathbf{F}_x \, dA_x + \mathbf{F}_y \, dA_y + \mathbf{F}_z \, dA_z\right)_i = 0 \quad (3)$$

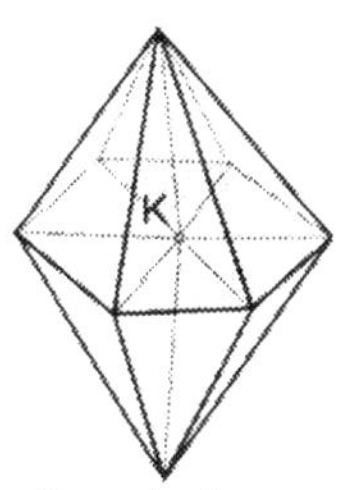

Figure1. A typical vertex-based finite control volume in 3D.

where Ω is the control volume associated with the node K, $\mathbf{Q}$ is the solution vector at node K. $\mathbf{F}_x$, $\mathbf{F}_y$, and $\mathbf{F}_z$ are the Cartesian components of the flux vector and dA_x, dA_y and dA_z are surface area of face i of the control volume, Nface is the total number of external faces which surround the control volume. The discretized equations contain only the first differences between the flow variables, thus are non-dissipative. This means that any errors, such as discretization and round-off errors are not damped and oscillations may be present in the steady state solution. Also oscillations may be produced in regions of large pressure gradients such as near shocks, and will persist due to the non-dissipative nature of these schemes. In order to eliminate these oscillations, artificial dissipation terms, $\mathbf{D}_K$, are added to the Equation 3 at each point [8]. The system of this ordinary differential equations represents an initial value problem which must be solved in order to obtain the steady-state solution until the time derivatives all vanish. The integration in time to a steady state solution is performed using a multi-stage Runge-Kutta scheme. To accelerate convergence, residual averaging, local time stepping, and enthalpy damping techniques are used [9].

4. GRID ADAPTIVITY

For Euler equations, some of the main flow features of the solution can be shock waves, stagnation points, and vortices. Any indicator for adaptive sensor should accurately identify these flow characteristics. The method that has been chosen to identify these important flowfield features is a measure of the gradient of some dependent variable q, which may be density, pressure, total velocity, Cartesian components of velocity, total energy, and Mach number. For Euler equations, the following is used as adaptive sensor.

$$S = \left|\mathbf{U} \cdot \nabla q\right| \quad (4)$$

where S is the adaptive sensor values that will be used as a factor to refine or coarsen the grids, $\mathbf{U}$ is Cartesian velocity vector with components, u, v, w, and q is a weighted flow value. In this study, pressure is used as the flow variable for detection, thus $q = p$. A polynomial function is used to control expansion of grid density from high gradient regions to low gradient regions. The flowfields under analysis may contain unwanted transient features, for example, rough numerical data may have been supplied by the solution, or there may be very large localized variations of the adaptive sensor. To remove abrupt changes a low pass filter is applied. Details of adaptivity are given in [4].

5. PARALLELIZATION

Throughout this study, a partitioning program called the General Divider (GD), [6], was used to generate the unstructured grid partitions. Matching and one-point overlapping interfaces were used due to the simplicity of updating the solution variables. In such a case, the values of flow variables at interface boundary points are taken from the neighbors that are treated as one point interior by that neighbor. Therefore, it gives exactly the same value as in a regular interior point calculation.

Since an explicit time integration scheme was used in this study, parallelization of the solver and sensor are simple in implementation. Once the interface grid points of a partition are known, the only task is the construction of a communication routine. For the mesh generation, VGRID-POSTGRID [5], tools were used to obtain surface and volume grids. After generation of a single block mesh, then the GD was used to obtain partitions. Then, parallel flow solution was achieved. A flow-chart that summaries the developed parallel adaptive flow solution procedure is shown in Figure 2.

Exchange of information between the blocks was achieved in the multi-stage time stepping loop of the flow solver. In addition to the exchange of information at interface for the adaptive solver, maximum and minimum values of sensor at different stages of calculation should be exchanged so that each block works with global values.

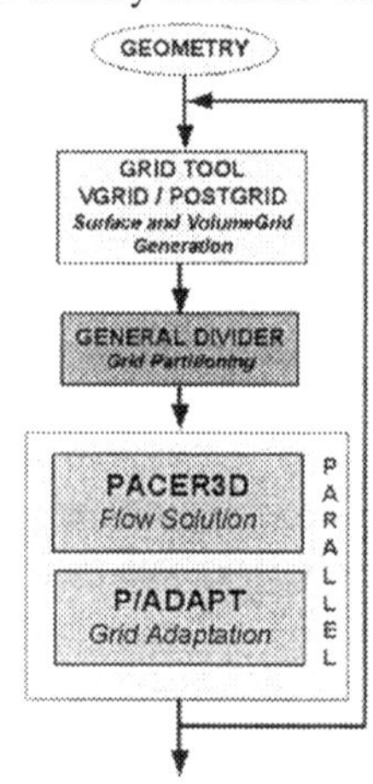

Figure 2. Flow chart of the developed parallel adaptive solution method.

6. CASE STUDIES

6.1 Parallel Adaptive Solution

The parallel adaptivity was applied on an Onera M6 wing geometry [10]. Four partitions were used and each partition was assigned to a processor. Three adaptation sequences seem to be enough to get fairly accurate results. One can also go further to get more precise contours. The remeshing sequence was done manually, since the grid generator, VGRID, runs on the SGI workstation platform that needs remote access. IBM RS6000 workstations in CFD Laboratory at IUPUI were used for the flow solver. Parallel adaptation step requires an elapsed time equal to almost one or two iteration of the flow solver. Three adaptation steps were achieved.

In Table 1, computation time and mesh size informations regarding the adaptation steps are given. Computation time corresponds to time spent in the flow solver, the adaptive sensor and VGRID. Only the program VGRID was run on the SGI O2. It is clear that mesh size was almost doubled after each iteration. An orthographic view of the adapted mesh is shown in Figure 3 and pressure contour is given in Figure 4. Mesh is also coarsened at some regions while refined along the shock structure.

Pressure coefficients at different spanwise sections of wing are given in Figure 5. Experimental data, [10], parallelly adapted mesh, and fine mesh computations were compared on the same graph. This figure shows the advantages of adaptation very clearly. When compared to fine mesh solution with the same number of iterations on a single computer, its size is less than two and half times, as given in Table 1, whereas the cost of adaptation was recorded as four times less. Moreover, pressure coefficients match the experimental data better than the fine mesh results.

Table 1
Total computation time of parallel adaptation steps for Onera M6 wing case (four partitions).

Steps	*# of cells*	*Flow Solver (sec)*	*Sensor (sec)*	*VGRID sec)*	*# of timesteps*
0th	129,798	1,158	NA	NA	3000
1st	246,825	2,469	3	172	3000
2nd	417,998	3,911	5	319	3000
3rd	702,462	19,636	10	556	7000
Total	NA	27,174	23	1047	NA

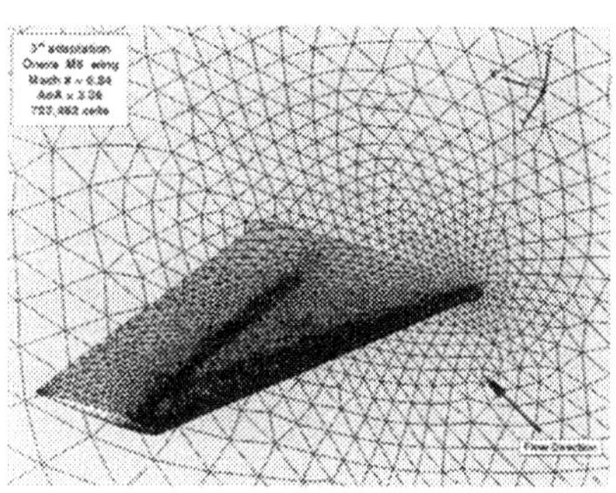

Figure 3. Mesh at 3rd parallel adaptation step for Onera M6 wing case.

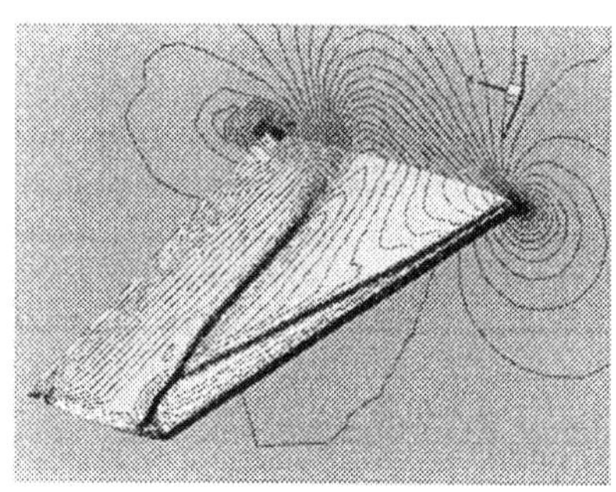

Figure 4. Pressure contours at 3rd parallel adaptation step for Onera M6 wing case.

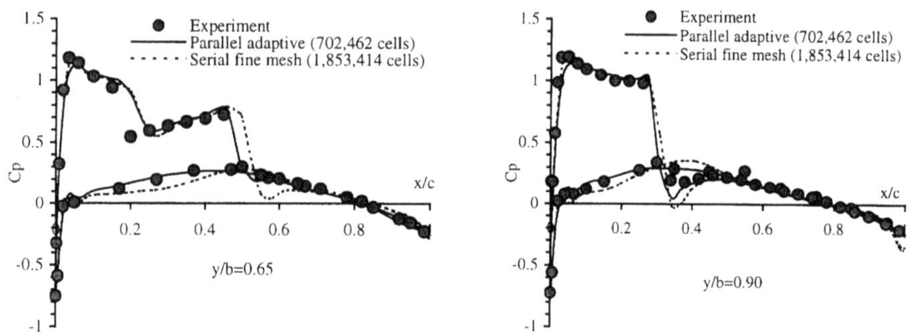

Figure 5. Pressure coefficient comparison of parallel adaptive solution at 3rd step and fine grid with experiment for Onera M6 wing case at two spanwise sections.

6.2 Parallel Performance Studies

Finally, speedup of present parallel flow solver was tested on the LINUX PC clusters of NASA Glenn Research Center. The geometry is a generic aircraft configuration that brings more complexity to geometry as well as the flow field. Computer system has 32 nodes that each node has 512MB of memory, 512KB of L2 Cache, and two Pentium II 400Mhz processors. Two, four, eight, 16, and 32 partitions were used for the speedup study. All speedup values were based on 1000 of the explicit time integration. Shown in Figure 6 is the mesh with 16 partitions. The same mesh is used for the other partitions. Note that there is no mesh adaptation in the parallel performance study. Because of the partitioning algorithm interface sizes are not exactly equal. This will cause some overhead if no load balancing is done [11]. Since load balancing is out of the scope of this study it was not taken into account.

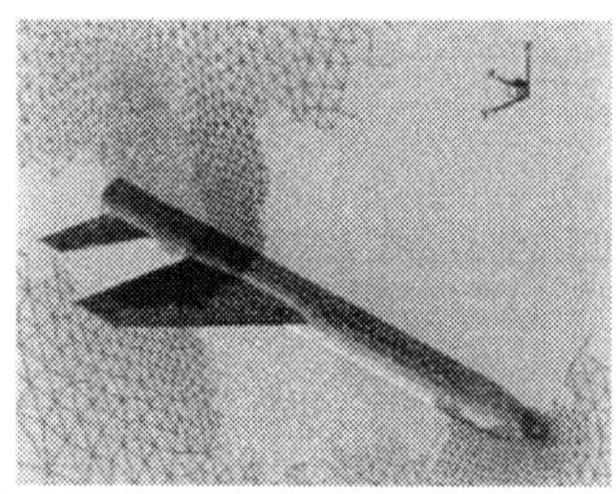

FIGURE 6. Mesh with 16 partitions for a generic aircraft case.

Parallel run time performance of present study was compared with a similar study by Uzun [12], where the same test case was solved. In that study, the NASA Langley based program USM3D [13], is parallelized and run for the same geometry. Also, in that study, solutions are obtained with 200 implicit time integration steps. This gives almost three orders of residual convergence. In that study, parallel communication is done with PVM. The present study uses MPI for communication and runs for 3000 of explicit time integration steps to get close to three orders of decrease in residuals. Figure 7 gives comparison of the elapsed time of both studies. As the number of partitions increases, the difference in elapsed time becomes smaller. Even for 32 partitions the total CPU time of the present explicit solution is smaller than the implicit parallel USM3D solution.

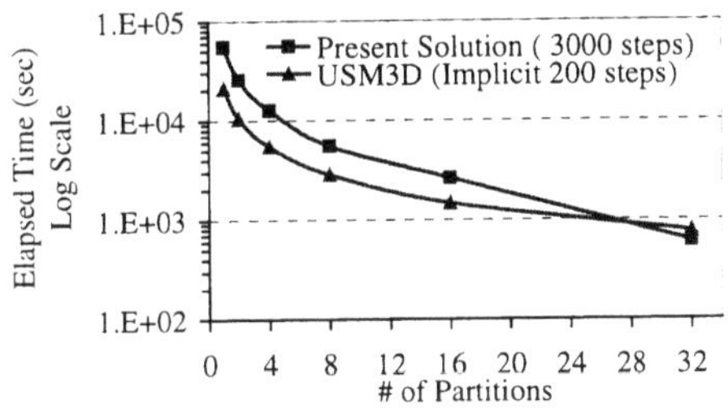

Figure 7. Comparison of elapsed time to steady state for generic aircraft case.

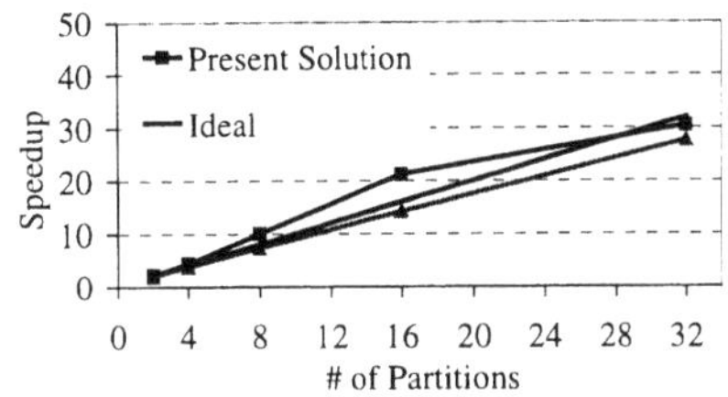

Figure 8. Speedup comparison for generic aircraft case.

The speedup values of both codes were compared in Figure 8. It is interesting to note that the present solution has higher values than USM3D solutions and ideal. The reasons for this are the differences in solution algorithm, memory requirement, and the paging size between memory and cache in each code.

7. CONCLUSIONS

It has been shown that efficient flow solutions can be achieved by grid adaptation with the parallel implementations of a flow solver and a solution adaptive sensor for grids. As opposed to going through modifications in the flow solver for the parallel adaptive refinement, the present study proved the applicability of parallel adaptivity with mesh *regeneration* and *repartition* with an affordable computational cost.

The adaptive flow solutions have better agreement with experiments even better than that of uniformly distributed fine meshes around high gradient regions. One of the important aspects of the grid adaptivity is to choose adequate flow features for the gradient calculation. The gradient of pressure seems to be good enough for very localized mesh refinements in the Euler flow analysis. In addition, a combination of flow variables can be constructed for any particular flow problems. Some parameters to control mesh size, number of mesh and adaptation locality were defined to give more degrees of freedom to the user. It was seen that 3-4 adaptation sequences are enough for an efficient adaptive solution.

The parallel performance of present flow solution algorithm has shown that the speedup is higher than that of a cell-centered implementation and ideal case. It was concluded that this is mainly related to the problem size and cache size of a particular computing environment and memory requirement of programs.

ACKNOWLEDGEMENTS

We would like to thank to The Scientific and Technical Research Council of Turkey (TUBITAK) for supporting the work of the first author through the NATO A2 scholarship. We extend our appreciation to several graduate students and staff at the IUPUI CFD Laboratory. We thank Dr. S.Z. Pirzadeh of the NASA Langley Research Center for his kind advice on the use of the grid generation program VGRID. Our appreciation also goes to the NASA Glenn Research Center for the use their parallel Linux cluster.

REFERENCES

1. Oliker, L. and Biswas, R., *PLUM: Parallel Load Balancing for Adaptive Unstructured Meshes,* NASA, NAS Reports, NAS-97-020, (1997).
2. Ozturan, C., de Cougny, H.L., Shephard, M.S., and Flaherty, J.E., *Parallel Adaptive Mesh Refinement and Redistribution on Distributed Memory Computers*, NASA, NAS Reports, NAS-96-011, (1996).
3. Coupez, T., Digonnet, H., Clinckemaillie, J., Thierry, G., Maerten, B., Roose, D., Basermann, A., Fingberg, J., Lonsdale, G., and Ducloux, R., "Dynamic Re-Allocation of Meshes for Parallel Finite Element Applications," *4th ECCOMAS Computational Fluid Dynamics Conference*, Athens, Greece, (1998).
4. Yilmaz, E., *A Three Dimensional Parallel and Adaptive Euler Flow Solver For Unstructured Grids*, PhD Thesis, Middle East Technical University, Ankara, Turkey, (2000).
5. Pirzadeh, S.Z., "Three-Dimensional Unstructured Viscous Grids by the Advancing-Layers Methods," *AIAA Journal*, Vol. 34, No. 1, (1996), pp. 43-49.
6. Bronnenberg, C.E., *An Unstructured Grid Partitioning Program for Parallel Computational Fluid Dynamics*, MS Thesis, Indiana University-Purdue University Indianapolis, Indianapolis IN, USA, (1999).
7. Karypis, G. and Kumar, V.A, *Fast and High Quality Multilevel Scheme for Partitioning Irregular Graphs*, University of Minnesota, Department of Computer Science, TR 95-035, (1995).
8. Jameson, A., Schmidt, W., and Turkel, E., "Numerical Solutions of the Euler Equations by Finite Volume Methods Using Runge-Kutta Time Stepping Schemes, " *AIAA Paper 81-1259,* (1986).
9. Mavriplis, D., *Solution of Two Dimensional Euler Equations on Unstructured Triangular Meshes*, Ph.D. Thesis, Princeton University, (1987).
10. Schmitt, V. and Charpin, F., *Pressure Distributions on the ONERA-M6-Wing at Transonic Mach Numbers. Experimental Data Base for Computer Program Assessment*, Report of the Fluid Dynamics Panel Working Group 04, AGARD AR 138, (1979).
11. Chien, Y.P., Ecer, A., Akay, H.U., Carpenter, F., and Blech, R.A., "Dynamic Load Balancing on a Network of Workstations for Solving Computational Fluid Dynamics Problems," *Computer Methods in Applied Mechanics and Engineering*, 119, pp. 17-33, 1994.
12. Uzun, A., *Parallel Computation of Unsteady Euler Equations on Dynamically Deforming Unstructured Grids*, MS Thesis, Indiana University-Purdue University Indianapolis, Indianapolis IN, USA, (1999).
13. Frink, T.F., Parikh, P., and Pirzadeh, S.Z., "A Fast Upwind Solver for the Euler Equations on Three Dimensional Unstructured Meshes," *AIAA Paper* 91-0102, (1991).

12. Multiphase and Reacting Flows

Parallel Computational Fluid Dynamics – Trends and Applications
C.B. Jenssen et al. (Editors)

Interaction Between Reaction Kinetics and Flow Structure in Bubble Column Reactors

H. A. Jakobsen[a], I. Bourg[a], K. W. Hjarbo[b] and H. F. Svendsen[a]

[a]Department of Chemical Engineering, The Norwegian University of Science and Technology, NTNU, Sem Sælands vei 4, NO-7491 Trondheim, Norway.

[b]SINTEF Applied Chemistry, NO-7491 Trondheim, Norway.

This paper evaluate the capabilities of steady multifluid flow models to predict the behaviour of a chemical reacting system. The system investigated is CO_2 absorption into methyldiethanolamine (MDEA) solution in a bubble column. The model predictions are compared with experimental data obtained in our laboratory.

1. INTRODUCTION

Bubble column reactors are widely used in the process and biotechnology industries owing to their simplicity of construction and effectiveness in heat and mass transfer. Although simple to build and operate, the behavior of the gas-liquid flow is complicated and seemingly unpredictable. In recent years attempts have been made to improve the understanding of the phenomena of circulation and bubble movement in bubble columns.

Status on the fluid dynamic flow modeling of bubble columns was discussed by [1]. The intent of this paper is to evaluate the capabilities of the CFD based reactor models when predicting reactive systems, as most papers found in the literature concern the flow variables only. The complex CFD reactor models are of limited value for chemical engineers if they are unable to predict reasonable conversions, concentrations and heat distributions in multiphase reactors.

The system investigated is CO_2 absorption in a methyldiethanolamine ($MDEA$) solution. Under the experimental conditions used, this reaction takes place in the fast reaction regime, making the overall absorption an interface reaction with a rate independent of the liquid phase mass transfer coefficient. This alleviates the need at this point, for introducing an empirical correlation for the liquid side mass transfer coefficient (e.g. [2]). In addition to the well known industrial importance of carbon dioxide removal in gas treating processes, this system offers the advantage of allowing an individual validation of how the specific interfacial area and the liquid-phase mass transfer coefficient (without reaction) influence the interfacial mass transfer rate. It is thus possible to evaluate the modeling of the interfacial area separately. On the other hand, by altering the amine concentration in the solution, thereby moving into another reaction regime, the influence of the liquid-phase mass transfer coefficient increases. In this work only the interfacial area modeling is investigated.

2. MODELING

The basic multifluid flow model used in this work is the same as given by [1,3–5]. The time averaged variables calculated by the model are volume fractions of the gas and liquid phases, the vertically and horizontally varying bubble size distributions and velocities for both phases, turbulent kinetic energy, turbulent energy dissipation and pressure of the liquid phase. Furthermore, the chemical species mass fractions in both phases are calculated. The interfacial contact area is determined by the bubble size and shape distribution in the system and this will have a significant impact on the mass transfer rate. A good description of the bubble size and shape distributions is also needed in order to improve the calculation of the interfacial drag. The bubble size distribution model of [4] has been used in this work.

The original model is formulated based on a volume after time averaging operator approach. In the time averaging part of this procedure two different approaches have been used.

In model version 1, in line with the previous model versions, a Reynolds like averaging procedure has been applied. This procedure gives rise to additional covariance terms in the transport equations as the turbulent mass diffusion terms in the continuity equations and in the species transport equations. Before the time averaged correlations in the fluid continuity equations and the equivalent equations for the dispersed phase can be modeled, the physical mechanisms included in these terms must be interpreted and understood. The volume fraction-velocity covariances describe the turbulent mass diffusion of gas, i.e. a bubble distribution effect, and is modeled based on a gradient hypothesis. The mass fraction-velocity covariances in the species transport equations describe turbulent diffusion of the components. In the continuous phase this term accounts for transport of components due to the eddies motion in the fluid. In the dispersed phase the internal turbulent diffusion inside individual bubbles is not described, but rather the mixing of bubbles having different composition and residence times because of movement at different velocities and trajectories. The gas phase mixing effect caused by bubble coalescence is also lumped into this, and not separately modeled. This term is also modeled based on the gradient hypothesis. The mass fraction-volume fraction covariances describe the species transport due to fluctuations in the volume fraction. The physics included in these terms is not yet understood. The mass fraction-volume fraction correlations are therefore neglected in this work. In addition, correlation terms of order higher than two are neglected.

A Favre like averaging procedure has been applied in model version 2. In this formulation, some of the covariances found in the first model apparently disappear, e.g. the turbulent mass diffusion terms in the continuity equations and the corresponding terms in the species transport equations vanish. The physical mechanisms described by these terms now occur as additional forces in the momentum equations, as described by [6,5]. In the latter model version, the turbulent dispersion force occurring in the momentum equations is described applying the model suggested by [6]. In both model formulations the inter-phase momentum transfer due to inter-phase mass transfer is neglected.

As discussed by [7,8], considering the Reynolds averaged model formulation, the turbulent mass diffusion terms may have several undesirable properties when considering reactive flows and should thus be avoided. First, the physics reflected by these terms is questionable as they may predict mass transfer of species due to volume fraction gradients in the multiphase system. Second, the total continuity set requirements on the phase dispersion coefficients that may not be physically realistic. Third, they may also introduce numerical instabilities in the system. The magnitudes of the possible shortcomings are analyzed in this work.

For chemistry modeling the reaction mechanism given by [2] is used. It is assumed that the two main reactions are parallel pseudo-first-order and can be expressed by:

$$CO_2 + R_3N + H_2O = HCO_3^- + R_3NH^+ \text{ and } CO_2 + OH^- = HCO_3^-$$

The interfacial carbon dioxide mass flux is given by

$$M_{CO_2} = Mw_{CO_2} E k_{L,CO_2} a (C^*_{CO_2} - C_{CO_2}) \approx Mw_{CO_2} H_{CO_2} (k_2 C_{Am} D_{CO_2})^{\frac{1}{2}} a (p_{CO_2})$$

When the alkanolamine concentration can be assumed constant through the diffusion boundary layer, the enhancement factor, E, is equal to the Hatta number, Ha. At low CO_2 loadings the back pressure of CO_2 can be neglected. The second-order rate constant has a value of approximately $k_2 = 5.0\ m^3/(\text{kmol s})$ at 20 °C.

The liquid density is assumed constant, while the gas density is a function of the total pressure according to the ideal gas law. The reactor is operated at isothermal conditions.

The model equations are solved using an extended version of the SIMPLE method (see [8] and references therein). The simulation of reactive bubble driven flows involves the solution of two sets of conservation equations for mass and momentum, plus the chemical species transport equations for the two phases. The convergence rate of the iterative procedure is severely limited by the interfacial mass balance requirements in these kinds of simulations.

3. EXPERIMENTAL

The bubble column used was 4.25 m tall, $ID = 0.144\ m$. The media used were $MDEA$ ($2.89M$) and $CO_2/N_2(g)$, ($y_{CO_2} \approx 0.08$). Concentration measurements were taken at three internal axial and radial positions, in addition to the in- and outlets. Bubble velocity, size distribution and gas fraction were measured at level $2m$ above the inlet using the five point conductivity method [9]. Temperature was $\sim 20°C$, pressure ambient, and the inlet gas and liquid velocities were 8.3 cm/s and 0.28 cm/s respectively. The physical properties of the MDEA solution were according to data given by [10,11].

4. RESULTS AND DISCUSSION

The experimental results from the bubble column are compared with simulated results for the radial distributions of gas void fraction, axial gas velocity and bubble size at level $2m$ above the bottom of the column. The basis for the fluid dynamics is a flow model with model parameters (i.e. transversal force-, bubble induced turbulence production - and bubble size model coefficients) tuned to experimental results for the air/water system in the heterogeneous flow regime. In this work the multifluid model has been used to model the absorption of CO_2 into aqueous $MDEA$ solutions, including the same set of model parameters.

Previous model predictions on non-reactive systems (e.g. [3,4,1]) often show reasonable agreement for bubble size distribution, gas phase axial velocity and gas void fraction.

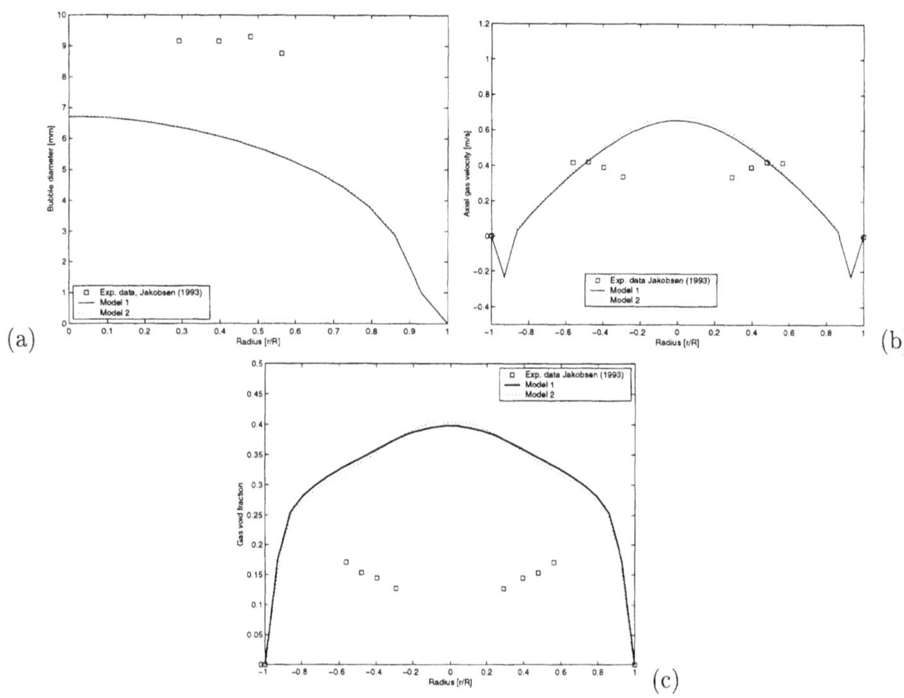

Figure 1. Results from the 2D steady Euler/Euler model versus measured data. The graphs in the figures indicate: a) Sauter Mean Diameter (SMD). b) Local axial gas velocity. c) Gas volume fraction.

Figure (1a) shows a comparison between predicted and measured profiles of the Sauter Mean Diameter, SMD in the reactive system. The radial distribution of the measured data points is too narrow to draw any firm conclusion, but for the experimental range the experimental and predicted profiles are almost flat. The simulated profiles based on parameters tuned to the air/water system thus seem to predict the qualitative form of the profile. The magnitude of the predicted SMD's is however about $\sim 20\%$ lower than the experimental data.

The measured gas phase velocity profiles in the reactive system, shown in figure (1b), are only in magnitude in agreement with the predicted gas phase velocity profiles. The measured profile is nearly flat with a weak increase towards the wall, whereas the simulated ones show a distinct maximum in the center region. The experimental velocity profile is thus consistent with the bubble size distribution shown in figure (1a). Based on single bubble data, one would expect that the gas velocity should be almost flat because of the flat bubble size profile. The predicted profiles show an increasing gas phase velocity towards the center of the column.

These discrepancies may partly be due to uncertainties in the experimental results. The electro conductivity method used accepts or rejects bubbles according to criteria related to their sphericity. Deformed bubbles will thus be rejected and not used in the velocity and size measurements. The rejected fraction of bubbles is for all positions as high as $\sim 25\%$. Considering that large bubbles are the most likely candidates for rejection, this may lead to too low measured average velocities. In addition, the fluid properties and the interfacial structure may also affect the steady drag coefficient. An experimental validation of this parameter in the MDEA system is needed.

The measured gas volume fraction profile, shown in figure (1c), is much lower than the predicted profile. In addition, the measured gas volume fraction profile shows a tendency of intermediate or wall peaking, in contrast to the air/water system which has a clear core peak (not shown). The measured data for SMD and axial gas velocity may therefore be consistent with an intermediate or core peaking system.

In this context it is important to note that previous measurements of the gas volume fractions show that the experimental techniques used have a tendency to underpredict the local values (e.g. [12,3]). However, the total averaged void fraction calculated from the local data are in reasonable agreement with total void fraction measurements performed in parallel (not shown). It is therefore reasonable to assume that the model predicts far too high local gas fraction profiles for this system. This discrepancy must also be seen in light of the fact that the experimental method also takes into account rejected bubbles when evaluating the gas fraction. However, the coalescence properties of the MDEA system is clearly different from the air/water system used for parameter tuning, and as shown previously ([4]) this may lead to dramatic changes in gas fraction. This is also consistently reflected by large discrepancies in the bubble size distribution.

The phase distribution phenomena are, however, not sufficiently understood yet, as discussed by [5]. The present modeling of the parameters in the underlying models for steady drag, lateral bubble movement, bubble induced turbulence production and bubble size and shape are not yet accurate enough to predict the effects of large changes in interfacial structure, turbulent interaction and system and coalescence properties. It is however believed that the discrepancy between the measured and the predicted profiles are to some extent caused by the uncertainties in the measurements and the errors in the treatment of the measured data.

When comparing the resulting profiles obtained by the fluid dynamic flow variables using the two model versions in figure (1) (i.e. model 1 is based on the Reynolds averaging procedure, whereas model 2 is based on the Favre averaging procedure), it can be seen that the predicted profiles for all the flow variables (bubble diameter, axial gas velocity and gas void fraction) are hardly distinguishable. These findings have been further discussed by [5].

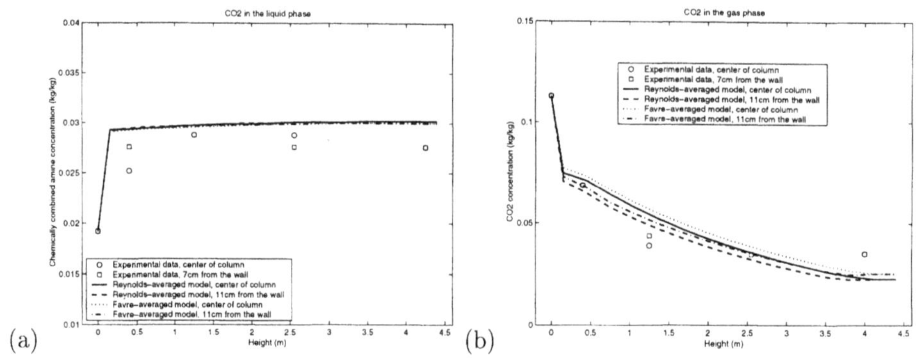

Figure 2. Results from the 2D steady Euler/Euler model versus measured data. The graphs in the figures indicate: a) Chemically combined amine concentration in liquid phase. b) CO_2 concentration in gas phase.

In the following the interaction between the chemical processes and flow structure will be discussed. As seen in figure (2a) the predicted conversion based on the liquid phase is higher than that experimentally observed. The accuracy level of the rate constant can explain part of the observed discrepancy. It is however believed that most of the deviation is caused by the bubble size description. The chemical conversion is very sensitive to the local bubble size distribution.

The experimental gas conversion data (figure(2b)) indicate that most of the reaction takes place in the bottom $1.5m$ of the column. In fact the conversion seems to go down towards the outlet. This last observation must be seen in connection with the sampling method used for the internal positions. The gas/liquid mixtures were withdrawn from the column through a tube during which reaction could continue before gas liquid separation could take place. Thus the measured internal conversion levels are thought to be slightly on the high side. The same conclusion can be drawn from the liquid phase concentration measurements that also show a reduction in conversion towards the reactor outlet. The predicted gas concentration profiles are however fairly smooth and falls off evenly towards the outlet. The sharp decrease in gas phase concentration shown also by the predicted curves is associated with the large changes in bubble size distribution taking place in the region close to the inlet, and backmixing of both gas and liquid. In the middle part of the column the experimental absorption rate is close to the predicted one, indicating that the axial mixing of the gas phase is well described.

The axial mixing of phases predicted by the model is however to some extend determined by the boundary conditions formulated. Applying the traditional reactor modeling approach, the Danckwerts boundary conditions [13] determine the dispersion level at the inlet (i.e. splitting the prescribed convective inlet flux into convective and diffusive flux components at the reactor inlet plane). In CFD codes it seems to be common practice to neglect the dispersion at the inlet plane to simplify the mass balance calculations. This simplification induces a zero concentration gradient and a no axial mixing boundary condition at $z = 0$. According to [13], this is not in agreement with experimental RTD observations in many chemical reactors. Experimental data is therefore needed to determine and validate proper boundary conditions for fluid dynamic bubble column models. In this work the usual CFD approach neglecting the dispersion term at the inlet plane has been used.

A spatial comparison with local CO_2 concentration measurements has also been performed indicating that the radial liquid and gas phase transport is somewhat overpredicted by the model.

Even though previous work ([3,4]) indicate that the Reynolds averaged model used is sufficient to obtain a satisfactory description of the overall flow pattern in the column for the non-reactive air/water system, it has been found that the capabilities of the same model describing an 'unknown' reactive system is limited. The chemical reaction system studied has a conversion rate solely dependent on the specific area and the partial pressure of CO_2. It is therefore believed that poor interfacial area estimates in the bottom and wall areas of the reactor is the main reason for the overprediction found. The chemical conversion was overpredicted due to the lack of a reliable model enabling an accurate local description of bubble size and shape. Thus, to be able to predict the reacting system it is necessary to develop an improved bubble size and shape model based on detailed knowledge of bubble coalescence and breakage coupled with the influence of turbulence effects and physical properties of the system. A model describing such phenomena has been presented by [14]. For model validation the bubble size and shape distributions, especially close to the gas distributor, need to be measured.

On the other hand, the interfacial mass transfer in the chemical reacting system seems not to have affected the flow structure in the reactor. This is in accordance with the small amount of CO_2 removed. The simulated liquid concentration profiles are consistent with the gas phase data.

Comparing the results obtained with the two model versions (i.e. the Reynolds and Favre formulations) for the prediction of chemical species concentration, some deviations can be observed in the gas phase. In the liquid phase the various concentration profiles are nearly equal. The deviations found in the gas phase concentration profiles are related to the formulation of the turbulent dispersion force. A Favre like averaging procedure may be recommended due to the above mentioned shortcomings of the Reynolds procedure.

5. CONCLUSIONS

A multifluid model, fluid dynamically tuned to the air/water system has been used to model the absorption of CO_2 into aqueous MDEA solutions.

The model predictions of the flow variables are found to deviate considerably from the measured data. This is due to the limited accuracy reflected by the parameters in the underlying models for steady drag, lateral bubble movement, turbulence and bubble size. To improve on the models applied, the complex physics lumped into these parameters has to be resolved.

Considering the interaction between the chemical processes and flow structure, a spatial comparison between predicted and measured local CO_2 concentration profiles has been performed. The chemical conversion was somewhat overpredicted due to the lack of an accurate description of bubble size and shape distributions in the reactor.

To enable good predictions of reacting systems it is necessary to develop an improved bubble size and shape model based on detailed knowledge on bubble coalescence and breakage coupled with the influence of turbulence effects and physical properties of the system.

The boundary conditions usually applied in fluid dynamic bubble column models should be validated.

Comparing the results obtained with the Reynolds and Favre averaged model versions, the predicted flow variables are hardly distinguishable. For the prediction of chemical species concentration some deviations can be observed in the gas phase, whereas in the liquid phase the various concentration profiles are nearly equal. A Favre like averaging procedure is recommended.

6. ACKNOWLEDGEMENTS

We gratefully acknowledge the financial support of the Research Council of Norway (Programme of Super-computing) through a grant of computing time.

REFERENCES

1. Jakobsen, H. A., Sannæs, Grevskott, S. and Svendsen, H. F. *Ind. Eng. Chem. Res.*, **36** (10), (1997) 4052-4074.
2. Littel, R.J., Van Swaaij, W.P.M. and Versteeg, G.F. *AIChE J.*, **36**, (1990) 1633-1640.
3. Svendsen, H. F., Jakobsen, H. A. and Torvik, R. *Chem. Eng. Sci.*, **47**, (13/14), (1992) 3297-3304.
4. Jakobsen, H. A., Svendsen, H. F. and Hjarbo, K. W. *Comp. Chem. Eng.*, **17S**, (1993) S531-S536.
5. Jakobsen, H. A., 2000. Phase Distribution Phenomena in Two-Phase Bubble Column Reactors. Submitted to *ISCRE-16* and *Chem. Eng. Sci.*.
6. Laux, H. *Modeling of Dilute and Dense Dispersed Fluid-Particle Flow.* Dr. ing. Thesis, NTNU, Trondheim, Norway, 1998.
7. Gray, W. G. *Chem. Eng. Sci.*, **30**, (1975) 229-233.
8. Jakobsen, H.A. *On the Modelling and Simulation of Bubble Column Reactors Using a Two-Fluid Model.* Dr.ing. Thesis, NTH, Trondheim, Norway, 1993.
9. Steinemann, J. and Buchholz, R. *Part. Charact.*, **1**, (1984) 102-107.
10. Haimour, N., Bidarioan, A. and Sandall, O.C. *Chem. Eng. Sci.*, **42**, (1987) 1393-1398.
11. BASF, Datenblatt-Data Sheet, April, 1988, *Methyldiethanolamine*, D 092 d, e.
12. Menzel, T., *Die Reynolds-Schub-Spannung als Wesentlicher Parameter zur Modellierung der strömungs- structur in Blasensäulen*, VDI Verlag, Düsseldorf, 1990.
13. Danckwerts, P. V. *Chem. Eng. Sci.*, **2** (1), (1953) 1-13.
14. Hagesæther, L., Jakobsen, H. A. and Svendsen, H. F. *Computer-Aided Chemical Engineering*, **8**, (2000) 367-372.

Parallel Computational Fluid Dynamics – Trends and Applications
C.B. Jenssen et al. (Editors)

Parallel DNS of Autoignition Processes with Adaptive Computation of Chemical Source Terms

Marc Lange [a]

[a] High-Performance Computing Center Stuttgart (HLRS), Stuttgart University
Allmandring 30, D-70550, Germany
E-mail: lange@hlrs.de

Direct numerical simulation (DNS) has become an important tool to study turbulent combustion processes. Especially in the case of using detailed models for chemical reaction kinetics, computation time still severely limits the range of applications accessible for DNS. The computation of the chemical source terms is one of the most time-consuming parts in such DNS. Using an adaptive evaluation of the chemical source terms can strongly reduce this time without a significant loss in accuracy which is shown for DNS of autoignition in a turbulent mixing layer. A dynamic load-balancing scheme is used to maintain a high efficiency in the parallel adaptive computations.

1. INTRODUCTION

Combustion processes are important for a wide range of applications like automotive engines, electrical power generation, and heating. In most applications the reactive system is turbulent and the reaction progress is influenced by turbulent fluctuations and mixing in the flow. The optimization of combustion processes, e.g. the minimization of pollutant formation, requires accurate numerical simulations. Better and more generally applicable models for turbulent combustion are needed to be able to perform such simulations. The coupling between the chemical kinetics and fluid dynamics constitutes one central problem in turbulent combustion modeling [1].

During the last few years, direct numerical simulations (DNS), i.e. the computation of time-dependent solutions of the Navier-Stokes equations (see Sect. 2), have become one of the most important tools to study turbulent combustion. Due to the broad range of occuring length and time scales, such DNS are far from being applicable to most technical configurations, but they can provide detailed information about turbulence-chemistry-interactions and thus aid in the development and validation of turbulent combustion models. However, many of the DNS carried out so far have used simple one-step reaction mechanisms. Some important effects cannot be captured by simulations with such over-simplified chemistry models [2,3]. By making efficient use of the computational power provided by parallel computers, it is possible to perform DNS of reactive flows using detailed chemical reaction mechanisms at least in two spatial dimensions [4,5]. Nevertheless, computation time is still the main limiting factor for the DNS of reacting flows, especially in the case of using detailed chemical schemes.

2. GOVERNING EQUATIONS FOR DETAILED CHEMISTRY DNS

Multicomponent reacting ideal-gas mixtures can be described by a set of coupled partial differential equations expressing the conservation of total mass

$$\frac{\partial \varrho}{\partial t} + \mathrm{div}(\varrho \vec{u}) = 0 \,, \tag{1}$$

momentum

$$\frac{\partial (\varrho \vec{u})}{\partial t} + \mathrm{div}(\varrho \vec{u} \otimes \vec{u}) = -\mathrm{grad}\, p + \mathrm{div}\, \mathsf{T} \,, \tag{2}$$

energy

$$\frac{\partial e_\mathrm{t}}{\partial t} + \mathrm{div}((e_\mathrm{t} + p)\vec{u}) = \mathrm{div}(\mathsf{T}\, \vec{u}) - \mathrm{div}\, \vec{q} \,, \tag{3}$$

and the masses

$$\frac{\partial (\varrho Y_\alpha)}{\partial t} + \mathrm{div}(\varrho Y_\alpha \vec{u}) = M_\alpha \dot{c}_\alpha - \mathrm{div} \vec{j}_\alpha \tag{4}$$

of the N_S chemical species α [6,7]. Herein ϱ denotes the density and $\vec{u}$ the velocity, Y_α, $\vec{j}_\alpha$ and M_α are the mass fraction, diffusion flux and molar mass of the chemical species α, T denotes the viscous stress tensor and p the pressure, $\vec{q}$ is the heat flux and e_t is the total energy given by

$$e_\mathrm{t} = \varrho \left(\frac{\vec{u}^2}{2} + \sum_{\alpha=1}^{N_\mathrm{s}} h_\alpha Y_\alpha \right) - p \,, \tag{5}$$

where h_α is the specific enthalpy of the species α. The computation of the chemical source terms on the right-hand-sides of the species mass equations (4) is one of the most time-consuming parts in such DNS. The production rate $\dot{c}_\alpha$ of the chemical species α is given as the sum over the formation rate equations for all N_R elementary reactions,

$$\dot{c}_\alpha = \sum_{\lambda=1}^{N_\mathrm{R}} k_\lambda (\nu_{\alpha\lambda}^{(\mathrm{p})} - \nu_{\alpha\lambda}^{(\mathrm{r})}) \prod_{\alpha=1}^{N_\mathrm{S}} c_\alpha^{\nu_{\alpha\lambda}^{(\mathrm{r})}} \,, \tag{6}$$

where $\nu_{\alpha\lambda}^{(\mathrm{r})}$ and $\nu_{\alpha\lambda}^{(\mathrm{p})}$ denote the stoichiometric coefficients of reactants and products respectively, and c_α is the concentration of the species α. The rate coefficient k_λ of an elementary reaction is given by a modified Arrhenius law

$$k_\lambda = A_\lambda T^{b_\lambda} \exp\left(-\frac{E_{\mathrm{a}\lambda}}{RT} \right) \,. \tag{7}$$

The chemical reaction mechanism for the $H_2/O_2/N_2$ system which has been used in the simulations presented in Sect. 5 contains $N_\mathrm{S} = 9$ species and $N_\mathrm{R} = 37$ elementary reactions [8].

This system of equations is closed by the state equation of an ideal gas

$$p = \frac{\varrho}{\overline{M}} RT \tag{8}$$

with R being the gas constant and $\overline{M}$ the mean molar mass of the mixture.

3. PERFORMANCE OF THE PARALLEL DNS-CODE

We have developed a code for the DNS of reactive flows using chemical mechanisms of the type described above on parallel computers with distributed memory [9,10]. Besides the computation of the reaction kinetics, detailed models are also utilized for the computation of the thermodynamical properties, the viscosity and the molecular and thermal diffusion velocities. The spatial discretization is performed using a finite-difference scheme with sixth-order central-derivatives, avoiding numerical dissipation and leading to high accuracy. The integration in time is carried out using a fourth-order fully explicit Runge-Kutta method with adaptive timestep control. The parallelization strategy is based on a regular two-dimensional domain decomposition with "halo" elements at the domain boundaries. Our main production platform is the Cray T3E on which we implemented a version of the code using PVM as well as one using MPI for the communication. During the normal integration in time, the performance difference between both versions is less than 1% CPU-time, whereas for the parts of the simulation in which values of the output variables from all subdomains are gathered for I/O, the MPI-version clearly outperforms the PVM-version [3]. In these parts, messages are sent with sizes scaling with the number of grid points per sub-domain, whereas during the rest of the temporal integration the message-sizes scale with the number of grid-points along the sub-domain boundaries. Due to the fact, that on the Cray T3E MPI delivers a higher communication bandwidth than PVM, but has also a higher latency [11], the MPI-version performs better with increasing message-sizes.

Below, we present performance results for the Cray T3E optimized implementation of our code using MPI as message-passing library. All computations have been performed on Cray T3E-900 systems, i.e. 450 MHz clock speed and stream buffers enabled. Having access to a node with 512 MB RAM allows us to perform a one-processor reference computation for the $H_2/O_2/N_2$ system with 544 × 544 grid points, a problem size which corresponds to some real production runs. The achieved speedups and efficiencies for this benchmark are given in Table 1. An average rate of 86.3 MFlop/s per PE is achieved in the computation using 64 processors.

Table 1
Scaling on a Cray T3E for a DNS with 9 species and 37 reactions on a 544^2 points grid

processors	1	4	8	16	32	64	128	256	512
speedup	1	4.3	8.1	15.9	30.5	57.9	108.7	189.0	293.6
efficiency	100.0	106.6	100.7	99.2	95.3	90.4	84.9	73.8	57.4

4. AUTOIGNITION IN A TURBULENT MIXING LAYER

Autoignition takes place in combustion systems like Diesel engines, in which fuel ignites after being released into a turbulent oxidant of elevated temperature. The influence of the turbulent flow field on the ignition-delay time and the spatial distribution of ignition spots is studied in a model configuration shown in Fig. 1. A cold fuel stream and an

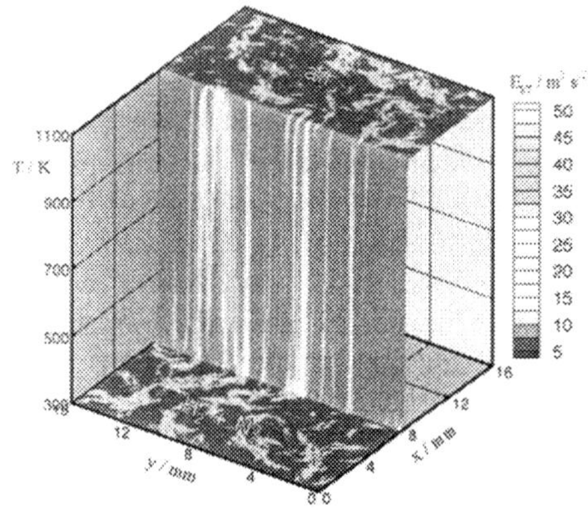

Figure 1. Configuration for the DNS of turbulent mixing of cold fuel with hot oxydizer

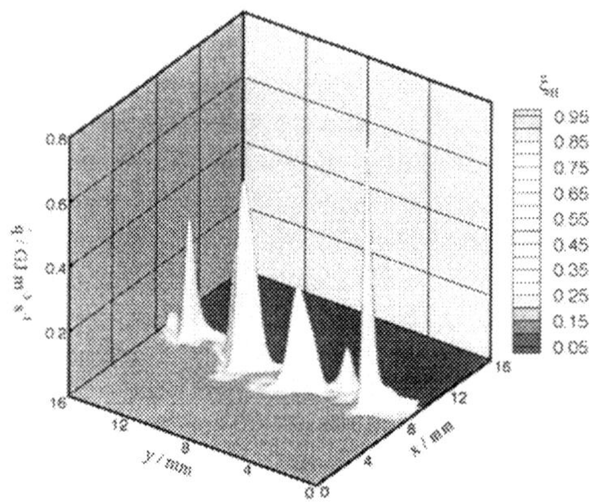

Figure 2. Ignition spots in an autoigniting turbulent mixing layer

air stream with an elevated initial temperature are superimposed with a turbulent flow field computed by inverse FFT from a von-Kármán-Pao-spectrum with randomly chosen phases. Non-reflecting outflow conditions [12] are used at the boundaries in x-direction and periodic boundary conditions are used in y-direction. After a specific temporal delay, which depends on the compositions and temperatures of the two streams as well as on the characteristics of the turbulent flow field, ignition spots occur as shown in Fig. 2. (More details on the spatial distribution of the ignition spots in such systems can be found in [3].)

5. ADAPTIVE CHEMISTRY AND DYNAMIC LOAD-BALANCING

The configuration described above has some features which are typical for many DNS studies of turbulent reacting flows: A very fine grid is used to resolve the smallest turbulent length-scales everywhere in the computational domain. In a fully coupled simulation the complex chemistry model is normally computed on every point of the same grid although in big parts of the domain no or almost no reactions occur. Thus, computation time can be saved by computing the chemical source terms using the detailed chemical mechanism as described in Sect. 2 only in those regions where reaction-rates are non-negligible. Criteria which can be fast evaluated are then needed to decide if a grid point belongs to such a region. For non-premixed and partially premixed reactive systems, the mixture fraction ξ, i.e. the (normalized) local ratio of fuel and oxidizer element mass fractions, is such a criterion. In a system of two streams (denoted by the indices 1 and 2), it can easily be written as

$$\xi_e = \frac{Z_e - Z_{e2}}{Z_{e1} - Z_{e2}} \tag{9}$$

where Z_e is the element mass fraction of the chemical element e. If equal diffusivities are assumed, ξ_e is a conserved scalar which is independent from the chosen e. In this case, the criterion for the necessity of performing the full chemistry computation can be expressed as $\epsilon_a < \xi < 1 - \epsilon_b$ with sufficently small values ϵ_a and ϵ_b. As preferential diffusion is known to be important for systems like those investigated here [3], a detailed transport model is used in our simulations. Therefore, the local element mass fractions of fuel and oxidizer are checked independently. If one of these two is nearly zero, no reactions will occur.

As a test for this method of adaptively evaluating the chemical source terms, simulations of autoignition in a turbulent mixing layer, i.e. with initial conditions as shown in Fig. 1, have been performed. The initial temperature of the fuel stream consisting of 10% H_2 and 90% N_2 (mole fractions) was $T_1 = 298\,\mathrm{K}$ and the initial temperature of the air stream was $T_2 = 1398\,\mathrm{K}$. The turbulent Reynolds number based on the integral length-scale was $Re_\Lambda = \mathrm{r.m.s.}(\vec{u}') \cdot \Lambda/\nu = 238$, and the computational grid has 800×800 points. The temporal evolution of maximum heat release rate and heat release integrated over the computational domain for this DNS and the corresponding laminar case are shown in Fig. 3.

In the adaptive computation for every timestep the chemical source terms have been set to $\dot{c}_\alpha = 0$ for all chemical species α at those points at which $Z_\mathrm{H} < \epsilon_a$ or $Z_\mathrm{O} < \epsilon_b$ with $\epsilon_a = \epsilon_b = 1 \cdot 10^{-5}$. The limiting values $\epsilon_{a,b}$ have been estimated from the results of a similar one-dimensional simulation. Another possibility would be the computation of a library of production-rates in homogeneous mixtures depending on temperature and

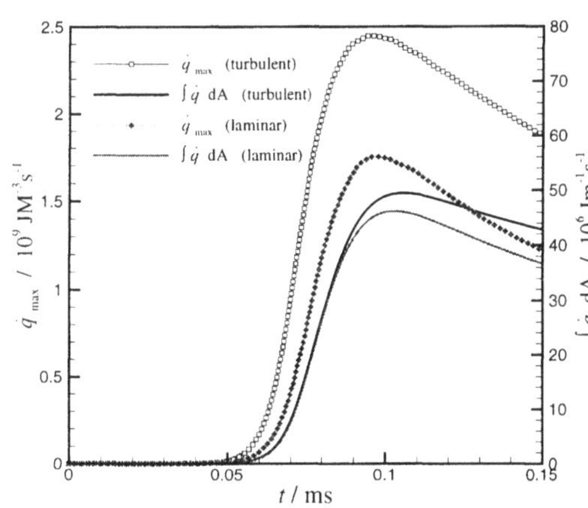

Figure 3. Temporal evolution of maximum heat release rate and heat release rate integrated over the computational domain in a laminar and a turbulent mixing layer

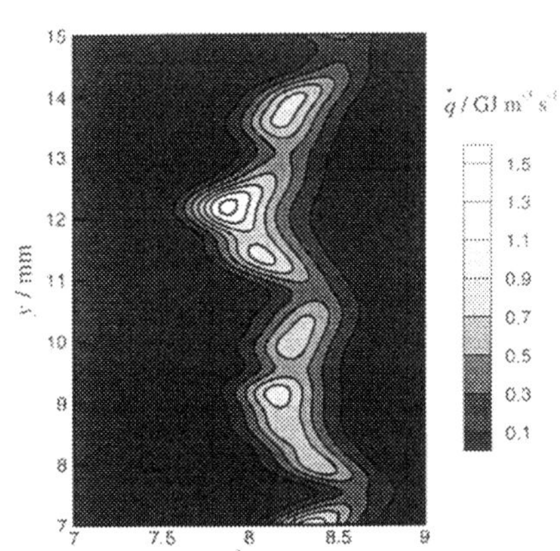

Figure 4. Comparison of heat release in an autoigniting turbulent mixing layer for adaptive (lines) and full (gray levels) chemistry at $t = 75\,\mu\mathrm{s}$

pressure, from which limiting values for a broad class of applications could be determined.

Figure 4 shows a snapshot of heat release rate in the turbulent mixing layer at a time near the strongest increase of maximum heat release rate obtained with adaptive (denoted by the contour-lines) and full (filled areas) computation of chemical source terms. (To be able to clearly identify some details, only a part of the computational domain is shown.) No differences between both computations are visible. The results of a more rigor error analysis are given in Table 2 which lists the maximum

$$\delta_X^{\max}(t) = \max_{x,y} \left(\delta_X(x, y, t)\right) \tag{10}$$

of the relative error

$$\delta_X(x, y, t) = \frac{|X_{\text{full}}(x, y, t) - X_{\text{adapt}}(x, y, t)|}{\max_{x,y}(X_{\text{full}}(x, y, t))} \tag{11}$$

of the quanitity X. Values of $\delta_X^{\max}(t)$ are given at times $t = n \cdot 25\,\mu\text{s}$ ($n = 1, \ldots, 6$ for the following physical variables X: x- and y-components of velocity u and v, pressure p, density ϱ, and temperature T, the heat release rate $\dot{q}$, and the mass fractions Y_α of all chemical species α. It can clearly be seen that no significant loss in accuracy is introduced by the adaptive computation of the chemical source terms.

In the one-dimensional simulation of the corresponding laminar situation the biggest maximum relative errors found in the same set of variables (except v) and times as listed in Table 2 are $\delta_{Y_{\text{HO}_2}}^{\max}(25\,\mu\text{s}) = 0.15\%$ for the HO_2 mass fraction and $\delta_{\dot{q}}^{\max}(25\,\mu\text{s}) = 0.18\%$

Table 2
Maximum relative errors $\delta_X^{\max}(t)$ for the adaptive chemistry computation

t/μs	25	50	75	100	125	150
u	$2.4 \cdot 10^{-9}$	$2.0 \cdot 10^{-7}$	$1.5 \cdot 10^{-5}$	$1.1 \cdot 10^{-5}$	$8.0 \cdot 10^{-6}$	$5.1 \cdot 10^{-6}$
v	$2.3 \cdot 10^{-9}$	$1.3 \cdot 10^{-7}$	$4.5 \cdot 10^{-6}$	$1.9 \cdot 10^{-5}$	$3.3 \cdot 10^{-5}$	$1.8 \cdot 10^{-5}$
p	$8.4 \cdot 10^{-11}$	$9.3 \cdot 10^{-9}$	$2.9 \cdot 10^{-7}$	$3.1 \cdot 10^{-7}$	$2.3 \cdot 10^{-7}$	$8.7 \cdot 10^{-8}$
ϱ	$5.3 \cdot 10^{-10}$	$2.3 \cdot 10^{-8}$	$4.0 \cdot 10^{-6}$	$1.1 \cdot 10^{-5}$	$9.9 \cdot 10^{-6}$	$8.2 \cdot 10^{-6}$
T	$7.9 \cdot 10^{-10}$	$7.4 \cdot 10^{-8}$	$7.8 \cdot 10^{-6}$	$1.2 \cdot 10^{-5}$	$8.2 \cdot 10^{-6}$	$5.9 \cdot 10^{-6}$
$\dot{q}$	$1.8 \cdot 10^{-3}$	$7.2 \cdot 10^{-4}$	$1.6 \cdot 10^{-3}$	$1.2 \cdot 10^{-3}$	$1.4 \cdot 10^{-3}$	$1.4 \cdot 10^{-3}$
Y_{H_2}	$6.8 \cdot 10^{-9}$	$1.3 \cdot 10^{-6}$	$3.3 \cdot 10^{-5}$	$2.2 \cdot 10^{-5}$	$1.3 \cdot 10^{-5}$	$8.3 \cdot 10^{-6}$
Y_{O_2}	$3.3 \cdot 10^{-9}$	$5.8 \cdot 10^{-7}$	$1.4 \cdot 10^{-5}$	$1.2 \cdot 10^{-5}$	$1.1 \cdot 10^{-5}$	$9.8 \cdot 10^{-6}$
Y_{H_2O}	$6.1 \cdot 10^{-4}$	$6.0 \cdot 10^{-4}$	$2.4 \cdot 10^{-4}$	$8.4 \cdot 10^{-5}$	$4.4 \cdot 10^{-5}$	$2.8 \cdot 10^{-5}$
$Y_{\text{H}_2O_2}$	$3.7 \cdot 10^{-4}$	$9.7 \cdot 10^{-4}$	$2.8 \cdot 10^{-4}$	$1.6 \cdot 10^{-4}$	$2.0 \cdot 10^{-4}$	$1.5 \cdot 10^{-4}$
Y_{HO_2}	$1.5 \cdot 10^{-4}$	$4.6 \cdot 10^{-4}$	$2.6 \cdot 10^{-4}$	$1.7 \cdot 10^{-3}$	$4.4 \cdot 10^{-3}$	$4.1 \cdot 10^{-3}$
Y_{OH}	$6.7 \cdot 10^{-4}$	$6.0 \cdot 10^{-4}$	$1.7 \cdot 10^{-4}$	$8.6 \cdot 10^{-5}$	$1.7 \cdot 10^{-4}$	$2.0 \cdot 10^{-4}$
Y_{H}	$6.0 \cdot 10^{-4}$	$5.9 \cdot 10^{-4}$	$2.0 \cdot 10^{-4}$	$4.5 \cdot 10^{-5}$	$3.3 \cdot 10^{-5}$	$3.0 \cdot 10^{-5}$
Y_{O}	$6.0 \cdot 10^{-4}$	$6.0 \cdot 10^{-4}$	$2.0 \cdot 10^{-4}$	$3.5 \cdot 10^{-5}$	$9.4 \cdot 10^{-5}$	$1.4 \cdot 10^{-4}$
Y_{N_2}	$3.6 \cdot 10^{-10}$	$4.4 \cdot 10^{-9}$	$6.7 \cdot 10^{-7}$	$1.5 \cdot 10^{-6}$	$1.0 \cdot 10^{-6}$	$9.8 \cdot 10^{-7}$

for the heat release rate which is closely related to HO_2-concentration at the initial phase of autoignition. The time needed for the computation of the chemical source terms is reduced by a factor of 5.6 in this simulation using one processor of a Cray T3E-1200. As the adaptive evaluation of chemical source terms leads to different CPU-times needed per grid point, a dynamic load-balancing has to be used to maintain a high efficiency in the parallel case. The implemented dynamic load-balancing algorithm relies on the transfer of boundary points between neighbouring processors [10]. At regular intervals during the run, the computation time needed by each node to carry out an integration step is measured. These local times are then averaged along rows and columns and the global mean value is computed. If the relative discrepancies between the measured times are larger than a given tolerance value, a grid-point redistribution is performed. In this redistribution process, grid-lines are transfered from the nodes in columns or rows currently needing more computing time than the average to their neighbours on the side with the smallest average load per processor. The number of exchanged grid points is approximately proportional to the additional computing time needed in comparison with the average value. This procedure turns out to be quite efficient. Starting with an equal number of grid points per processor, a nearly perfect load-distribution is achieved within a few redistributions. After this initial phase, load-changes introduced by the adaptivity occur slowly compared to the size of the timestep for the simulation. Therefore, the necessity of a load-balancing is only checked in every nth timestep. For small n, a redistribution of points is not necessary in every timestep and if it becomes necessary, typically only one or two grid-lines have to be migrated to the neighbouring column or row of processors. For the described DNS using 64 processors of a Cray T3E-1200, the time needed for checking the need of a redistribution and transfering one grid-line to the neighbouring processors is less than 10% of the time needed for a full computation of the chemical source terms for one timestep. Thus, the overhead for the dynamic load-balancing is small compared to the computation-time reduction due to the adaptive chemistry.

The presented technique of adaptively evaluating the chemical source terms can be extended to premixed systems, e.g. by using the value of some kind of reaction progress variable like $\operatorname{grad} T$ as the evaluation criterion.

6. CONCLUSION

A method for the adaptive evaluation of chemical source terms in detailed chemistry DNS has been presented, which strongly reduces the time needed for the computation of the chemical source terms without a significant loss in accuracy. This method has been implemented into a parallel code for the DNS of turbulent reacting flows. This code is primarily used on Cray T3E systems, on which high parallel speedups and performance rates are achieved. Dynamic load-balancing is performed to maintain high parallel efficiency in the adaptive chemistry case. Adaptivity and parallelism are the key paradigms to further enlarge the domain of configurations accessible for DNS.

ACKNOWLEDGEMENT

The author would like to thank the High Performance Computing Center at Stuttgart (HLRS) and the John von Neumann Institute for Computing at Jülich (NIC) for granting him access to their Cray T3E systems. The presented DNS would not have been possible without this support.

REFERENCES

1. J. Warnatz, U. Maas, R. W. Dibble, Combustion, 2nd Edition, Springer, Berlin, Heidelberg, New York, 1999.
2. T. Mantel, J.-M. Samaniego, Fundamental Mechanisms in Premixed Turbulent Flame Propagation via Vortex-Flame Interactions, Part II: Numerical Simulation, Combustion and Flame 118 (1999) 557–582.
3. M. Lange, J. Warnatz, Investigation of Chemistry-Turbulence Interactions Using DNS on the Cray T3E, in: E. Krause, W. Jäger (Eds.), High Performance Computing in Science and Engineering '99, Springer, Berlin, Heidelberg, New York, 2000, pp. 333–343.
4. M. Baum, Performing DNS of Turbulent Combustion with Detailed Chemistry on Parallel Computers, in: E. D'Hollander, G. Joubert, F. Peters, U. Trottenberg (Eds.), Parallel Computing: Fundamentals, Applications and New Directions, no. 12 in Advances in Parallel Computing, Elsevier Science, Amsterdam, 1998, pp. 145–153.
5. M. Lange, U. Riedel, J. Warnatz, Parallel DNS of Turbulent Flames with Detailed Reaction Schemes, AIAA Paper 98-2979 (1998).
6. R. B. Bird, W. E. Stewart, E. N. Lightfoot, Transport Phenomena, Wiley, New York, 1960.
7. F. A. Williams, Combustion Theory, Addison-Wesley, Reading, Mass., 1965.
8. U. Maas, J. Warnatz, Ignition Processes in Hydrogen-Oxygen Mixtures, Combustion and Flame 74 (1988) 53–69.
9. D. Thévenin, F. Behrendt, U. Maas, B. Przywara, J. Warnatz, Development of a Parallel Direct Simulation Code to Investigate Reactive Flows, Computers and Fluids 25 (5) (1996) 485–496.
10. M. Lange, D. Thévenin, U. Riedel, J. Warnatz, Direct Numerical Simulation of Turbulent Reactive Flows Using Massively Parallel Computers, in: E. D'Hollander, G. Joubert, F. Peters, U. Trottenberg (Eds.), Parallel Computing: Fundamentals, Applications and New Directions, no. 12 in Advances in Parallel Computing, Elsevier Science, Amsterdam, 1998, pp. 287–296.
11. E. Anderson, J. Brooks, C. Grassl, S. Scott, Performance of the Cray T3E Multiprocessor., in: Proc. of SC97: High Performance Networking & Computing, http://www.supercomp.org/sc97/, 1997.
12. M. Baum, T. J. Poinsot, D. Thévenin, Accurate Boundary Conditions for Multicomponent Reactive Flows, Journal of Computational Physics 116 (1995) 247–261.

Parallel Computational Fluid Dynamics – Trends and Applications
C.B. Jenssen et al. (Editors)

Application of swirling flow in nozzle for CC process

Shinichiro Yokoya[1], Shigeo Takagi[1], Manabu Iguchi[2], Katukiyo Marukawa[3], Shigeta Hara[4]

1) Department of Mechanical Engineering, Nippon Institute of Technology, Miyashiro, Saitama, 345-8501, Japan,
2) Division of Materials Science and Engineering, Hokkaido University, North 13, West 8, Kita-ku, Sapporo, 060-8628,
3) Sumitomo Metal Industries, Ltd, 16-1, Sunayama, Hazakichou, Kasimagun, Ibaraki, 314-02, Japan,
4) Dept. Materials Science and Processing, Suita, Osaka University, Yamadaoka, Suita, Osaka-fu, 565-0871, Japan

A numerical and water model are used to study the flow pattern in an immersion nozzle of a continuous casting mold and mold region with a novel injection concept using swirling flow in the pouring tube, to control the heat and mass transfer in the continuous casting mold. The maximum velocity at the outlet of the nozzle with swirl is reduced significantly in comparison with that without swirl. Heat and mass transfer near the meniscus can be remarkably activated compared with a conventional straight type immersion nozzle without swirl.

1.INTRODUCTION

In the continuous casting process, it is well known that fluid flow pattern in the mold has a key effect both on the surface and the internal quality of the ingots, because the superheat dissipation induced by the flow pattern has a great influence on the growth of the solidifying shell as well as on the resulting development on the micro-structure. Accordingly, numerous efforts have been expended to control the fluid flow in the mold region. There are many ideas proposed for the controlling using electromagnetic force and some of them have been used in practice until now. Application of the electromagnetic braking and stirring to the fluid in the mold region are typical example. All these electromagnetic installations require quite costly equipment, especially, in the case of mold stirring the electromagnetic field has to penetrate the copper mold.

In this work, we show how to control the outlet flow in the immersion nozzle and the metal flow in the mold region by imparting a swirling motion to the inlet stream of a divergent nozzle.

2. OUTLET FLOW PATTERN OF IMMERSION NOZZLE WITH IMPARTING SWIRLING MOTION

The purpose of this section shows an alternative, potentially cost effective way to obtain a low velocity and uniform dispersion of the hot metal stream as it enters the mold region by imparting a swirling motion to the stream inside a submerged entry nozzle with divergent outlet.

Let us consider flow in an axisymmetric divergent nozzle that is stirred by a swirling blade as shown in Fig. 1. Calculations were performed for these systems with and without swirl flow.

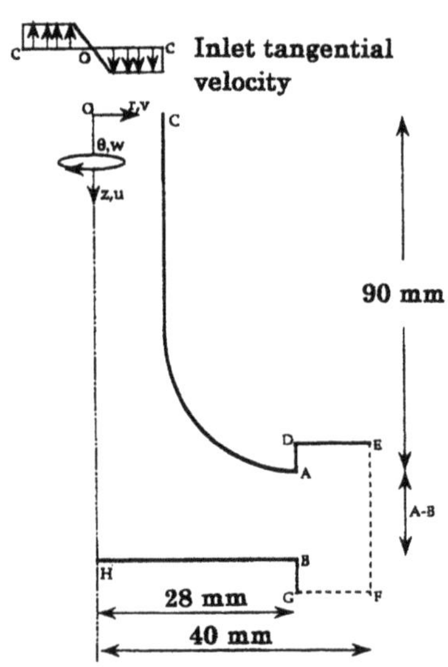

Figure 1. Schematic of divergent nozzle having swirling flow at the entrance impinging on an "opposite face" which is placed to turn the flow radially outward at the nozzle exit. Only one side of the axi-symmetric nozzle is shown.

2.1. Governing equation

Governing equation to be considered are the time averaged continuity and momentum equations for an incompressible single-phase Newtonian fluid. An eddy viscosity model is used to account for the effect of turbulence. The mode chosen is the standard k- ε model.

Boundary condition

The boundary conditions prescribed at the various types of boundaries are in general quite standard. At the solid surfaces the semi-empirical "wall functions" [1] are used to approximate the shear stress due to the no-slip condition at the wall. The value of k and ε are those derived from an assumption of an equilibrium boundary layer. At the exit boundaries, a constant pressure boundaries is assumed. The equations are solved using the finite volume, fully implicit procedure embodied in the FLUENT computer code [2].

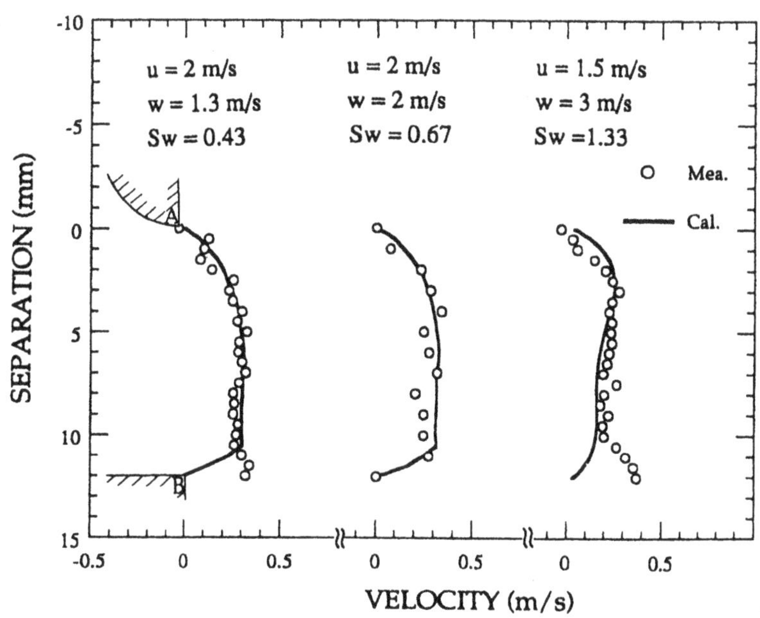

Figure 2. Comparison between the calculated (solid line) and experimentally measured (symbols) profiles of the radial velocity at the nozzle outlet, with the swirl strength denoted by the inlet tangential velocity, w for the case: u, mean velocity through the tube and Sw, swirl number. (A-B=12mm)

2.2. Water model.

Figure 2 shows the comparison between the calculated (solid line) and measured axial profile of the radial velocity at the nozzle outlet region (separation 12mm), for the cases; the entrance axial mean velocity, 2 m/s and various swirl number, Sw defined by 2w/3u. Those figures clearly show that by using the divergent nozzle with the swirling, an uniform and low outlet flow profile can be obtained. Figure 3 shows the change of the calculated radial component of velocity at the nozzle outlet (separation 12mm) when the strength of the swirl at the entrance of nozzle is gradually increased. The velocity distribution changes significantly with the increase of swirl velocity from 0.5 to 1.17 m/s, but the change of velocity distribution over swirl strength 1.17m/s was small even with an increase of the swirl strength of seven times. [3)]

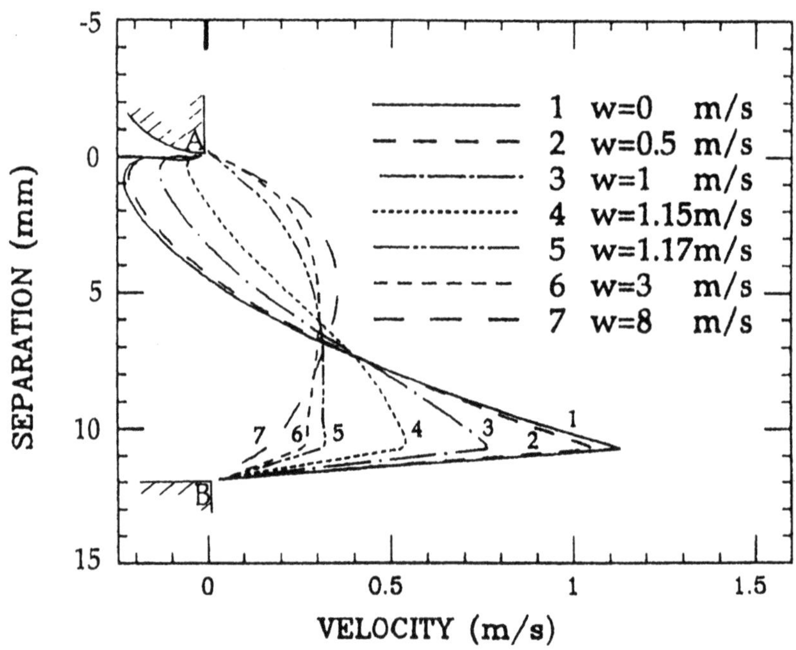

Figure 3. Calculated profiles of the radial velocity for the several different swirl strength, denoted by the inlet tangential velocity, w with an inlet mean axial velocity, 2m/s. (A-B=12mm).

3. SWIRLING EFFECT IN HEAT AND MASS TRANSPORT IN BILLET CC

In previous section, we discuss that the application of swirling motion to a liquid flow in the immersion nozzle works effectively to control the nozzle outlet flow pattern at the outlet. In this section, we show the results used for a billet CC mold on water model experiments.[4]

The dimensions of the mold are 150mm on the diameter as shown in Fig.4. The axial velocity of 2m/s was chosen at the nozzle entrance. When the swirling velocity of 1.7m/s was imposed to the entrance flow, following issues was observed:

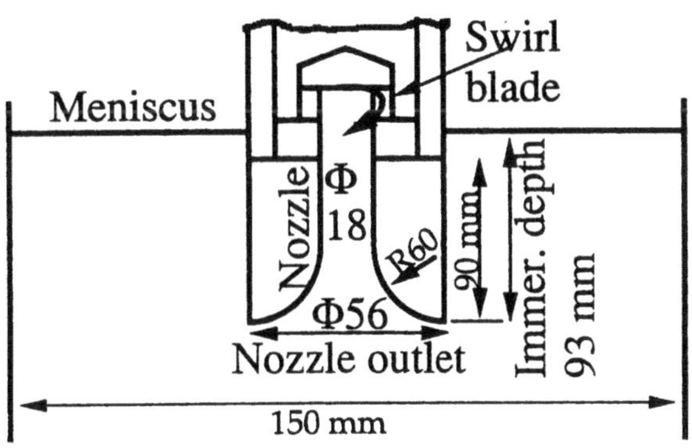

Figure 4. Schematic diagram of water model mold, showing the meniscus, immersion depth, nozzle, nozzle outlet and swirl blade.

Figure 5 shows the experimental radial profile of the axial component velocity for the cases both with and without swirl downwards from the outlet of the immersion nozzle.

It can be seen that, for the case without swirl, the flow has a maximum, which is very high on the centerline and considerable velocity fluctuation because of separation of boundary layer. In contrast, for the case with swirl, maximum velocity reduce at 25% of that without swirl, and velocity profile becomes both very uniform and calm within very short distance downward from the nozzle outlet. The results of the calculation show the same tendency as the experimental results.

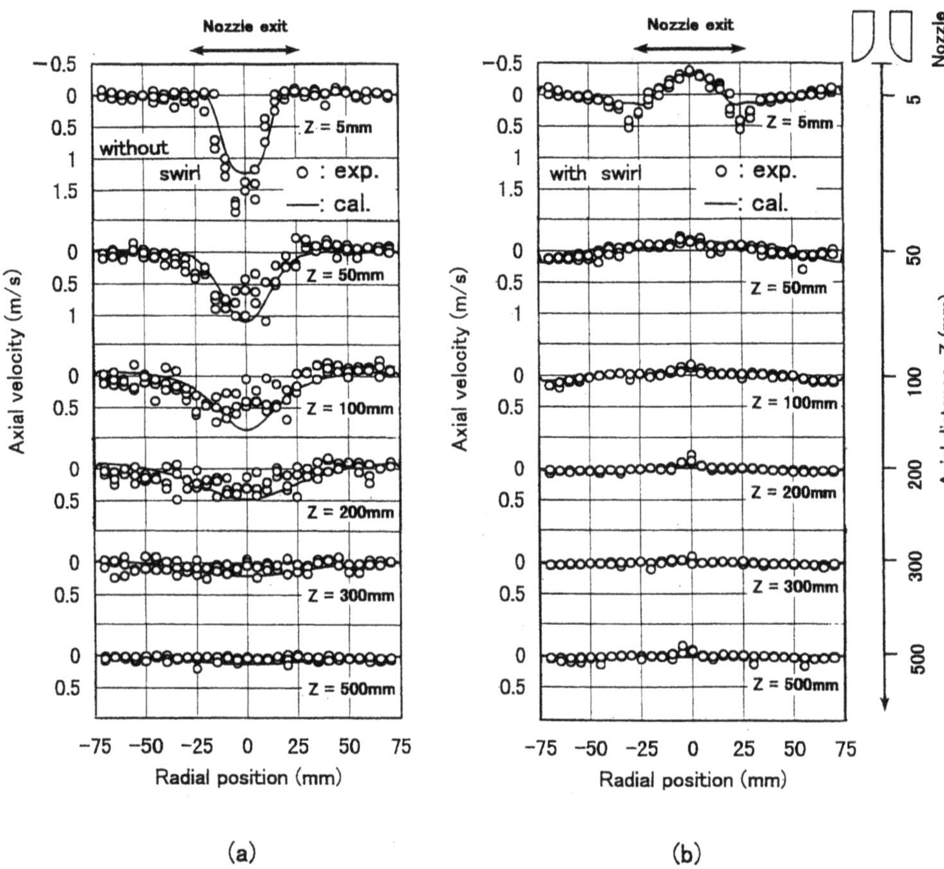

Figure 5. Radial profiles of the axial velocity vs. various axial position at the several axial positions from the outlet of the immersion nozzle for the cases; without (a) and with swirl velocity 1.72 m/s (b), entrance mean axial velocity 2 m/s.

Effect of Immersion Depth of Nozzle on Maximum Temperature and Velocity at Meniscus (Molten steel)

We examine the effect of swirl on the flow pattern in the mold region and the resulting effect on the temperature.

In no swirl case, shown in Fig. 6 (a), (inlet axial velocity 2 m/s, inlet temperature 1773 K, mold wall temperature 1748K), the flow passes straight down into the mold and the maximum velocity is always on the centerline.

In the case of high swirl, shown in Fig. 6 (b), (inlet axial velocity 2 m/s, inlet swirl velocity 1.72 m/s, inlet temperature 1773 K, mold wall temperature 1748K), the flow passes parallel to the curved wall, impinging on the wall of the mold with a low velocity, then splits vertically to create upper and lower recirculation region. The axial velocity components are maximum near the mold wall, however those maximum velocities are only within 0.066 m/s.

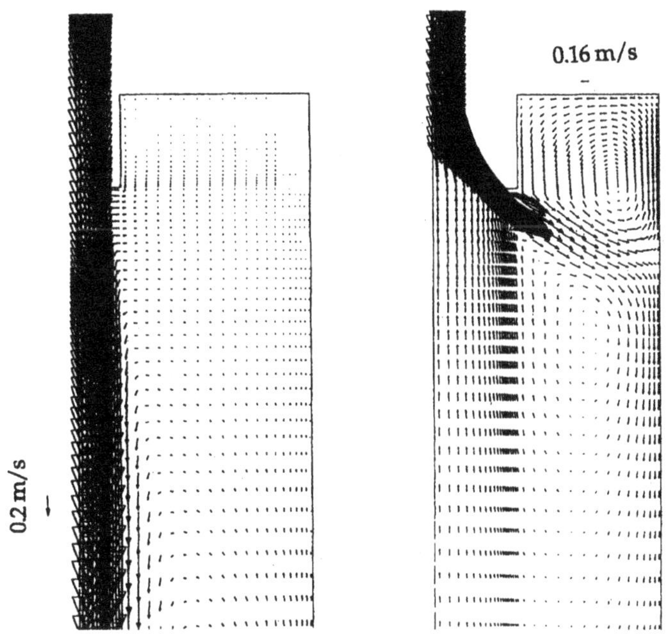

(a) Straight nozzle without swirl

(b) Divergent nozzle with swirl, w, 1.72 m/s

Figure 6. Velocity fields in the mold with the entrance mean axial velocity 2 m/s both for the cases; (a) straight nozzle without swirl, (b) divergent nozzle with swirl, 1.72 m/s.

The maximum radial component velocities at the meniscus are shown for the various immersion depths in Fig. 7. For the case of divergent nozzle with swirl, maximum surface velocity decreases with increasing immersion depth from 0 to 93 mm, while, for the case of conventional straight nozzle with no swirl, that remains almost nearly 0 m/s at any immersion depths. It was cleared that using the divergent nozzle with swirl, mass transfer near the meniscus can be considerably activated. Figure 8 shows a maximum temperature at the meniscus vs. immersion depth. For the case of divergent nozzle with swirl, maximum surface temperature decreases with increasing immersion depth, while, for the case of conventional straight nozzle with no swirl, that is always fixed to almost nearly the melting point at any immersion depths. It was cleared that using the divergent nozzle with swirl, heat transfer near the meniscus can be considerably activated.

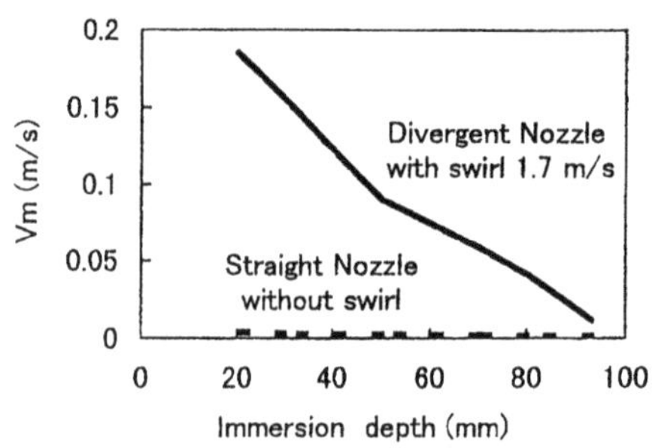

Figure 7. Maximum radial velocity at the meniscus.

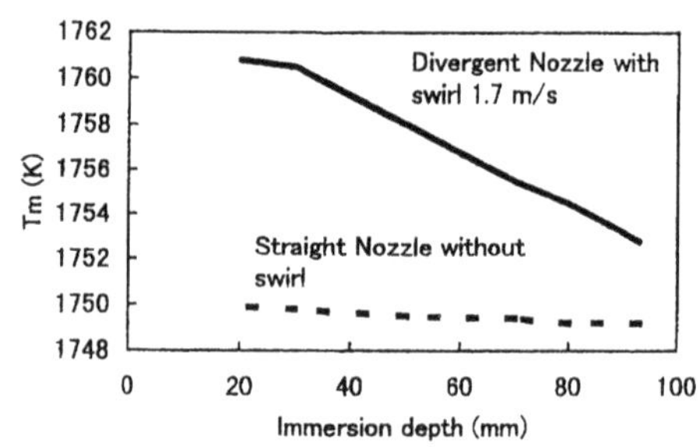

Figure 8. Maximum temperature at the meniscus.

4. CONCLUDING REMARK

This paper has demonstrated a number of possibilities presented by a novel immersion nozzle. These may be summarized as follows.

(1) By changing swirl strength, it is easy to control the flow pattern as well as the direction of the flow.

(2) Heat and mass transfer near the meniscus can be remarkably activated compared with a conventional straight type immersion nozzle without swirl.

(3) Uniform velocity distribution can be obtained within a very short distance from the outlet of the nozzle.

REFERENCES

1) B.E. Launder and D. B. Spalding: Comput. Method. Appl. Mech. Eng., 3(1974), 269.
2) Fluent User's Manual Version 4.4, ed. By Fluent Inc., August, (1996)
3) S. Yokoya, R. Westoff, Y. Asako, S. Hara and J. Szekely: ISIJ Int. , 34(1994), No.11, 889.
4) Shinichiro. Yokoya, Sigeo Takagi, Manabu Iguchi, Yutaka Asako, R. Westoff, Sigeta. Hara : ISIJ Int. , 38(1998), No.8, 827.

13. Unsteady Flows

Parallel Computational Fluid Dynamics – Trends and Applications
C.B. Jenssen et al. (Editors)

Computational Fluid Dynamic (CFD) Modelling of the Ventilation of the Upper Part of the Tracheobronchial Network

A. E. Holdø[a], A. D. Jolliffe[a], J. Kurujareon[a], K. Sørli[b], C. B. Jenssen[c]

[a]University of Hertfordshire, Hatfield, Herts., AL10 9AB, England

[b]SINTEF Applied Mathematics, 7465 Trondheim, Norway

[c]Statoil, 7005 Trondheim, Norway

Simulations of respiratory airflow in a three-dimensional asymmetrical single bifurcation were performed. Two breathing conditions, normal breathing condition and high-frequency ventilation (HFV), were selected for the present study. A parallelised CFD code based the finite volume method (FVM) together with the implicit scheme was utilised. The technique of multi block method was applied to the three-dimensional asymmetric single bifurcation. The multi block structured grids for the bifurcation model were applied with an object-oriented code for geometric modelling and grid generation.

The simulation results obtained from the present study were in a good agreement with the previous experiments. It was found that the results for the normal breathing were similar to the steady state airflow. Whereas the result obtained from the HFV condition were strongly influenced by the unsteadiness effect.

1. INTRODUCTION

The understanding of the ventilation airflow in the human lung is important. It is thought that diseases such as Asthma could be linked to the particle deposition in the lung and the effects of air pollution also requires an enhanced understanding of particle deposition in the lung. Another use of knowledge of particle deposition is medication through the lung using inhaler system. The beneficiaries of such systems could be sufferers of diseases such as Diabetes where continued hypodermic injections of the necessary drugs become problematic with time. Medication through the lung could bring real benefits for such patients.

In order to understand the particle deposition, it is essential to be able to correctly model the airflow in the tracheobronchial network. There have been numbers of investigations studying such flow in CFD models [1–5]. Those studies, however, were based on steady state simulations within a symmetrical single bifurcation model. The present work shows that results can be misleading and that transient, time dependent simulations are essential for the full description of the airflow patterns resulting in the ventilation of the lung. The more realistic asymmetric bifurcation model based on anatomy detail of Horsfield et al. [6] was taken into account (Figure 1). The results also suggest that breathing patterns in terms of peak flow rates and ventilation of flow rate with time are strong

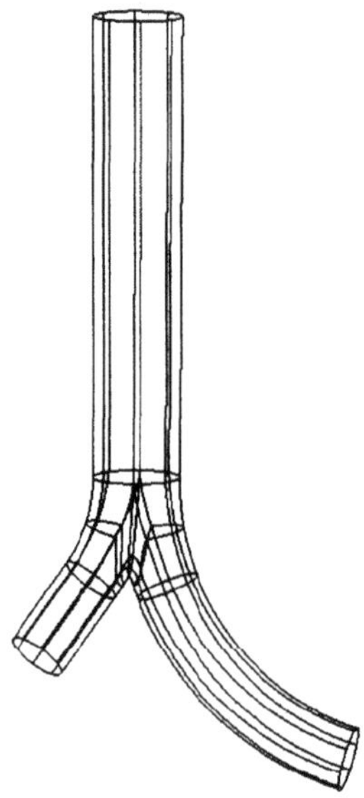

Figure 1. An asymmetric single bifurcation model of the central airway in the lung.

contributors to the resulting airflow patterns inside the lung. These flow patterns will strongly affect the particle deposition within the lung. Preliminary work also indicates that in many circumstances it is necessary to model more than one bifurcation as shown in Figure 2.

The resulting CFD models become necessarily very large in terms of node numbers and geometry complexity. Consequently it has become necessary to use parallel methods. The present work employed a three-dimensional Navier-Stokes solver based on the FVM using an implicit scheme. The multi-block structured grids were used and simulated on parallel computers using the PVM message passing system.

2. NUMERICAL BOUNDARY CONDITION AND MESH MODELLING

A three-dimensional asymmetric single bifurcation model (Figure 1) was selected for the respiratory airflow in the present study. The airway geometry and dimensions are based on the anatomic details of the central airway given by Horsfield et al [6]. The technique of the multi block technique was applied to airway model. The mesh model consisted of 84 blocks with 157,820 node points of the hexahedral mesh cells. The multi block structured grids for the single bifurcation model (Figure 3) were applied with an object-oriented code for geometric modelling and grid generation.

Two breathing conditions under the normal breathing condition (Re =1.7103, f = 0.2Hz)

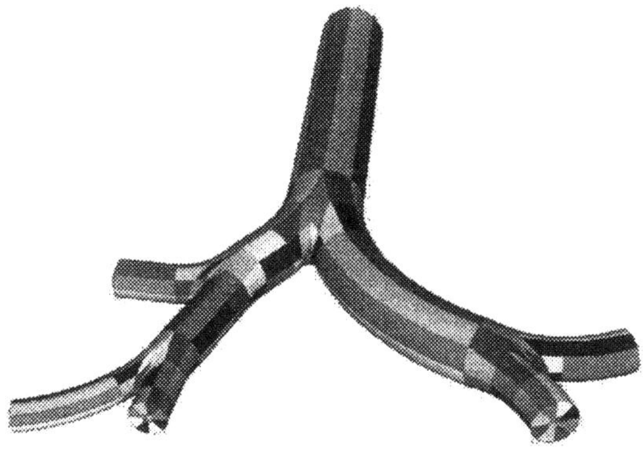

Figure 2. Multiple bifurcation model of the central airway including trachea and five lobar bronchi.

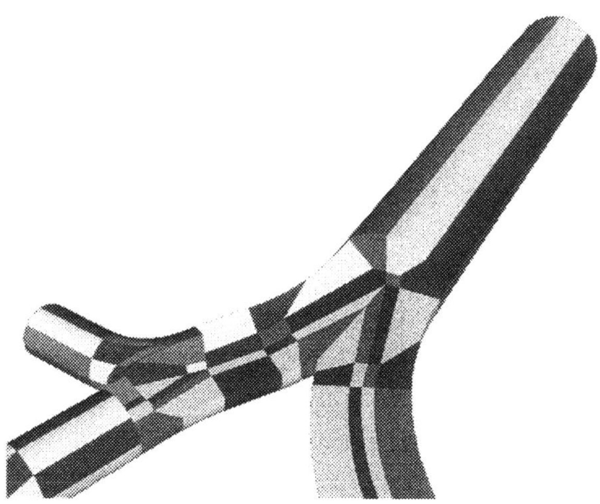

Figure 3. Multi block technique employed into the multi bifurcation geometry.

and the HFV condition (Re = 4.3103, f = 5Hz) were selected. The velocity boundary conditions were imposed at the inflow/out flow boundaries varying with respect to time as a sinusoidal time function to regulate the oscillatory airflow.

The numerical method to discretise the Navier-Stokes equations used in the present study was based the FVM using concurrent block Jaconi (CBJ) with the implicit scheme [7]. In the CBJ solver, the use of implicit multi block method is also available for the flow calculation on parallel processors. The solutions in each block is solved separately applying explicit boundary conditions at block boundaries. A coarse grid correction scheme [8] is then applied to link between blocks and speed up convergence. This approach has been shown to work well for time-accurate simulations, ensuring both fast convergence and high parallel speed up for the 84 blocks used here.

3. RESULTS AND DISCUSSIONS

The results obtained from the normal breathing and the HFV conditions are shown in Figure 4a and Figure 4b. For the normal breathing condition at the peak flow rate (Figure 4a), the results are similar to the steady state respiratory airflow studied by many investigators. [9–11,4,5] Menon et al [12] and Jolliffe [13], who studied the oscillatory flow in the model of multi generations of central airways model, also obtained the similar results that the flow pattern at peak inspiration were resemble to those steady state study. These observations can be explained by that the velocity gradient, , at peak flow of the respiratory cycle is near zero. Hence the unsteadiness effect can be neglected at the peak of the respiratory cycle. The axial flows are skewed towards the inner walls of the bifurcation (outer wall of the bend). The secondary flow motions were obtained in both right and left daughter airways on the inspiration and in parent airway on the expiration. This conforms to the steady flow in curved pipe [14]. However the unsteadiness effect becomes significant for the other flow rate during the respiratory cycle.

In comparison between the resulting respiratory airflow simulation under the normal breathing condition within the single bifurcation model in the present study and the multi bifurcation model of Jolliffe [13], the flow fields were well similar on the inspiration phase. Within the inspiration the particle deposition is most influenced while for the expiration the particle deposition is not significant. Hence the single bifurcation model is sufficient in considering the inspiratory airflow patterns effect on the particle deposition.

For the HFV condition (Figure 4b), the secondary motions were not observed. The axial flow, therefore, was not distorted. The axial flow under the HFV condition was different from those observed under the normal breathing condition. The axial velocity profiles throughout the bifurcation model for the HFV condition are in the same patterns with no change of boundary layer thickness. This indicates that geometry is not the significant effect on the respiratory flow for this breathing condition.

4. CONCLUSIONS

The CFD model of the respiratory flow in the present study gives realistic results respiratory airflow that agree well with experiments. As a result of this study, new information about high-frequency ventilation condition has been found Without parallel computing it would have been virtually impossible to simulate and solve such geometrically complex

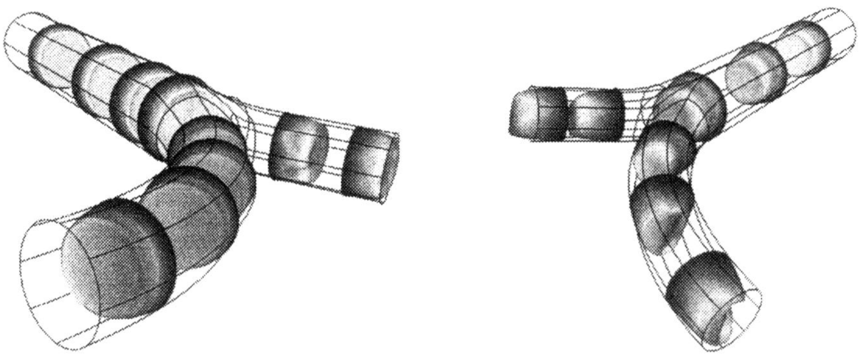

(a) Normal breathing condition (b) HFV condition

Figure 4. Peak inspiratory flow during the normal breathing condition (a) and HFV condition (b).

problem of the airway network in the respiratory system.

REFERENCES

1. Gatlin, B., Cuicchi, C., Hammersley, J., Olson, D.E., Reddy, R. and Burnside, G. Computation of coverging and diverging flow through an asymmetric tubular bifurcation. ASME FEDSM97-3429 (1997) 1-7.
2. Gatlin, B., Cuicchi, C., Hammersley, J., Olson, D.E., Reddy, R. and Burnside, G. Paticle paths and wall deposition patterns in laminar flow through a bifurcation. ASME FEDSM97-3434 (1997) 1-7.
3. Gatlin, B., Cuicchi, C., Hammersley, J., Olson, D.E., Reddy, R. and Burnside, G. Computational simulation of steady and oscillating flow in branching tubes. ASME Bio-Medical Fluids Engineering FED-Vol.212 (1195) 1-8.
4. Zhao, Y. and Lieber, B.B. Steady expiratory flow in a model symmetric bifurcation. ASME Journal of Biomechanical Engineering 116 (1994a) 318-323.
5. Zhao, Y. and Lieber, B.B. Steady inspiratory flow in a model symmetric bifurcation. ASME Journal of Biomechanical Engineering 116 (1994a) 488-496.
6. Horsfield, K., Dart, G., Olson, D.E., Filley, G.F. and Cumming, G. Models of the human bronchial tree. J.Appl.Physiol. 31 (1971) 207-217.
7. Jenssen, C. B. Implicit Multi Block Euler and Navier-Stokes Calculations AIAA Journal Vol. 32 (1994) No. 9.
8. Jenssen, C. B and Weinerfelt P.A. Parallel Implicit Time-Accurate Navier-Stokes Computations Using Coarse Grid Correction, AIAA Journal Vol. 36 (1998) No. 6.
9. Schroter, R.C. and Sudlow, M.F. Flow patterns in model of the human bronchial airways. Respir.Physiol. 7 (1969) 341-355.
10. Chang, H.K. and El Masry, O.A., A model study of flow dynamics in human central

airways. Part I:Axial velocity profiles. Respir.Physiol. 49 (1982) 75-95.
11. Isabey, D. and Chang, H.K. A model study of flow dynamics in human central airways. Part II:Secondary flow velocities. Respir.Physiol. 49 (1982) 97-113.
12. Menon, A.S., Weber, M.E. and Chang, H.K. Model study of flow dynamics in human central airways. Part III:Oscillatory velocity profiles. Respir.Physiol. 55 (1984) 255-275.
13. Jolliffe, A.D. Respiratory airflow dynamics. PhD thesis, University of Hertfordshire (2000) pp.1-500.
14. Snyder, B., Hammersley, J.R. and Olson, D.E. The axial skew of flow in curved pipes. J.Fluid Mech. 161 (1985) 281-294.

Parallel Computational Fluid Dynamics – Trends and Applications
C.B. Jenssen et al. (Editors)

Parallel Computing of an Oblique Vortex Shedding Mode

T. Kinoshita[a] **and** **O. Inoue**[b]

[a]Scalable Systems Technology Center, SGI Japan, Ltd.,
P.O. Box 5011, Yebisu Garden Place, 4-20-3 Yebisu, Shibuya-ku, Tokyo, 150-6031 Japan

[b]Institute of Fluid Science, Tohoku University,
2-1-1 Katahira, Aoba-ku, Sendai, Miyagi, 980-8577 Japan

The parallelizing strategy for a numerical simulation of an oblique vortex shedding mode for a free-ended circular cylinder at the low Reynolds number will be presented. A hybrid method of distributed memory parallelization and shared memory parallelization is employed. The transition from a parallel shedding mode to an oblique shedding mode will be also discussed.

1. INTRODUCTION

It has been observed in experiments that vortices are shed at oblique angles in the low-Reynolds-number circular cylinder wake [1-3]. When a circular cylinder is located in a towing tank and it is traveled from the starting position, vortices are shed parallel to the cylinder across almost the whole span initially. After the cylinder travels for a while, the wake vortices near the ends begin to be shed at oblique angles. Finally, the oblique vortex shedding mode takes over across the whole span, and the vortex cores form a 'chevron'-shaped pattern, that is symmetrical with respect to the center span.

The phenomenon is obviously caused by boundary conditions at the cylinder ends. Williamson [1] suggests that the end effects have a direct influence over a region of the cylinder span of the order of 10-20 diameters in length. Their influence over the rest of the span is of an indirect nature. An oblique front gradually travels inwards along the span from each end, bringing behind it a region of oblique shedding. He also suggests that the presence of oblique shedding does not require a difference in the two end conditions.

The authors have had much interests if the oblique shedding mode can be observed in a symmetrical numerical simulation, that is free from flow-induced cylinder vibrations and flow non-uniformity. According to Williamson, the cylinder must travel of the order of 500 diameters for the wake to reach its oblique shedding asymptotic form, and it means that the simulation time should be long enough to see the phenomenon. It is believed that the careful consideration should be also directed to the number of grid points. The enough number of grid points must be employed both around the cylinder surface and on the ends in order to simulate this kind of flow phenomenon. The numerical simulation of an oblique vortex shedding mode, therefore, is very time consuming, and its parallelization becomes an essential part of the analysis. A hybrid method of distributed memory parallelization and shared memory parallelization was employed.

2. NUMERICAL METHODS

The incompressible three-dimensional Navier-Stokes equations are solved on the general curvilinear coordinates. The equations are discretized in a finite difference formulation, where a third-order QUICK scheme is applied. The length-to-diameter ratio (*L/D*) of the circular cylinder is 107 in the simulation. The symmetrical boundary condition is given at the center span, and only a half of the computational domain is computed. The computational domain extends 30*D* upstream, 30*D* in the upper and lower directions from the cylinder surface, also 30*D* outwards form the free end, and 150*D* downstream, where *D* indicates the cylinder diameter. An O-type grid system is employed around the cylinder, and an orthogonal grid system is connected downstream of the O-type grids. In addition to it, curvilinear coordinate grids are inserted into the O-type grids for flow fields on the outside of the free end. The grid space in the radial direction in the nearest layer from the cylinder is 0.002*D*, and the space in the span direction near the cylinder free end is also 0.002*D*. They are increased in a region off the cylinder surface or the free end. The total number of grid points in the whole computational domain is about forty-seven million. The boundary conditions consist of uniform inflow velocity, zero-normal-velocity and zero-shear-stress at the lateral boundaries and the outflow boundary, and no-slip on the cylinder. Reynolds number of the simulation is 150, and the time step is set to 0.025.

3. PARALLELIZING STRATEGY

Because the numerical simulation of oblique vortex shedding requires huge computing resources, it was planned that clustered Origin2000 systems would be used for the simulation in order to reduce the computing time. When only a single system is available, the same code would be used for parallel processing on the single system. Though, the domain decomposition method would be straightforward for parallelizing a structured-grids-code for both clustered multi-processor systems and a single multi-processor system, a hybrid method of distributed memory parallelization and shared memory parallelization was employed in the present work. Grid systems for the flow simulation are composed of three zones. These zones are computed in parallel using a distributed memory parallel approach. Velocity and pressure data at overlapped grid points in each zone is exchanged between the tasks with MPI or SHMEM message passing library. MPI is used when the tasks are executed over clustered systems, and SHMEM is used when all the tasks are executed on a single system. Furthermore, loop level parallelization is carried out in each zone (i.e. each task) using a shared memory parallel approach. The so-called multi-level parallelization enables parallel computing on clustered multi-processor systems with minimum programming efforts. Even if all the tasks corresponding to each zone are run on a single multi-processor system, the multi-level parallelization contributes to making the work granularity large compared to parallelizing each loop with all the available processors.

4. NUMERICAL RESULTS

Figure 1 shows vorticity isosurfaces for $|\boldsymbol{\omega}| = 0.375$ at $t = 125$ and $t = 400$, where $t = 0$ indicates the time when the uniform flow reaches the leading edge of the cylinder. It corresponds to the time when the cylinder starts traveling in the towing tank at experiments. At $t = 125$, oblique vortex shedding is observed near the cylinder ends, but the parallel shedding mode is dominant over the rest of the span. The oblique front gradually travels inwards along the span, and the whole span sheds oblique vortices at around $t = 400$. It is noted that an oblique angle near the cylinder end becomes approximately 45° at its maximum during the oblique front stays near the end. The oblique angle gradually becomes smaller again as the oblique front travels inwards, and the vortices are shed at about 22° across the whole span finally. Figure 2 shows power spectra of velocity at a point $10D$ downstream from the cylinder center and $2D$ outwards from the center span (i.e. the symmetrical plane). Figure 2 (a) shows spectra from $t = 100$ to $t = 200$, and Fig. 2 (b) shows spectra from $t = 400$ to $t = 500$. This point is located in the parallel shedding mode for the period of Fig. 2 (a), and it is entered in the oblique shedding mode for the period of Fig. 2 (b). The Strouhal number decreases from 0.18 (S_0) to 0.165 (S_θ) due to the transition of the mode, and it indicates that the following relationship holds:

$$S_\theta = S_0 \cdot \cos\theta \tag{1}$$

(a) $t = 125$

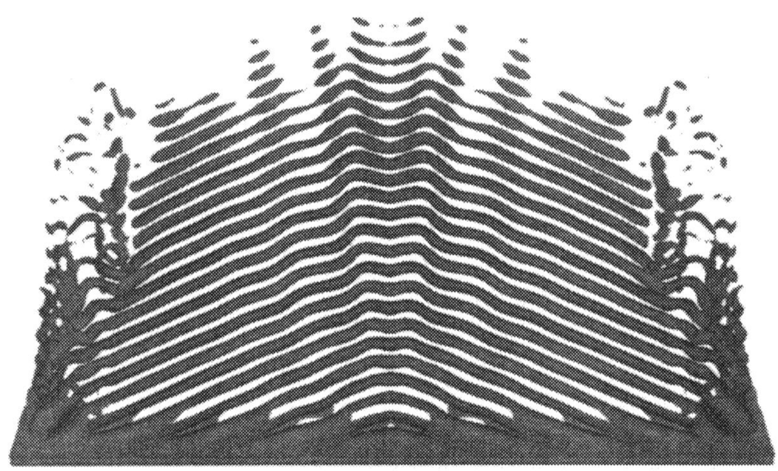

(b) $t = 400$

Fig.1 Vorticity isosurfaces for $|\boldsymbol{\omega}| = 0.375$

where θ is the oblique angle for the period of Fig. 2 (b). The oblique angle of $\theta = 22^\circ$ in the final chevron-shaped pattern is much larger than the result of $\theta = 13^\circ$ in Williamson's experiments for the same Reynolds number. Williamson used endplates whose diameter is 10 to 30 times of the cylinder diameter as the end conditions, while the end conditions in the present work are free ends. The different end conditions may bring about the distinct oblique angles.

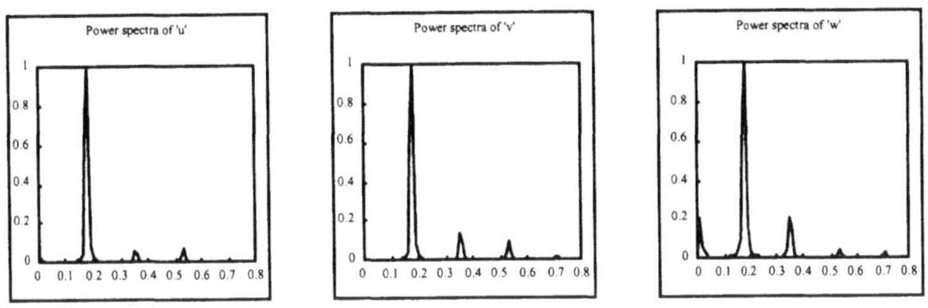

(a) $t = 100$ to $t = 200$

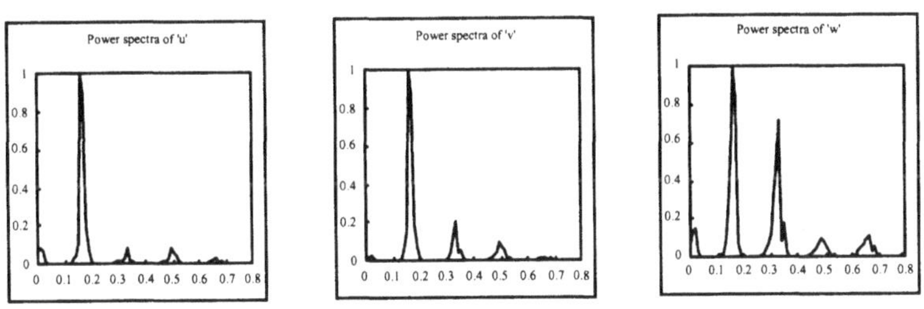

(b) $t = 400$ to $t = 500$

Fig. 2 Power spectra of 'u', 'v', and 'w' at a point $x = 10.0D, y = 1.0D, z = 2.0D$

5. CONCLUSIONS

The parallelizing strategy of mixing OpenMP and MPI was employed, and it has shown relatively good parallel performance for numerical simulations of an oblique vortex shedding mode. The computed results indicate that the whole circular cylinder span sheds oblique vortices even if the conditions are perfectly symmetrical. The present work found that the oblique front does not keep a constant oblique angle when it travels inwards from the end. The angle becomes far larger than the final oblique angle in a region near the cylinder ends, and then it comes to a smaller degree as the oblique front travels inwards.

ACKNOWLEDGEMENTS

All the computations in this work were carried out on SGI Origin2000 in Institute of Fluid Science, Tohoku University.

REFERENCES

1. C.H.K. Williamson, Oblique and parallel modes of vortex shedding in the wake of a circular cylinder at low Reynolds numbers, J. Fluid Mech. Vol. 206 (1989) 579.

2. F.R. Hama, Three-dimensional vortex pattern behind a circular cylinder, J. Aerosp. Sci. Vol. 24 (1957) 156.

3. E. Berger, Transition of the laminar vortex flow to the turbulent state of the Karman vortex street behind an oscillating cylinder at low Reynolds number, Jahrbuch 1964 de Wiss. Gess. L. R. (1964) 164.

Parallel Computational Fluid Dynamics – Trends and Applications
C.B. Jenssen et al. (Editors)

Three-dimensional numerical simulation of laminar flow past a tapered circular cylinder

Brice Vallès[a] * and Carl B. Jenssen[b] † and Helge I. Andersson[a] *

[a]Department of Applied Mechanics, Thermodynamics and Fluid Dynamics, Norwegian University of Science and Technology, 7491 Trondheim, Norway

[b]Statoil R&D Centre, 7005 Trondheim, Norway

1. Introduction

Since the earliest investigations of Tritton [1] and Gaster [2] many studies of the flow around bluff bodies, such as cylinders or cones, have been made. The majority were experimental works, and numerical simulations of vortex dynamics phenomena mainly appeared over the last decade (cf Williamson [3,4] for more details). Surprisingly, only few numerical studies have been concerned with the flow behavior behind tapered cylinders, despite the fact that the complex vortex shedding which occurs in the wake of a tapered cylinder is a subject of substantial interest to engineers (factory chimney or support of offshore platform, for example). Hence, three-dimensional numerical simulations of the flow field past a tapered circular cylinder, at low Reynolds number, have been conducted using an implicit multiblock Navier-Stokes solver. Firstly, two-dimensional investigations were carried out to ensure the feasibility of these type of simulations with this solver. The results showed the effect of the mesh and some parameters such as the time step on the accuracy of the solutions. Moreover, the primary results of the two-dimensional simulations were successfully compared with a large variety of earlier works. Secondly, detailed results of three-dimensional simulations, which aimed at reproducing the previous laboratory experiments by Piccirillo and Van Atta [5], are presented. The main features of the computed flow fields seem to agree well with experiments as well as other numerical simulation [6]. It is concluded that the Concurrent Block Jacobi solver has proved to perform fully satisfactorily both for two and three-dimensional laminar flow computations.

2. Numerical method

Three-dimensional wakes behind tapered cylinders are physically very complex and computer simulations inevitably require a large amount of CPU time. Therefore a parallel Navier-Stokes solver running on a Cray T3E was used. The adopted solver, called "CBJ"(Concurrent Block Jacobi), cf Jenssen [7] and Jenssen and Weinerfelt [8,9], is a par-

*brice.valles@mtf.ntnu.no helge.i.andersson@mtf.ntnu.no
†CABJ@statoil.com

allel implicit multiblock time-accurate Navier-Stokes solver, using a coarse grid correction scheme (CGCS), based on the Finite Volume Method. An implicit scheme is chosen in order to accelerate the convergence of solutions of the Navier-Stokes equations in which different time scales are present. Moreover, a multiblock technique was chosen mainly for two reasons: firstly, when using a parallel computer, different processors can work on different blocks thereby achieving a high level of parallelism; secondly, the three-dimensional implicit computations, performed in the present work, require a storage too large to fit in the computer's central memory. By splitting the domain into multiple blocks, it is sufficient to allocate temporary storage for the number of blocks being solved at a given time. The governing equations, written in integral form, are solved on a structured multiblock grid. The convective part of the fluxes is discretized with a third-order, upwind-biased method based on Roe's scheme. The viscous fluxes are obtained using central differencing. Derivatives of second-order accuracy are first calculated with respect to the grid indices and then transformed to derivatives with respect to the physical spatial coordinates. Implicit and second-order-accurate time stepping is achieved by a three-point, A-stable, linear multistep method:

$$\tfrac{3}{2}(V_i/\Delta t)U_i^{n+1} - 2(V_i/\Delta t)U_i^n + \tfrac{1}{2}(V_i/\Delta t)U_i^{n-1} = R(U_i^{n+1}) \tag{1}$$

$R(U_i^{n+1})$ denotes the sum of the flux into volume V_i of the grid cell i, Δt is the time step and n refers to the time level. Equation (1) is solved by an approximate Newton iteration technique. Using l as the iteration index, it is customary to introduce a modified residual

$$R^*(U_i^{n+1}) = R(U_i^{n+1}) - \tfrac{3}{2}(V_i/\Delta t)U_i^{n+1} + 2(V_i/\Delta t)U_i^n - \tfrac{1}{2}(V_i/\Delta t)U_i^{n-1} \tag{2}$$

which is to be driven to zero at each time step by the iterative procedure

$$\left(\frac{\partial R}{\partial U} + \frac{3}{2}\frac{V_i}{\Delta t}\right)\Delta U^l = R^*\left(\left[U_i^{n+1}\right]^l\right) \tag{3}$$

updating $\left[U_i^{n+1}\right]^{l+1} = \left[U_i^{n+1}\right]^l + \Delta U_i$ at each iteration.

The Newton iteration process is approximate because some approximations inevitably are used in the linearisation of the flux vectors and also because an iterative solver is used to solve the resulting linear system. In particular, a first-order approximation for the implicit operator is used. For each Newton iteration procedure, a septadiagonal linear system of equations is solved in each block. By ignoring the block interface conditions, this system is solved concurrently in each block using a line Jacobi procedure [7]. Then, for each iteration of this line Jacobi procedure, a tridiagonal system is solved along lines in each spatial direction. A Coarse Grid Correction Scheme [8] is used to compensate for ignoring the block interface conditions by adding global influence to the solution. The coarse mesh is obtained by dividing the fine mesh into different cells by removing grid lines. Then, the coarse grid system is solved using a Jacobi-type iterative solver that inverts only the diagonal coefficients in the coarse grid system at each of the 25 iterations.

Table 1
Comparison of predicted Strouhal number and the total cpu-time required per shedding period for three different meshes (two-dimensional test cases). The Strouhal number (St) is defined as $(f.D)/U$, where f is the vortex shedding frequency, D the diameter and U the speed of the incoming flow. Note: cc=convergence criterion for the Newton iteration, ts=dimentionless time step.

Mesh (cc=0.01, ts=0.1)	St	Total cpu-time (min) per period
100×100	0.1926	36
200×200	0.1933	162
400×400	0.1956	781

3. Two-dimensional simulations

Firstly, two-dimensional simulations, were carried out to estimate optimum values of different parameters such as time step, convergence criterion and so on, which lead to the best accuracy/cpu-time ratio. The convergence criterion is the accuracy required for each Newton iteration. For each time step, the code iterates on the Newton iterations until the convergence criterion is satisfied. These simulations were performed at $Re = 200$; the Reynolds number is defined by $Re = (U.D)/\nu$ where U is the uniform speed of the incoming flow, D the cylinder diameter and ν the kinematic viscosity of the incompressible fluid. Table 1 shows the effect of the mesh size on the total cpu time, which is the sum of the cpu-time used by all the different processors employed in the simulation. Note that the more nodes, the more iterations per time step are performed and, consequently, higher cpu-time consumption per grid node is required.

The best accuracy/cpu-time ratio was found for the 200×200 mesh. Figure 1 furthermore suggests that the best compromise is for a convergence criterion equal to 0.001 with a time step equal to 0.1.

To demonstrate the validity of the present results, a comparison with a variety of other studies was made, as can be seen in Table 2.

CBJ refers to the present simulation on the 200×200 mesh (cc=10^{-3} and ts=10^{-1}).

Table 2
Comparison of drag coefficient (Cd), lift coefficient (Cl) and Strouhal number (St) for two-dimensional vortex shedding at Re=200.
[a] Pressure forces only.

Reference	Cd	Cl	St
CBJ	1.411	± 0.684	0.1952
Multigrid	1.2[a]	± 0.68[a]	0.195
Belov et al.	1.232	± 0.64	0.193
Braza et al.	1.3	± 0.775	0.20
Williamson	—	—	0.197
Roshko	—	—	0.18–0.20

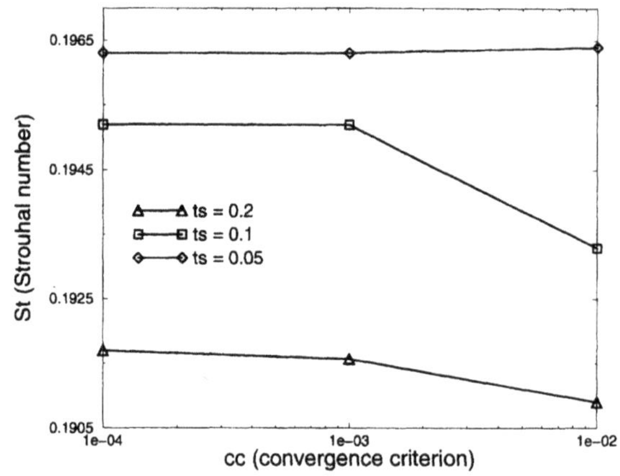

Figure 1. Predicted Strouhal number versus convergence criterion for three time steps. Total cpu-time (min) per shedding period (from the right to the left): △ 87, 133, 215; □ 162, 220, 335; ◇ 261, 392, 557

Multigrid refers to a simulation made by Jenssen and Weinerfelt [9] with a multigrid code based on the Jameson scheme, whilst Belov et al. [10] refers to another multigrid algorithm using pseudo-time stepping. Braza et al. [11] performed two-dimensional numerical simulations of flow behind a circular cylinder using a SMAC (Simplified Marker-and-Cell) method with Finite Volume approximations. Besides these numerical investigations, the results of two other works are seen. Williamson [12] established a mathematical relationship between the Strouhal number and the Reynolds number leading to the definition of a universal curve. At last, the range of values of the Strouhal number, corresponding to $Re = 200$, measured by Roshko in 1954 is given. Despite the large variety of the earlier investigations (experimental, theoretical and numerical), all the results are in good agreement. The differences noticed, especially in the drag coefficient, could result from the actual meshes, the different discretization schemes and their accuracy (second-order accurate only for [11]). Moreover, care should be taken when comparisons are made with the Williamson's relation and Roshko's experiments. Since their results are for three-dimensional straight cylinders which encounter a transition regime close to $Re = 200$.

4. Three-dimensional simulations

For the three-dimensional simulations, two different meshes of 256 000 nodes divided into 28 blocks were constructed, corresponding to two tapered cylinders tested by Piccirillo & Van Atta: Figure 2 shows how the mesh was constructed: 6 fine blocks surrounded the

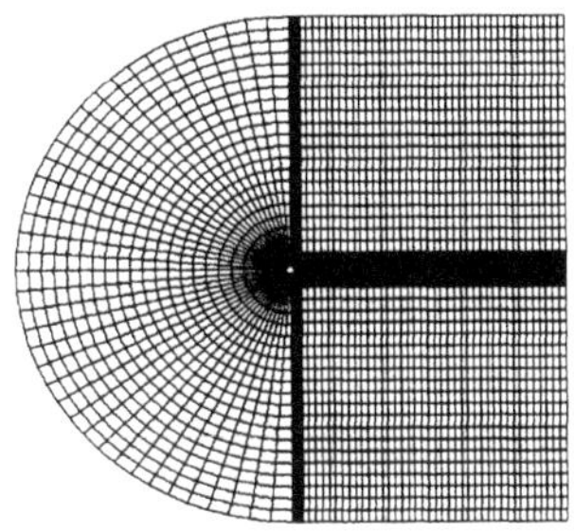

Figure 2. Three-dimensional mesh: view perpendicular to the axis of the cylinder

cylinder with 8 coarser blocks surrounding the first ring in the x-y plane (cylinder cross-section). Two subdivisions were made in the z direction (parallel to the cylinder axis). The CBJ code ran on 8 processors on the Cray T3E such that each processor handled 32 000 points. The time step was fixed at 0.1, the convergence criterion was 0.001 with a maximum of 20 Newton iterations per time step. The two tapered cylinders studied, A and B, had a taper ratio equal to 100 and 75, respectively. The definition of the taper ratio used is: $R_T = \frac{l}{d_2 - d_1}$, where l is the length of the cylinder, d_2 the diameter at the widest end and d_1 at the narrowest end. To eliminate some of the end effects Neumann-type boundary conditions were imposed on the $x - y$ planes at the two ends of the cylinder. After 500 time steps, the total cpu-time for all 8 processors was approximately 424 hours for "Case A" and 429 for "Case B", that means an average of 6 sec per processor and per grid point. The wall-clock time was approximately 62 hours. In these cases, the use of 8 processors running in parallel on Cray T3E allowed us to obtain the results 7 times quicker than on a computer with only one single processor. The simulations aimed at reproducing the experiments of Piccirillo and Van Atta [5] of different tapered cylinders in crosswise flow at low Reynolds number. In particular, the simulations "Case A" and "Case B", reproducing what they called "run14" and "run23" respectively, have a Reynolds number *Re* (based on the wide diameter) equal to 178 and 163, respectively.

The results showed the same type of flow behavior behind the body as the experiments did, especially the oblique shedding angle of the vortices occurring along the span of the cylinder, from the smallest diameter to the largest one. Moreover, the same number of shedding vortex cells was found. The span of the cylinder can be divided into a set of cells. Each cell (or vortex cell) shed vortices with one typical frequency. Only results from "Case A" are presented herein, whereas results for "Case B" and deeper results analysis are shown in an accompanying paper [13].

Figure 3 shows the pressure in a plane through the cylinder axis and parallel with the incoming flow and Figure 4 shows the iso-contour surface with the non-dimensional

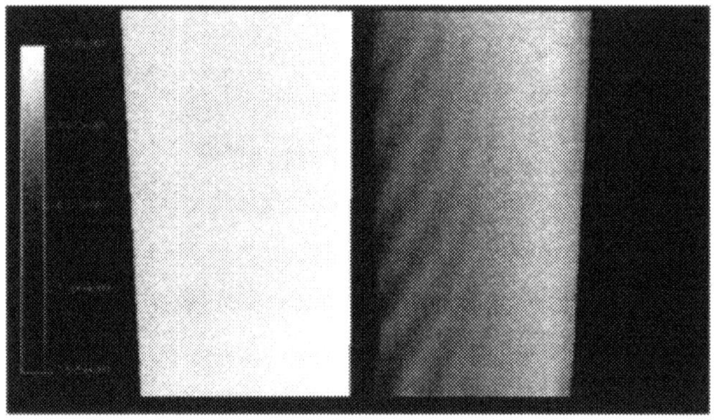

Figure 3. Pressure in the stream direction ("Case A")

pressure equal to -1 in this plane. The different shedded vortices can easily be identified along the spanwise direction.

In 1991, Jespersen and Levit [6] conducted similar three-dimensional simulations for a tapered cylinder with taper ratio $R_T = 100$ in a Reynolds number range from 90 to 145, i.e. somewhat lower than those considered in the present work. They implemented a parallel implicit, approximate-factorization central-difference code for the Navier-Stokes equations with no thin-layer assumption and they used central-difference discretization in space and a three-point implicit time-stepping method. Their code was developed on a VAX machine and ran on a single-instruction, multiple data parallel computer. Their mesh had 131 072 nodes.

Their results showed qualitatively the same type of flow behavior as the experiments [5] (i.e. velocity-time trace, vortex shedding) but the quantitative comparison in Figure 5 is not satisfactory. This figure compares the $St(Re)$ results of the simulations with the results of the experiments made by Piccirillo & Van Atta [5]. The curve fit they employed was $St_c = 0.195 - 5.0/Re$, where St_c is the Strouhal number associated with an individual shedding vortex cell. The $St(Re)$ relation deduced from the present simulation is in good agreement with the experimental curve for Reynolds numbers below 150, whereas the two curves diverge for $Re > 150$. This is mainly due to the fact that the latter curve is a fit on Strouhal number values taken at the center of each vortex cell only, and not values taken at each spanwise location. The fact that the spanwise boundary conditions used are not fully consistent with the experiment end conditions may also cause some deviations. For the sake of completeness, the universal Strouhal-Reynolds number curve for straight circular cylinders $St = -3.3265/Re + 0.1816 + 1.6E^{-4}Re$ due to Williamson [12], is also plotted. As reported from the experiments by many authors [2,5], the simulations showed that the Strouhal number for tapered cylinders is lower than those for straight cylinders at the same Reynolds number.

Figure 4. Isopressure surfaces, p=-1 ("Case A")

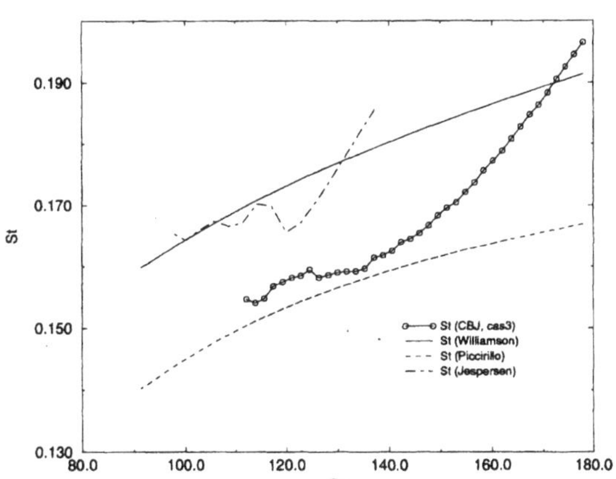

Figure 5. Strouhal number (St) versus Reynolds number (Re). CBJ and Jespersen refer to simulations for cylinders with $R_T = 100$. Piccirillo refers to curve fitting of all results reported in [5]. Williamson refers to the universal St-Re curve for straight circular cylinders [12].

5. Conclusion

The present results compare favorably with other simulations and experimental data, both in two-dimensional and three-dimensional cases. The CBJ code has been proved to perform satisfactorily for simulations of the complex laminar flow behind tapered cylinders and reproduce the dominant flow phenomena observed experimentally. The next stage would be to simulate the turbulent flow behavior for flow past a tapered cylinder at higher Reynolds numbers, typically $Re > 1000$. This will be accomplished by means of large-eddy simulations in which parts of the turbulent fluctuations are accounted for by a sub-grid-scale model.

REFERENCES

1. D. J. Tritton. Experiments on the flow past a circular cylinder at low Reynolds numbers. *J. Fluid Mech.*, 6:547–567, 1959.
2. M. Gaster. Vortex shedding from slender cones at low Reynolds numbers. *J. Fluid Mech.*, 38:565–576, 1969.
3. C. H. K. Williamson. Oblique and parallel modes of vortex shedding in the wake of a circular cylinder at low Reynolds numbers. *J. Fluid Mech.*, 206:579–627, 1989.
4. C. H. K. Williamson. Vortex dynamics in the cylinder wake. *Annu. Rev. Fluid Mech.*, 28:477–539, 1996.
5. P. S. Piccirillo and C. W. Van Atta. An experimental study of vortex shedding behind linearly tapered cylinders at low Reynolds number. *J. Fluid Mech.*, 246:163–195, 1993.
6. D. C. Jespersen and C. Levit. Numerical simulation of flow past a tapered cylinder. 29th Aerospace Sciences Meeting, Reno, NV, jan 7–10 1991.
7. C. B. Jenssen. Implicit multiblock Euler and Navier-Stokes calculations. *AIAA J.*, 32(9):1808–1814, 1994.
8. C. B. Jenssen and P. Å. Weinerfelt. Coarse grid correction scheme for implicit multi-block Euler calculations. *AIAA J.*, 33(10):1816–1821, 1995.
9. C. B. Jenssen and P. Å. Weinerfelt. Parallel implicit time-accurate Navier-Stokes computations using coarse grid correction. *AIAA J.*, 36(6):946–951, 1998.
10. A. Belov, L. Martinelli, and A. Jameson. A new implicit algorithm with multigrid for unsteady incompressible flow calculations. *AIAA J.*, 95–0049:Jan, 1995.
11. M. Braza, P. Chassaing, and H. Ha Minh. Numerical study and physical analysis of the pressure and velocity fields in the near wake of a circular cylinder. *J. Fluid Mech.*, 165:79–130, 1986.
12. C. H. K. Williamson. Defining a universal and continuous Strouhal-Reynolds number relationship for the laminar vortex shedding of a circular cylinder. *Phys. Fluids*, 31:2742–2744, 1988.
13. H. I. Andersson, C. B. Jenssen, and B. Vallès. Oblique vortex shedding behind tapered cylinders. Presented at IUTAM Symp. on Bluff Body Wakes and Vortex-Induced Vibrations, jun 13–16 2000.

www.ingramcontent.com/pod-product-compliance
Ingram Content Group UK Ltd.
Pitfield, Milton Keynes, MK11 3LW, UK
UKHW050949290926
15470UKWH00005B/1735

9 780444 506733